International Conference on Advances in Renewable Energy and Technologies (ICARET 2016)

Sustaining a Green Future

IOP Conference Series: Earth and Environmental Science
Volume 32

Putrajaya, Malaysia
23 – 25 February 2016

Editors:

Abd Halim Shamsuddin
Adlansyah Abd Rahman
Halina Misran

ISBN: 978-1-5108-2273-3
ISSN: 1755-1307

Printed from e-media with permission by:

Curran Associates, Inc.
57 Morehouse Lane
Red Hook, NY 12571

Some format issues inherent in the e-media version may also appear in this print version.

Printed by Curran Associates, Inc. (2016)

For permission requests, please contact the Institute of Physics
at the address below.

Institute of Physics
Dirac House, Temple Back
Bristol BS1 6BE UK

Phone: 44 1 17 929 7481
Fax: 44 1 17 920 0979

techtracking@iop.org

Additional copies of this publication are available from:

Curran Associates, Inc.
57 Morehouse Lane
Red Hook, NY 12571 USA
Phone: 845-758-0400
Fax: 845-758-2633
Email: curran@proceedings.com
Web: www.proceedings.com

Table of contents

Volume 32

International Conference on Advances in Renewable Energy and Technologies (ICARET 2016)23–25 February 2016, Putrajaya, Malaysia

Accepted papers received: 15 March 2016Published online: 19 April 2016Preface

Papers

TENAGA NASIONAL
CENTRE FOR RENEWABLE ENERGY
UNIVERSITI TENAGA NASIONAL
The National Energy University
PROGRAMME BOOK
ICARET 2016
2nd INTERNATIONAL CONFERENCE ON ADVANCES IN RENEWABLE ENERGY TECHNOLOGIES
" SUSTAINING A GREEN ENERGY FUTURE "
Co-sponsor
energy institute
Supported by
KeTTHA
KEMENTERIAN TENAGA, TEKNOLOGI HIJAU DAN AIR
Co-organized by
YAYASAN CANSELOR
IEPRe
iRMC
Power Engineering Centre

ICARET 2016 2nd International Conference on Advances in Renewable Energy Technologies

ORGANISING COMMITTEE

MAIN ORGANIZING COMMITTEE

Chair: Prof. Ir. Dr. Abd Halim Shamsuddin
Co-Chair: Prof. Ir. Dr. Au Mau Teng
Prof. Ir. Dr. Tuan Ab Rashid Bin Tuan Abdullah
Prof. Ir. Dr. Lariyah Bte. Mohd Sidek
Secretary: Prof. Dr. Saifuddin Bin Haji Nomanbhay

TECHNICAL AND PUBLICATIONS SUB-COMMITTEE

Chair: Prof. Dr. Mujibur Rahman
Co-Chair: Prof. Dr. Ibrahim Bin Ahmad
Secretary: Dr. Adlansyah Abd Rahman
Publications: Assoc. Prof. Dr. Jagadeesh Pasupuleti
Paper Submission: Assoc. Prof. Dr. Halina Misran
Media: Dr. Abreeza Manap
Track Chairs: Prof. Dr. T. M. Indra Mahlia
Prof. Ir. Dr. Tiong Sieh Kiong
Dr. Norhayati Bte Mat Husin
Prof. Ir. Dr. Lariyah Mohd Sidek
Assoc. Prof. Dr. Salman Yussof

FINANCE AND LOGISTICS SUB-COMMITTEE

Chair: Hj. Mohd Sakarnor Bin Deris
Co-Chair: Ir. Dr. Kumaran Palanisamy
Zulkifli Bin Ahmad
Members: Nora Bte Dato' Yahya

IOP Conf. Series: Earth and Environmental Science **32** (2016) 011003 doi:10.1088/1755-1315/32/1/011003

ICARET 2016 2nd INTERNATIONAL CONFERENCE ON ADVANCES IN RENEWABLE ENERGY TECHNOLOGIES

WELCOME REMARKS

PROF. IR. DR. KAMAL NASHARUDDIN MUSTAPHA

Vice Chancellor, Universiti Tenaga Nasional

At Universiti Tenaga Nasional (UNITEN), we take pride in expanding new ideas and promoting lifelong knowledge. UNITEN has an academic environment where exceptional students and scholars worldwide can be inspired to excel in their fields of study and research. The BOLD 2020 strategic plan for Universiti Tenaga Nasional represents a commitment to improving quality, and an aspiration to be recognised among the leading institutions of higher learning in Malaysia and in the rest of the world.

Sustainable research in the energy sector is one of the key agenda of UNITEN. This is clearly worded in its mission statement as; "We strive to advance knowledge and learning experience through research and innovation that will best serve human society". BOLD 2020 has also set a goal for Advancing Research in the Energy Industry, in line with UNITEN as "The National Energy University". UNITEN's commitment in promoting research activities is evident from the amount of research undertaken that continues to increase year after year.

The Centre for Renewable Energy, CRE, was established by UNITEN, as a centre of excellence for research and development in Renewable Energy. The core of CRE's strategic direction is in research and development programmes mainly in understanding renewable resources for energy, to the conversion of these resources to renewable electricity and fuels. CRE has established its core strength through continuous, innovative and excellent research achievements. CRE provides a sustainable platform for strategic partnerships between the authorities, industry and academia into greater achievements in national renewable energy development with the vision to be the leading centre of excellence and regional knowledge hub in renewable energy. The centre's pool of scientists and researchers support critical market objectives to accelerate research from scientific innovations to market-viable alternative energy solutions. The centre thereby directly contributes to our nation's goal for finding new renewable ways to power our homes, industries, businesses, and transportation.

My heartiest congratulations to CRE and the ICARET 2016 Organizing Committee for putting together this important event, providing an intellectual forum for exchange of knowledge and information in the development of renewable energy (RE), and its related enabling technologies. The outcomes and proceedings of the conference will be an important input to the Nation's effort on Green Technology and Renewable Energy implementation.

I would like to take this opportunity to welcome all participants and wish all of you the very best in your deliberations and discussions on this important topic of renewable energy.

Thank you.

IOP Conf. Series: Earth and Environmental Science **32** (2016) 011004 doi:10.1088/1755-1315/32/1/011004

PROF. DR. IR. ABD HALIM SHAMSUDDIN

Chairman, ICARET 2016 Organizing Committee

At the recent Paris Climate Summit in December 2015, global leaders forged a landmark deal with the ultimate goal of reaching a new global climate change agreement. This new agreement will include commitments to reduce GHG emissions, adaptation to the impacts of climate change and will also provide assistance to countries that need it. Countries worldwide have taken initiatives in developing alternative energy scenarios and strategies aimed at a more sustainable and cleaner energy future. Renewable energy (RE) has been accepted as a promising and important alternative in the energy and generation mix in our carbon-constrained world and the development worldwide has been very encouraging.

Centre for Renewable Energy, CRE, at Universiti Tenaga Nasional (The National Energy University) has been actively spearheading R&D in RE since its inception in 2008. CRE directly contributes to our nation's goal to fine new renewables as alternative to the conventional fossil energy sources. CRE's mission and strategy are focused on advancing our nation's energy goals, in line with the 5 Fuel Strategy of the National Energy Policy and the subsequent enactment of the RE Act 2011.

This conference is organized by CRE, in collaboration with UNITEN's Power Engineering Centre (PEC) and the Institute of Energy Policy Research (IEPRe), to provide a platform for researchers, industries and stakeholders to analyse and explore the future options that are available through advancements in research and enabling technologies for application of renewable energy as a viable and cost-effective energy supply in the future. This conference is expected to benefit the researchers, relevant authorities and industries aspiring to provide an effective solution for the future of sustainable and clean energy.

On behalf of the organizing committee, I wish to thank the Ministry of Energy, Green Technology and Water Malaysia, UNITEN, Tenaga Nasional Berhad and the Energy Institute United Kingdom on the support given for this conference. I wish to acknowledge the efforts of the keynote speakers and all the authors of papers for their valuable contributions to the conference, and session chairs for organization and execution of the technical sessions. A warm welcome is especially extended to the overseas participants for their interest in this conference, and their time and effort spent in particpating.

Finally, I extend special words of thanks to the members of the organizing committee, and all those who have contributed for this conference. A remarkable amount of effort was expended by everyone to ensure that this conference was a success.

Thank you.

IOP Conf. Series: Earth and Environmental Science **32** (2016) 011005 doi:10.1088/1755-1315/32/1/011005

AREAS OF INTEREST

The areas of interest covered in ICARET 2016 are as follows:-

1. ***Sustainable Clean Energy***

   The objective of the Sustainable Clean Energy track is to sustain provision of sustainable clean energy development that meets the needs of the present without compromising the ability of future generations to meet their needs. This track focuses on the reduction of energy and water consumption and the use of environmentally friendly energy sources. The program under the track includes renewable energy, energy conservation, sustainable transport and sustainability as well as those aspects that are a prerequisite for finding intelligent and pragmatic solutions for a sustainable clean energy future.

2. ***Power and Energy Services Convergence***

   The aim of the Power and Energy Service Convergence track is to provide a platform for engineers, academics and researchers to disseminate, discuss and exchange innovative ideas, leading edge concepts, new technologies and devices, ongoing R&D efforts, prototype and demonstration projects, and visions of the future related to the theme.

3. ***Energy Social Sustainability***

   The Energy and Social Sustainability cluster track aims to highlight the increase in research capacity in energy, the green economy, policy, and social transformation towards sustainable communities. In line with the New Economy Model (NEM), this track reports the social, environmental and economic (SEE) impact of renewable energy, the energy sector and its regulatory framework and development of energy reporting/audit guidelines. It also focuses on energy security, energy policy related to energy production, distribution, consumption and energy economics on energy demand, energy efficiency and potential savings. In addition, the track illustrates economic and business structural changes related to energy and environmental impact due to the energy production, energy conversion and energy consumption and other fugitive uses of energy resources towards sustainability.

ICARET 2016 2nd INTERNATIONAL CONFERENCE ON ADVANCES IN RENEWABLE ENERGY TECHNOLOGIES

INTRODUCTION

4. ***Energy Infrastructure***

   The transition from a traditional economy to a modern high-technology urban-based economy often stimulates profit-seeking development, with a time-lag before provision of infrastructure and methods of coping with hazards are seen as necessary and viable. The recent natural disaster events such as floods and landslides have introduced the potential threat to energy infrastructure from geohazards. Therefore, the Energy Infrastructure track aims to provide a platform for researchers and all sectors of society to discuss their latest research findings and developments in addressing the engineering challenges to ensure environmentally sound, hazard and risk free power generation and delivery. The combination of environmentally friendly and hazard-free risk leads to advanced knowledge and technologies that enable engineers and researchers to design and build advanced systems to reduce the vulnerability of the people and areas most at risk from natural hazards.

5. ***Information and Communication Technology (ICT)***

   The Information and Communication Technology (ICT) track aims to provide an avenue for researchers to present their latest research findings and development related to the use of ICT in the energy sector, especially related to renewable energy. Given that many power companies are moving towards smart grids, ICT has become a necessity in the energy industry. However, there are still many issues and problems related to applications, protocols, interoperability and security that are yet to be resolved for the potential of smart grids to be fully realized. Furthermore, the use of computer science techniques and algorithms may also open a way for a more efficient operation that can provide cleaner energy for society.

IOP Conf. Series: Earth and Environmental Science **32** (2016) 011006 doi:10.1088/1755-1315/32/1/011006

ICARET 2016 2nd INTERNATIONAL CONFERENCE ON ADVANCES IN RENEWABLE ENERGY TECHNOLOGIES

PROGRAMME SCHEDULE

Day 1: 23rd February 2016 (Tuesday)

Time	Programme	Venue
0800 – 0830	Registration	Grand Ballroom Foyer
0830 – 1000	Opening Ceremony	Grand Ballroom
1000 – 1030	Morning Break	Matahari
1030 – 1115	Keynote Address 1: ***"Renewable Energy in the Electricity Industry in Malaysia"*** **Dr. Ahmad Jaafar Abd Halim** *TNB Energy Services / Tenaga Nasional Berhad*	Grand Ballroom
1115 – 1200	Keynote Address 2: ***"Redesigning Anaerobic Digestion Technology – Resource Recovery, Energy Production and Storage, Chemicals and Materials Production"*** **Prof. Sandra Esteves** *University of South Wales, UNITED KINGDOM*	Grand Ballroom
1200 – 1245	Keynote Address 3: ***"Adsorption of Fuel Gases on MOFs Characterised by High-Pressure Adsorption Microcalorimetry"*** **Prof. Akihiko Matsumoto** *Toyohashi University of Technology, JAPAN*	Grand Ballroom
1245 – 1400	Lunch	Matahari
1400 – 1430	Invited Speaker 1: ***"Updates in Biogas industry in Germany"*** **Prof. Dr.-Ing. Isabel Kuperjans** *FH Aachen - University of Applied Sciences, GERMANY*	Grand Ballroom
1430 – 1500	Invited Speaker 2: ***"Innovative Japanese Technologies for Waste-To-Green Product Conversion"*** **Prof. Dr. Kunio Yoshikawa** *Tokyo Institute of Technology, JAPAN*	Grand Ballroom
1500 – 1515	Afternoon Break	Grand Ballroom Foyer
1515 – 1715	Energy Institute Malaysia Forum: : ***"Challenges and Opportunities in the Financing of Renewable Energy Projects in Malaysia"***	Grand Ballroom
1715 - 1800	Energy Institute Malaysia Drink Reception	Matahari

IOP Conf. Series: Earth and Environmental Science **32** (2016) 011006 doi:10.1088/1755-1315/32/1/011006

ICARET 2016 2nd INTERNATIONAL CONFERENCE ON ADVANCES IN RENEWABLE ENERGY TECHNOLOGIES

PROGRAMME SCHEDULE

Day 2: 24th February 2016 (Wednesday)

Time	Programme	Venue
0800 – 0830	Registration	Ballroom Foyer
0830 – 1000	Parallel Sessions 1 – 5	Grand Ballroom, Melati 1, Melati 2, Melati 3 & Raya
1000 – 1015	Morning Break	Grand Ballroom Foyer
1015 – 1045	Invited Speaker 3: "***Challenges in Implementing RE connections to TNB Distribution Network***" **Mr. Saharuddin Sulaiman** *Tenaga Nasional Berhad*	Grand Ballroom
1045 – 1115	Invited Speaker 4: ***"Development of Biogas Industry in Thailand"*** **Prof. Dr. Pruk Aggarangsi** *Energy Research and Development Institute-Nakornping Chiang Mai University, THAILAND*	Grand Ballroom
1115 – 1145	Invited Speaker 5: ***"Employment and Career Development in Renewable Energy and Energy Efficiency Industry"*** **Sarah Beacock** *Energy Institute, UNITED KINGDOM*	Grand Ballroom
1145 – 1215	Invited Speaker 6: ***"Cold Spray Fabricated Dye-Sensitized Solar Cell"*** **Prof. Kazuhiro Ogawa** *Tohoku University, JAPAN*	Grand Ballroom
1215 – 1245	Invited Speaker 7: ***"Integrated Solution for Municipal Solid Waste Treatment – The Zero Waste Park"*** **Mr. William Tan** *SP Multitech Renewable Energy Sdn. Bhd.*	Grand Ballroom
1245 – 1400	Lunch	Matahari
1400 – 1530	Parallel Sessions 6 – 10	Grand Ballroom, Melati 1, Melati 2, Melati 3 & Raya
1530 – 1545	Afternoon Break	Grand Ballroom Foyer
1545 – 1715	Parallel Sessions 11 – 15	Grand Ballroom, Melati 1, Melati 2, Melati 3 & Raya
1715 – 1745	Closing of ICARET 2016	Grand Ballroom

ICARET 2016 2nd International Conference on Advances in Renewable Energy Technologies

ACKNOWLEDGEMENTS

EDITORS

Prof. Ir. Dr. Abd Halim Shamsuddin
Dr. Adlansyah Abd Rahman
Assoc. Prof. Dr. Halina Misran

PAPER REVIEWERS

Dr. Bakhtiar Alrazi
Dr. Adlansyah Abd Rahman
Ir. Hasril Hasini
Dr. Agusril Syamsir
Dr. Lim Chin Wai
Prof. Dr.Md. Mujibur Rahman
Dr. Abreeza Noorlina Bte. Abd. Manap
Dr. Gasim Hayder Ahmed Salih
Herda Yati Katman
Dr. Norhayati Mat Husin
Dr. Nur Irfah Mohd Pauzi
Mohd Zakwan Ramli
Wan Mohammad Taufik Wan Abdullah
Assoc. Prof. Dr. Khairul Salleh Mohamed Sahari
Assoc. Prof. Dr. Ungku Anisa Ungku Amirulddin
Assoc. Prof. Dr. Jawaid I. Inayat-Hussain
Dr. Ing. Zarina Itam
Prof. Dr. Saifuddin Hj. M. Normanbay
Dr. Weria Khaksar
Dr. Norlela Ismail
Prof. Ir. Dr. Lariyah Mohd Sidek
Dr. Tan Ching Sin
Assoc. Prof. Dr. Ng Khai Ching
Prof. Dr. Ong Hang See
Dr. A.k.m. Parvez Iqbal
Dr. Md Ashraful Alam
Dr. Omar Suliman Zaroog
Assoc. Prof. Dr. Shahida Begum
Assoc. Prof. Dr. Agileswari K. Ramasamy
Dr. Rabiah Eladwiah Abdul Rahim
Assoc. Prof. R. Jegatheesan V. Rajadurai
Assoc. Prof. Ir. Dr. Marlinda. Abdul Malek
Assoc. Prof. Dr. Yap Boon Kar
Assoc. Prof. Dr. Jagadeesh
Assoc. Prof. Dr. Salman Yussof
Dr. Aishah Mohd Isa
Dr. Nadia Tan Mei Lin
Dr. Grace Pua Fei Ling
Assoc. Prof. Dr. Abdul Aziz Mohamed
Assoc. Prof. Dr. Lee Choon Yong
Prof. Ir. Dr. Tiong Sieh Kiong
Assoc. Prof. Ir. Dr. Chong Kok Hen
Prof. Dr. T.M. Indra Mahlia
Assoc. Prof. Dr. Yap Keem Siah
Adzly Anuar
Assoc. Prof. Dr. Sivakumar Naganathan
Prof. Madya. Dr. Cheah Cheng Lai
Dr. Azrul Mohd Ariffin
Assoc. Prof. Rohayu Che Omar
Dr. Abdul Talip Zulkarnain
Assoc. Prof. Dr. Nasri A.Hamid
Dr. Zaipatimah Ali
Dr. Faizah Che Ros
Prof. Dr. Farrukh Hafiz Nagi
Prof. Ir. Dr. Au Mau Teng
Dr. Nurul Asyikin Mohamed Radzi
Prof. Ir. Dr. Abd Halim Shamsuddin
Dr. Leong Yeng Weng
Dr. Normy Norfiza Abdul Razak
Dr. Juraifa Jais
Assoc. Prof. Dr. Azizah Suliman
Dr. Azhana Ahmad
Dr. Alicia Tang Yee Chong
Nazirul Mubin Zahari

ICARET 2016 2nd INTERNATIONAL CONFERENCE ON ADVANCES IN RENEWABLE ENERGY TECHNOLOGIES

SPONSORS

MAIN SPONSOR

TENAGA NASIONAL

CO-SPONSOR

energy institute

detik aturan

SUPPORTED BY

KeTTHA
KEMENTERIAN TENAGA, TEKNOLOGI HIJAU DAN AIR

CO-ORGANISED BY

YAYASAN CANSELOR
UNIVERSITI TENAGA NASIONAL

iRMC
INNOVATION & RESEARCH MANAGEMENT CENTRE
UNIVERSITI TENAGA NASIONAL

IEPRe
Institute of Energy Policy and Research

Power Engineering Centre

Peer review statement

All papers published in this volume of *IOP Conference Series: Earth and Environmental Science* have been peer reviewed through processes administered by the proceedings Editors. Reviews were conducted by expert referees to the professional and scientific standards expected of a proceedings journal published by IOP Publishing.

IOP Conf. Series: Earth and Environmental Science **32** (2016) 012001 doi:10.1088/1755-1315/32/1/012001

Gravitational search algorithm based tuning of a PI speed controller for an induction motor drive

Jamal Abd Ali[1,2], M A Hannan[1] and Azah Mohamed[1]
[1]Department of Electrical, Electronic and Systems Engineering, Faculty of Engineering and Built Environment, Universiti Kebangsaan Malaysia, Malaysia.
[2]General company of electricity production middle region, Ministry of electricity, Iraq.

E-mail: eng_jhy@yahoo.com, hannan@eng.ukm.my, azah@eng.ukm.my.

Abstract. Proportional-integral (PI)-controller is very useful for controlling speed and mechanical load variables for the three-phase induction motor (TIM) operation. However, the conventional PI-controller has a very exhaustive trial and error procedure for obtaining it is parameters. In this paper, PI speed controller has been improved in it is design technique to suite TIM by utilizing a gravitational search algorithm (GSA) optimization technique. The mean absolute error (MAE) of the speed response has been used as an objective function. An optimal GSA based PI speed controller (GSA-PI) objective function is also employed to tune and minimize the MAE for developing the performance of the TIM in terms of changes speed and mechanical load. This experiment use space vector pulse width modulation (SVPWM) technique to create pulse width modulation for switching devices for three phase bridge inverter. Results obtained from the GSA-PI speed controller are compared with those obtained through particle swarm optimization (PSO) to validate the developed controller. Then it has been proved that the robustness of the GSA-PI speed controller is far better than that of the PSO controller in all tested cases in terms of damping capability and transient response under different mechanical loads and speeds.

Introduction

Three-phase induction motors (TIMs) are used widely in the industrial applications because of their simple structure, easy maintenance, low cost, robustness, and ruggedness [1]. Scalar controller is one of the controller methods used for induction motor, it is used in the most of controller design of induction motor because of easy design, simple structure and low cost [2]. Three phase bridge inverter is also plays an important role to control the variable speed of the induction motor such as fans and pumps, and controlling on three phase bridge inverter via pulse width modulation (PWM) techniques. There are many PWM techniques are used to control inverters SVPWM method is one of the best methods this is because has a great capability to minimize harmonic distortion [3].
PI-controller is one of these speed control schemes. It is considered very good controller technique that uses a numerous applications. These applications including scalar and vector techniques due to it is easy design, low cost, and simple structure [1],[4]. Moreover, PI-controller used to regulate for the main variables as voltages, currents, speed, torque, and rotor flux in the induction motors. Therefore, the disadvantages is difficulty to get on suitable coefficients for PI-controller (Proportional gain (kp), integral gain (ki)). These parameters are plays a role in sensitivity and stability to model control [1],[4],[5]. There are many methods to search for coefficient PI controller as Ziegler-Nichols method, Cohen-Coon method, and Lambda tuning method, etc. These methods have some problems as process upset, some trial-and-error, needed for some calculation operation, and needed for the mathematical

model [6],[7]. There are number of researchers has used different way of optimization to solve these problems involved with PI-controller such as [6] use genetic algorithm (GA), [7] use PSO algorithm to find the best values for PI-controller parameters and etc. In this paper, the GSA optimization is developed to improve the performance of the TIM speed controller by tuning the free parameters and selecting the best parameters for PI speed controller. The results obtained from the developed speed controller and the PSO optimization algorithm has been compared in terms of the control process and robustness under specific condition such as sudden change in the speed and the mechanical load. On the other hand, the high-performance membership functions are also obtained by minimizing the error function using MAE of the system. In addition, employing the SVPWM switching algorithm to develop the GSA optimization-based PI speed controller is useful for TIM drives.

Design of tuning GSA based PI speed controller

Primary GSA is an evolutionary computation technique developed by E. Rashediet et al. at 2009, inspired by the law of gravity and mass interactions introduced in [8]. The numerical values of the parameters are difficult to calculate in PI speed controller design. Therefore, in this study, the PI speed controller was fixed with the parameters tuned by GSA to obtain the optimal numerical values of the parameters limits. The objective function is used to minimize the speed error (i.e., reduce overshoot, steady-state error, and settling time) while enhancing or maintaining the driving performance characteristics of TIM. The developed GSA-PI speed controller simulation model for the TIM drive model is shown in Figure 1. SVPWM switching technique was utilized to control the TIM drive with the V/f control PI-speed controller. The induction motor model is a stationary reference frame in which rotor speed is measured depending on the feedback signal. Therefore, the sensor speed needs to be measured. The developed PI speed controller was also utilized to tune the parameters and determine the best values for the parameters of PI speed controller through GSA optimization. The GSA-PI speed controller was used for the speed controller system for simplicity. The task of this controller is to allow the actual speed to track the reference speed. The speed controller provides the required slip speed to reach the reference speed (synchronous speed). The slip speed is then added to the feedback signal is the rotor speed to produce the required frequency to the inverter then to TIM. A V/f relationship is required to fix the machine at its rated flux linkage.

The procedure for implementing the GSA-PI speed controller is given by the following pseudo-codes:

```
Initialize the population size (N), the positions and the dimensions of the population (D):
For t = 1 to Maximum Iteration T
    Run the simulation with PI speed controller for each position
    Calculate objective function (MAE = Σ_{i=1}^{n} |error| /n) for each population (Xij)
    Update forces, velocity, and position using GSA equations
end
Output the PI speed controller with optimal parameters Xij
```

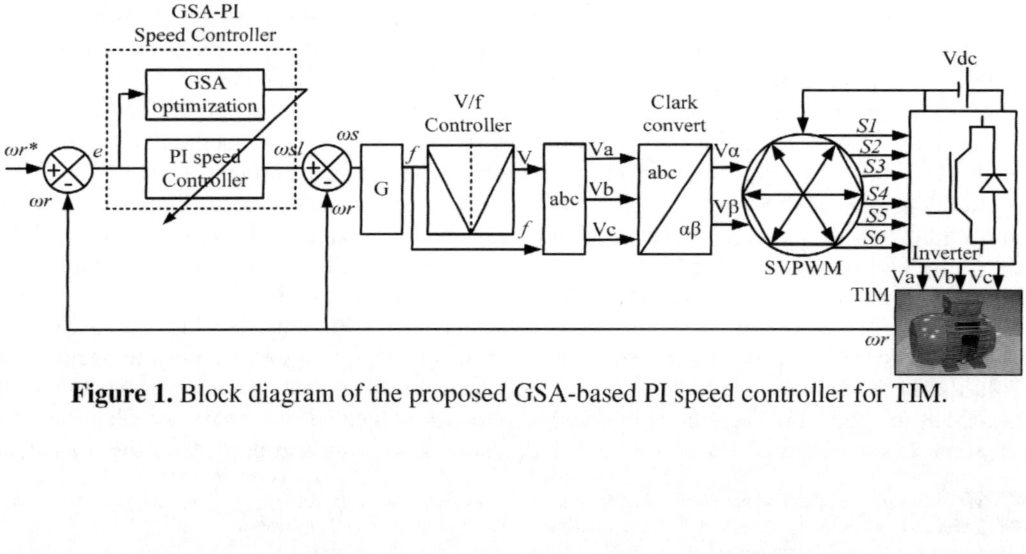

Figure 1. Block diagram of the proposed GSA-based PI speed controller for TIM.

Results and discussion

The results were obtained by execution the control scheme to validate the proposed GSA-PI speed controller and the performance of the overall system. The switching frequency used was 20 kHz in the SVPWM technique. The PI speed controller parameters were tuned by GSA methods in order to search for the optimal parameters. Using GSA-PI speed controller and PSO-PI for optimal parameter speed controller for the purpose of comparison and to evaluate the GSA optimization robustness. Figure 2 shows the relationship between the objective function curve and the number of iterations performed by the two optimization methods (GSA and PSO). For fair comparison, the optimization methods take the same number of population and iteration. The simulation results are presented in two test cases to determine the effectiveness and robustness of the proposed PI speed controller. In this work, the two tests carried out in the system were variation of speed and mechanical load.
The first test involves increasing or decreasing the reference speed with the mechanical load constant. This case study aims to evaluate the performance of the proposed method of GSA-PI speed controller and estimate the reference speed variation with the mechanical load constant of TIM controlled by the V/f ratio. The reference speed variation was applied on induction motor in short times as shown in Figure 3 also shows the speed responses of GSA-PI and PSO-PI aped controller with speed change. The second test for induction motor involves constant speed and mechanical torque variations. This test aims to determine the system performance and robustness. The results were indicated by the speed response and its zoomed locations as in Figure 4, the estimated speed is consistent with the actual speed with good accuracy. GSA-PI speed controller achieved better results than PSO-PI controller by minimizing the overshoot, settling time, steady-state error, and damping ratio after change and exhibiting rapid stability after each speed change.

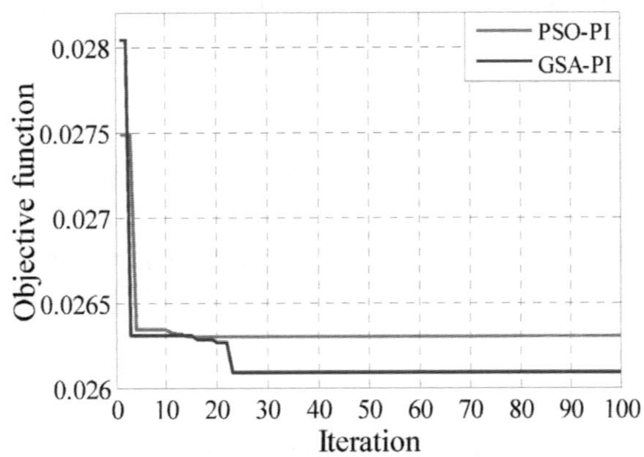

Figure 2. Objective function curve.

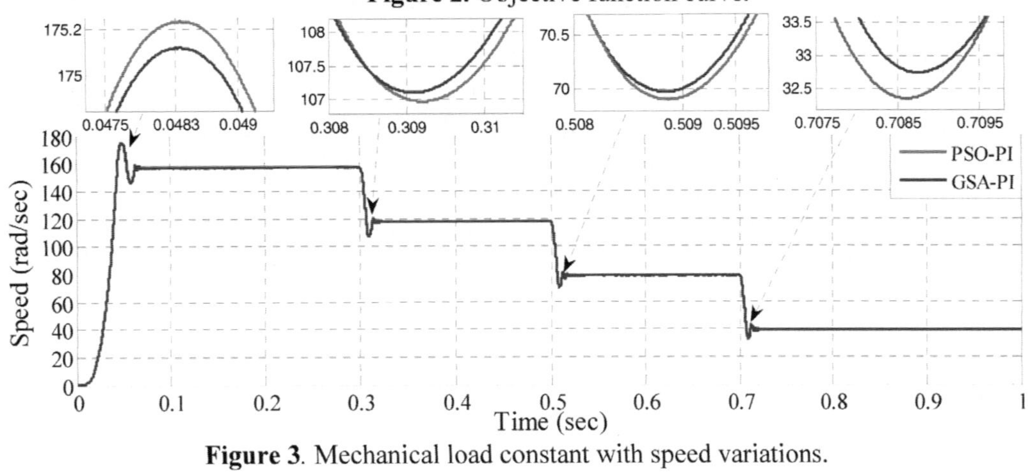

Figure 3. Mechanical load constant with speed variations.

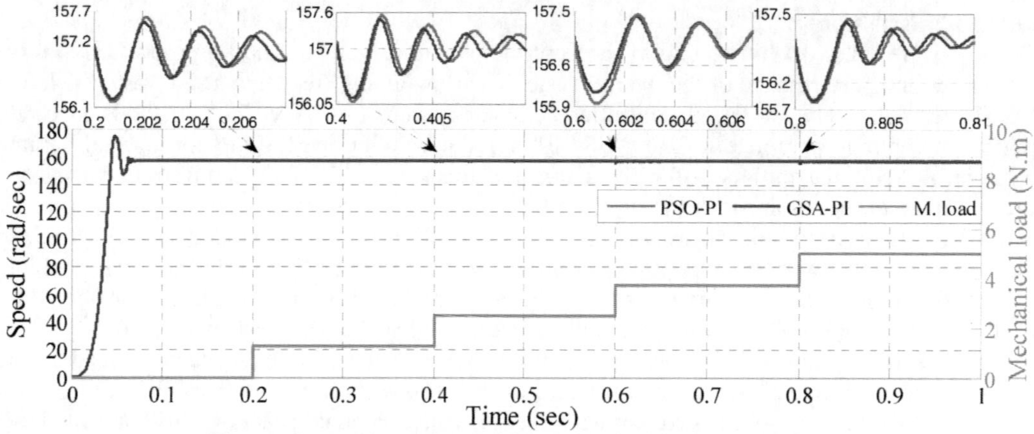

Figure 4. speed constant with mechanical load variations.

Conclusion

This paper presents PI speed controller-based optimization approach for induction motor using the GSA. The proposed method is formulated to automatically change the parameters of the PI speed controller. To effectively tune the parameters of the proposed PI speed controller used in the TIM applications, a suitable objective function was developed to minimize the MAE of the speed response. The developed GSA-PI optimization method helped to avoid traditional trial and error procedure for obtaining the best parameters. The performances of GSA-PI and PSO-PI were compared and the results show that the GSA-PI speed controller is far better than that of the PSO-PI speed controller in terms of robustness, damping capability, and improvement of the transient responses.

Acknowledgment

The authors are grateful to Ministry of Science, Technology and Innovation, Malaysia for supporting this research financially under grant 06-01-02-SF1060.

References

[1] A. Hazzab, I. Bousserhane, M. Zerbo, and P. Sicard 2006 Real Time Implementation of Fuzzy Gain Scheduling of PI Controller for Induction Motor Machine Control," *Neural Processing Letters* 24 203-215.

[2] T. Santosa, A. Goedtelb, S. Silvab, and M. Suetakec 2014 Scalar control of an induction motor using a neural sensorless technique," *Electric Power Systems Research* 108 322-330.

[3] S. Nageswari, and V. Kumar 2014 Field programmable gate array implementation of variable common mode injection PWM for three-level inverters," *Computers and Electrical Engineering* 40 1238-1252.

[4] M. Jovanović, M. Simonović, D. Zorić, S. Lukić, N. Stupar, and S. Ilić 2013 Experimental studies on active vibration control of a smart composite beam using a PID controller," *IOP Publishing* 22.

[5] P. D. Ngo, and Y. C. Shin 2015 Gain estimation of nonlinear dynamic systems modeled by an FBFN and the maximum output scaling factor of a self-tuning PI fuzzy controller," *Engineering ApplicationsofArtificial Intelligence* 42 1-15.

[6] C. Elmas and T. Yigit 2007 Genetic Algorithm Based On-line Tuning of a PI Controller for a Switched Reluctance Motor Drive," *Electric Power Components and Systems* 35 675-691.

[7] J. A. Ali, M. A. Hannan, and A. Mohamed 2014 PSO Algorithm for Three Phase Induction Motor Drive with SVPWM Switching and V/f Control," *2014 International conference Power & Energy* 250-254.

[8] E. Rashedi, H. Nezamabadi-pour, and S. Saryazdi, 2009 GSA: A Gravitational Search Algorithm," *Information Sciences* 179 2232-2248.

A new fuzzy self-tuning PD load frequency controller for micro-hydropower system

M Reyasudin Basir Khan[1], Razali Jidin[1] and Jagadeesh Pasupuleti[1]
[1] College of Engineering, Universiti Tenaga Nasional, Jalan IKRAM-UNITEN, 43000 Kajang, Selangor, Malaysia

E-mail: {Reyasudin, Jagadeesh, Razali}@uniten.edu.my

Abstract. This paper presents a new approach for controlling the secondary load bank of a micro-hydropower system using a fuzzy self-tuning proportional-derivative (PD) controller. This technology is designed in order to optimize the micro-hydropower system in a resort island located in the South China Sea. Thus, this technology will be able to mitigate the diesel fuel consumption and cost of electricity supply on the island. The optimal hydropower generation for this system depends on the available stream flow at the potential sites. At low stream flow, both the micro-hydropower system and the currently installed diesel generators are required to feed the load. However, when the hydropower generation exceeds the load demand, the diesel generator is shut down. Meanwhile, the system frequency is controlled by a secondary load bank that absorbs the hydropower which exceeds the consumer demand. The fuzzy rules were designed to automatically tune the PD gains under dynamic frequency variations. Performances of the fuzzy self-tuning PD controller were compared with the conventional PD controller. The result of the controller implementation shows the viability of the proposed new controller in achieving a higher performance and more robust load frequency control than the conventional PD controller.

1. Introduction

Load frequency controller (LFC) is a very important component of an electrical power system's operation and control. The objective of a LFC is to regulate the voltage and frequency of the system to the specified nominal values, in case of fluctuations [1]. A well designed and operated power system must able to handle changes in the load and system disturbances. Furthermore, it should provide a high level of power quality while maintaining both voltage and frequency within acceptable limits.

Many investigations have been reported relating to load frequency controller. There are researchers proposed control strategies for LFC, based on classical linear control theory [2]. However, due to the inherent characteristics of the varying loads, the operating point of a power system varies continuously throughout a daily cycle. Therefore, a fixed controller may no longer be suitable under all operating conditions. Many advanced control methods utilizing soft computing approaches have been reported, such as Artificial neural network (ANN) [3], Genetic algorithms (GA) [4], Particle swarm optimization (PSO) [5], Fuzzy logic [6] and Bacterial foraging optimization algorithm (BFOA) [7]. These controllers have been extensively used in LFC and have shown to provide more robust control compared to classical control techniques. However, these methods require the system's plant model, which is very challenging to obtain for a large electrical power system.

The artificial intelligence controller also tends to be complex and specific to a particular installation, thus making them expensive to construct and difficult to maintain and reconfigure. Hence, these particular types of controllers are not suitable to be implemented in micro-grid systems for remote areas that require low cost and easy-to-maintain controllers.

A new load frequency controller, based on fuzzy gain scheduling of PD controller is proposed in this paper. Fuzzy rules and reasoning has been utilized in order to determine the PD controller parameters. Hence, the controller has the capability to modify its parameters when the system dynamics or the disturbance characteristics vary. Furthermore, this method does not require a precise model of the system dynamics under control. In this system, a fast response secondary load system is required in order to control the frequency of the micro-grid, to allow the diesel generators to be shut down during periods of high stream flows. Therefore, the control strategy must be adaptive and robust.

2. Fuzzy self-tuning PD load frequency controller

In this work, a new fuzzy self-tuning PD controller has been implemented to achieve a higher performance of the discrete frequency regulator than the conventional PD controller. Typically, PD gains are determined by the Ziegler-Nichols method based on the transient response characteristics of a plant. This method may produce initial estimations of the PD gains. However, a fine-tuning of the PD gains is required to obtain the most optimal dynamic frequency control under various load conditions. There is a trade-off between the system performance and robustness with fixed PD gain. Therefore, a systematic real-time PD gain tuning is required in order to increase the performance of a fixed gain PD controller. To achieve the performance targets in dynamic responses, the PD gains are tuned as in equation (1) and equation (2).

$$K_P = K_P^{'} + \int \Delta K_P \tag{1}$$

$$K_D = K_D^{'} + \int \Delta K_D \tag{2}$$

where K'_P and K'_D are the initial estimations of the proportional and derivative gains, whereas ΔK_P and ΔK_D are the adjustments of the PD gains. The block diagram of the new fuzzy self-tuning controller design is shown in figure 1. The fuzzy self-tuning scheme uses error, e and rate of change in error, $\dot{e}$ in making decisions for tuning the PD gains. The fuzzification, fuzzy rule base, max-min inference mechanism and defuzzification for tuning the PD gain adjustments ΔK_P and ΔK_D, are based on the Mamdani model.

2.1. Fuzzification

The list The error and derivative of error are fuzzified from the crisp values to fuzzy variables with seven linguistic variables: Zero (ZE), Positive Small (PS), Positive Medium (PM), Positive Big (PB), Negative Small (NS), Negative Medium (NM) and Negative Big (NB) with each linguistic variable having a fuzzy membership value.

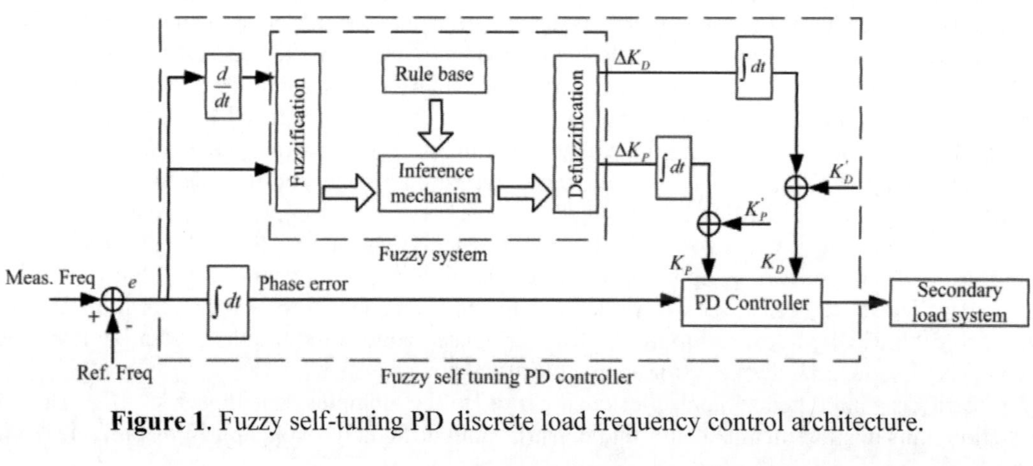

Figure 1. Fuzzy self-tuning PD discrete load frequency control architecture.

2.2. Inference mechanism

The inference mechanism takes the fuzzy input from the fuzzifier in form of membership values matrix, and uses the fuzzy rule base to make decisions on the fuzzy value of the output. The max-min inference mechanism is interpreted as follows:

For ΔK_P,

Rule i: If e is A_e^i and $\dot{e}$ is $A_{\dot{e}}^i$ then ΔK_P is B_P^i ; $i = 1,\ldots,49$.

For ΔK_D,

Rule j: If e is A_e^j and $\dot{e}$ is $A_{\dot{e}}^j$ then ΔK_D is B_D^j ; $j = 1,\ldots,49$.

where i and j are the indexes, A_e and $A_{\dot{e}}$ are the input linguistic variables of the error and the rate of change in error respectively, whereas B_P and B_D are the output linguistic variables of the proportional and derivative gain adjustments, respectively.

2.3. Defuzzification

The center of gravity method has been used to convert the fuzzy output of PD gain adjustments into a crisp value.

3. Performance analysis

The performance of the proposed controller was verified by considering two different scenarios: an additional 50 kW load at t=1 second, and a 50 kW load drop at t=1 second. The system's frequency output for both the new controller design and the conventional PD controller were compared and shown in figure 2.

To demonstrate the performance and robustness of the new proposed controller, performance indices such as overshoot, undershoot, settling time (at 0.05 %), steady state error (SSE), integral of the absolute error (IAE), integral of the squared error (ISE) and integral of time multiplied by the absolute error (ITAE) were considered in the analysis. The performance indices of both fuzzy self-tuning PD controller and the conventional PD controller are shown in table 1.

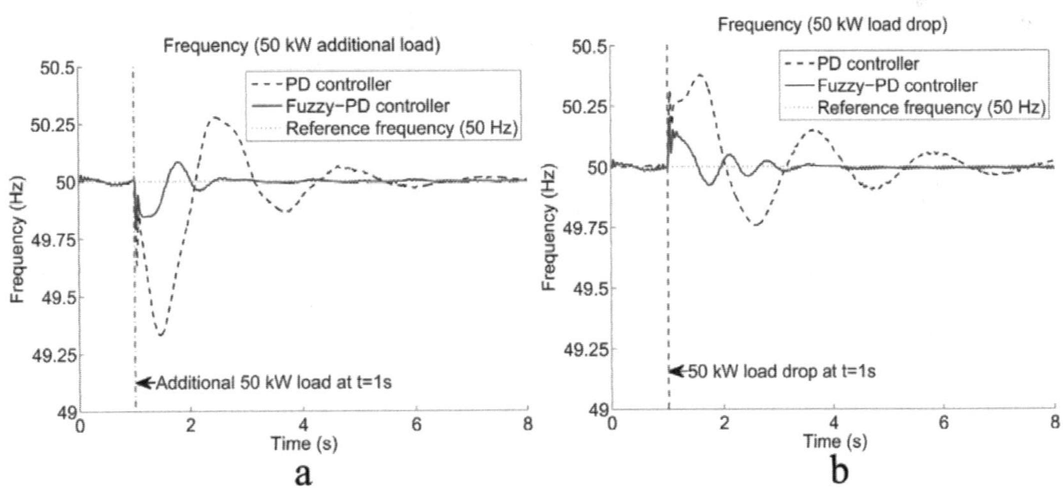

Figure 2. Frequency comparison between conventional PD and fuzzy-PD controller for a) 50 kW additional load b) 50 kW load drop.

Table 1. Performance indicators of the load frequency controllers.

Performance indicators	Conventional PD controller		Fuzzy self-tuning PD controller	
	50 kW additional load	50 kW load drop	50 kW additional load	50 kW load drop
Overshoot (%)	0.554	0.756	0.166	0.404
Undershoot (%)	1.333	0.484	0.409	0.150
Settling time (s)	5.132	6.316	1.299	1.545
SSE (Hz)	0.004	0.023	0.003	0.010
IAE	2.430	2.404	0.981	0.977
ISE	0.641	0.630	0.185	0.183
ITAE	9.583	9.487	3.811	3.815

4. Conclusion

A new fuzzy self-tuning PD control architecture was proposed in this paper in order to achieve the dynamic performance targets. The fuzzy rules were generated based on expert knowledge to efficiently tune the PD controller gains. The results of the new controller implementation on the system were compared with the PD controller without fuzzy self-tuning mechanism. It is observed that the fuzzy self-tuning mechanism considerably decreases the overshoot, settling time, SSE, IAE, ISE and ITAE in the frequency control.

Acknowledgements

The author would like to thank Ministry of education (MOE), Malaysia for funding this research under FRGS research grant (20150214FRGS). The author would also like to thank MOE for funding the author (M. Reyasudin Basir Khan) doctoral studies through MyBrain15 (MyPhD) program.

References

[1] Pandey S K, Mohanty S R and Kishor N 2013 A literature survey on load–frequency control for conventional and distribution generation power systems *Renewable and Sustainable Energy Reviews*. **25** 318-34

[2] Bose A and Atiyyah I 1980 Regulation error in load frequency control *IEEE Transactions on Power Apparatus and Systems*. **2** 650-7

[3] Kassem A M 2010 Neural predictive controller of a two-area load frequency control for interconnected power system *Ain Shams Engineering Journal*. **1** 49-58

[4] Toulabi M R, Shiroei M and Ranjbar A M 2014 Robust analysis and design of power system load frequency control using the Kharitonov's theorem *International Journal of Electrical Power & Energy Systems*. **55** 51-8

[5] Shayeghi H, Jalili A and Shayanfar H A 2008 Multi-stage fuzzy load frequency control using PSO *Energy Conversion and Management*. **49** 2570-80

[6] Yousef H 2015 Adaptive fuzzy logic load frequency control of multi-area power system *International Journal of Electrical Power & Energy Systems*. **68** 384-95

[7] Ali E S and Abd-Elazim S M 2011 Bacteria foraging optimization algorithm based load frequency controller for interconnected power system *International Journal of Electrical Power & Energy Systems*. **33** 633-8

A micro-hydropower system model with PD load frequency controller for Resort Islands in the South China Sea

M Reyasudin Basir Khan[1], **Jagadeesh Pasupuleti**[1] **and Razali Jidin**[1]
[1] College of Engineering, Universiti Tenaga Nasional, Jalan IKRAM-UNITEN, 43000 Kajang, Selangor, Malaysia

E-mail: {Reyasudin,Jagadeesh,Razali}@uniten.edu.my

Abstract. A model of high-penetration micro-hydropower system with no storage is presented in this paper. This technology is designed in order to reduce the diesel fuel consumption and cost of electricity supply in a resort island located in the South China Sea. The optimal hydropower generation for this system depends on the available stream flow at the potential sites. At low stream flow, both the micro-hydropower system and the currently installed diesel generators are required to feed the load. However, when the hydropower generation exceeds the load demand, the diesel generator is shut down. Meanwhile, the system frequency is controlled by a secondary load bank that absorbs the hydropower which exceeds the consumer demand. This paper also presents a discrete frequency control system using proportional-derivative (PD) controller. The controller is employed in order to manipulate the system frequency by controlling the secondary load system. The simulation results indicate that a variety of load conditions can be satisfactorily controlled by the PD controller. Hence, this particular type of controller is suitable to be implemented in micro-grid systems for remote areas that require low cost and easy-to- maintain controllers.

1. Introduction

The resort island selected for this study is Tioman, as it represents the typical energy requirements of many resort islands in the South China Sea. It is situated at 2°47'47"N latitude and 104°10'24"E longitude. The island's geographical conditions, such as its hilly landscape and abundant stream flows from highland areas permits the installation of run-of-river hydropower system [1]. The renewable energy (RE) assessment for Tioman Island had been previously assessed by several researchers. Ashourian et al. proposed an optimal combination of solar energy and wind energy for the Juara village of Tioman Island [2]. Meanwhile, Chik et al. performed sustainability indicators to determine sustainability degree for solar, wind and hydro resources on the island [3]. Both of the studies performed only RE assessments and did not consider the system level modeling of an RE system for the island based on actual assessed RE resources. This paper will provide a detailed system level model of a run-of-river hydropower that has the potential to be installed on the island, based on the available hydropower resources of the island.

In previous work [1], energy audit and the surveys of available hydro resource potentials in Tioman Island were conducted. The results show that, based on annual average hydropower estimation, a total of 10 potential sites have been identified to have run-of-river hydropower potential from 26 investigated sites on Tioman Island. Therefore, it is suitable to develop a run-of-river hydropower system in order to mitigate the diesel fuel consumption on the island.

Secondary load controller (SLC) is required in the micro-grid systems to dissipate the excess hydropower and maintain the system's stability. The controller maintains a constant generator output by providing a secondary load. There are several advantages of SLC [4], such as the incorporation of a cheaper and simpler turbine with less moving parts in the run-of-river hydropower system architecture. The SLC also provides high reliability, low maintenance, and a simple operating system that can be installed anyplace in the electrical system. Therefore, SLC is the prominent solution for regulating the induction generator output voltage and frequency for run-of-river hydropower system supplying rural load with cost-effectiveness as the primary factor.

2. System level run-of-river hydropower model

The run-of-river hydropower Simulink SimPowerSystems model is shown in figure 1. The hydro turbine model uses general hydropower equation as in equation 1 to compute the turbine mechanical power (P_m).

$$P_m = Q \times h \times g \times e_f \tag{1}$$

where P_m is theoretical power output from turbine (kW), h is gross head height (m), g is gravitational constant (9.81 m/s^2), and e_f is the efficiency factor (assumed as 0.7 [1]). The turbine mechanical power is then converted to turbine torque output (T_m) as an input to the asynchronous generator.

The grid voltage is controlled at its nominal value by the synchronous condenser and its excitation system. The secondary load block is designed to absorb the excess power generated by the hydropower system. The secondary load system architecture comprised of eight sets of three-phase resistors connected in series with Gate turn-off thyristor (GTO) switches. The nominal power of each set follows a binary progression in order for the load to be varied from 0 to 446.25 kW by steps of 1.75 kW. Meanwhile, the system's frequency is controlled by the discrete frequency regulator block.

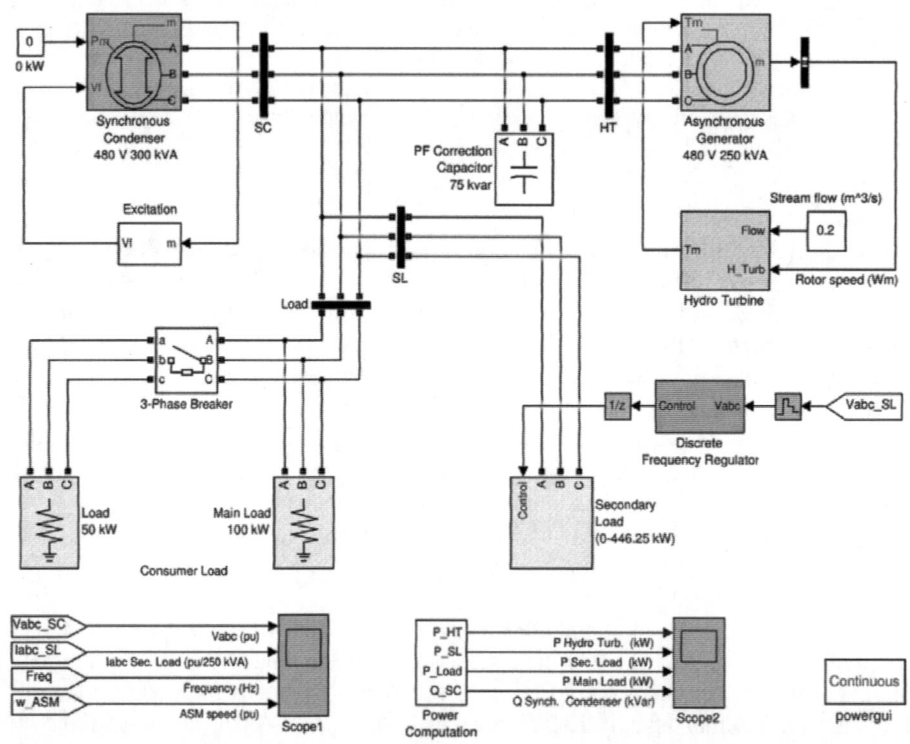

Figure 1. Simulink model of run-of-river hydropower system with load frequency control.

3. System level run-of-river hydropower model

The block diagram of the system control design is presented in figure 2. The controller is employed in order to manipulate the system frequency by controlling the secondary load system. Three Phase locked loop (PLL) systems are used to measure the system frequency. Then, the system frequency is compared to the reference frequency (50 Hz) in order to obtain the frequency error. This error will be integrated in order to get the phase error, which will be used by the controller to produce an output signal representing the required secondary load power. This signal is converted to an 8-bit digital signal for controlling the switching of the three phase secondary loads. All switching is performed at zero crossing voltage in order to minimize the voltage disturbances.

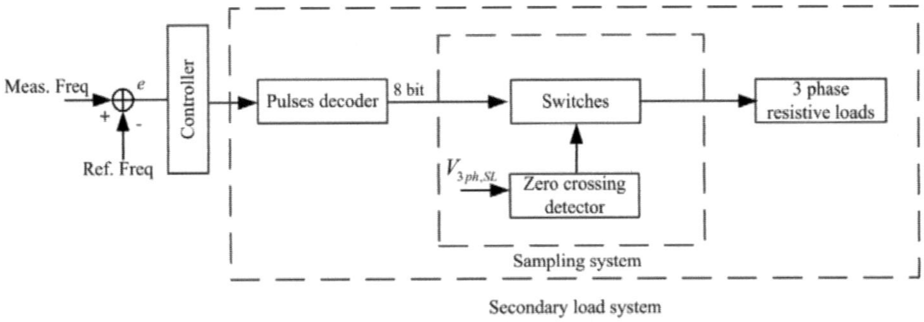

Figure 2. Secondary load system block diagram.

3.1. PD controller

The PD controller has an error signal e(t) as its input, which is obtained from desired output r(t) and actual output y(t), as seen in equation (2). The error signal used by the controller in order to produce a control signal u(t) as its output is seen in equation (3).

$$e(t) = r(t) - y(t) \tag{2}$$

$$u(t) = K_P e + K_D \frac{de}{dt} \tag{3}$$

4. Simulation results and discussions

The model simulated under two different scenarios: an additional 50 kW load at t=1 second, and a 50 kW load drop at t=1 second. The system's frequency output for the PD controller is shown in figure 3.

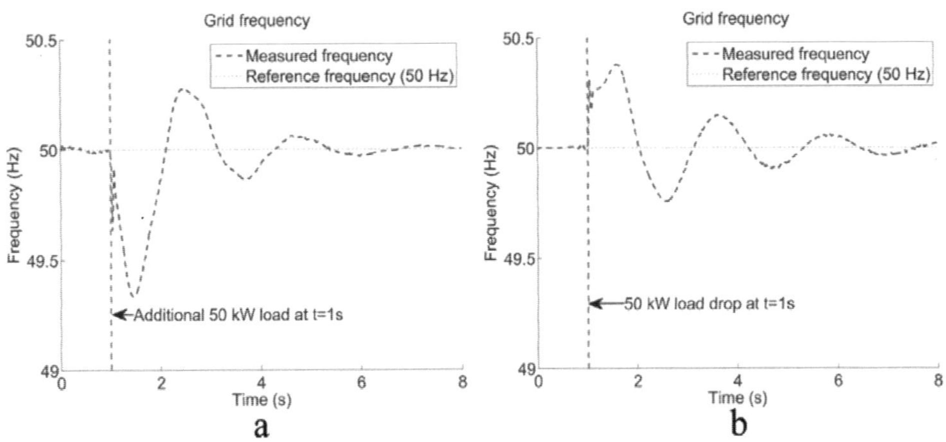

Figure 3. Simulation results of grid frequency for a) 50 kW additional load b) 50 kW load drop.

To demonstrate the performance and robustness of the proposed controller, performance indices such as overshoot, undershoot, settling time (at 0.05 %), steady state error (SSE) integral of the absolute error (IAE), integral of the squared error (ISE) and integral of time multiplied by the absolute error (ITAE) were considered in the analysis. The performance indices of the PD controller are shown in table 1.

Table 1. Performance indicators of the PD load frequency controller.

Performance indicators	Load conditions	
	50 kW additional load	50 kW load drop
Overshoot (%)	0.554	0.756
Undershoot (%)	1.333	0.484
Settling time (s)	5.132	6.316
SSE (Hz)	0.004	0.023
IAE	2.430	2.404
ISE	0.641	0.630
ITAE	9.583	9.487

5. Conclusion

In this paper, a system level of a run-of-river hydropower system model with PD secondary load controller was presented to describe the system's behaviour under different load conditions. The system frequency was regulated by a secondary load bank that absorbs the hydropower, which exceeds the consumer's demand. The simulation results indicate that a variety of load conditions can be satisfactorily controlled by the PD controller since it exhibits small overshoot, settling time, SSE, IAE, ISE and ITAE in the frequency control.

Acknowledgements

The author would like to thank Ministry of education (MOE), Malaysia for funding this research under FRGS research grant (20150214FRGS). The author would also like to thank MOE for funding the author (M. Reyasudin Basir Khan) doctoral studies through MyBrain15 (MyPhD) program.

References

[1] Basir Khan M R, Jidin R, Pasupuleti J and Shaaya S A 2015 Optimal combination of solar, wind, micro-hydro and diesel systems based on actual seasonal load profiles for a resort island in the South China Sea *Energy*. **82** 80-97

[2] Ashourian M H, Cherati S M, Zin A A M, Niknam N, Mokhtar A S and Anwari M 2013 Optimal green energy management for island resorts in Malaysia *Renewable Energy*. **51** 36-45.

[3] Chik M N, Othman A B, Rashid M A, Syed Abdul Hamid S A F and Azit A H 2010 In Search of a Sustainable Energy for Pulau Tioman. *Asia Pacific Regional Conference and Exhibition on Energy Efficiency*

[4] Kusakana K 2014 A survey of innovative technologies increasing the viability of micro-hydropower as a cost effective rural electrification option in South Africa *Renewable and Sustainable Energy Reviews*. **37** 370-9

IOP Conf. Series: Earth and Environmental Science **32** (2016) 012004 doi:10.1088/1755-1315/32/1/012004

Influence of Continuous Flow Microwave Pre-Treatment on Anaerobic Digestion of Secondary Thickened Sludge for Sustainable Energy Recovery in Sewage Treatment Plant

D Hephzibah[1]**, P Kumaran**[2]**, N M Saifuddin**[2]

[1] Postgraduate Student, Centre for Renewable Energy, Universiti Tenaga Nasional, 43000, Kajang, Selangor

[2] Principal Researcher, Centre for Renewable Energy, Universiti Tenaga Nasional, 43000 Kajang, Selangor

E-mail: hephzibahdavid89@gmail.com

Abstract. This work elucidates the effects of pre-treatment of secondary thickened sludge (STS) for enhancement of biogas production that has great potential to generate energy for the utilization of the sewage treatment plant (STP) itself. Microwave pre-treatment has been adopted for this study. Experiment works have been designed and conducted to examine the effectiveness of continuous flow microwave pre-treatment on the solubility of STS, digestibility of STS and biogas production at a power level of 80 W for 5, 10 and 15 minutes. A few characteristics of the sewage sludge were monitored daily to identify the effect of pre-treatment on the sludge. The soluble chemical oxygen demand (SCOD)/total chemical oxygen demand (TCOD) ratio increased by 0.1, 1.0 and 1.8%, while the volatile fatty acids (VFA) concentration of the pre-treated sludge improved by 4.4, 5.1, 5.9% at the irradiation time of 5, 10 and 15 minutes, respectively at a microwave power level of 80 W. Besides that, the digestate also indicates that the pre-treated sludge undergoes efficient VS removal and TCOD removal after anaerobic digestion compared to the untreated sludge. Moreover, the biogas quantity increased by an average of 19.2, 24.1 and 32.2% in 5, 10 and 15 minutes irradiation time respectively compared to the untreated sludge. The additional quantity of biogas generated has shown a great potential for sustainable energy generation that can be utilized internally by the STP.

1 Introduction

The ambitious goal of Malaysia to introduce renewable energy to the national electricity supply mix has motivated the adoption of renewable resources in various sectors such as the wastewater treatment sector.

Currently, anaerobic digesters have been installed in few modern mechanized STPs in Malaysia to generate biogas from the anaerobic treatment of sewage sludge. Presently, in a typical modern STP that deploys anaerobic digesters to treat 250,000 population equivalent (PE) sewage solids the average biogas production rate is about 1,150 m^3 daily whereby the electricity plausible to be generated after

This research is supported financially in part by the Malaysian Electricity Supply Industries Trust Account (MESITA), better known as Akaun Amanah Industri Bekalan Elektrik (AAIBE).

gas engine conversion is approximately 2,900 kWh/day. However, the average plant electricity consumption is 43,503 kWh/day [1]. Thus, the recovery of energy from biogas will replace 6.6% of the electricity consumption during the treatment of the wastewater itself. However, it is not economically feasible for commercial implementation due to the quantity of biogas produced presently. Hence, the methane gas produced is flared in order to mitigate greenhouse gas emission. Therefore, the valuable electricity energy potential of the methane remains untapped.

In order to enhance the biogas production, the limiting factors of the anaerobic process which includes the hydrolysis stage needs to be overcome. The substrates can be very slow to be degraded due to the molecular structure of the substrates such as high crystalline structure or low surface area, which are poorly accessible to the microorganism and their enzymes [2]. Recent publications have suggested that pre-treatment of substrates prior to anaerobic digestion can be used to overcome this problem [2,3,4]. Among the types of pre-treatment is mechanical pre-treatment, thermal pre-treatment, ultrasound, pre-treatment, microwave pre-treatment acid pre-treatment and others. In this study, microwave pre-treatment will be explored.

Previously, a study on batch microwave pre-treatment has been conducted and the quantity of biogas produced with and without pre-treatment has been evaluated and compared whereby the microwave pre-treated sludge for 5 minutes and 15 minutes enhances the anaerobic digestibility rate and biogas production by 11.9% and 38.5% respectively compared to untreated sewage sludge [5]. Although, batch microwave pre-treatment shows optimistic effect on the biogas production, it is not feasible to be implemented in the STP. Therefore, a feasible microwave pre-treatment system that can be put into operation in the STP need to be deliberated. Literatures have suggested that STS from STP is the most widely studied substrates on pre-treatment application due to low biodegradability [4]. The aim of this study is to investigate the influence of continuous flow microwave pre-treatment of STS for sustainable energy recovery in STP.

2 Materials and Methods

Two Sartorius Stedim BIOSTAT Aplus bioreactors were filled with 4.3 L of digested sludge from Pantai Dalam STP as inoculum and purged with nitrogen gas to ensure that the digesters are in anaerobic condition. The digesters have been allowed to stabilise for 5 days. After this period the experiment was started for 30 days under mesophilic condition at 35°C and stirred continuously at 200 rpm. The primary thickened sludge (PTS) and STS samples collected from Pantai Dalam STP were standardized to 4% concentration prior being fed into the digester. A domestic SHARP R-298H (S) microwave oven with maximum power output of 800 W was modified to accommodate the continuous flow heat exchanger, was used in this study. A 150 ml of digestate is drawn out from the digesters daily at a fixed time and fed with 150 ml of PTS for the first 15 days until the quantity of biogas generated start to be in steady state. Subsequently, from day 16 onwards STS was fed into the digesters daily by gradually increasing the volume by 10% in which one digester was fed with untreated STS while the other received pre-treated STS. The untreated, pre-treated and digested sludge was characterised daily for total solid (TS), volatile solid (VS), total suspended solid (TSS), volatile suspended solid (VSS), TCOD, SCOD, VFA and alkalinity. Meanwhile, the quantity and quality of biogas generated was measured by using a graduated eudiometer and COMBIMASS GA-m portable chemical sensor gas analyser.

3 Results and Discussion

3.1 Sludge Solubilization

The SCOD/TCOD ratio was used to evaluate the efficiency of sludge solubilization of STS. The SCOD/TCOD ratio for the untreated STS shows no significant differences, leading to conclusion that no substantial fluctuations in the STS characteristics although the samples were collected at different time. Thus, any observed enhancement of anaerobic digestion will be due to the alteration in the structure of the STS and not due to the difference in the amount of organics fed into the digester. The microwave pre-treated STS significantly change the averaged SCOD/TCOD ratio as depicted in Table

1. The results show that the SCOD/TCOD ratio increased by 0.1, 1.0 and 1.8% at the irradiation time of 5, 10 and 15 minutes, respectively, at a microwave power of 80 W compared to the untreated. The increase in SCOD/TCOD ratio is the result of the disruption of the complex STS floc structure and the release of the extracellular and intracellular materials from the cells into aqueous phase [6,7]. However, the extent of sludge floc disruption by microwave pre-treatment is influenced by the irradiation time as presented in Table 1. At a shorter irradiation time, the structural change was minimal enough, while when the contact time increases, the floc structure were disrupted completely [8].

Table 1: SCOD/TCOD ratio of untreated and pre-treated STS

SCOD/ TCOD Ratio (%)	80 W @ 5 min	80 W @ 10 min	80 W @ 15 min
Untreated STS	0.6827	0.7129	0.6944
Pre-treated STS	0.6836	0.7201	0.7069

Table 2: VFA concentration of untreated and pre-treated STS

VFA Concentration (mg/L)	80 W @ 5 min	80 W @ 10 min	80 W @ 15 min
Untreated STS	142.9402	142.6435	142.7030
Pre-treated STS	149.2104	149.9053	151.1545

Besides that, total VFA also represents the essential organics in anaerobic digestion. The averaged total VFA that was analysed is included in Table 2 for the untreated and microwave pre-treated STS. The concentration of the measured total VFA increased significantly due to the microwave pre-treatment [6]. The total VFA concentration increased by 4.4, 5.1, 5.9% at the irradiation time of 5, 10 and 15 minutes, respectively, at a microwave power of 80 W. The VFA concentration is also affected by the irradiation time of the microwave pre-treatment [8].

3.2 Sludge Digestibility

Figure 1 represents the removal efficiency of VS in digesters fed with untreated and microwave pre-treated STS at different pre-treatment condition which is affected by microwave pre-treatment. The VS removal efficiencies of the untreated STS for each condition were averaged for comparison purpose. The VS removal efficiency increased to the range of 41.2% and 43.4% gradually in the first 15 days of the digestion time due to the effect of PTS sample which is high in strength compared to STS. From day 16 onwards, it can be observed that the VS removal efficiency decreases steadily as the volume of STS fed into the digester increases. However, the VS removal efficiency of the microwave pre-treated STS from day 16 onwards was higher compared to the untreated STS. The VS removal efficiency also increases as the irradiation time increases [9]. The average VS removal efficiency, when STS was fed 100% into the digester from day 25 onwards, at the irradiation time of 5, 10 and 15 minutes are 14.9, 17.1, and 18.9%, respectively.

Meanwhile, Figure 2 illustrates the TCOD removal efficiency of the untreated and microwave pre-treated STS at different pre-treatment condition. The pattern of TCOD removal efficiency is almost similar to the VS removal efficiency. For the first 15 days, the TCOD removal efficiency increases due to the PTS sample being fed into the digester. However, from day 16 onwards, the TCOD removal efficiency decreases slowly. On the other hand, the pre-treated STS shows greater TCOD removal compared to the untreated STS. Meanwhile, the TCOD removal efficiency is also affected by the pre-treatment condition. As the irradiation time increases, the TCOD removal efficiency increases too [8]. The average TCOD removal efficiency for 5, 10 and 15 minutes irradiation time at the last 5 days when 100% STS was fed into the digester is 7.2, 7.9 and 8.1%, respectively. The improved VS and TCOD removal compared to the untreated strongly indicates that solubilzation by microwave pre-treatment converted a portion of organic materials which are difficult to be biodegraded into materials which are more easily biodegradable [9].

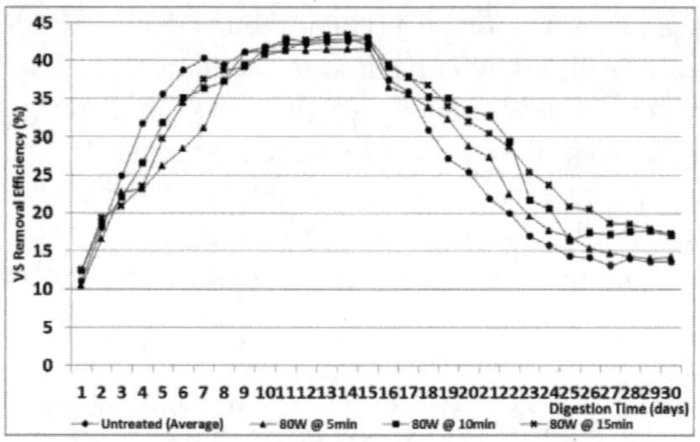

Figure 1: VS removal efficiency of digesters fed with untreated and microwave pre-treated STS at different pre-treatment condition

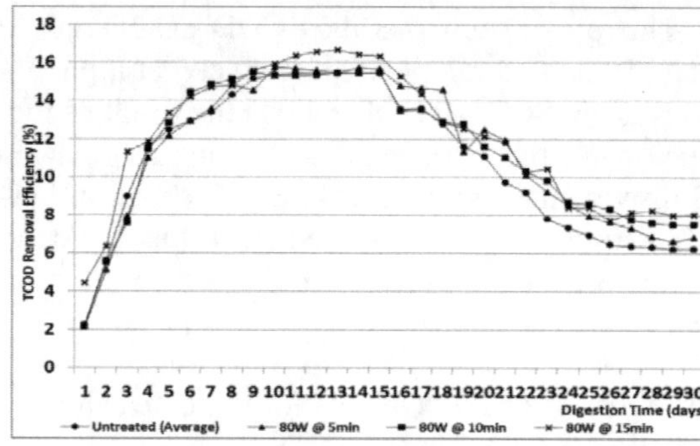

Figure 2: TCOD removal efficiency of digesters fed with untreated and microwave pre-treated STS at different pre-treatment condition

3.3 Biogas Production

The total biogas production in the digesters is as illustrated in Figure 3. It was seen that the biogas production in the microwave pre-treated digester was significantly higher compared to the untreated digester. The biogas quantity increased by an average of 97, 112 and 172 ml when the digesters were fed with 100% STS for 5, 10 and 15 minutes irradiation time respectively compared to the untreated.

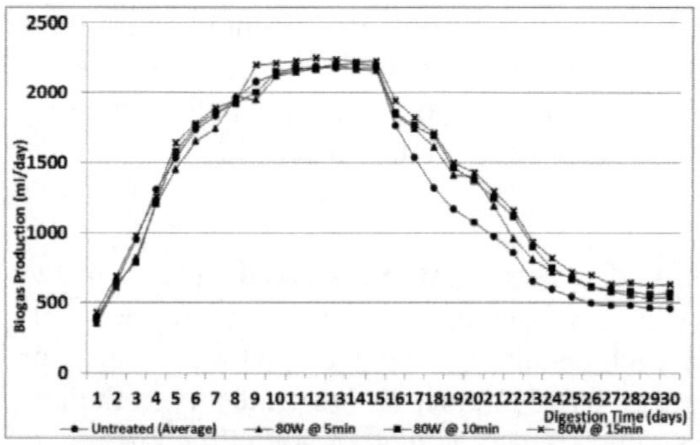

Figure 3: Daily biogas production of the untreated and microwave pre-treated STS at different pre-treatment condition

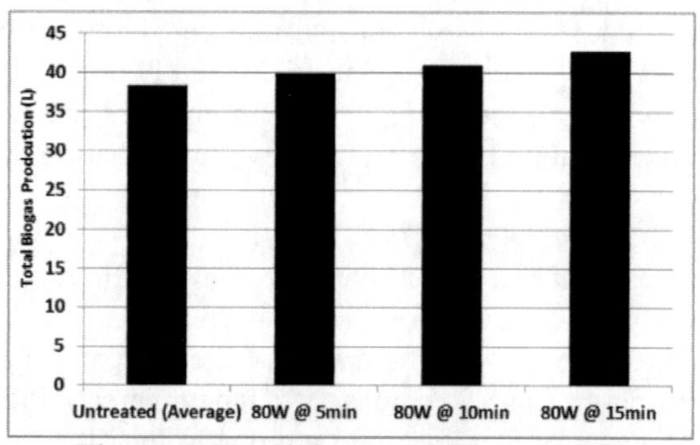

Figure 4: Total biogas production of the untreated and microwave pre-treated STS at different pre-treatment condition

Besides that, the total biogas production of the untreated and pre-treated STS at different pre-treatment condition is illustrated in Figure 4. The averaged total biogas generated by the untreated sample is approximately 38.3 L. Meanwhile, for the pre-treated samples, the total biogas generated is around 39.9, 41.0 and 42.6 L for 5, 10 and 15 minutes irradiation time respectively. The results show an averaged increment in total biogas production by 19.2, 24.1, and 32.2% for 5, 10 and 15 minutes irradiation time respectively compared to the untreated.

Besides that, the methane content of the produced biogas was in the range of 65-69%, which is suitable for electricity generation by using a gas engine. The additional quantity of biogas which is highly potential to be generated after microwave pre-treatment will be able to increase the potential of energy recovery from the biogas. Therefore, the biogas production pattern suggests that the pre-treatment increased the amount of material available for biodegradation as discussed earlier [6,7]. Moreover, it can be observed that as the irradiation time increases, the enhancement of biogas production is greater [8,9].

4 Conclusions

In this study the effect of continuous flow microwave pre-treatment was investigated on the STS sludge solubilization, digestibility and biogas production after anaerobic digestion. The results show that microwave pre-treatment is an efficient and rapid technique to disintegrate the sludge to enhance the digestibility of the sludge which leads to increase in biogas generation. The biogas production from STP is considered as one of the most promising methods to generate renewable energy from the wastewater which can sustain the electricity consumption of the STP after converting the gas to electricity. However, further research is recommended to identify the optimum microwave power and irradiation time for the pre-treatment and to perform cost-benefit analysis.

Acknowledgments
The authors would like to acknowledge the contributions of Indah Water Konsortium (IWK) management for providing us their support in completing this research. We would like to convey our sincere gratitude to all the research assistants and laboratory staffs of the Department of Mechanical Engineering in Universiti Tenaga Nasional (UNITEN), Department of Civil Engineering in UNITEN and the management of UNITEN for their support and direct involvement in this study.

References

[1] Yew, L. (2012) *Energy and Resource Baseline Analysis of a Sewage Treatment Plant*. Undergraduate. Universiti Tenaga Nasional.
[2] Montgomery, L. and Bochmann, G. (2014). *Pretreatment of Feedstock for Enhanced Biogas Procution*. Energy from Biogas. IEA Bioenergy.
[3] Appels, L., Baeyens, J., Degrève, J. and Dewil, R. (2008). Principles and Potential of the Anaerobic Digestion of Waste-activated Sludge. *Progress in Energy and Combustion Science*, 34 (6), pp. 755-781.
[4] Carlsson, M., Lagerkvist, A. and Morgan-Sagastume, F. (2012). The Effects of Substrate Pre-treatment on Anaerobic Digestion Systems: A Review. *Waste Management*, 32(9), pp.1634-1650.
[5] Hephzibah, D., Kumaran, P. and Saifuddin, N. (2014). Pre-Treatment of Sewage Sludge to Enhance Biogas Production to Generate Green Energy for Reduction of Carbon Footprint in Sewage Treatment Plant (STP). In: *Green Energy for Sustainable Development (ICUE), 2014 International Conference and Utility Exhibition*. [online] Pattaya City: IEEE, pp.1-5. Available at: http://ieeexplore.ieee.org/stamp/stamp.jsp?tp=&arnumber=6829007&isnumber=6828886 [Accessed 15 Jun. 2015].
[6] Tyagi, V. and Lo, S. (2013). Microwave irradiation: A sustainable way for sludge treatment and resource recovery. *Renewable and Sustainable Energy Reviews*, 18, pp.288-305.
[7] Appels, L., Houtmeyers, S., Degrève, J., Van Impe, J. and Dewil, R. (2013). Influence of microwave pre-treatment on sludge solubilization and pilot scale semi-continuous anaerobic digestion.*Bioresource Technology*, 128, pp.598-603.
[8] Yu, Q., Lei, H., Yu, G., Feng, X., Li, Z. and Wu, Z. (2009). Influence of microwave irradiation on sludge dewaterability. *Chemical Engineering Journal*, 155(1-2), pp.88-93.
[9] Park, W. and Ahn, J. (2011). Effects of microwave pretreatment on mesophilic anaerobic digestion for mixture of primary and secondary sludges compared with thermal pretretament. *Environmental Engineering Research*, 16(2), pp.103-109.

IOP Conf. Series: Earth and Environmental Science **32** (2016) 012005 doi:10.1088/1755-1315/32/1/012005

Application of Moving Bed Biofilm Reactor (MBBR) and Integrated Fixed Activated Sludge (IFAS) for Biological River Water Purification System: A Short Review

M S Lariyah [1], H A Mohiyaden [1], G Hayder[1], A Hussein[1], H Basri[1], A F Sabri[2], MN Noh[2]
[1]Centre for Sustainable Technology and Environment, Universiti Tenaga Nasional, Jalan IKRAM-Uniten, 43000 Selangor, Malaysia.
[2]River Basin Division, Department of Irrigation and Drainage, Jalan Sultan Salahuddin, 50626, Kuala Lumpur

E-mail: hairun@uniten.edy.my; lariyah@uniten.edu.my

Abstract. This review paper present the MBBR and IFAS technology for urban river water purification including both conventional methods and new emerging technologies. The aim of this paper is to present the MBBR and IFAS technology as an alternative and successful method for treating different kinds of effluents under different condition. There are still current treatment technologies being researched and the outcomes maybe available in a while. The review also includes many relevant researches carried out at the laboratory and pilot scales. This review covers the important processes on MBBR and IFAS basic treatment process, affecting of carrier type and influent types. However, the research concluded so far are compiled herein and reported for the first time to acquire a better perspective and insight on the subject with a view of meeting the news approach. The research concluded so far are compiled herein and reported for the first time to acquire a better perspective and insight on the subject with a view of meeting the news approach. To this end, the most feasible technology could be the combination of advanced biological process (bioreactor systems) including MBBR and IFAS system.

1. Introduction

Several methods can be used to purify the polluted river. Through implementation of proper analysis and environmental control, polluted river can be treated biologically. Among the selected methods of biological processes are using Moving Bed Biofilm Reactor (MBBR) and Integrated Fixed-Film Activated Sludge (IFAS). This is considered as attached growth secondary biological treatment in the water treatment process. Polluted river can be treated biologically, provided with proper analysis and environmental control. However, it is essential to understand the characteristic of each biological process to ensure the proper environment is produced and controlled effectively.[1]

MBBR process uses biological carrier as a media with a density close to water so that it can be kept in suspension with minimum mixing energy provided by aeration or mechanical mixing [2]. Biological carrier are manufactured in various shapes and are sufficiently large in specific surface area so that suspended support media can be retained in the reactor by screen or wire wedges.

This type of treatment process will be adapted for river purification as it increases biomass or biofilm concentration and treatment ability [3]. According to Briones & Raskin (2003), biofilm is defined as a microbial engineered community system which may comprise any type of microorganism including algae, fungi, bacteria, archaea and protozoa/ metazoan [4]. These multiverse microorganisms require the presence of a surface to adhere to and grow on, and biological carrier in MBBR and IFAS system will support as a media for them to grow.

Biofilm is one of the microbial attached growth processes in water/ wastewater treatment process. Meanwhile, another concept is suspended culture process. The most common suspended growth process used for municipal wastewater treatment is the conventional activated sludge (CAS) process. . Overall performance proves that MBBR has better rate of constituent removal efficiency compared to CAS in the laboratory basis study[5].

2. Moving Bed Biofilm Reactor (MBBR)

In the past, conventional activated sludge is widely used as biological treatment due to cost-friendly factor. Nowadays, a lot of modifications and upgrading works have been made to meet strict rules and regulations requirement for discharging treated wastewater into the natural water bodies Compact wastewater treatment plants that produce an effluent of high standard in the presence of smaller footprint and minimize waste is increasingly become worldwide concern particularly in the densely populated areas where limited space is available for the treatment plants.

Biological processes particularly MBBR and IFAS is one of the biological treatment processes in wastewater treatment which offer compact treatment plant design to overcome the drawbacks of CAS process and produce higher quality effluent even in smaller foot print.

2.1. MBBR Configurations

MBBR was developed by Kaldnes Miljøteknologi in cooperation with SINTEF, a Norwegian research organization and this system was firstly installed in 1990 [1]. In the MBBR process, small plastic carrier elements support the growth of biofilm in a continuously mixed reactor [6].

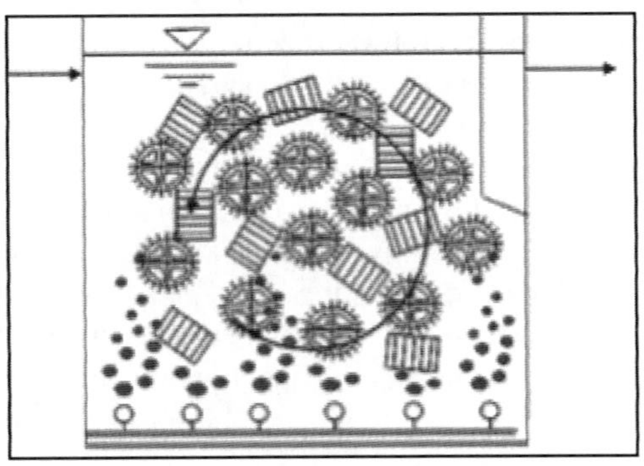

Figure 1. Aerobic MBBR reactor system[1].

MBBR systems are mainly based on the aeration rate and reactors filled with the specially designed biological carriers to provide a surface to colonize by bacteria. When the suspended porous biofilm carriers are continuously mixed in operated aeration tank as illustrated in Figure 1, active biomass will grows into biofilm on the surface of these carriers which having a density slightly less than the water [7]. MBBR system is the efficient method to retain slow growing microorganisms such as nitrifiers in the form of biofilm.

Field study had been conducted by Zimmerman (2005) prove that MBBR process is very flexible and can be retrofit into almost any size or shape of tank [8]. In general, the reactors are straight forward to install and maintain, requiring only a tank of adequate size and a bank of reactors

[9]. Ideally, the reactor is totally mixed with no dead space. The MBBR can be used to upgrade overloaded activated sludge, trickling filter, or other processes. A MBBR operates continuously and it is not affected by problem of clogging that may require backwashing or maintenance works. As compared to the fixed film system, the moving bed biofilm systems have much higher specific surface area for the biofilm [10].

2.2. Current Development of MBBR

The Wang (2006) carried out a laboratory scale test using biofilm carriers and a filling ratio of 50%. The coagulant used was Fe (II) Sulphate Heptahydrate solution. The performance of MBBR was studied at the DO level of 2 mg/L and HRT of 6 hours. The phosphorous removal efficiency was found to be greater than 75%. The highest removal efficiency was observed to be 92%. In the study by Wang, (2006), it was shown from the results of the average concentration of NH^{4+}, NO^{3-} and NO^{2-} and their removal efficiencies that the efficiency of nitrification could reach to more than 90% when the DO was kept greater than 2 mg/L. It was further stated that the average TN removal efficiency of 89.9% from domestic wastewater could be attained at DO of 2 mg/l. It has been considered that DO diffusion through the biofilm rates the determining step for media nitrification.

Chu & Wang (2011) investigated the performance of MBBR for the removal of organics and Nitrogen from wastewater with a low Carbon Nitrogen ratio using the two different materials as carrier for their research, namely PUF and biodegradable polymer PCL particles. The study demonstrated that the MBBR with PUF had good results in the TOC and ammonium removal, 90% and 65%, compared with 72% and 56% for reactor filled with PCL carriers at an average HRT of 14 h. The MBBR with biodegradable PCL carrier showed good performance in terms of TN removal.

Ødegaard (2000) analyzed the influence of carrier size and shape on the performance of moving bed biofilm process related to highly loaded working plants. It was concluded that the organic surface area loading rate (g $COD/m^2.d$) is the main component for the removal of organic matter using moving bed biofilm reactor. They also concluded that shape and size of the carrier do not seem to be significant as long as the effective surface area is the same.

Kermani (2008) evaluated MBBR filled with FLOCOR – RMP® in terms of organics and nutrient removal efficiency from synthetic wastewater which showed that the MBBR could be used as an ultimate and efficient option for the total nutrient removal from municipal wastewater. The experiment shows that the system is a very effective process for almost complete organic and nutrient removal, with average soluble COD, TN removal efficiencies of 96.9, 84.6% respectively during optimum operating conditions.

According to a study carried out by Shrestha (2013) on MBBR with polyethylene carriers, the DOC removal efficiency was found to be above 92% at all filling rates. The average COD removal efficiency was at 10, 20, 30 and 40% filling rates were 75.7, 91.1, 85.5 and 79.6% respectively. These results also showed that the MBBR system achieved higher DOC and COD removal efficiency at 20% PE carrier filling rate under the same condition of influent organic loading rate.

Rodgers (2003) monitored and evaluated one novel pilot plant, a new vertically MBBR system for treating municipal wastewater for eleven (11) months. The biological carrier applied was high surface area plastic media. On-site monitoring showed that DO was in the range of 1.5 to 5 mg/L. Removal rate of filtered COD was up to 35 g COD/(m^3 day) and the bulk fluid volumetric filtered COD removal rate was 2.62 kg COD/(m^3 day). No clogging was found in the biomedia. The power consumption was in the range of 0.09–0.25 kWh/m^3 wastewater flow, 0.40–2.19 kWh/kg COD removal and 1.24–1.74 kWh/kg BOD removal. The system of pilot plant has a vertical movement of the biofilm module which supplied sufficient oxygen for the removal of the organic carbon in the wastewater. The cost for power consumption was very low. Considering that the electric driving unit consumed 4.5 kWh/ day without loading, it may be assumed that only 30% power was consumed by the actual wastewater treatment. No performance chart has been analysed. The new biofilm system offers potential for reduced reactor volumes, energy saving, simple construction and easy operation since construction cost and operation cost are two important factors in evaluating a wastewater treatment process.

Zhang (2014) conducted a trial to study nitrification kinetics of a pilot scale MBBR for treating polluted raw water. A pilot-scale MBBR with an effective reactor volume of 4.4 m^3 (1.1 m × 1.0 m × 4.25 m) was constructed at the Chongshan water treatment plant. Two factors contributed to the performance of NH_4-N removal which are temperature and nutrient loads. Oxidation rates of NH_4-N and nitrite increased with temperature, although the biomass concentrations decreased from 348.7 to 188.9 mg Volatile Solid (VS)/L as temperature increased from 16.1 to 28.3 °C. NH_4-N and Nitrite oxidation were very highly dependent on temperature, while microbial population in the biofilm was more affected by NH_4-N loading than temperature. Based on this study, the authors suggest that Polysaccharides (PS) and phospholipids (PL) and protein (PN) correlate well with VS and can be used to estimate the attached biomass in MBBR systems. In the study, nitrification performance in the MBBR was severely affected when water temperature dropped to as low as 5.0 °C.

Previous studies have proven that the MBBR has established itself as a robust and compact reactor for wastewater treatment. The efficiency of the reactor has been demonstrated in many process combinations, both for BOD removal and nutrient removal. The primary advantage of the process as compared to activated sludge reactors refers to its compactness. Also, it does not require sludge recirculation. The advantage over other biofilm processes is its flexibility, and a further study on this bioreactor can yield more results. However, all the previously mentioned methods suffer from some serious limitations. Further, most studies in the field of MBBR have only focussed on domestic, industrial and aquaculture wastewater. No previous study has been conducted focusing on multiples of biological carrier particularly for river water purification neither in laboratory nor pilot plant basis.

3. Integrated Fixed-Film Activated Sludge (IFAS)

In the 1980's and 1990's, IFAS application system work began in the United State of America (USA) especially on the integration and modification of fixed film and activated sludge technologies [13]. In proportion to today's problem like more stringent effluent requirements, high cost of reactor tank expansion and lower fund options, IFAS technology is being introduced to represent an attractive solution for wastewater applications.

Recent developments in IFAS have highlighted a few advantages for this system. One of the most common operational problems in conventional activated sludge wastewater treatment systems around the world is poor settling of biological solids although it is supported with sedimentation tank [17]. This can lead to an increase in the cost for exclusive sludge treatment, an increase in TSS concentration in effluent, and increased risks to downstream ecosystems and public health. However, based on findings reported by Kim (2010), IFAS and non-IFAS systems were likely related to the observed differences in density and settleability for the suspended phases because stable and efficient removal of the biological solids produced in biological reactors is critical to the operation of biological wastewater treatment systems for the production of high quality effluent [17].

IFAS application has been increasingly recognized as a cost effective solution for wastewater treatment process. However, the implementation, operation and performance of such systems in full scale are less reported.

3.1. Current Development of IFAS

IFAS sometimes expressed as Fixed Bed Biological Reactor (FBBR) since it media was installed with fixed-in the reactor. Basically, biomedia in the MBBR system was freely in the reactor and in the meantime, biomedia in FBBR system is installed fixed attached operating mode in the reactor (Ye, Iii & Mcdowell 2010 in McDowell and Hubbell, 2000). An attempt have been made to compare performance between FBBR and MBBR and CAS for example Choi (2012) and G. Andreottola, R Foladori (2000). The foremost difference between the MBBR and IFAS systems is the presence of a return activated sludge flow that remains to the IFAS process.

According to Choi (2012), MBBR and FBBR were compared for biological phosphorus removal and denitrification. Results indicated that all nutrients were removed by the FBBR process compared with the MBBR process: 19.8% (total COD), 35.5% (filtered COD), 27.6% (BOD_5), 62.2% (acetate), 78.5% (PO_4-P), and 54.2% (NO_3-N) in MBBR; 49.7% (total COD), 54.0% (filtered COD), 63.2% (BOD_5), 99.6% (acetate), 98.6% (PO_4-P), and 75.9% (NO_3-N) in FBBR. The phosphate uptake and NO_3-N decomposition in the FBBR process during the denitrification phase were much higher than for the MBBR process despite being of shorter duration. Results obtained from this study are helpful in elucidating the practical implications of using MBBR and FBBR for the removal of nutrients and denitrification from wastewater.

Ye, Chestna (2010) in their case study in IFAS wastewater treatment Hopedale, US able to maintain DO level in range 4-6 mg/L to provide sufficient aeration process for both process and mixing requirement. Types of aeration used are tapered aeration. After three month start-up operation, Effluent NH_3-N is consistently below the limit. However, the NH_3-N removal capacity provided by the attached growth on the media was able to keep the plant within the NH_3-N limit even with a concurrent wastewater temperature as low as 7-9°C. The recovery of the system from the peak conditions appeared to be instantaneous, which had attributed to the attached biomass inventory on the media. In addition, the installation of the structured sheet media in the aeration basins also improved the process stability due to the increased biomass inventory. Effluent BOD and TSS concentrations were averaging about 5 to 7 mg/L during the IFAS period, as opposed to 15-20 mg/L for the previous CAS process. Implementation of the IFAS system appeared to improve the overall settleability of the suspended solids. An average SVI of approximately 100 mL/g was observed in the IFAS system, as compared to 150 mL/g for the prior CAS process.

Jianchang (2010) study about biofilm performance of high surface area density vertical flow structured sheet media. The study demonstrated that the Vertical Flow (VF) media combined with the proprietary distribution media is capable of achieving complete nitrification and high-rate BOD removal for both IFAS and FBBR applications. As an essential element in the VF media system, the distribution media not only maximized the air and wastewater distribution over the entire surface area of the media, but also optimized the airlift pumping through the VF media for sufficient mixing and effective biomass control. Favourable kinetic rates for example tertiary ammonia rates up to 1.4 g NH_3-N/m^2-day at 15°C, SCOD removal rate of 30 g SCOD/m2-day at a SCOD load of 45 g SCOD/m^2-day, and pre- denitrification rates of 1.0-2.0 g NO_3-N/m^2-day have been consistently observed in the VF structured sheet media system, mainly due to the intimate contact between thin biofilm and substrates/oxygen as promoted by the dedicated aeration associated with the media towers.

Previous studies reported that the integrated fixed-film activated sludge (IFAS) with a media of extruded high density polyethylene demonstrated higher removal efficiencies of 90%, 90% and 85% for COD, TP and ammonia, respectively, with a solids residence time (SRT) of 8 days. Recently, performance of a hybrid activated sludge/biofilm process for wastewater treatment in a cold climate region. One of the biofilm support media for IFAS is generally made of plastic. They use various forms/types of plastic, such as polyethylene (PE) and polypropylene (PP) [22], [23]. However all the previously mention case study suffer from some serious drawbacks. Data on the effects of IFAS systems on biomass settleability are scarce and inconsistent[17]. Authors suggests that optimization of nutrient removal process should be taken into consideration for good settling character in designing IFAS system. The potential effects of IFAS on settleability in non-nutrient removal systems should be the subject of future research.

3.2. Experimental Study on IFAS and MBBR

It is essential to know about previous experimental methodology on extensive biological carrier support in MBBR and IFAS system. In order to achieve the study objectives, the methodology must be done in laboratory basis. A few factors need to take into consideration for laboratory experiment in water and wastewater treatment. The important factors are design Hydraulic Retention Time (HRT), reactor size, biomedia size, influent type and types of biological carrier. As for biological performance in biomedia, biofilm and biomass analysis and observation from previous studies also will be highlighted in this section.

3.2.1. Design HRT

Hydraulic loading is important factor. On the one hand, hydraulic loading directly determine HRT of the reactor, further influence the size of volume, on the other hand, hydraulic loading directly affect the removal effect in biofilm process [24]. So, choosing appropriate hydraulic loading is necessary. Table shows HRT and Feeding Value from previous MBBR and IFAS studies. Many of studies tested various HRT for their experimental analysis. Most of their laboratory operated in parallel and fed from a common feed tank by a multichannel peristaltic pump.

Table 1: HRT and Feeding Value from previous MBBR and IFAS study

Process	**HRT (hrs)**	**Feeding value (mg/l)**	**Location**	**Reference**
MBBR	5,10 & 15	COD:120-150	Spain	(Calderón , 2012) [25]
SBR	4	COD: 641	Iran	(Moghaddam & Sargolzaei, 2014)[26]
Fixed Media Submerged Biofilters	1.5	COD : 270-300 NH_4-N : 30-35 TN : 35-40	China	(Guohui, 2012) [24]
Hybrid MBBR	12	COD : 600	Sweden	(Falås, 2013) [27]
MBBR	24	NH_4-N : 40-50	Italy	(Bertino, 2010) [10]
Lab MBBR	4-8	COD : 500, 1000, 2000, 4000 & 8000	Turkey	(Aygun, Nas, & Berktay, 2008) [28]
Real MBBR	9 & 12	NH_3-N: 600	Spain	(González-Martínez et al., 2013) [29]
Lab IFAS	1.9-13.6	Total Kjeldahl Nitrogen (TKN) : 25-70	USA	(Regmi et al., 2011) [30]
Lab IFAS & MBBR	1-2	NH_3-N : 30-35	USA	(Ye, Kulick III, et al., 2010) [18]

Low strength feeding water test normally have low HRT in range 3 to 12 hrs. High strength feeding water and anaerobic treatment acquire longer period of HRT. However, HRT setup is be subject to researcher's objective and experimental design factor.

3.2.2. Types of Influent

Influent concentration for experimental test normally are depends on the objective and expecting results. It is under a controlled range of parameter like pH, BOD, COD, TN and other parameter. Table 2 show the types of influent sources for MBBR and IFAS study.

Table 2: Influent sources for MBBR and IFAS experimental and pilot plant studies

No	Author	Location	Influent Source	Influent (mg/l)	Output
1	(Leyva-Díaz et al., 2014) [31]	Spain	Urban WW	BOD_5: 100-200 COD: 200-400	The hybrid MBBR–MBR showed the best performance of COD and BOD_5 removal as compare to pure MBBR–MBR.
2	(Gong et al., 2012) [32]	China	Rural domestic WW	COD: 150-300 TN: 60-90	TN was removed averagely by 69.3% and under internal recycling ratio of 200% and less proportions of biomass assimilation (less than 3%).
3	(Nguyen et al., 2014) [22]	Korea	Domestic WW	TN : 30-50 COD: 150-300 TSS: 175-460	NH_4-N was completely eliminated and T-P removal efficiency was also up to 100%. It was found that increasing in internal recycling ratio could improve the nitrate and nitrogen removal efficiencies. TSS after treatment was < 5 mg/L
4	(R. C. Wang, Wen, & Qian, 2005) [33]	China	Synthetic WW	Glucose: 200	The optimum carrier concentration was about 50% with the average COD and ammonia removal rate about 70% and 30%, respectively
5	(Shore, M'Coy, Gunsch, & Deshusses, 2012) [34]	USA	Municipal and Synthetic WW	NH_3-N: 10-15 TN : 10-15	Experiment able to remove more than 90% of the influent ammonia (more than 19 mg/L NH_3–N) in both the synthetic and industrial wastewater.
6	(Pfeiffer & Wills, 2011) [35]	USA	Hatchery Aquaculture WW	pH: 7.3-7.5 Nitrite: 0.2-0.5	The TAN removal rates for the MB3 media was the highest of the three media types at both the low and high feed load rates averaging 12.3% and 14.4%, respectively.
7	(Clifford, Forde, McNamara, Rodgers, & O'Reilly, 2013) [36]	Ireland	Synthetic WW	COD: 0.6 TN: 0.2, and NH_4-N: 0.11 in $g/m^2/day$	The average removal rates 92.4%, 34.8&, and 98.5% for COD, TN and NH_4-N, respectively

Overall, average value from Table 2 shows all the influents from municipal wastewater. Average suitable influent values for experiment study are COD in range 200-250 mg/L and BOD 100 mg/L. The role of influent value is not only to provide feeding load for biomedia to treat the wastewater but also to load nutrient for attached microorganism to form a biofilm layers.

Ødegaard and Rusten (1995) gathered data from various small full-scale wastewater treatment plants and the MBBR systems started to develop. However, later, organic matter removal MBBR treatment systems were developed. Currently, MBBR systems are used as stand-alone treatment solutions and in tandem with other treatment processes including AS and membrane bioreactors for high strength organic wastewaters MBBR processes.

Brinkley John in Borkar (2013) investigated processes that would treat variable high strength wastewater in a small footprint and provided provisions for future expansion. He selected the MBBR process due to the success the process had for treating high strength wastewater for comparable pharmaceutical applications. The 0.5 million gallon per day (mgd) MBBR process consisted of two reactors operated in series designed to treat an influent and effluent BOD_5 of 3,197 mg/L and less than 75 mg/L, respectively.

The media provides increased surface area for the biological microorganisms to attach to and grow in the aeration tanks. The increased surface area reduces the footprint of the tanks required to treat the wastewater. according to [38] attached microorganism based process like MBBR and IFAS are suitable for high strength wastewater. however there are still study to test workability of MBBR in river water [39]. However in water purification is the process of removing undesirable chemicals, biological contaminants, suspended solids and gases from contaminated river water[40].

3.2.3. Biological Carriers in MBBR and IFAS

The media on which the biofilm develops are carefully designed with high internal surface area having density slightly less than the water so that it can easily float. The most commonly used solid surface for attached growth processes are; stones, clinker, sand, activated charcoal, ceramic, metals, plastic sheets, and foams. There are different types of media which can be used as a media for the microbial growth[7]. Specific surface area (SSA) is the main factors that contribute the biomass characteristic in water treatment and as well as the total performance of IFAS and MBBR system. Table shows the types of carriers available in the markets and being apply in laboratory and pilot plant study for both MBBR and IFAS system.

Table 3: MBBR and IFAS biological carriers that has been used in various study

Authors	**Country**	**Types**	**SSA (m^2/m^3)**	**Volume concentratio n**	**Figure**
Brentwood, 2009 [41]	USA	PVC Structured Sheet Media	160	75-100%	 
Brentwood, 2009 [41]	USA	Fabric Web	100	75-100%	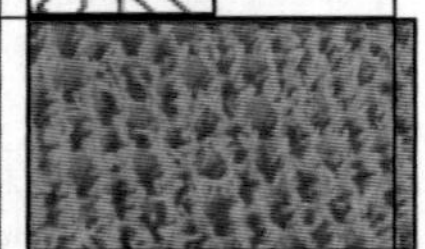 
Kim., 2010; Shore., 2012 [17], [34]	USA	Bioportz™	576		
Azimi, Hooshyari, Mehrdadi, & Bidhendi, 2007 [42]	Iran	Bee-Cell 2000	650		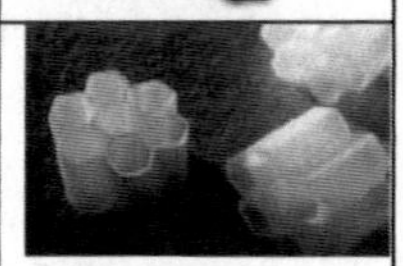
Borkar et al., 2013 [37]	India	Chips	260		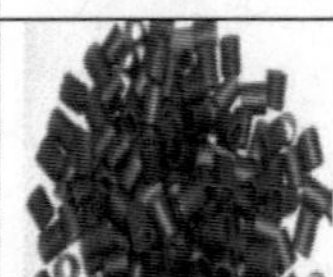
Zhang et al., 2013 [16], [43]	China	Yuhuan	230		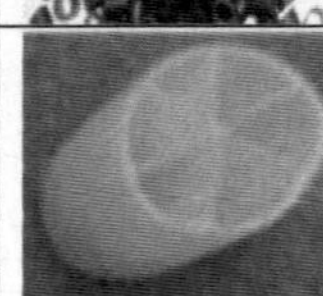
Bio, 2005[44]	USA	ActiveCell™	400	50-70%	

Continue Table 3..

Authors	Country	Types	SSA (m^2/m^3)	Volume concentration	Figure
Aygun et al., 2008; Bertino, 2010; Calderón et al., 2012; Jaroszynski, Cicek, Sparling, & Oleszkiewicz, 2011; Leyva-Díaz et al., 2013, 2014; Pfeiffer & Wills, 2011; Daniele Di Trapani, Mannina, Torregrossa, & Viviani, 2010 [10], [25], [28], [31], [35], [45]–[47]	USA, Turkey, Canada, Spain & Italy	Anodkalness™ K_1	500	50-70%	
Hoang et al., 2014; Regmi et al., 2011; Bjorn Rusten, Kolkinn, & Odegaard, 1997 [30], [48], [49] [12][30], [48], [49]	USA	Anodkalness™ K_3	500	50%	
Chu & Wang, 2011 [12]	China	Polyurethane foam (PU)	900	30%	
Xie et al., 2005; Zhu, Yu, Wu, & Yao, 2014 [39], [50]	China	Bio-ceramic	2500	50%	
Chu & Wang, 2011 [12]	China	Polymer polycaprolactone (PCU)	346	70%	

Continue Table 3..

Authors	Country	Types	SSA (m2/m3)	Volume concentration	Figure
Pfeiffer & Wills, 2011 [49]	USA	MB3	604	-	
(P. Wang, Wang, Ai, & Yang, 2011 [51] [13]	China	Cageball	300	40-50	
Kermani et al., 2008 [50],	Iran	Flocor RMP-HPS®	277	70	
Aloysius, 1999; Hussain, Tat, & Idris, 2014 [52], [53]	Malaysia	Cosmoball ™	160	100	
Falås et al., 2013 [27]	Sweden	Biofilm Chip M ™	500	-	

3.2.4. Aeration rate

In MBBR and IFAS system, aeration is used to provide DO for the biomass and also to create the cross flow velocity to scour the biomedia surface [8]. The high crossflow activities that are generated at the biomedia surface tend to shear off the deposited materials and thus reduce the hydraulics resistance of the fouling layer. An optimum value of air flow rate was identified beyond which further increases had no effect. The similar phenomenon was also verified by some researchers [32], [38], [54]. Wang (2012) investigated the effect of hydrodynamic on biomedia and concluded that the treatment filtration rate sharply increased at aeration intensity below the critical aeration intensity.

4. Conclusion

There are varieties of biomedia used in river water purification technologies which are available in the local and international water industry market nowadays, and the types of biomedia can be characterized by considering the types of treatment system. Different types of biomedia have their unique performance based on certain treatment conditions. Apart from biomedia removal efficiency and physical characteristics at the allocated design flow, other important aspects that affect the decision in selection of biological carriers for MBBR and IFAS are frequency of maintenance and cost. Biomedia should be able to provide a habitation for the growth of the attached microorganism which is called biofilm layer, either fixed or freely moving inside the river purification system. Based on previous studies, Biomedia in MBBR and IFAS have been widely tested only for domestic sewage or industrial wastewater including aquaculture wastewater. However, approaches of this kind carry with them various well known limitations. There have been no controlled studies which compare differences between biomedia physic-chemical and biological performance for river water treatment plant in local tropical river condition. However, the research concluded so far are compiled herein and reported for the first time to acquire a better perspective and insight on the subject with a view of meeting the news approach. To this end, the most feasible technology could be the combination of advanced biological process (bioreactor systems) including Moving Bed Biofilm Reactor (MBBR) and Integrated Fixed Activated Sludge (IFAS) system.

Acknowledgments
This research is a Fundamental Research Grant Scheme (FRGS/1/2014/STWN01/UNITEN/03/1) supported by Ministry of Higher Education Malaysia under the title "Gross Pollutant Characteristic Using Economic Analysis For Stormwater Treatment Alternatives Decision Making in Tropical Climate Country" with the collaboration of several Higher Educations Institutions and Government Sectors.

References
[1] H. Odegaard, B. Rusten, and T. Westrum, "A new moving bed biofilm reactor - applications and results," in *Water Science and Technology*, 1994, vol. 29, no. 10–11, pp. 157–165.
[2] M. Henze, M. C. M. Van Loosdrecht, G. A. Ekama, and D. Brdjanovic, *Biological wastewater treatment : principles, modelling and design*, 2008th ed., vol. 96. London, UK.: IWA Publishing, 2008.
[3] A. Æsøy, H. Ødegaard, and G. Bentzen, "The effect of sulphide and organic matter on the nitrification activity in a biofilm process," in *Water Science and Technology*, 1998, vol. 37, no. 1, pp. 115–122.
[4] A. Briones and L. Raskin, "Diversity and dynamics of microbial communities in engineered environments and their implications for process stability," *Current Opinion in Biotechnology*, vol. 14, no. 3. pp. 270–276, 2003.
[5] L. Mohd Sidek, H. A. Mohiyaden, H. Basri, G. H. A. Salih, A. H. Birima, Z. Ali, A. F. Mohd Sabri, and M. N. Md. Noh, "Experimental Comparison between Moving Bed Biofilm Reactor (MBBR) and Conventional Activated Sludge (CAS) for River Purification Treatment Plant," *Adv. Mater. Res.*, vol. 1113, pp. 806–811, 2015.
[6] B. Rusten, C. H. Johnson, S. Devall, D. Davoren, and B. S. Cashion, "Biological pretreatment of a chemical plant wastewater in high-rate moving bed biofilm reactors," in *Water Science and Technology*, 1999, vol. 39, no. 10–11, pp. 257–264.
[7] A. Shrestha, "Specific Moving Bed Biofilm Reactor in Nutrient Removal from Municipal Wastewater," University of Sydney, 2013.
[8] R. a. Zimmerman, D. Richard, S. Lynne, and W. Lin, "Is Your Moving Bed Biofilm Reactor (MBBR) Running on All Cylinders?," in *Proceedings of the Water Environment Federation*, 2005, vol. 2005, no. 9, pp. 6238–6265.
[9] Y. Qiqi, H. Qiang, and T. I. Husham, "Review on Moving Bed Biofilm Process," *Pakistan J. Nutr.*, vol. 11, no. 9, pp. 804–811, 2012.
[10] A. Bertino, "Study on one-stage Partial Nitritation-Anammox process in Moving Bed Biofilm Reactors: a sustainable nitrogen removal," Royal Institute of Technology, 2010.

[11] X. J. Wang, S. Q. Xia, L. Chen, J. F. Zhao, N. J. Renault, and J. M. Chovelon, "Nutrients removal from municipal wastewater by chemical precipitation in a moving bed biofilm reactor," *Process Biochem.*, vol. 41, no. 2/3, pp. 824–828, 2006.

[12] L. Chu and J. Wang, "Comparison of polyurethane foam and biodegradable polymer as carriers in moving bed biofilm reactor for treating wastewater with a low C/N ratio.," *Chemosphere*, vol. 83, no. 1, pp. 63–68, Mar. 2011.

[13] H. Ødegaard, B. Gisvold, and J. Strickland, "The influence of carrier size and shape in the moving bed biofilm process," in *Water Science and Technology*, 2000, vol. 41, no. 4–5, pp. 383–391.

[14] M. Kermani, B. Bina, H. Movahedian, M. M. Amin, and M. Nikaein, "Application of Moving Bed Biofilm Process for Biological Organics and Nutrients from Municipal Wastewater," *Am. J. Environ. Sci.*, vol. 4, no. 6, pp. 675–682, Jun. 2008.

[15] M. Rodgers, X.-M. Zhan, and B. Gallagher, "A pilot plant study using a vertically moving biofilm process to treat municipal wastewater," *Bioresour. Technol.*, vol. 89, no. 2, pp. 139–143, Sep. 2003.

[16] S. Zhang, Y. Wang, W. He, M. Wu, M. Xing, J. Yang, N. Gao, and M. Pan, "Impacts of temperature and nitrifying community on nitrification kinetics in a moving-bed biofilm reactor treating polluted raw water," *Chem. Eng. J.*, vol. 236, pp. 242–250, Jan. 2014.

[17] H.-S. Kim, J. W. Gellner, J. P. Boltz, R. G. Freudenberg, C. K. Gunsch, and A. J. Schuler, "Effects of integrated fixed film activated sludge media on activated sludge settling in biological nutrient removal systems.," *Water Res.*, vol. 44, no. 5, pp. 1553–61, Mar. 2010.

[18] J. Ye, F. M. Kulick III, C. Mcdowell, and K. L. Chestna, "Biofilm Performance of High Surface Area Density Vertical-Flow Structured Sheet Media for IFAS and Fixed Bed Biofilm Reactor (FBBR) Applications," *83rd Annu. Water Environ. Fed. Tech. Exhib. Conf.*, pp. 1–20, 2010.

[19] H. J. Choi, A. H. Lee, and S. M. Lee, "Comparison between a moving bed bioreactor and a fixed bed bioreactor for biological phosphate removal and denitrification.," *Water Sci. Technol.*, vol. 65, no. 10, pp. 1834–8, Jan. 2012.

[20] M. R. and F. T. G. Andreottola, R Foladori, "Experimental comparison between MBBR and activated sludge system for the treatment of municipal wastewater," *J. Water Sci. Technol.*, p. 6, 2000.

[21] J. Ye, K. L. Chestna, F. M. K. III, and B. Rothermel, "Full Scale Implementation , Operation , and Performance of a Structured Sheet Media IFAS Systemia IFAS System," *83rd Annu. Water Environ. Fed. Tech. Exhib. Conf.*, 2010.

[22] D. D. Nguyen, H. H. Ngo, and Y. S. Yoon, "A new hybrid treatment system of bioreactors and electrocoagulation for superior removal of organic and nutrient pollutants from municipal wastewater.," *Bioresour. Technol.*, vol. 153, pp. 116–25, Mar. 2014.

[23] D. Di Trapani, G. Mannina, M. Torregrossa, and G. Viviani, "Hybrid moving bed biofilm reactors: a pilot plant experiment," *Water Science & Technology*, vol. 57, no. 10. p. 1539, 2008.

[24] Y. Guohui, J. Chundong, X. Zhi, F. Qingzhi, and G. Haiyan, "Treatment of Synthetic Wastewater in a Pre-Denitrification Biofilm Reactor Packed with Polyurethane Media," *Energy Procedia*, vol. 16, pp. 1642–1649, 2012.

[25] K. Calderón, J. Martín-Pascual, J. M. Poyatos, B. Rodelas, A. González-Martínez, and J. González-López, "Comparative analysis of the bacterial diversity in a lab-scale moving bed biofilm reactor (MBBR) applied to treat urban wastewater under different operational conditions.," *Bioresour. Technol.*, vol. 121, pp. 119–26, Oct. 2012.

[26] A. H. Moghaddam and J. Sargolzaei, "Biofilm development on normal and modified surface in a hybrid SBR-based bioreactor," *J. Taiwan Inst. Chem. Eng.*, Dec. 2014.

[27] P. Falås, P. Longrée, J. la Cour Jansen, H. Siegrist, J. Hollender, and A. Joss, "Micropollutant removal by attached and suspended growth in a hybrid biofilm-activated sludge process.,"

Water Res., vol. 47, no. 13, pp. 4498–506, Sep. 2013.

[28] A. Aygun, B. Nas, and A. Berktay, "Influence of High Organic Loading Rates on COD Removal and Sludge Production in Moving Bed Biofilm Reactor," *Environ. Eng. Sci.*, vol. 25, no. 9, pp. 1311–1316, Nov. 2008.

[29] a González-Martínez, K. Calderón, A. Albuquerque, E. Hontoria, J. González-López, I. M. Guisado, and F. Osorio, "Biological and technical study of a partial-SHARON reactor at laboratory scale: effect of hydraulic retention time.," *Bioprocess Biosyst. Eng.*, vol. 36, no. 2, pp. 173–84, Feb. 2013.

[30] P. Regmi, W. Thomas, G. Schafran, C. Bott, B. Rutherford, and D. Waltrip, "Nitrogen removal assessment through nitrification rates and media biofilm accumulation in an IFAS process demonstration study.," *Water Res.*, vol. 45, no. 20, pp. 6699–708, Dec. 2011.

[31] J. C. Leyva-Díaz, J. Martín-Pascual, M. M. Muñío, J. González-López, E. Hontoria, and J. M. Poyatos, "Comparative kinetics of hybrid and pure moving bed reactor-membrane bioreactors," *Ecol. Eng.*, vol. 70, pp. 227–234, Sep. 2014.

[32] L. Gong, L. Jun, Q. Yang, S. Wang, B. Ma, and Y. Peng, "Biomass characteristics and simultaneous nitrification–denitrification under long sludge retention time in an integrated reactor treating rural domestic sewage," *Bioresour. Technol.*, vol. 119, pp. 277–284, Sep. 2012.

[33] R. C. Wang, X. H. Wen, and Y. Qian, "Influence of carrier concentration on the performance and microbial characteristics of a suspended carrier biofilm reactor," *Process Biochem.*, vol. 40, no. 9, pp. 2992–3001, Sep. 2005.

[34] J. L. Shore, W. S. M'Coy, C. K. Gunsch, and M. A. Deshusses, "Application of a moving bed biofilm reactor for tertiary ammonia treatment in high temperature industrial wastewater.," *Bioresour. Technol.*, vol. 112, pp. 51–60, May 2012.

[35] T. J. Pfeiffer and P. S. Wills, "Evaluation of three types of structured floating plastic media in moving bed biofilters for total ammonia nitrogen removal in a low salinity hatchery recirculating aquaculture system," *Aquac. Eng.*, vol. 45, no. 2, pp. 51–59, Sep. 2011.

[36] E. Clifford, P. Forde, S. McNamara, M. Rodgers, and E. O'Reilly, "Performance of Air Suction Flow Biofilm Reactor in Treating Municipal-Strength Wastewater," *J. Environ. Eng.*, vol. 139, no. 6, pp. 864–872, Jun. 2013.

[37] R. P. Borkar, M. L. Gulhane, and A. J. Kotangale, "Moving Bed Biofilm Reactor – A New Perspective in Wastewater Treatment," *IOSR J. Environ. Sci. Toxicol. Food Technol.*, vol. 6, no. 6, pp. 15–21, 2013.

[38] S. L. (Nanyang T. U. Khor, "Membrane Bioreactor For of High Strength Wastewater," Nanyang Technological University, 2008.

[39] S. G. Xie, X. Y. Tang, W. Z. Wu, D. H. Wen, and Z. S. Wang, "Biological pretreatment of Yellow River water," *J. Environ. Sci.*, vol. 17, no. 4, pp. 557–561, 2005.

[40] H. A. Mohiyaden, L. M. Sidek, H. Basri, and G. Hayder, "Biological Carrier Performance Evaluation for Artificial River Purification," in *13th International Conference on Urban Drainage, Sarawak, Malaysia*, 2014, no. September, pp. 7–12.

[41] I. Brentwood, "Integrated Fixed Film / Activated Sludge (IFAS) Technology," *Water Technology Group*, vol. 5, no. June. Water Technology Group, p. 6, 2009.

[42] A. A. Azimi, B. Hooshyari, N. Mehrdadi, and G. H. N. Bidhendi, "Enhanced COD and nutrient removal efficiency in a hybrid integrated fixed film activated sludge process," *Iran. J. Sci. Technol.*, vol. 31, pp. 523–533, 2007.

[43] S. Zhang, Y. Wang, W. He, M. Wu, M. Xing, J. Yang, N. Gao, and D. Yin, "Responses of biofilm characteristics to variations in temperature and NH4(+)-N loading in a moving-bed biofilm reactor treating micro-polluted raw water.," *Bioresour. Technol.*, vol. 131, pp. 365–73, Mar. 2013.

[44] H. Bio, "Applications of Technologies Wastewater Technologies." Houston, Texas, p. 40, 2005.

[45] J. C. Leyva-Díaz, K. Calderón, F. a. Rodríguez, J. González-López, E. Hontoria, and J. M.

Poyatos, “Comparative kinetic study between moving bed biofilm reactor-membrane bioreactor and membrane bioreactor systems and their influence on organic matter and nutrients removal,” *Biochem. Eng. J.*, vol. 77, pp. 28–40, Aug. 2013.

[46] D. Di Trapani, G. Mannina, M. Torregrossa, and G. Viviani, “Quantification of kinetic parameters for heterotrophic bacteria via respirometry in a hybrid reactor.,” *Water Sci. Technol.*, vol. 61, no. 7, pp. 1757–66, Jan. 2010.

[47] L. W. Jaroszynski, N. Cicek, R. Sparling, and J. a Oleszkiewicz, “Importance of the operating pH in maintaining the stability of anoxic ammonium oxidation (anammox) activity in moving bed biofilm reactors.,” *Bioresour. Technol.*, vol. 102, no. 14, pp. 7051–6, Jul. 2011.

[48] V. Hoang, R. Delatolla, T. Abujamel, W. Mottawea, A. Gadbois, E. Laflamme, and A. Stintzi, “Nitrifying moving bed biofilm reactor (MBBR) biofilm and biomass response to long term exposure to 1 °C.,” *Water Res.*, vol. 49, pp. 215–24, Feb. 2014.

[49] B. Rusten, O. Kolkinn, and H. Odegaard, “Moving bed biofilm reactors and chemical precipitation for high efficiency treatment of wastewater from small communities,” *Water Sci. Technol.*, vol. 35, no. 6, pp. 71–79, 1997.

[50] X. L. Zhu, J. Y. Yu, Z. X. Wu, and L. Yao, “The Preparation of Iron Ore Tailings Bio Ceramic Filter Material and Domestic Sewage Treatment,” in *Advanced Materials Research*, 2014, vol. 997, pp. 877–881.

[51] P. Wang, C. Wang, X.-Y. Ai, and C.-Q. Yang, “Biofilm characteristics of globular biofilter in the ecological spur-dike and water quality improvement effect on Wangyu River,” in *2011 International Conference on Electronics, Communications and Control*, 2011, no. 9–11, pp. 3598–3603.

[52] L. M. Y. Aloysius, “A Full-Scale Combined Anaerobic-Aerobic System For Treatment Of Melamine Formaldehyde Wastewater,” University Putra Malaysia, 1999.

[53] S. A. Hussain, T. H. Tat, and A. Idris, “Numerical study of fluid flow across multiple Cosmo ball,” *Int. J. Nano Biomater.*, vol. 5, no. 1, p. 89, 2014.

[54] J. S. Weiss, M. Alvarez, C. Tang, R. W. Horvath, and J. F. Stahl, “Evaluation Of Moving Bed Biofilm Reactor Technology For Enhancing Nitrogen Removal In A Stabilization Pond Treatment Plant,” in *Water Environment Federation*, 2005, pp. 2085–2102.

[55] J. Wang, X. D. Liu, and J. Lu, “Urban River Pollution Control and Remediation,” *Procedia Environ. Sci.*, vol. 13, pp. 1856–1862, 2012.

Performance evaluation of cognitive radio in advanced metering infrastructure communication

Yik-Kuan Hiew, Norazizah Mohd Aripin, and Norashidah Md Din
Universiti Tenaga Nasional
Kajang, Malaysia

E-mail: yikkuanhiew@gmail.com; {norazizahm, norashidah}@uniten.edu.my

Abstract. Smart grid is an intelligent electricity grid system. A reliable two-way communication system is required to transmit both critical and non-critical smart grid data. However, it is difficult to locate a huge chunk of dedicated spectrum for smart grid communications. Hence, cognitive radio based communication is applied. Cognitive radio allows smart grid users to access licensed spectrums opportunistically with the constraint of not causing harmful interference to licensed users. In this paper, a cognitive radio based smart grid communication framework is proposed. Smart grid framework consists of Home Area Network (HAN) and Advanced Metering Infrastructure (AMI), while AMI is made up of Neighborhood Area Network (NAN) and Wide Area Network (WAN). In this paper, the authors only report the findings for AMI communication. AMI is smart grid domain that comprises smart meters, data aggregator unit, and billing center. Meter data are collected by smart meters and transmitted to data aggregator unit by using cognitive 802.11 technique; data aggregator unit then relays the data to billing center using cognitive WiMAX and TV white space. The performance of cognitive radio in AMI communication is investigated using Network Simulator 2. Simulation results show that cognitive radio improves the latency and throughput performances of AMI. Besides, cognitive radio also improves spectrum utilization efficiency of WiMAX band from 5.92% to 9.24% and duty cycle of TV band from 6.6% to 10.77%.

1. Introduction

Smart grid is an intelligent electricity system that integrates sensors and intelligent electronic devices for automatic measuring, controlling, and monitoring. These sensors and devices measures and reports grid status frequently to ensure smart grid is in stable status. A reliable communication link is needed to transmit these smart grid data. However, there are ubiquitous smart grid components and these components are scattered in wide geographical area. It is difficult to locate a huge chunk of dedicated spectrum for smart grid communication. It is also expensive to implement wired communications in smart grid. The third option is to implement cognitive radio based communication in smart grid.

Spectrum measurements have been conducted worldwide and these occupancy results indicated that spectrum bands are poorly utilized [1]-[9]. Static-assigned licensed spectrum bands are not fully utilized while unlicensed spectrum bands are facing overcrowded issue. Cognitive radio improves spectrum utilization efficiency and mitigates the issue of spectrum scarcity by allowing unlicensed users to utilize idle licensed spectrum bands. In order to protect licensed users from potential harmful interference, unlicensed users can only access these idle spectrum when licensed users are not using these spectrum.

The benefits of using cognitive radio in smart grid communication are cost-efficient, energy-efficient, low interference level, and increases interoperability. As cognitive radio let smart grid users to access spectrum bands opportunistically, spectrum leasing is not required. Besides, one of the features of cognitive radio is transmit power control. Hence, it reduces energy consumption and keeps interference level low by computing optimal transmit power for smart grid users. Furthermore, it is expected that multiple communication techniques will be implemented in smart grid. Cognitive radio can interoperate in heterogeneous communication networks.

A cognitive radio based smart grid communication framework which consists of Home Area Network (HAN) and Advanced Metering Infrastructure (AMI) is proposed. AMI comprises communication in neighborhood area network (NAN) and wide area network (WAN). HAN supports communication between home electrical appliances and smart meter. AMI sends meter data from HAN to billing center and deals with automatic meter reading and dynamic respond. The components in AMI include smart meters, data aggregator units, and billing center. In this paper, only the performance of cognitive radio in AMI is presented.

2. Related works

Conventional communication techniques such as WiMAX, GSM, and GPRS are implemented for smart grid communications. These technologies offer long range propagation and licensed spectrum bands provide good quality channel for smart grid communication. However, they require spectrum leasing cost. Besides, channel assignment and priority are given to the licensed users of these bands. For instance, priority are given to mobile users because they are legal GSM subscribers. Depends on the service level agreement, smart grid data might or might not be assigned a good quality channel. On the other hand, wireless mesh network is implemented in smart grid. In [10], smart meter data are transmitted through RF mesh network. RF mesh networks offers the benefits of extended coverage, redundant links to cope with link failure and varying channel conditions, scalable, and low lost. However, RF mesh networks have high routing complexity. Besides, number of hops might incur huge latency in packet delivering.

In [11], a hybrid spectrum access in cognitive smart grid communication is proposed. Smart grid users are allowed to access both licensed and unlicensed spectrum bands opportunistically. If idle licensed bands are available, smart grid users can switch to idle licensed spectrum bands because licensed spectrum bands offer better quality of service than unlicensed spectrum bands. Several channels from both licensed and unlicensed spectrum bands are reserved as guard bands for handoff purpose. In [12], Omid Fatemieh and his team studies the feasibility of using TV white spaces for AMI communications. The proposed architecture suggests that NAN uses White-Fi and WAN uses 802.22 communication technologies. Both White-Fi and 802.22 standards transmit data by using cognitive radio in low duty cycle TV spectrum bands. White-Fi and 802.22 in different in terms of channel assignment and channel access. Omid Fatemieh and his team did not provide any justification to their framework.

3. Cognitive radio based smart grid communication framework

Figure 1 shows the proposed cognitive radio based smart grid communication framework. The framework comprises HAN, NAN, and WAN. In this paper, the performance of cognitive radio in NAN and WAN are presented.

There are three types of meter data, which are residential meter data, commercial meter data, and industrial meter data. In AMI, smart meter sends residential meter data to billing center four to six times a day. The latency is less than four hours. The interval for industrial and commercial meter data is 12 to 24 times a day and the latency is less than two hours. The packet size of these scheduled meter data is approximately 1600 to 2400 bytes. If on-demand meter data is required, smart meter sends data to data aggregator unit immediately. The packet size of on-demand data is 100 bytes and its latency is 15 seconds. AMI requires reliability of more than 98% [13].

In this proposed framework, AMI is separated into NAN and WAN. In NAN, meter data are sent from smart meters to data aggregator units. Data aggregator unit performs data compression, error checking, and data aggregation. In WAN, data aggregator unit forwards all the meter data to billing center. In this paper, WAN communication does not refer to core network that involves grid measuring, monitoring, controlling, and protection.

NAN communication is carried out using cognitive 802.11. In conventional 802.11, access points select one channel out of 11 channels from ISM spectrum bands. Wireless users that are attached to this access point will contend for this channel only for data transmission. On the other hand, cognitive users scan all six channels in ISM bands and utilize idle channels[1]. In NAN, smart meters are equipped with cognitive radio abilities, which allow them to sense multiple channels to detect idle channel. Cognitive users access channels through CSMA/CA.

The distance between data aggregator units and billing center is relatively larger than the distance between smart meters and data aggregator units. So, a long-range communication technique is required for WAN communication. Cognitive WiMAX and TV white spaces are used for WAN communication. WiMAX has a coverage area of 4km (NLOS) and 50km (LOS). TV white spaces refer to idle channels/ spectrum holes in TV bands. TV bands have excellent propagation characteristics. They can travel long distance and penetrate obstacles. WiMAX and TV white spaces are licensed spectrum bands. Hence, cognitive users must be able to detect the presence of licensed users in order to avoid causing harmful interference to licensed users. In [14], three methods are proposed to protect licensed users which are beacons, periodical spectrum sensing, and geo-location database. In this paper, geo-location database is used in WAN to protect licensed users and assign idle channels to cognitive users.

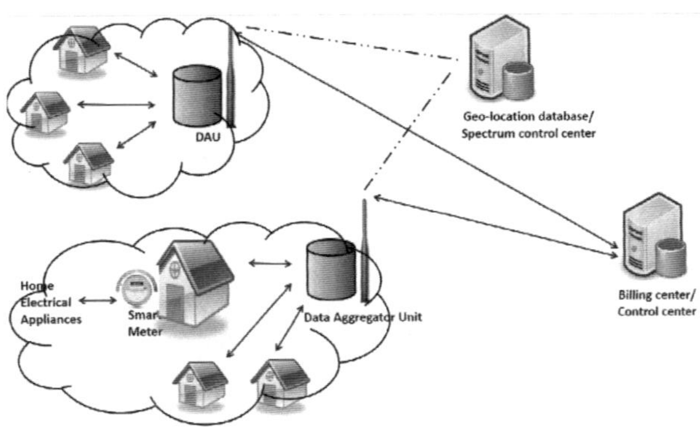

Figure 1. Cognitive smart grid communication framework.

4. Simulation

Simulation is carried out using Network Simulator 2 (NS-2). NS-2 is an open source event-driven simulator. Cognitive Radio Cognitive Network (CRCN) is patched into NS-2 to provide multi-interface multi-channel functions. Several modifications are done to support cognitive radio and multi-channel in 802.11 coding. Simulation parameters for NAN and WAN are listed in Table 1.

[1] There are 11 channels in ISM spectrum bands. However, five channels are used for routing purpose. Hence, there are six channels left for contention and data transmission.

Table 1. Simulation parameters.

Area	NAN: 500m × 500m WAN: 2km × 2km
Routing protocol	AODV
MAC protocol	NAN: 802.11 WAN: WiMAX, 802.22
Simulation time	240 seconds
Number of pairs of licensed users	2
PU CBR traffic	512 bytes
SU Traffic type	Poisson
SU Packet size	NAN: 1600 bytes WAN: 2400 bytes

The packet sizes are set according to [13]. WiMAX MAC protocol is implemented by patching 802.16 coding. By right, a 802.22 standard should be patched to simulate cognitive radio based communication in TV white spaces. However, there is no complete 802.22 coding by the time simulation is conducted. According to [15], there are many similarities between 802.16 and 802.22 MAC protocols. Hence, 802.16 MAC protocol is used to simulate 802.22 MAC.

Primary users are set to use certain channels during simulation. For example, the first pair of primary users use Channel 0 and the second pair of primary users use Channel 1 for data transmission. Cognitive users in NAN detect idle channels through CSMA/CA and cognitive users in WAN acquire idle channels through consulting geo-location database.

5. Results

The performance metric are delivery ratio, latency, and throughput. The quality of service for AMI are (i) delivery ratio more than 98%, (ii) latency less than 2000 milliseconds, and (iii) throughput of 14 to 100 kbps.

5.1 NAN

When the number of users increases, more nodes are contending for channels. The probability of collisions is higher. Nodes that encounter collisions have to back off and wait for a random period to contend for channel again. The waiting period incurs delay. In spite of the high probability of collisions and delay, the consistent performance of NAN delivery ratio is shown in Figure 2. Both conventional 802.11 and cognitive radio (no. of channel=1) have 100% delivery ratio while cognitive radio (no. of channel =2-6) has slightly lower delivery ratio.

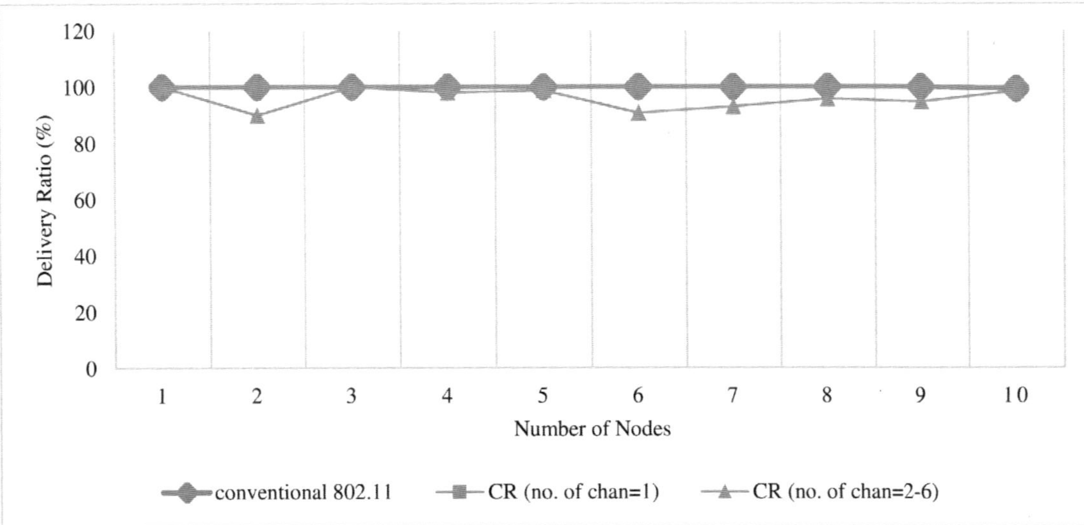

Figure 2. Delivery ratio for NAN.

Figure 3 shows the delay performance in NAN. When the number of channel is one, the delay of cognitive 802.11 outperforms conventional 802.11 by 3.32%. Once collision occurs, nodes cannot switch to other channel because there is only one channel in conventional 802.11. Nodes have to back off and wait for the channel to be idle again. For cognitive 802.11, there are two pairs of licensed users transmitting in the network. Sensing multiple channels and competing with two pairs of primary users cause its delay higher than the delay of conventional 802.11. In case of sudden appearance of primary users, cognitive users do not need to wait for primary users to finish data transmission and contend channels. Cognitive users can switch to other available channels.

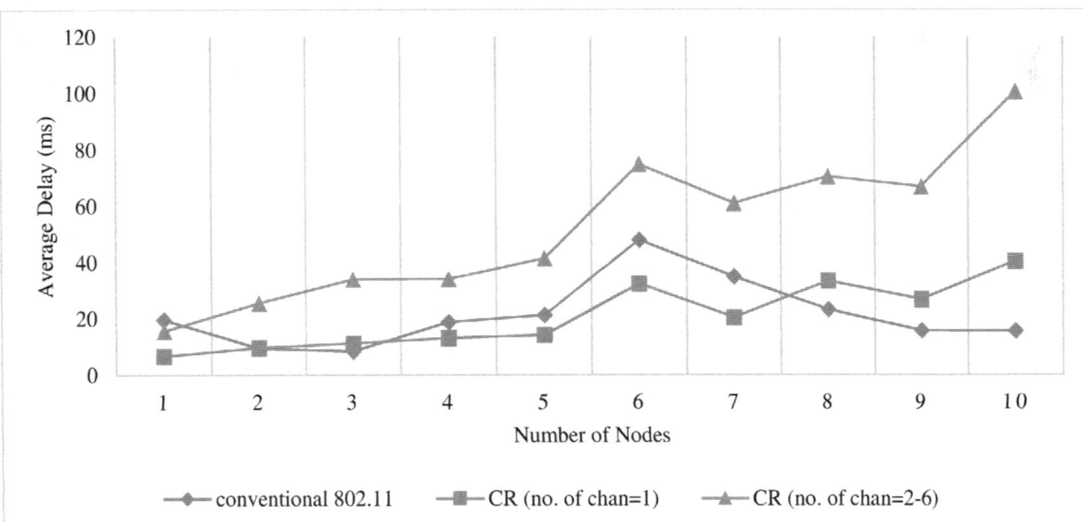

Figure 3. Average delay for NAN.

Figure 4 shows that the throughput value for conventional and cognitive 802.11 are comparable. The throughput of cognitive 802.11 is higher by 2.32%. Throughput is proportional to delivery ratio. In both cases, all packets sent are received successfully. Plus, the simulation time is same in both cases. Thus, both cases have comparable throughput.

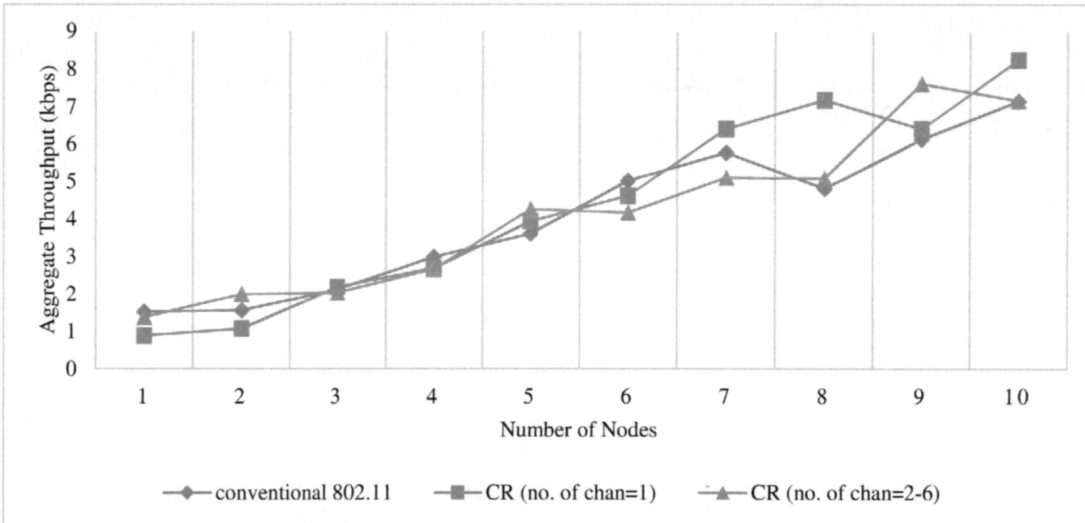

Figure 4. Aggregate throughput for NAN.

5.2 WAN

Figure 5 to 7 show the performance of WAN when using WiMAX band and TV white spaces opportunistically. Different from ISM bands, WiMAX and TV white spaces are licensed spectrum bands. If smart grid users wish to access WiMAX and TV bands legally, they have to pay for spectrum leasing. Although they are licensed users (since they pay for spectrum leasing), they still have to obey the policy of tele-operators in accessing spectrum. In this paper, the policy issue is not in the scope of work. Hence, instead of comparing the performance of non-cognitive radio and cognitive radio in WiMAX and TV bands, simulation is run to compare the performance of cognitive radio in WiMAX and TV white spaces. Figure 5 shows that the delivery ratio of cognitive radio in both WiMAX and TV white spaces is good. Although there is a slight drop of delivery ratio when number of nodes equals three, this packet drop is due to end of simulation time. The packet is generated near the end of simulation. When simulation times up, the packet is simply dropped.

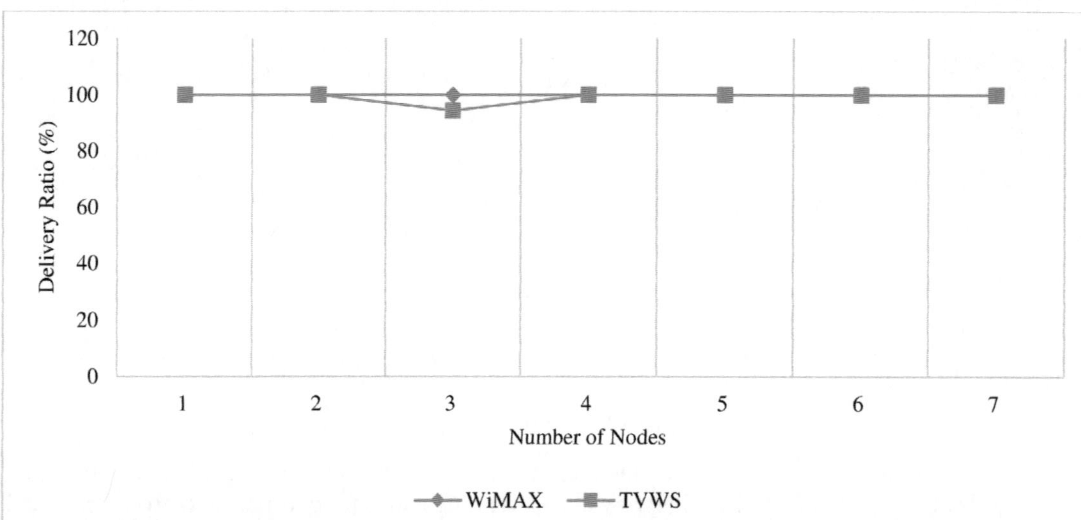

Figure 5. Delivery ratio in WAN.

Figure 6 shows that cognitive WiMAX offers lower delay than cognitive TV white space. The difference of delay is 36.83 %. Cognitive WiMAX allows users to access channels faster. This is because TV bands can propagate long distance. However, long distance propagation causes its round trip time higher. Despite the delay of TV bands is higher, the advantage of using TV white spaces is that TV frequency band can cover wider coverage area compared to WiMAX band. In other words, one 802.22 base station can support more cognitive users. Therefore, in a fix coverage area, less number of 802.22 base stations is required compare with the number of WiMAX base station.

In random access MAC, delay increases with the number of users. This is because nodes contends channels through CSMA/ CA. However, the MAC protocol of 802.16 (WiMAX) and 802.22 (TVWS) is time-slotted. Users are allocated timeslot for data transmission. TDMA performs better than CSMA/ CA when number of user is high. When there is less users (i.e. no. of nodes < 4), there are many timeslots that are not assigned to any users. Users cannot use unassigned channel and have to wait for their timeslot. These unused timeslot become wasted. The number of user increases until a certain point where all timeslots are assigned to users (i.e. no, of nodes >= 4). When there are four or more than four users, all timeslots are assigned to users. There will be no wasted timeslot. More packets can be sent and therefore the average delay is lower. Hence, average delay for all nodes become smaller when there more than four users in the system

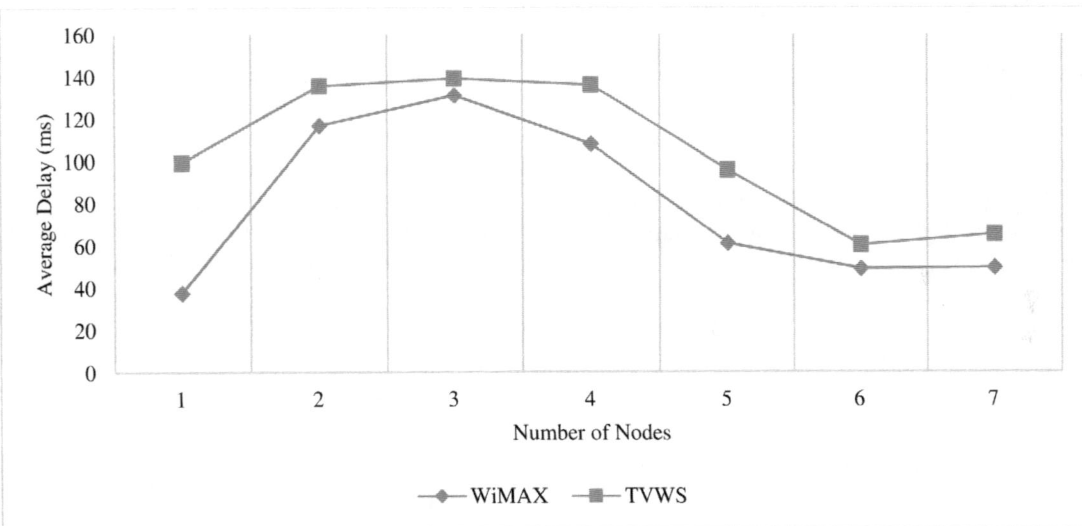

Figure 6. Average delay in WAN.

Figure 7 shows that the throughput performance for cognitive WiMAX and cognitive TV white space. The throughput of cognitive WiMAX is higher by 55.33%. This is because cognitive WiMAX has higher capacity than cognitive TV white space. WiMAX is a high frequency communication spectrum bands that provide high speed capacity for subscribers while TV spectrum bands are low frequency bands with low capacity data transmission.

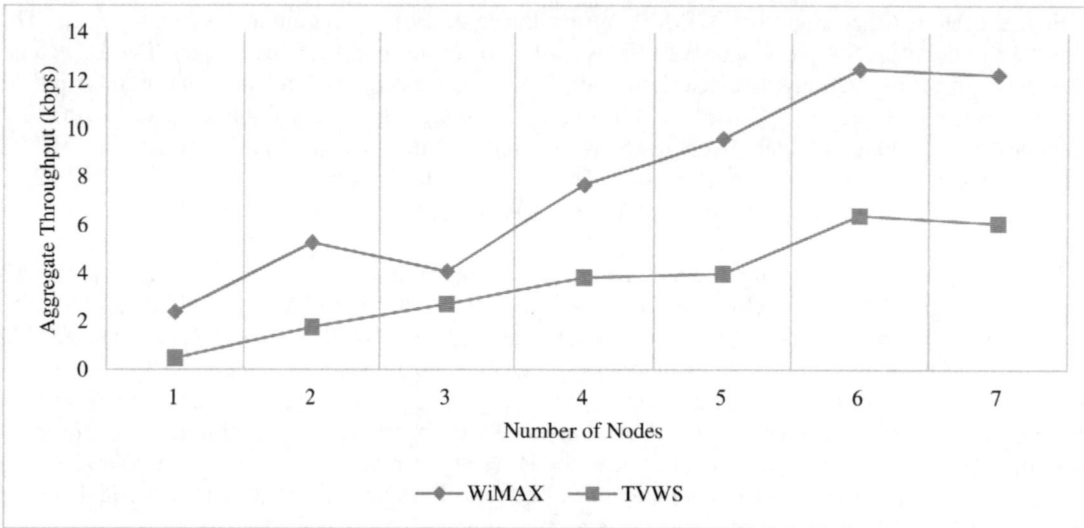

Figure 7. Aggregate throughput in WAN.

6. Spectrum Utilization Efficiency

As mentioned in Introduction, cognitive radio is able to improve spectrum utilization efficiency of spectrum bands. Hence, the new duty cycles of WiMAX band and TV band are calculated to investigate the percentage of improvement in duty cycle. Table 2 shows that allowing smart grid users to use spectrum band opportunistically greatly improve the spectrum utilization efficiency. On the left column of the table is the duty cycles of each spectrum band as reported in [9]. On the right column of the table, the duty cycle is calculated by measuring the total active time nodes sending data by using that particular spectrum and divides by total simulation by allowing smart grid users to use spectrum band opportunistically. It is observed that the duty cycles of each spectrum band have been improved

Table 2. Spectrum utilization efficiency.

Spectrum band	Spectrum utilization efficiency		Improvement
	Non-cognitive radio	Cognitive radio	
WiMAX band	5.92%	9.24%	3.32%
TV band	6.6%	10.77%	4.17%

7. Conclusion

The simulation results clearly show that cognitive radio outperforms non-cognitive radio. In NAN, cognitive radio outperforms non-cognitive radio by 3.32% and 2.32% in latency and throughput, respectively. In WAN, the performance of using cognitive radio in WiMAX and TVWS is promising. Both spectrum bands can offer high channel quality, low latency, and high throughput for WAN communication. The latency in NAN and WAN are kept well below the maximum latency of AMI requirement, which is 2000 milliseconds. This is due to the ability of sensing multiple channels and utilizing idle channels opportunistically decrease the probability of collision in random access, which in turn reduces the queuing delay of each packet. Besides, cognitive radio also improves spectrum utilization efficiency. Future work include a setting up a test-bed to test the practicability of cognitive radio based smart grid communication in real life.

8. References

[1] Patil K, Prasad R, and Skouby K 2011 A survey of worldwide spectrum occupancy measurement campaigns for cognitive radio *Int. Conf. Devices Commun.* 1-5

[2] Mehdawi M, Riley N, Ammar M, and Zolfaghari M. 2012 Comparing historical and current spectrum occupancy measurements in the context of cognitive radio *20th Telecommun. Forum* 623-6

[3] S D Barnes, P A Jansen van Vuuren, and B T Maharaj 2013 Spectrum occupancy investigation: measurements in South Africa *Measurement* **46** 3098-12

[4] Han Y, Wen Y, Tang W, and Li S 2012 Spectrum occupancy measurement: focus on the TV frequency *2nd Int. Conf. Signal Process. Syst.* Pp V2–490–V2–494

[5] Islam M H, Koh C L, Oh S W, Qing X, Lai Y Y, Wang C, et al. 2008 Spectrum survey in singapore: occupancy measurements and analyses *3rd Int. Conf. Cogn. Radio Oriented Wirel. Networks Commun. (CrownCom 2008)* 1–7

[6] Vo N Q B, Le Q C, Le Q P, Tran D T, Nguyen T Q, and Lam M T 2011 Vietnam spectrum occupancy measurements and analysis for cognitive radio applications *Int. Conf. Adv. Technol. Commun. (ATC 2011)* 135–43

[7] S Jayavalan 2014 Measurements and analysis of spectrum occupancy in the cellular and TV bands *Lecture Notes on Software Engineering* **2** 133-138

[8] W. A. Abood, N Din, A Ismail, and H Mohamad 2013 Electromagnetic spectrum survey of the environment in a locality in Kuala Lumpur, Malaysia *Conf. Ser. Earth Environ.* **16** 1-4

[9] Hiew Y K, N M Aripin, S Jayavalan, and N M Din 2013 Spectrum band for smart grid implementation in Malaysia *IEEE Student Conf. Res. Dev.* 26–30

[10] Lichtensteig B, Bjelajac B, Mueller C, and Wietfeld C 2010 RF mesh systems for smart metering: system architecture and performance *1st IEEE Int. Conf. Smart Grid Commun* 379–84

[11] Yu R, Zhang C, Zhang X, Zhou L, and Yang K 2014 Hybrid spectrum access in cognitive-radio-based smart-grid communications systems *IEEE Syst. J.* **8** 577-87

[12] Fatemieh O, Chandra R, abd Gunter CA 2010 Low cost and secure smart meter communications using the TV white spaces *3rd Int. Symp. Resilient Control Syst* 37–42

[13] Kuzlu M, Pipattanasomporn M, and Rahman S 2014 Communication network requirements for major smart grid applications in HAN, NAN and WAN *Computer. Networks* **67** 74–88.

[14] Nekovee M 2011 Current trends in regulation of secondary access to tv white spaces using cognitive radio *IEEE Glob. Telecommun. Conf. - GLOBECOM 2011* 1–6

[15] Sherman M, Mody A, Martinez R, Rodriguez C, and Reddy R 2008 IEEE standards supporting cognitive radio and networks, dynamic spectrum access, and coexistence *IEEE Commun. Mag.* **46** 72-79

Acknowledgements

The authors would like to express gratitude to Universiti Tenaga Nasional and Ministry of Higher Education (MOHE). This project is funded under MOHE (20130114 FRGS).

International Conference on Advances in Renewable Energy and Technologies (ICARET 2016) IOP Publishing
IOP Conf. Series: Earth and Environmental Science **32** (2016) 012007 doi:10.1088/1755-1315/32/1/012007

Artificial Intelligence Application in Power Generation Industry: Initial considerations

Rahmat Izaizi B. Ismail[1], **Firas B. Ismail Alnaimi**[1]*, **Haidar F. AL-Qrimli**[2]

E-mail: Firas@uniten.edu.my

Abstract. With increased competitiveness in power generation industries, more resources are directed in optimizing plant operation, including fault detection and diagnosis. One of the most powerful tools in faults detection and diagnosis is artificial intelligence (AI). Faults should be detected early so correct mitigation measures can be taken, whilst false alarms should be eschewed to avoid unnecessary interruption and downtime. For the last few decades there has been major interest towards intelligent condition monitoring system (ICMS) application in power plant especially with AI development particularly in artificial neural network (ANN). ANN is based on quite simple principles, but takes advantage of their mathematical nature, non-linear iteration to demonstrate powerful problem solving ability. With massive possibility and room for improvement in AI, the inspiration for researching them are apparent, and literally, hundreds of papers have been published, discussing the findings of hybrid AI for condition monitoring purposes. In this paper, the studies of ANN and genetic algorithm (GA) application will be presented.

1. Introduction

1.1. Artificial Neural Networks

Artificial Neural Network (ANN) is a fast-growing soft computing method, which has been used in different type of industries recently. ANN is a computational model inspired by natural neurons. ANN imitates the characteristic of a natural neurons by several functions, namely inputs (like synapses), which are multiplied by weights (strength of signals) and then computed by mathematical function, determining the activation of neuron. Another function will compute the output, which will sometime depend on a certain threshold. A neural network model is made up of interconnected artificial units (neurons). Neurons are arranged in different layers, including input layer, hidden layer(s), and output layer. The number of neurons and layers depends on the type of problems need to be solved and the complexity of the system to be modelled. Figure 1 shows a simple structure of a typical ANN with 4 inputs, first hidden layer with 5 neurons, second hidden layer with 3 neurons, and one output.

[1] Power Generation Research Centre, University Tenaga Nasional, 43000 Kajang, Malaysi.
[2] Department of Mechanical Engineering, Curtin University Sarawak, 98009 Miri, Malaysia.

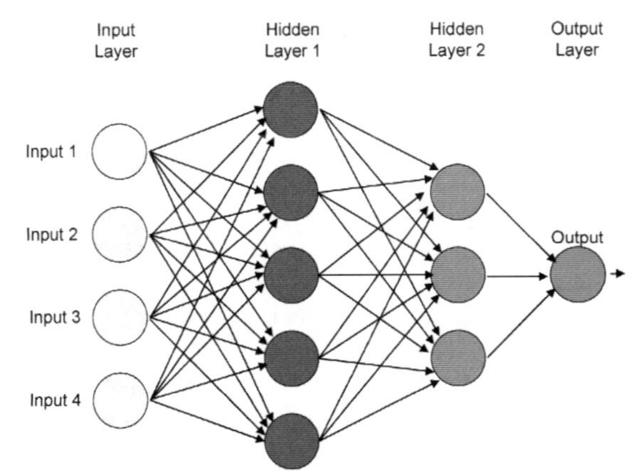

Figure 1. Simple ANN structure

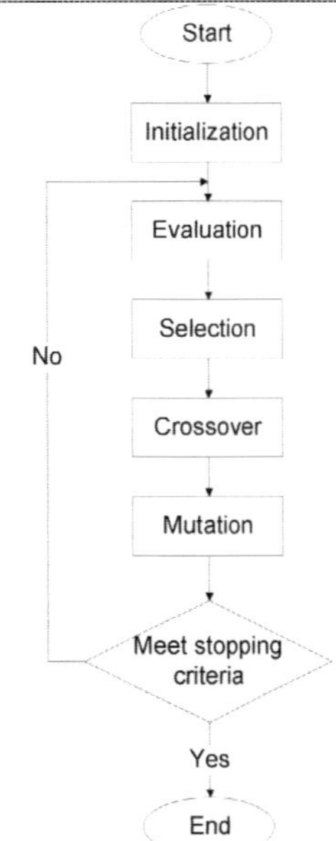

Figure 2. Genetic algorithms concept flowchart

Firas B. Ismail Alnaimi *et al.* [1] presented a detailed data preparation procedure for steam boiler fault detection and diagnosis (FDD) analysis, where real data of steam boiler are captured, identified, clustered, and sampled. The boiler behaviour was studied, and the most influencing parameters were decided. For fault detection and diagnosis neural network (FDDNN) model training-validation phase, feed-forward neural networks were used. The FDDNN model can detect and diagnose the super heater low temperature quickly and accurately, hence resulting in satisfactory performance.

Thomas Palmé *et al.* [2] demonstrated a solution for sensor fault detection, isolation, and accommodation by employing ANN as a classifier. Thomas Palmé *et al.* [3] also used nonlinear Principal Component Analysis (PCA) for early warning of gas turbine failure implemented through the use of Auto-Associative Neural Network (AANN). In this case study, the use of AANNs for early detection of abnormal engine behaviour could warn the operator a few days prior to full failure. The pros and cons of applying NN monitoring approaches are summarized in Table 1.

Table 1. The pros and cons of applying NN monitoring approaches.

Pros	• No detailed physical information about the GT is needed. • Only operational data is required. • NN-calculation is fast and can be used for online use • The interpretation is easy to understand. • It can establish relationships between performance parameters and operational conditions that are difficult to model
Cons	• As in all statistical models, data covering the entire operation range is needed for training. • Any new operational condition requires a retraining.

1.2. Genetic Algorithm

Genetic algorithms (GA) are a class of probabilistic optimization algorithms pioneered by John Holland in the 1970's, and became popular in the late 1980's. They are based on ideas from Darwinian Evolution inspired by the biological evolution process, as shown in Figure 2. GA is a way of solving problems by mimicking the natural processes combination of selection, recombination and mutation to evolve a solution to a problem. GA exploits historical information to direct the search into better performance within the search parameters.

B.Kishore *et al.* [4] proposed an application of Adaptive GA for fault detection in machinery. The results confirm that the networks correctly diagnose faults and guarantee good performances in terms of sensitive data obtained. A. Azadeh *et al.* [5] proposed a flexible algorithm based on SVM, GA and particle swarm optimization (PSO) for centrifugal pumps fault diagnosis. It was also applied to noisy data to show the robustness of the proposed algorithm in noisy environments. The result showed that support vector classifier improves when hybridized with GA and PSO.

2. Case Studies

2.1. Gas Turbine Sensor Validation through Classification with Artificial Neural Networks [2]

In this study, Thomas Palmé *et al.* focus on detecting single sensor faults. Steam turbine data such as pressures and temperatures, were produced with a Siemens performance deck. The data from sensors is divided into 3 classes; healthy, positive sensor drift, and negative sensor drift. Based on the sensor readings, the result would be translated into '1' if true and '0' if false as shown in Table 2. This approach has two advantages, namely information about the direction of the drift is provided, and the NN classification capabilities are improved.

Table 2. Example of Data Preparation For ANN Classification Training [2]

Sensor 1 reading	Sensor 2 reading	Class 1: healthy	Class 2: S1 too high	Class 3: S2 too high	Class 4: S1 too low	Class 5: S2 too low
1	2	1	0	0	0	0
5	6	1	0	0	0	0
1.1	2	0	1	0	0	0
5.5	6	0	1	0	0	0
1	2.4	0	0	1	0	0
5	7.2	0	0	1	0	0
0.9	2	0	0	0	1	0
4.5	6	0	0	0	1	0
1	1.6	0	0	0	0	1
5	4.8	0	0	0	0	1

Once a failing sensor has been identified by the classification neural network, a recovered value can be reproduced using remaining healthy sensors. This is achieved through employing individual regression neural networks using healthy sensors as inputs, and the desired parameter as output. This is further clarified when the author introduced white noise to create errors in the sensor, but the ANN predictions are near as good as those of a healthy sensor.

2.2. Using genetic algorithms to improve the thermodynamic efficiency of gas turbines [6]

Jose M. Chaquet *et al.* presented a method for optimizing thermodynamic efficiency of an aeronautical gas turbine. This method is based in the transformation of the original constrained optimization problem. Several tools are involved in the traditional design process of a gas turbine, which are efficiency estimator module and ThroughFlow codes. Efficiency estimator module computes the thermodynamic efficiency of a turbine design, rendering it capable of calculating the fitness function. ThroughFlow is a specific CFD code for turbo-machinery design that computes the flow variables along all the surface plane.

GA communicates with the ThroughFlow by two interfaces called export and import, which have been specifically designed to carry out our implementation. In the export process, all the data needed by the GA is generated, particularly all the fluid variables of the reference individual for computing the efficiency. After the run of the GA, the efficiency of best individual in the last generation is compared with the efficiency of the ThroughFlow model. Only three iterations are needed to obtain the optimal configuration, as shown in Table 3. For each iteration, the thermodynamic efficiency and the total number of airfoils are given for the ThroughFlow model and the best individual in the last generation of the GA. We can check that the total number of airfoils in ThroughFlow models is similar to the one in the GA in the previous iteration. The iterative process is halted when the GA does not change the input data from the ThroughFlow. Comparing the first and last iteration, we can see that the GA has increased the efficiency by 0.36%, and reduced the total number of airfoils by 10.56%.

Table 3. Iterative process between *throughflow* and GA.

Iteration	ThroughFlow		GA	
	$\eta-\eta_0$	*Num of airfoils*	Efficiency	*Num of airfoils*
1	0	1486	0.003924	1334
2	0.003253	1334	0.003480	1329
3	0.003408	1329	0.003408	1329

3. Discussion and Conclusion

ANNs were shown to be robust and reliable tools. They have been utilized to solve many operational problems, especially problems of complex systems with nonlinear dynamics. In this paper, a brief overview for applications of ANNs and GA for fault detection and diagnosis was presented. The studies done in this paper are to pave the way for the author to model an intelligent condition monitoring system for steam turbine application as the next initiative. As a final note, AI will continue to develop and play an increasingly important role in the area of fault detection and diagnosis for power generation industry in the coming future.

Acknowledgment

The author would like to express acknowledgement to Universiti Tenaga Nasional (UNITEN) and Ministry of Higher Education (MOHE), Malaysia for the opportunity and sponsorship of this corresponding research under the FRGS research fund.

References

[1] Firas B. Ismail Alnaimi and Hussain H. AL-Kayiem *Artificial Intelligent System for Steam Boiler Diagnosis Based On Superheater Monitoring* Journal of Applied Science 11 (9): 1566 – 1572, 2011.
[2] Thomas Palmé, Magnus Fast and Marcus Thern *Gas turbine sensor validation through classification with artificial neural networks* Applied Energy 88 (2011) 3898–3904.
[3] Thomas Palmé, Peter Breuhaus, Mohsen Assadi, Albert Klein, Minkyo Kim *Early Warning Of Gas Turbine Failure By Nonlinear Feature Extraction Using An Auto-Associative Neural Network Approach* Proceedings of ASME Turbo Expo 2011.
[4] B.Kishore, M.R.S.Satyanarayana, and K.Sujatha *Adaptive Genetic Algorithm with Neural Network for Machinery Fault Detection* International Journal of Advances in Engineering & Technology, Sept. 2013.
[5] A. Azadeh, M. Saberi, A. Kazem, V. Ebrahimipour, A. Nourmohammadzadeh, Z. Saberi *A flexible algorithm for fault diagnosis in a centrifugal pump with corrupted data and noise based on ANN and support vector machine with hyper-parameters optimization* Applied Soft Computing 13 (2013) 1478–1485
[6] Jose M. Chaquet, Enrique J. Carmona, Roque Corral *Using genetic algorithms to improve the thermodynamic efficiency of gas turbines designed by traditional methods* Applied Soft Computing 12 (2012) 3627–3635.
[7] Michael Negnevitsky *Artificial Intelligence, A Guide to Intelligent Systems* 3rd edition, University of Tasmania, 2011, pp. 1–25.

Biogas Production Modelling: A Control System Engineering Approach

D Stollenwerk[1], C Rieke[1], M Dahmen[1], M Pieper[1]

[1] Institut NOWUM-Energy, FH Aachen - University of Applied Science, Heinrich-Mussmannstr. 1, 52428 Jülich, DE

E-mail:info@nowum-energy.com

Abstract. Due to the Renewable Energy Act, in Germany it is planned to increase the amount of renewable energy carriers up to 60 %. One of the main problems is the fluctuating supply of wind and solar energy. Here biogas plants provide a solution, because a demand-driven supply is possible. Before running such a plant, it is necessary to simulate and optimize the process feeding strategy. Current simulation models are either very detailed like the ADM 1, which leads to very long optimization runtimes or not accurate enough to handle the biogas production kinetics. Therefore this paper provides a new model of a biogas plant, which is easy to parametrize but also has the needed accuracy for the output prediction. It is based on the control system approach of system identification and validated with laboratory results of a real biogas production testing facility.

1. Introduction

The restructuring of the energy sector as a result of the European climate targets is a big challenge in Germany. The greenhouse gas emissions should be reduced to at least 50 % below 1990 levels by 2050 [1,2]. As a result, in Germany the dispatch of renewable energy sources increased up to 25 % in 2013 caused by the Renewable Energy Act [3].

The security of supply can be increased by many options, as an improvement in weather forecasting, development of long and short-term storages and grid expansion [4,5]. Since biogas plants are weather-independent and can be used locally, they are an alternative and offer a high potential for grid stabilization. Today more than 7850 biogas plants operate with an installed capacity of about 3.5 GW [6]. With the introduction of direct marketing of the produced energy and the flexibility bonus it becomes economically interesting to run these plants in grid stabilizing configuration. In order to achieve the highest possible return on the exchange, the production must be accurately predicted. Consequently a simple, fast and reliable model for the biogas production is required.

Today the standard model of the anaerobic fermentation process is the Anaerobic Digestion Model No.1 (ADM1) [7]. It describes the degradation of biomass with 42 Parameters and 19 ordinary differential equations (ODEs). Additional models like the agriADM1 simplify and aggregate the input variables, but the basic model concept remains.

In this paper, in order to reduce the number of free parameters, we characterize the degradation process with a new approach. We show that it is possible to depict the process with only two parameters and one differential equation which can be determined by single additional feeding and response characterization. On the one hand, this eliminates complex and expensive substrate analysis. Due to the

IOP Conf. Series: Earth and Environmental Science **32** (2016) 012008 doi:10.1088/1755-1315/32/1/012008

small number of parameters, the new model can be simple adjusted for various substrates. On the other hand, we can show that the output prediction accuracy is sufficient to predict the gas amount of the fermenter.

2. Model

The widely accepted standard model of the anaerobic digestion process is the ADM1. Unfortunately it requires numerous parameters, which have to be measured. Especially if a new substrate should be fed to an existing plant for every substrate the model has to be parameterized again which is time consuming, not economically and often not possible, due to the numerous measurements required for the parameters. Thus we developed a new model, which is reduced to the most important parameters.
To define the degradation characteristics, some assumptions were applied: First the biogas process was considered as linear, similar to ADM1. Second the coefficients of the equations were assumed to be time independent. Furthermore, an inhibition up to the maximum amount of substrate was excluded. Consequently the degradation of biomass could be described by one simple ODE, represented as transfer function in the Laplace domain.
The fundamental idea behind the model is that the process can be simplified by using model reductions and aggregations. So we considered the fermenter as grey box, summarized in Figure 1

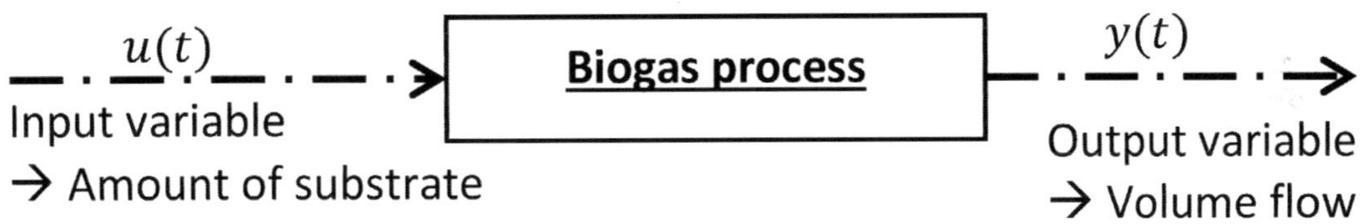

Figure 1. Block diagram of the biogas process.

The feed substrate is used as input variable u. The output variable represents the volume flow $\dot{V}_{total}$. As demonstrated in independent experiments [8,9], it is possible to model the volume flow $\dot{V}_{total}$ by means of ordinary differential equations of first order. The corresponding transfer function in the Laplace domain is:

$$G(s) = \frac{K}{1 + s\,T} \quad (1)$$

This result in

$$\dot{V}_{total} = G(s) \cdot U \quad (2)$$

The dynamic parameters K and T are determined by a conventional method in control theory, the system identification.
Since the feeding period is small compared to the expected large time constants, the feeding can be regarded as Dirac Delta pulse. Due to the integrative measurement and the linear behaviour the response of the system was characterized by typical step response.
The objective is now to determine the parameters of the transfer function as seen in equation (1). The required parameters are K and T. The gain value K represents the biogas amount. T is the corresponding time constants, which determines the time response.
Therefore an additional feeding was carried out at an arbitrary time t. The feeding amount corresponded to *6 kg*.

3. Methods and Experimental Setup

The experiments were performed on a test biogas plant (depicted in Figure 2) with a fermenter volume of 1.5 m³. The volume is split into a 1 m³ working part and 0.5 m³ gas storage. The pH value of the test facility can be assumed on average as neutral.
The reactor temperature was managed by a control system and kept constant at 40°C. Thus, the process was maintained under mesophilic conditions. Furthermore, the installed stirrer operated with an on-off ratio of 10 minutes to 20 minutes.

The gas measurement was sampled every 30 minutes with an accuracy of 10 litres, so 48 values could be measured per day. The measurement was performed as soon as the installed pressure valve had reached a pressure difference of 2 mbar.

Figure 2. Lab-Scale Fermenter
Source: PlanET Biogastechnik GmbH Documentation.

4. Experimental Results

The additional feeding was carried out at an arbitrary time *t*. The feeding amount corresponded to *6kg*. The response of the biogas fermenter is depicted in Figure 3.

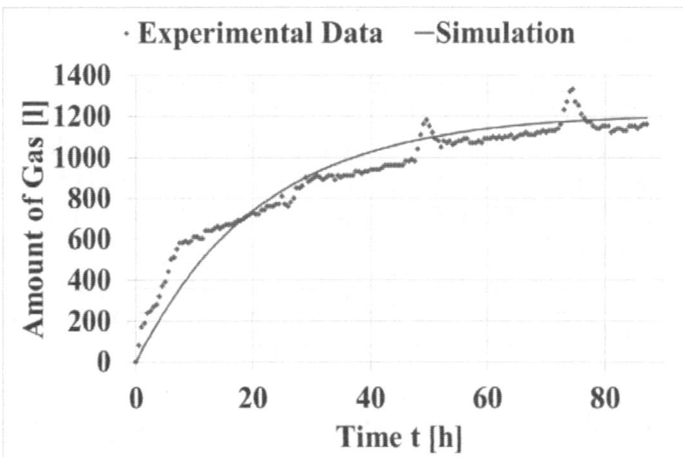

Figure 3. Comparison Experimental versus Simulation Data

The parameters K and T are determined by a curve fitting based on the method of least squares in MATLAB. On that account, the square deviation between the set point and actual value was calculated and used as the objective function. As solver method "fmincon" was selected. The boundary conditions of the parameter fitting can be noted as follows, the time constants have to be bigger than zero. In addition the sum of the K Value should be close to the gas generation potential of maize silage. The literature value was specified with 586.1 N l/kg substrate. This corresponds to an amount of gas of 202 litres/kg [10,11].

The result of the parameter fitting can be summarized in equation (3) and (4)

$$G(s) = \frac{202\ l/kg}{21.3\ h\ s + 1} \tag{3}$$

$$\dot{V}_{total} = \frac{202\ l/kg}{21.3\ h\ s + 1} \cdot U \tag{4}$$

5. Discussion and Conclusion

The results show that it is possible to model the biogas production with one differential equation of 1st order. Thus, the process can be described by means of two parameters, one transfer coefficient and the associated time constant. The parameters can be determined for each substrate by an additional feeding at an arbitrary time t. This allows the determination during operation. It should be noted, that the fast response kinetics in the beginning cannot be modelled with one first order transfer function.
The validation shows that the model works qualitatively correct. Due to various factors, in particular the water content in the substrate, the result is quantitatively not matching with the set point specification. As part of future studies, the parameters and the interaction of different substrates should be investigated. In addition, the model should be extended to more than one transfer function and to an extent, so it is possible to divide the gas production in methane and carbon dioxide parts.

Acknowledgements
The authors are grateful to the company PlanET GmbH for providing the test facility.
The research is partially financed by the Federal Ministry of Education and Research (BMBF). Contract No. 17N1011

6. References

[1] European Union. DECISION No 406/2009/EC OF THE EUROPEAN PARLIAMENT AND OF THE COUNCIL of 23 April 2009 on the effort of Member States to reduce their greenhouse gas emissions to meet the Community's greenhouse gas emission reduction commitments up to 2020: 406/2009/EC; 2009.
[2] European Union. DIRECTIVE 2009/28/EC OF THE EUROPEAN PARLIAMENT AND OF THE COUNCIL of 23 April 2009 on the promotion of the use of energy from renewable sources and amending and subsequently repealing Directives 2001/77/EC and 2003/30/EC: 2009/28/EC; 2009.
[3] BDEW. Erneuerbare Energien und das EEG: Zahlen, Fakten, Grafiken (2014) Anlagen, installierte Leistung, Stromerzeugung, EEG-Auszahlungen, Marktintegration der Erneuerbaren Energien und regionale Verteilung der EEG-induzierten Zahlungsströme 2014.
[4] Szarka N, Scholwin F, et al. A novel role for bioenergy: A flexible, demand-oriented power supply. **Energy** 2013;**61**:18–26.
[5] Hahn H, Ganagin W, et al. Cost analysis of concepts for a demand oriented biogas supply for flexible power generation. **Bioresource technology** 2014;**170**:211–20.
[6] Fachverband Biogas e.V. German Biogas Association. *Branchenzahlen*. [April 17, 2015]; Available from: http://www.biogas.org/edcom/webfvb.nsf/id/DE_Branchenzahlen.
[7] Batstone D J, Keller J, et al. (2002). Anaerobic Digestion Model No. 1. *IWA Task Group for Mathematical Modelling of Anaerobic Digestion Processes, Scientific and Technical Report No.* **13**, IWA Publishing
[8] Weinreich S, *Praxisnahne Modellierung von Biogasanlagen*. Leipzig; 2014.
[9] Engler N, Nelles, M, *Messungen zur Abbaukinetik von Einzelsubstraten und Substratmischungen*. Leipzig; 2014.
[10] Quaschning V. *Regenerative Energiesysteme: Technologie - Berechnung - Simulation*. 8th ed. München: Hanser; 2013.
[11] *Biogasausbeuten verschiedener Substrate - Programm Berechnung* - Bayerische Landesanstalt für Landwirtschaft (LfL). [May 02, 2015]

Biogas Production on Demand Regulated by Butyric Acid Addition

K Kasper[1], J Schiffels[3], S Krafft[1], I Kuperjans[1], G Elbers[2], T Selmer[3]

[1] Institute NOWUM-Energy, FH Aachen - University of Applied Sciences, Heinrich-Mussmann-Str. 1, 52428 Juelich, DE; [2] FH Aachen - University of Applied Sciences, Heinrich-Mussmann-Str. 1, 52428 Juelich, DE; [3] Instituts of Nano- und Biotechnologies (INB), FH Aachen - University of Applied Sciences, Heinrich-Mussmann-Str. 1, 52428 Juelich, DE

E-mail: selmer@fh-aachen.de

Abstract. Investigating effects of volatile fatty acids on the biogas process it was observed that butyric acid can be used for transient stimulation of the methane production in biogas plants operating with low energy substrates like cattle manure. Upon addition of butyrate the methane output of the reactors doubled within 24 h and reached almost 3-times higher methane yields within 3-4 days. Butyrate was quantitatively eliminated and the reactors returned to the original productivity state within 3 days when application of butyrate was stopped. The opportunity to use butyrate feeding for increased biogas production on demand is discussed.

1. Introduction

Limited availability of fossil fuels and global warming, which is associated with their use, represent major challenges for the future of modern societies. A sustainable energy supply requires the consistent use of wind, water and solar energy, but also the use of biomass for energy production. Only the latter is able to compensate for power shortfalls that may occur in adverse weather conditions.

Anoxic degradation of biomass yields carbon dioxide and methane as final products. Methane can be used as an energy source and could be provided by consistent use of waste streams from households, agriculture and food production industry. The formation of biogas requires a complex syntrophic interaction of biomass-degrading bacteria and methanogenic archaea [1-3]. The former group of organisms degrades biopolymers and thrives on fermentation of the monomers to yield complex mixtures of organic acids, alcohols, carbon dioxide and small amounts of hydrogen. The organic intermediates are further oxidized when methanogenic archaea are present and use hydrogen to form methane. This permits anoxic oxidation of organic intermediates to increase the yields of methanogenic substrates, e.g. acetate, hydrogen and carbon dioxide. In stable biogas processes the hydrolysis of biopolymers regulates the whole process, so that no more intermediates are formed, than the methanogens are capable to consume.

The production of methane is positively influenced by increasing biomass feeding only within narrow limits: the increased availability of fermentable substrates quickly leads to acidification and thus eliminates the methanogens from the system. Nevertheless it is well known that volatile fatty acids formed at the end of the fermentative process are readily converted into methane [4, 5]. These findings encouraged us to study the effect of large doses of butyrate on biogas quantity and quality. The data presented herein suggest that butyrate effectively increases the formation of biogas reversibly in short time and permits compensation of power shortfalls caused by adverse weather conditions.

2. Methods and Experimental Setup

Experiments were performed in continuously fed lab scale fermenters of 1 L volume with a content volume of approx. 0.8 L which was continuously stirred at 37 °C. The fermenters were fed multiple times per day to yield a total of 10 g organic matter/(L*d) with cattle manure (3.75 % organic dry matter). Neutralized sodium butyrate solution (pH 7) was added in 48 h intervals (80 mmol butyrate per injection). The fatty acid concentrations in samples were determined by HPLC as recently described [5]. Gas quantity and quality (CH_4- and CO_2-content) were monitored online with near infrared gas sensors (BlueSense).

3. Results

The laboratory fermenters constantly fed with cattle manure produced about 44 mL of gas per hour yielding approximately 2.1 L of biogas in 48 h. The methane concentration during these control periods was 67.5 % (Fig. 1). These values correspond to 63 mmol of methane formed per day.

Following the addition of 80 mmol butyrate, the rate of gas formation gradually rose from 44 mL/h to 86 mL/h. The biogas yield increased to 3.4 L over a time period of 48 h (162 % of control). Likewise, the gas quality improved yielding a methane concentration of up to 72 %. Repetition of butyrate additions in time intervals of 48 h further increased the maximum gas formation rate to about 100 mL/h for the second addition up to about 120 mL in continuation. The maximum methane concentration slightly increased to 74.7%. The average biogas yield per 2 days rose up to 4.9 L in 2 days and the methane yield reached up to 161 mmol in 48 h. The reactors readily returned to the original productivity state and latest three days after the last addition of butyrate gas formation rates, quality and methane yield returned to the original values.

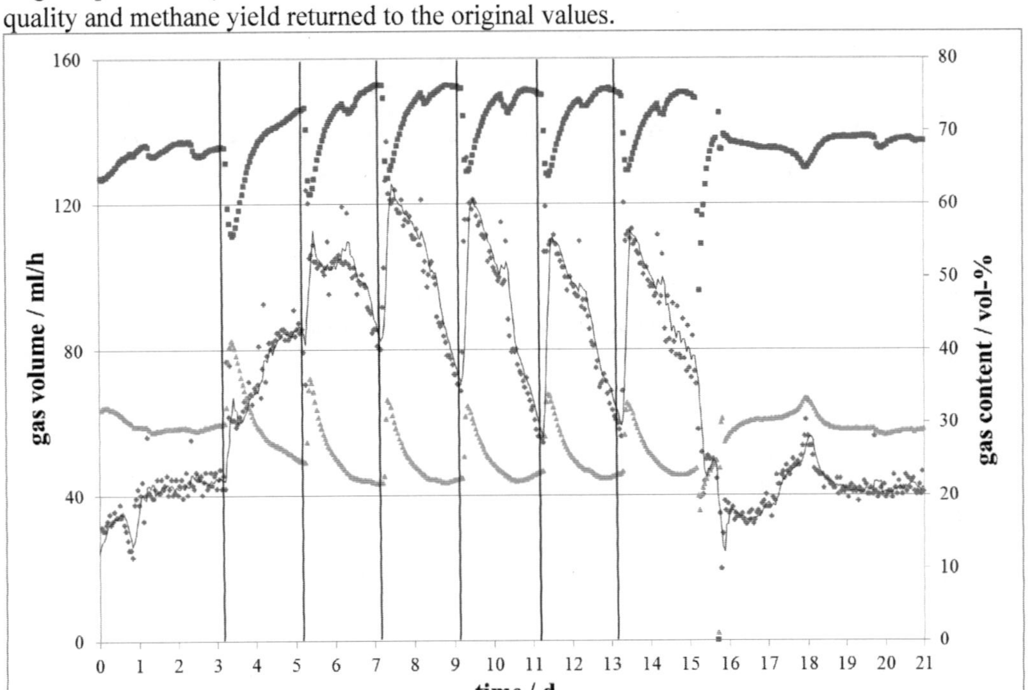

Figure 1. Continuous records of biogas formation rate and gas quality. The gas volume formed per h (blue diamonds) and the concentrations of methane (red squares, upper curve) and carbon dioxide (green triangles, lower curve) are shown. The points of butyrate addition (vertical lines) are indicated.

The stepwise increase in gas formation rates and methane yield in the beginning suggest a period of adaptation. This adaption was also evident when the fatty acids concentrations in the reactor were monitored (Fig. 2). When solely fed with cattle manure, the only fatty acid detectable in significant equilibrium concentrations (8 mM) was acetate. The concentrations of propionate (1.2 mM), butyrate (not detectable) and *iso*-butyrate (not detectable) were significantly lower. As expected, the addition of butyrate (100 mmol/L digestate, 80 mmol in total) caused an immediate increase of its concentration to about 60 mM. The addition of the next portion of butyrate caused further increase to 74 mM. The degradation of butyrate was evident in the intermediately formed acetate; its concentration increased up to 67 mM within 3 days. Beyond this point, however, the acetate concentration showed regular periodic ups and downs one day each displaced from the butyrate additions. It is clearly evident from the course of butyrate concentrations that the fermenter biocoenosis readily consumed butyrate once "trained". 24 hours after the first addition, 87 % of the originally measured butyrate could be detected. After the third addition, only 15 % of butyrate was left after 24 hours, compared to the initial concentration. Another interesting finding was the observation of increasing concentrations of the other fatty acids, which rose significantly to concentrations substantially above the initial levels.

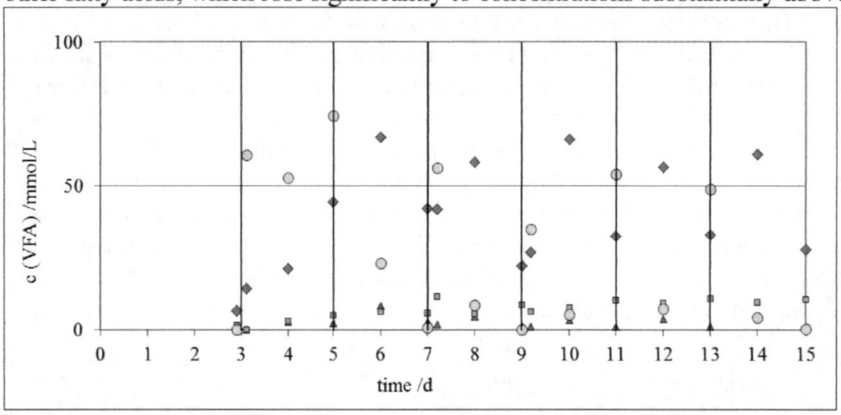

Figure 2. Concentrations of volatile fatty acids in the fermenter. The concentrations of acetate (magenta diamonds), propionate (green squares), butyrate (cyan dots) and iso-butyrate (red triangles) are shown.

4. Discussion

Previous studies suggested that about 1/3 of the methane produced from cattle manure is formed from short-chain fatty acids [5]. In particular acetate (~100 mM) and propionate (~25 mM) are present in significant amounts, whereas the concentrations of butyrate (~10 mM) and iso-butyrate (< 5 mM) are much lower. An increase of even-numbered fatty acids like butyrate or caproate was anticipated to account for significant increases in the carbon flux through the fermenter causing comparably smaller rates of acidification than acetate.

For theoretical yields it was calculated that 80 mmol of butyrate introduced should yield 200 mmol methane and 120 mol carbon dioxide, according to equation (1).

$$2\ \text{Butyrate}^- + 2\ H^+ + 2\ H_2O \rightarrow 5\ CH_4 + 3\ CO_2 \quad \Delta G'^\circ = -\ 88.5\ \text{kJ/mol butyrate (eq. 1)}$$

The observed productivity between 130 mmol to 161 mmol methane formed within 2 days, suggesting a conversion of butyrate to methane in the range of at least 40 %. This might be caused by immediate overflow after addition to the fermenter, as suggested by the measured butyrate concentrations. Evidently, the butyrate had an inhibitory effect on the formation of biogas from other VFAs. Thus, quantification of the methane formed from butyrate remains doubtful. However, for the measured maximum butyrate concentration of 60 mM, a conversion rate of about 68 % is theoretically reasonable. The observed gas yields, however, did not exceed 40% of the anticipated methane. It must be stated that part of the butyrate is consumed for biomass production and up to 25 % losses per day can be anticipated as wash-out of the fermenter caused by the ongoing background-feeding with cattle manure. Another interesting finding was the increasing gas quality, which is not readily explainable without further investigation.

The measured effects on gas production and VFA consumption turned out to be completely reversible: It is obvious that neither the butyrate oxidizing bacteria nor the methanogenic community in the reactor were significantly disturbed by addition of butyrate, although the biogas reactors organic load was increased from 10 g/(L*d) to 17 g/(L*d). This is reflected in the observation that gas volume and composition returned to the pre-experimental values within less than 72 hours after the last addition.

The response time to the following addition of butyrate was very short and within 24 h the amount of formed methane per hour roughly doubled. Although maximum performance with up to 3-fold increased methane yields was reached only after 4 days in the experimental setup, it must be stressed that gradually and controlled adjusted fermenters will likely reach the maximum performance of the microbial community within 3 days. The required period of 3 days to more than double the methane output of a given biogas plant is within the accuracy time of modern weather forecast. The speed of response provides, therefore, sufficient reaction time, to respond to bad conditions for the production of wind and solar energy.

This suggests that butyrate and probably also other even numbered volatile fatty acids are suitable for biogas process regulation and surplus production of larger quantities of biogas on demand: Fed with natural, comparable low yield substrates from waste streams, biogas plants can provide an important building block in a decentralized energy supply. The high energy substrate butyrate is easily formed from synthesis gas or hydrogen/CO_2 by acetogenic microbes. Thus, on windy, sunny days a surplus of produced energy can be stored in this compound. Likewise, butyrate is a common metabolic end product of lignocellolytic clostrial species, too. It is, therefore, easy to produce using biotechnological methods and can be stored save and easily at low costs.

To substantiate these findings in a technical context, the ongoing project "ACidestion" (funded by BMWi, grant no. 03KB084B) examines the effects of various ensiling strategies on the biogas process and the controllability via simulation by a simplified model.

Acknowledgements
The experimental work was part of the project "Bio-LAPS", funded by the BMELV (grant no. 22026407).

5. References

[1] Angelidaki, I., L. Ellegaard, and B.K. Ahring, *Applications of the anaerobic digestion process.* Adv Biochem Eng Biotechnol, 2003. **82**: p. 1-33.
[2] Blaut, M., V. Muller, and G. Gottschalk, *Energetics of methanogenesis studied in vesicular systems.* J Bioenerg Biomembr, 1992. **24**(6): p. 529-46.
[3] Ferry, J.G., *Fundamentals of methanogenic pathways that are key to the biomethanation of complex biomass.* Curr Opin Biotechnol, 2011. **22**(3): p. 351-7.
[4] Ahring, B.K. and P. Westermann, *Kinetics of butyrate, acetate, and hydrogen metabolism in a thermophilic, anaerobic, butyrate-degrading triculture.* Appl Environ Microbiol, 1987. **53**(2): p. 434-9.
[5] Schiffels, J., M.E. Baumann, and T. Selmer, *Facile analysis of short-chain fatty acids as 4-nitrophenyl esters in complex anaerobic fermentation samples by high performance liquid chromatography.* J Chromatogr A, 2011. **1218**(34): p. 5848-51.

IOP Conf. Series: Earth and Environmental Science **32** (2016) 012010 doi:10.1088/1755-1315/32/1/012010

Water Quality Improvement through Reductions of Pollutant Loads on Small Scale of Bioretention System

Norshafa Elyza Muha[1], Lariyah Mohd Sidek[2]Milad Jajarmizadeh[2]

[1]Faculty of Civil Engineering,
Universiti Teknologi MARA, Pulau Pinang.

[2]Centre for Sustainable Technology and Environment,
College of Engineering, Universiti Tenaga Nasional, Kajang, Selangor

E-mail: adeq_sha@yahoo.com

Abstract. Bioretention system is introduced as an important topic namely Urban Storm Water Management Manual for Malaysia (MSMA) by the Department of Irrigation and Drainage Malaysia (DID) in May 2012. The main objective of this paper is to evaluate the performance of water quality for small scale bioretention system under tropical climate via MUSIC model. Two bioretention systems 1 and 2 are observed based on the difference media depth. The result of bioretention system is compared with a reference model which has infrastructure with Urban Stormwater Improvement Conceptualisation (MUSIC) for pollutants load reduction and water quality results. Assessment of results via MUSIC software indicates a significant percentage of reduction for Total Suspended Solid (TSS), Total Phosphorus (TP) and Total Nitrogen (TN). The prediction of pollutant reduction via using MUSIC has the harmony for requirement in MSMA. TSS pollutant reduction is more than 80%, while for TP and TN more than 50%. The outcome of this study can be helpful for improvement of the existing MSMA guidelines for application of bioretention systems in Malaysia.

1. Introduction

Best Management Practices (BMPs) is introduced to help in planning storm water properly such as Low Impact Development (LID), Sustainable Urban Drainage (SUD), Water Sensitive Urban Design (WSUD) and Urban Storm Water Management Manual for Malaysia (MSMA) [1]. The main objective of stormwater management practices is to match the state of the development area that was altered in terms of its hydrologic condition due to urbanization to the level of condition before the area was developed. The common BMPs are to use the concept of detention and retention systems to attenuate the peak runoff and prevent or reduce sediments, nutrients and other pollutants from transport between land and surface water. Storm water management integrates to the aspect of quality, quantity and amenities. Best management practice should be seen as the best way in making the neighborhoods and natural habitats become greener with the improvement of environmental conditions in urbanized and surrounding areas [2].

In Malaysia, Urban Stormwater Manual (MSMA) has been introduced by Department of Irrigation and Drainage (DID) to provide guidance pertaining to the latest stormwater best management practices (BMPs). This technical guideline can assist decision makers such as engineers, developers and stakeholders to implement sustainable development in Malaysia. The first urban drainage manual, "Planning and Design Procedures No.1: Urban Drainage Design Standards and Procedures for Peninsular Malaysia" was published in 1975 and followed by "Urban Stormwater Management Manual (MSMA) published in year 2000 and was revised with the MSMA second edition in 2012. Every development project must follow and comply with the stormwater quantity and quality objectives to meet the pre-development condition.

A new chapter on bioretention system is introduced in Chapter 9 in Urban Stormwater Manual for Malaysia (MSMA) 2nd Edition. Bioretention system or also known as 'rain garden' is one of BMP's techniques that is used to control the water quantity at attributed sources and it can be the adaptable method applied throughout many regions. Bioretention system can retain and treat urban runoff by using vegetation or plants before it flows into the main storm drain system. Bioretention system is made of an excavated basin or landscape depression consisting of plants, ponding area, mulch layer, several layers of planting soil and underdrain (optional). Two types of bioretention system which are bioretention 1 and 2 with different media depth were monitored. This study focuses on the application of bioretention system as a micro scale of BMP's that works in reducing stormwater quality from entering the drainage system.

2. Site Description

A bioretention research and education site was constructed at Universiti Tenaga Nasional (UNITEN), Putrajaya Campus in August 2013. The bioretention systems (1 and 2) were designed as impermeable system and constructed according to the standard criteria based on Urban Stormwater Management Manual for Malaysia (MSMA) 2nd Edition. The site contains two parallel bioretention cells that capture and treat stormwater runoff from catchment of an asphalt surface parking area (approximately 0.08 hectare). Each bioretention cell (system) is rectangular in shape with the length of 12.0 m and width of 4.0 m resulting surface area is about 48m^2 for each cell. The standard materials in each cell consist of an engineered soil mix of 20 to 25 % top soil, 50 to 60 % medium sand and 20 to 25 % organic leaf compost. The total engineered soil depth in each cell is 1.0m and 0.45m thickness promoted stormwater treatment through infiltration process. A network of perforated pipes with 100mm in diameter in the drainage layer is conveyed the infiltrated stormwater to the outlet point.

3. Methodology

The stormwater from the catchment area flows through the drainage system to the bioretention system inlet where the stormwater is monitored. The monitoring stations are recorded on-site rainfall data, measured the inflow and outflow into the bioretention system and automatically captured the stormwater samples for selected rainfall events. All field data are recorded in a data logger that is kept inside the housing together with other apparatus such as automatic sampler and batteries. Water samples are transferred into sampling bottles and clearly labelled for future identification. During transportation, stormwater samples are stored in a cool box with ice cubes to maintain the required low temperature. The selections of parameters for analysis are Total Suspended Solid (TSS), Total Phosphorus (TP) and Total Nitrogen (TN) according to the targeting parameters for removal efficiency in Urban Stormwater Manual (MSMA) 2nd edition. The performance of bioretention systems at the study area has been modelled with MUSIC as conceptual design tool software. This software can estimate stormwater pollutant loads that are generated in a catchment and predict the performance of stormwater treatment measures. By stimulating the performance of stormwater quality improvement measures, MUSIC is able to determine whether if the proposed stormwater management systems can meet the specified water quality objectives.

IOP Conf. Series: Earth and Environmental Science **32** (2016) 012010 doi:10.1088/1755-1315/32/1/012010

4. Result and Discussion

The existing and future pollutant loads generated from catchment in the study area is computed by MUSIC using adopted pollutant concentrations. The rainfall data template for year 2014 is taken from UNITEN rainfall gauge station to provide more accurate simulation of study area condition. The entire cumulative pollutants load that coming from the upstream of the catchment has been considered to provide a clear sample of pollutions has come out from the catchment. Table 1 shows the results obtained from MUSIC simulation modelling with the application of bioretention systems. Bioretention systems are incorporated into the MUSIC modelling to study the performance water quality at the study area.

Table 1 MUSIC modelling result for bioretention 1 and 2

Parameter (kg/year)	Before Bioretention Implementation		After Bioretention Implementation		% of Reduction	
	Final Discharge				Bioretention 1	Bioretention 2
	Bioretention 1	Bioretention 2	Bioretention 1	Bioretention 2		
TSS	123	123	2.76	3.49	97.8	97.2
TP	0.328	0.328	0.047	0.052	85.5	84.2
TN	4.06	4.06	0.943	1.04	76.8	74.3

Based on the MUSIC modelling results with bioretention system, it can be seen that the water quality parameters of TSS, TP, TN and gross pollutants accomplished significant reduction with the implementation of bioretention system components (Table 1). The targeted reduction of TSS is 97.8% to 97.2% for bioretention 1 and bioretention 2 respectively. Moreover, reduction goal of TP is obtained 85.5% to 84.2% for bioretention 1 and 2 respectively. Finally, TN outputs are between 76.8% and 74.3% for bioretention 1 and 2. MUSIC has been used to predict the effectiveness of bioretention system in terms of water quality for selected storm event based on rainfall depth between 25.6 mm to 94.8 mm for year 2014 (Table 2). The attributed dates of selected rainfall events are 6th January 2014, 31st March 2014, 9th April 2014, 21st May 2014, 28th October 2014 and 15th November 2014. The ranges of recorded rainfalls intensity for the selected events are between 27.93 mm/hr and 66.92 mm/hr. Prediction at final outlet for all storms events, gross pollutants are predicted to reduce totally 100%. Based on the MUSIC result in Table 2, the prediction of removal efficiency for overall parameters shows that percentage of reduction is slightly higher for bioretention 1 compared to bioretention 2. In regard with the depth of filter media, which is different as 1.0m and 0.45m, the removal efficiency between two bioretention system are less than 1% for TSS and less than 15% for TP and TN based on selected events (Table 2).

Table 2 Percentage of differences on predicted removal efficiency by MUSIC Model between bioretention 1 and 2

Parameters	% Differences of Removal Efficiency					
	06/01/14	31/03/14	09/04/14	21/05/14	28/10/14	15/11/14
Rainfall intensity (mm/hr)	27.93	43.20	39.72	35.28	42.88	66.92
TSS	0.50	0.30	0.30	0.20	0.30	0.10
TP	6.10	4.91	4.91	3.02	3.23	1.90
TN	13.13	11.04	11.04	7.02	7.48	4.76

The result from bioretention system 1 and 2 is compared at the final discharge 1 and final discharge 2 (Table 3). It can be seen as an average, the percentage removal efficiency for monitoring samples and MUSIC modelling result are not much different. However, the predicted removal efficiency using MUSIC showed slightly higher percentage removal than monitoring results. For example, the TSS removal efficiency for monitoring bioretention 1 and 2 are 92.2 % and 92.4 % compared to MUSIC result 97.8 % and 97.2 %. For TP removal efficiency, ranges of monitoring result are 82.1% to 83.3% while MUSIC results are 85.5 % to 84.2%. Monitoring results are 61.4 % to 64.7 % for TN range in comparison with MUSIC result which are 74.3 % to 76.8 %.TN shows that it has highest overestimation by MUSIC model for two bioretention systems. Moreover, TP includes shortest overestimation by MUSIC model.

Table 3 Comparison of Percentage Removal Efficiency between MUSIC Modelling Results and Monitoring Results

Parameters	Average % Removal Efficiency from Observation		% Removal Efficiency from MUSIC Modelling		% Difference	
	Bioretention 1	Bioretention 2	Bioretention 1	Bioretention 2	Bioretention 1	Bioretention 2
TSS	92.2	92.4	97.8	97.2	5.7	4.9
TP	82.1	83.3	85.5	84.2	4.0	1.1
TN	61.4	64.7	76.8	74.3	20	13.0

This study can show the modelling of water quality using MUSIC is necessary during the planning and design process in order to predict the removal efficiency of applied bioretention system. MUSIC is a design tool which currently one of the popular models to predict the performance of various WSUD techniques applied in Australia [3].

5. Conclusion

Bioretention system on a local scale in Universiti Tenaga Nasional Malaysia is a pilot study that can serve as effective and great integrated solution to improve urban stormwater pollution. Results indicated that bioretention system have the ability to remove TSS, TP and TN significantly. The MUSIC model is able to simulate the performance of bioretention system with the optimum treatment train. In conclusion, the outcome of this study can contribute to support the design procedure and criteria of bioretention facilities as outlined in the MSMA.

Acknowledgement

The authors acknowledge the financial support provided by the Ministry of Higher Education of Malaysia under the LRGS grant scheme (Project No. 203/PKT/6720004).

References

[1] Lariyah M S Mohamed Roseli Z A Zulkefli M Amirah Hanim M P .2014. Application of porous pavement system for MSMA stormwater management ecoHydrology: comparison of asphalt, interlocking and turf pavement. 13th International Conference on Urban Drainage, 7-12 September , Sarawak, Malaysia.

[2] A. E. Barbosa, J. N. Fernandes, and L. M. David, "Key issues for sustainable urban stormwater management.," *Water Res.*, vol. 46, no. 20, pp. 6787–98, Dec. 2012.

[3] M. A. Imteaz, A. Ahsan, A. Rahman, and F. Mekanik, "Modelling stormwater treatment systems using MUSIC: Accuracy," *Resour. Conserv. Recycl.*, vol. 71, pp. 15–21, 2013.

IOP Conf. Series: Earth and Environmental Science **32** (2016) 012011 doi:10.1088/1755-1315/32/1/012011

Sustainable Approach of Reusing Open Dumping Area by Predicting Long-Term Settlement of Waste Soil

Ahmad Sulaimi Shariffuddin[1], Nur Irfah Mohd Pauzi[2] and Goh Yu Hao[3]

[1]Postgraduate Student, Civil Engineering Department, College of Engineering, Universiti Tenaga Nasional, Jalan UNITEN-IKRAM, Putrajaya Campus, 43009 Kajang, Selangor

[2]Senior Lecturer, Civil Engineering Department, College of Engineering, Universiti Tenaga Nasional, Jalan UNITEN-IKRAM, Putrajaya Campus, 43009 Kajang, Selangor

[3]Graduate Student, Civil Engineering Department, College of Engineering, Universiti Tenaga Nasional, Jalan UNITEN-IKRAM, Putrajaya Campus, 43009 Kajang, Selangor

Corresponding author: irfah@uniten.edu.my

Abstract

Sustainable approach towards an open dumping area can be determined by predicting the long-term settlement of waste soil. To predict the settlement that happens in closed open dumping area in Bukit Palong, Lukut in the state of Negeri Sembilan, Malaysia, the Power Creep Function is used. The waste soil were collected to determine their basic geotechnical properties by doing sieve analysis, Atterberg Limit, consolidation, compaction and direct shear. From the geotechnical properties of the waste soil, the Power Creep Function are be used to determine the long-term settlement of the reused dumping area. There are three case studies and one of the parameters is M' or reference compressibility that have the value from 1.6×10^{-5} to 5.8×10^{-5} kPa. There is also N' or rate of compression that have value from 0.50 to 0.67. In Case 1, M' and N' are constant which are 5.8 x 10-5 kPa and 0.67 respectively. Case 2 have the value of M' increased from 1.6 x 10-5 to 5.8 x 10-5 kPa with respect to age while N' constant at 0.5. Lastly the Case 3 shows the M' is constant at 1.6 x 10-5 kPa while the N' is increased ranged from 0.50 to 0.67 with respect to age. The area will reach a stable condition which means the settlement has stabilized after 50 years. From the three case studies, the analysis show that the settlement does not exceed the allowable settlement which is 0.05m every year.

1. Introduction

Nowadays the use of open dumping area does not attune with the rising of public awareness about environmental issues, including about sustainability. Open dumping requires the least development and operational cost requirement among other method of land disposal, and thus, are the most prevalent type of disposal facilities in most developing countries. And open dumping pose the greatest threat to the environment and public health from the others [1].

Open Dumping Area are unplanned particularly no respect to sitting consideration. Open dumping area are also haphazardly operated, no controls over waste inputs whether it in quantity or properties of the waste itself or both and have no controls over emission of pollutant released due to waste decomposition [2]. Closing, or upgrading open dumping area is a key issue and upgrading is an important procedure in decreasing future environmental and public health impacts, as well as averting future costs caused by the continuing transgression of waste, evident at open dumps [3]. To upgrade the open dumping area, the settlement occur at the area must be calculated first.

To calculate the settlement that occurs in the dumping area, there are a few models to predict. There is the soil mechanics approach [4], Rheological model [5], Empirical Function [6] and Bjangard and Edgers [7]models. Under the Empirical Functions contains three formula which are the Logarithmic Function [6] and Hyperbolic Function [6]. The other formula under the Empirical Function is used in this paper which is the Power Creep Function [6]. The other formula did specify on what is the time the waste was dumped onto the soil unlike the Empirical Functions. Among the Empirical Functions, this formula was chosen rather than Power Creep and Hyperbolic Function because it relate the settlement rate with time [8] and includes the compressive strength in the formula.

2. Methodology

Power Creep Function is a simple relation for time-dependent deformation under constant stress. The formula is expressed as follows [8]:

$$\Delta H = H0\Delta\sigma M'(\frac{t}{tr})^{N'}$$

where ΔH=settlement; H0=initial height of waste; Δσ=compressive stress depending upon waste height, density, and external loading; M'=reference compressibility (=1.6 x 10-5 to 5.8 x 10-5 kPa); N'=rate of compression (=0.50–0.67); t=time since load application; and tr =reference time is typically introduced to make time dimensionless (usually taken as 1 day). The parameter M' is site specific and N' is the rate of compression, which varies with respect to age and placement conditions of the waste. However, the variability of N' is less than that of M'.

From this formula, three samples were collected from a closed open dumping area at Bukit Palong, Lukut in the state of Negeri Sembilan of Malaysia from three different location and their geotechnical properties were determined. This site was chosen because some part of the site already closed and the studies requires waste soil from closed open dumping area. Basic soil properties experiments were done such as the sieve analysis, Atterberg limit, compaction test, consolidation test, and direct shear test.

After obtaining all the basic geotechnical properties, the compressive strength can be acquired by multiplying the density of sample with the height of waste soil. For all the samples, the initial height are all the same which are 2m. The case study is divided by three cases and divided by three more as there are 3 samples.

The first case is M' and N' are constant which is 5.8 x 10-5 kPa and 0.67 respectively, the 2nd case is that M' is increased in ranged value of 1.6 x 10-5 to 5.8 x 10-5 kPa with respect to age while N' constant at 0.5. Lastly the 3rd case shows the M' is constant at 1.6 x 10-5 kPa while the N' is increased ranged from 0.50 to 0.67 with respect to age. All the data must be done with a time frame within 50 years.

3. Analysis And Discussion

This study is to determine the prediction of settlement of waste soil. Based on Table 1, the basic geotechnical properties of the waste soil from the Bukit Palong site are shown. Figure 1, 2 and 3 shows the rate of settlement of the waste soil in Bukit Palong in a period of 50 years. The waste soil contains 4% – 26% of waste which such as plastic, gravel, twigs and undisposible foods.

Table 1 Geotechnical Properties of Waste Soil at Bukit Palong, Lukut

Properties	Ranged Values
Liquid Limit (%)	35 – 39
Plastic Limit (%)	32 – 42
Maximum Dry Density (Mg/m^3)	1.2 – 1.7
Optimum Water Content (%)	20 – 31
Cohesion (kN/m^2)	12 – 36
Friction Angle(^{0})	15 – 53

The settlement of waste soil shows a slight increase every year and for every sample, case 1 always have the highest reading followed by case 3 and then case 2.

For the first sample, the first case exhibit a rapid increase in settlement and the settlement is exceeding 2m but not more than 2.5m after 50 years because of the value chosen for the M' and N' are the highest within the range. For case 2, the settlement increase slightly within the 50 years which shows only goes from 0 to 0.5m. Case 3 shows the reading of the settlement similar to case 2 only to have the reading slightly over 0.5m. The settlement for case 2 and case 3 are almost similar because of the increasing value of M'.

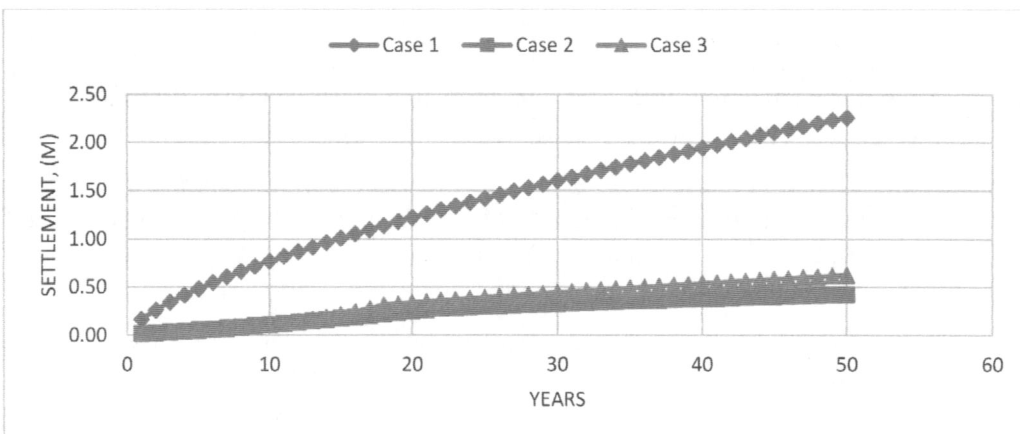

Figure 1: Settlement vs Years for Sample 1

For the second sample, the first case also shows a rapid increase in settlement and the settlement is exceeding 2.5m but not more than 3m after 50 years because of the value chosen for the M' and N' are the highest within the range. For both case 2 and case 3, the settlement increase slightly within the 50 years which shows only goes from 0 to slightly over 0.5m but not exceeding 1m. This due to the increasing value of M'.

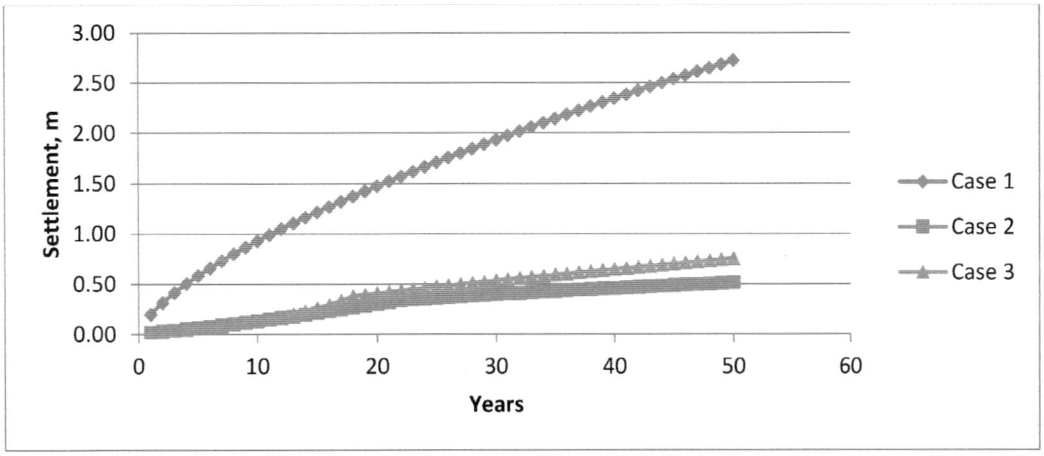

Figure 2: Settlement vs Years for Sample 2

For the third sample, it shows a reading which the first case also shows a rapid increase in settlement and the settlement is exceeding 2m but not more than 2.5m after 50 years because of the same reason for sample 1 and 2. This is because of the value chosen for the M' and N' are the highest within the range. The settlement for case 2 and case 3 increase slightly within the 50 years which shows only goes from 0 to 0.5m and from 0 to slightly over 0.5m respectively because of the increasing value of M' in case 2 and N' in case 3.

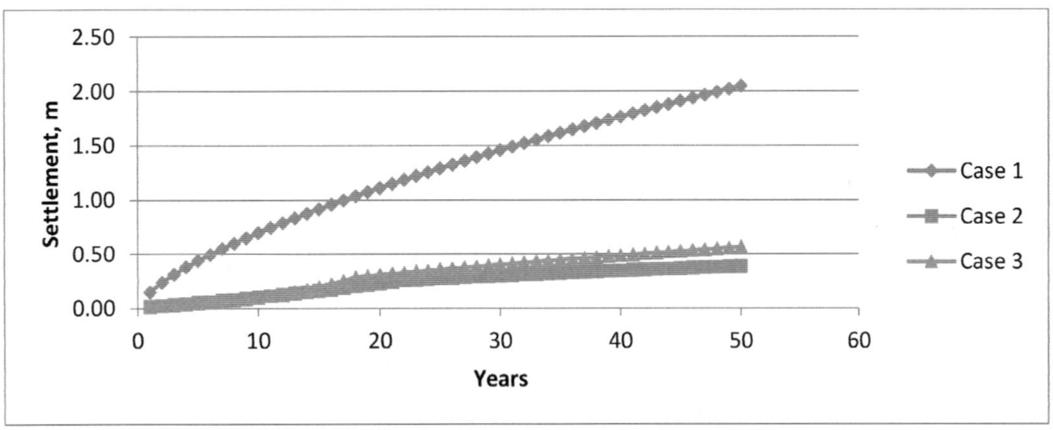

Figure 3: Settlement vs Years for Sample 3

4. Conclusion

As a conclusion, the waste soil collected at the open dump at Bukit Palong shows the settlement at the area will reach a stable condition after 50 years. For case 1, the area still shows an increasing in value after 50 years but does not exceed 0.05m between each year. For case 2 and 3, the waste soil does not exceed the 0.5m mark within the 50 years which means that the waste soil already reached it stable condition. The allowable settlement for the waste soil is 0.05m and in between years, all the soil sample for all cases does not exhibit increasing in settlement bigger than the allowable settlement. All of the sample will reach a stable condition after 50 years.

Acknowledgement

The author wish to acknowledge the Ministry of Education for supporting the research under grant number 20140110 FRGS and thank you to University Tenaga Nasional for the laboratory facilities and library.

References

[1] United Nations Environment Programme, Closing an Open Dumpsite and Shifting From Open Dumping to Controlled Dumping and to Sanitary Landfill, Pampanga, Phillipines: Department of Environment and Natural Resources, 2005.

[2] Department of Environmental and Natural Resources, *Closing an Open Dumpsite and Shifting from Open Dumping to Controlled Dumping and to Sanitary Land Filling*, Osaka: United Nation Environmental Programme's, 2005.

[3] D. Greedy and J. Thrane, "CLOSED FOR BUSINESS - A LOOK AT THE CLOSURE OF OPEN DUMPS," *Waste Management World,* vol. 9, no. 6, 2015.

[4] G. F. Sowers, "Settlement of waste disposal fills," in *Proceeding 8th international Conference on Soil Mechanics and Foundation Engineering*, Moscow, 1973.

[5] Gibson R.E. and Lo K.Y., A thoery of consolidation for soil exhibiting secondary compression, Norwegian Geotechnical Institute Publication, 1961.

[6] H. I. Ling, D. Leshchinsky and Y. Mohri, "Estimation of muncipal solid waste landfill settlement," *Journal Geotech. Geoenvironment.Eng,,* vol. 124(1), pp. 21-28, 1998.

[7] A. Bjangard and L. Edgers, "Settlement of municipal solid waste landfills," in *13th Annual Madison Waste Conference*, Wisconsin, 1990.

[8] T. Edil, V. Ranguette and a. W. Wuellner, "Settlement of municipal refuse,Geotechnics of wastefills: Theory and practice," *STP 1070,* pp. 225-239, 1990.

An Integrated Numerical Hydrodynamic Shallow Flow-Solute Transport Model for Urban Area

N A Alias[1], L Mohd Sidek[2]
[1]Faculty of Civil Engineering & Earth Resources. UMP, Lebuhraya Tun Razak, 26300 Kuantan, Pahang.
[2]School of Civil Engineering, College of Engineering. UNITEN, Jalan IKRAM-UNITEN, 43000 Kajang, Selangor.

E-mail: vlinders@rocketmail.com

Abstract. The rapidly changing on land profiles in the some urban areas in Malaysia led to the increasing of flood risk. Extensive developments on densely populated area and urbanization worsen the flood scenario. An early warning system is really important and the popular method is by numerically simulating the river and flood flows. There are lots of two-dimensional (2D) flood model predicting the flood level but in some circumstances, still it is difficult to resolve the river reach in a 2D manner. A systematic early warning system requires a precisely prediction of flow depth. Hence a reliable one-dimensional (1D) model that provides accurate description of the flow is essential. Research also aims to resolve some of raised issues such as the fate of pollutant in river reach by developing the integrated hydrodynamic shallow flow-solute transport model. Presented in this paper are results on flow prediction for Sungai Penchala and the convection-diffusion of solute transports simulated by the developed model.

1. Introduction

Numerous studies in modelling flow in open channel have been done, however there are still deficiencies in the models. As it is difficult to resolve the problematic river reach in a 2D manner, a 1D component with an ability to deal with the complex flow hydrodynamics in complex topography is essential. Enhancement of computer technology also raised the 2D models to be more prevalent, yet it is still computationally difficult to signify the small streams in a larger scale. This brought to the necessity of having an intensive 1D flow model when the 2D also tends to be more data demanding [1]. The computational cost of 1D model is also lower than the 2D models. For a large scale river network, still it consents closely to real time analysis [2]. However, there is still a subject to debate on the 1D model. Most of the 1D models were developed based on the solution to the approximation of the 1D shallow water formula ([3],[4]) and popular for its straightforwardness. This ignored the spatially hydraulics variables and caused inability to represent the full dynamics of a flow. Thus, for broader applications it is desired to have a high ability of 1D model in solving the more complex hydrodynamics flow pertaining to flood in urban areas.

[1] To whom any correspondence should be addressed.

2. Governing equations and the numerical scheme

It is common to assume that the 1D channel attached to a floodplain has a rectangular cross-section ([5],[6]). Dealing with complex geometry, the breadth variation is then included in order to accurately calculate the flow depth (h) and discharge (q=uh). The final equations can be written as;

$$\frac{\partial h}{\partial t}+\frac{\partial (uh)}{\partial x}=-\frac{uh}{b}\frac{\partial b}{\partial x} \quad (1)$$

$$\frac{\partial (uh)}{\partial t}+\frac{\partial (u^2 h)}{\partial x}+\frac{g}{2}\frac{\partial}{\partial x}\left(h^2\right)=-\frac{u^2 h}{b}\frac{\partial b}{\partial x}-C_f u|u|-gh\frac{\partial Z_b}{\partial x} \quad (2)$$

where u is velocity, Z_b is bed elvation, g is the gravity acceleration, b is channel width, $C_f = gn^2/h^{1/3}$ is the roughness coefficient of the river bed and n is the Mannings' coefficient. Equation (1) allows varying channel width and bottom, while the source terms in equation (2) includes breadth variations, bottom friction and bed slope. Adding the equation (3) into the model, solute transport and shallow waters flow is simulated simultaneously.

$$\frac{\partial (ch)}{\partial t}+\frac{\partial (uch)}{\partial x}=S_C \quad (3)$$

Godunov-type finite volume schemes in conjunction with Harten-Lax-vanLeer (HLL) approximate Reimann solver is chosen to solve the fully SWE's. Riemann solvers are of significant interest (e.g., [7],[8],[9],[10]) as widely implemented to modelling flows with abrupt and discontinuous gradients. The final expression for the scheme is;

$$\mathbf{u}_i^{k+1}=\mathbf{u}_i^k-\frac{\Delta t}{\Delta x}\left(\mathbf{f}_{i+1/2}-\mathbf{f}_{i-1/2}\right)+\Delta t\mathbf{s}_i \quad (4)$$

where i denotes the cell index, Δx and Δt are the cell size and time step respectively, $\mathbf{f}_{i+1/2}$ and $\mathbf{f}_{i-1/2}$ are the interface fluxes vector. The HLL approximate Riemann solver [11] is adopted for flux calculation as it facilitate the wetting and drying cases. The Riemann states obtained from the face values that are reconstructed from the cell-centred flow variables. Herein the MUSCL (Monotone Upstream-centred Schemes for Conservation Laws) slope limited linear reconstruction (van Leer, 1979) is adopted to rebuild the face values in order to achieve 2nd-order accuracy in space. Combining the Godunov-type finite volume method with the second-order Runge-Kutta method gives:

$$\mathbf{u}_i^{k+1}=\mathbf{u}_i^k+0.5\Delta t\left(\mathbf{K}_i(\mathbf{u}^k)+\mathbf{K}_i(\mathbf{u}^*)\right) \quad (5)$$

The $\mathbf{u}^*$ is the intermediate flow variables while superscript k in equation (4) and (5) represent the present time level. K in equation (5) is the Runge-Kutta coefficient at predictor and corrector steps defined by:

$$\mathbf{K}_i=-\frac{\left(\mathbf{f}_{i+1/2}-\mathbf{f}_{i-1/2}\right)}{\Delta x} \quad (6)$$

In reconstructing the Riemann states, the flow variables at the flux faces were first estimated from the central flow information using a slope limited linear interpolation to rebuild the face values to achieve second-order accuracy in space [12]. To preserves the accuracy and well-balance property, the source terms are split into two parts and discretized using central differences that are compatible to the flux calculation [13].

3. Model validation and implementation

The phenomenon of 1D surface flow is observed under various conditions experienced in open channel [13]. As the results are reliable, it is then implemented to the real case scenario. Sungai Penchala that is densely populated, surrounded by industrial and residential areas situated between Kuala Lumpur and Selangor is identified as the potential site studies. The river reach is divided into few sections and the readings of water level at predetermined locations were recorded. The time histories for flow depth along the Sungai Penchala were predicted numerically at 2000 m interval.

After validating its performance in open channels, the flow model was then integrated with the pollutant transport equation. Then, the model used to study the fate of pollutant.

4. Results and discussion

4.1. Flow hydrodynamic in Sungai Penchala

The proposed model has been validated against several benchmark tests and close agreement has been achieved. This confirms the effectiveness of the current 1D code [13]. The model used the channel geometry obtained from the field surveys done by Department of Irrigation and Drainage (DID). The bed profile is as plotted in figure 1.

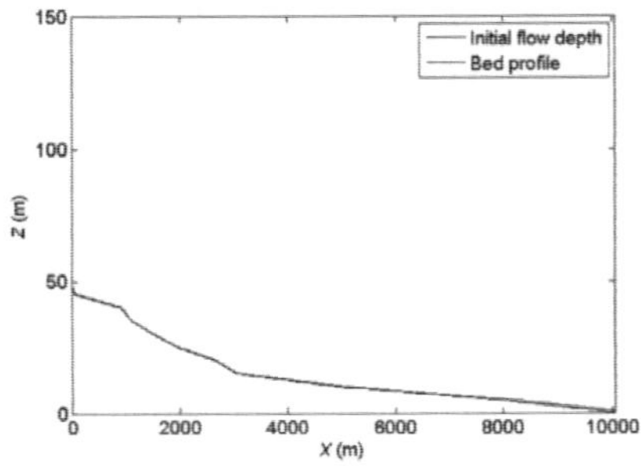

Figure 1. Bed profile of Sungai Penchala

From the sketched, the Penchala catchment is a hilly terrain. Almost 80% of Sungai Penchala is a concrete channel with bed frictional coefficient equals to 0.015 s/m$^{1/3}$. The imposed outflow depth is 0.15 m as recorded during the survey. Results of simulation as in figure 2 (a-d) which describe the time histories of flow depth at five different locations, x. Simulation was run for 7200 s on a uniform grid.

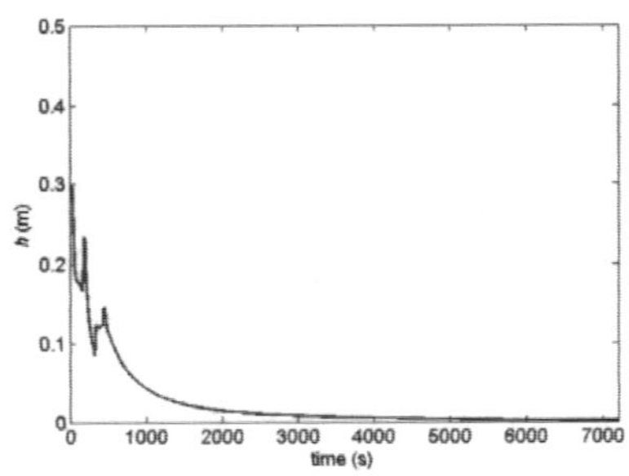

(a) $x = 2000$ m

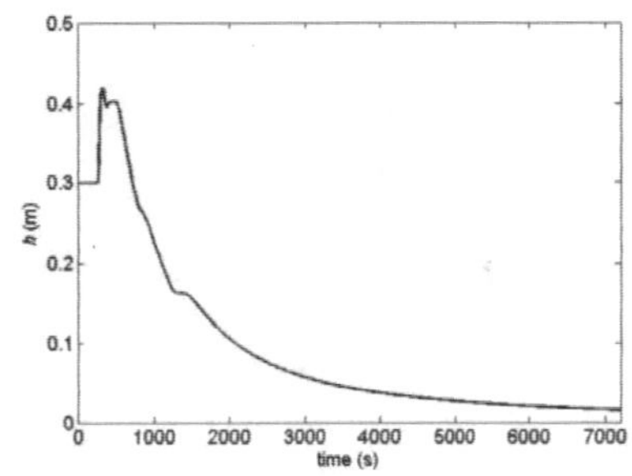

(b) $x = 4000$ m

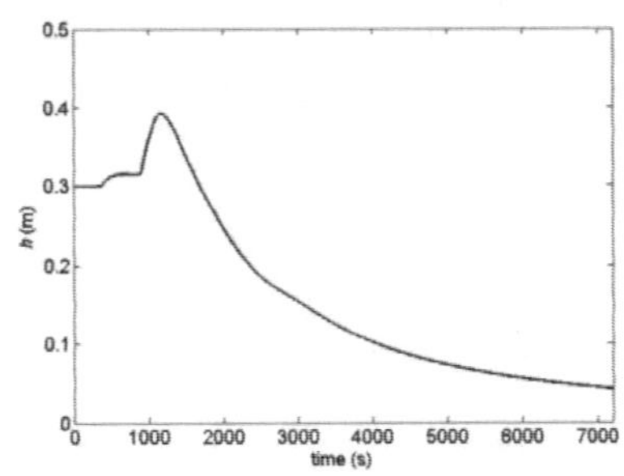

(c) $x = 6000$ m

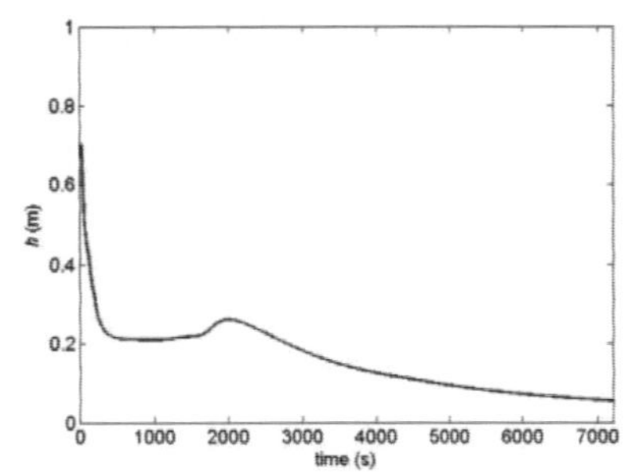

(d) $x = 8000$ m

Figure 2 (a-d). Time histories of water depth in five points.

The maximum flow depths at five locations numerically simulated and compared with the flow depth measured on site. Figure 3 illustrates the comparison between the maximum predicted and measured flow depth.

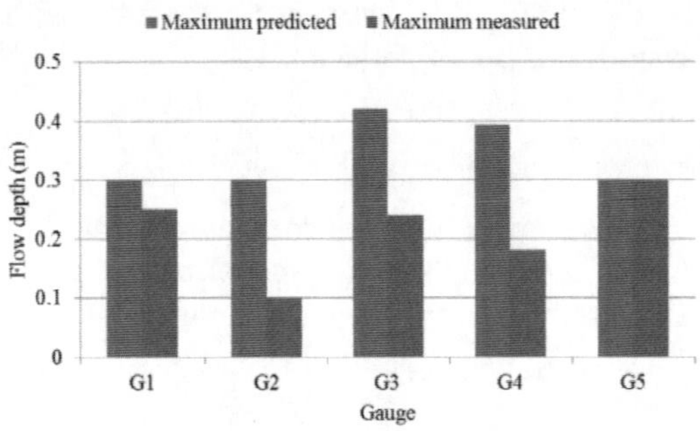

Figure 3. Comparison of numerical and measured maximum water depth along Sungai Penchala.

4.2. *Fate of pollutant*

The integrated model then used to study the fate of pollutant. The test involved a 1D convective – diffusion problem [14]. Figure 4 (a-b) exhibits the solute concentration at different simulation time.

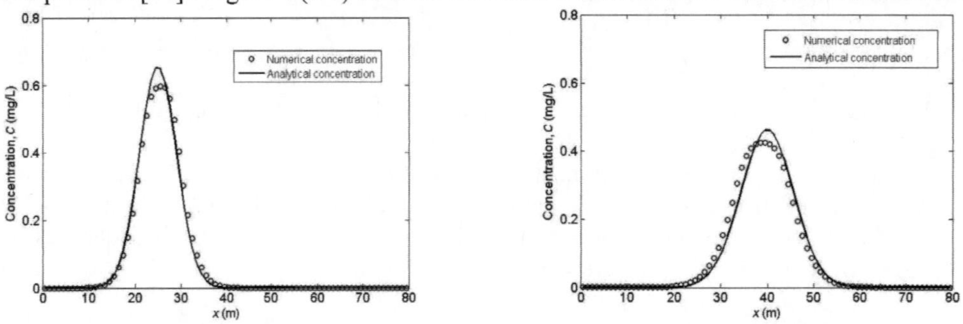

(a) $t = 20$s (b) $t = 40$s

Figure 4. Comparison of concentration C at different time histories.

The maximum concentration computed was 0.5982 mg/L and as the time of simulation increases, it reduced to 0.3537 mg/L. Results exhibit good accuracy in the convective-diffusion case with the global relative error, $R = 0.0949$ and $R = 0.10711$ for $t = 20$ s and 40 s respectively.

5. Conclusion and recommendations

The model has been validated by numbers of benchmark tests previously and good agreements were achieved. Model is then implemented on Sungai Penchala and gives sufficient results in flow prediction. This model is expected to be broadly tested on other rivers in urban area. Predictions on fate of pollution also exhibit good agreements. In future, it is suggested that the solver can include the advection-reaction-diffusion equation to model fate of pollutant as in reality the dynamics flow in urban areas are chemically reactive or biologically reproductive.

References

[1] Horritt M S 2000 *Int. J. Numer. Meth. Eng.* **47** 2019-37
[2] Gejadze I Y and Monnier J 2006 *European Conference on Computational Fluid Dynamics* (TU Delft Netherlands)
[3] Ahmed H and Norio T 2009 *J. Hydrodynamic* (Ser. B). **21** 564-72
[4] Wilson M D, Bates P D, Alsdorf D, Forsberg B, Horritt M, Melack J, Frappart F and Famglietti J S 2007 *Geophys. Res. Lett.* **34** L15404
[5] Bates P D and Roo A P J D 2000 *J. Hydro.* **236** 54-77
[6] Fan P, Li J C, Li Q and Singh V P 2008 *ASCE J. Hydraulic Eng.* **134** 1757–66
[7] Alcrudo F and García-Navarro P 1993 *Int. J. Numer. Meth. In Fluids.* **16** 489-505
[8] Anastasiou K and Chan C T 1997 *Int. J. Numer. Meth. In Fluids.* **24** 1225-45
[9] Rogers B D, Fujihara M and Borthwick A G L 2001 *Int. J. Numer. Meth. in Fluids.* **35** 247-80

[10] Toro E F 2001 *Shock-capturing methods for free-surface shallow flows*. (Chichester: John Wiley & Sons)
[11] Harten A, Lax P D and VanLeer B 1983 *Soc. for Industr. and Appl. Math. Rev.* **25** 35-61
[12] VanLeer B 1979 *J. Comput Phys.* **32** 101-36
[13] Nor A A and Lariyah M S 2015 *Int. Conference on Advances in Civil and Environmental Engineering* (UiTM Pulau Pinang)
[14] Li Y N and Huang P 2008 *J. Adv. Water Res.* **31** 1719-30

Acknowledgments
This work is supported in part by MOHE. This is a Project 3 Urban Ecohydrology for Resilient Environment through Long-term Research Grant Scheme (203/PKT/6720004). Authors wish to thank DID for assistance with data collection and fieldwork.

IOP Conf. Series: Earth and Environmental Science **32** (2016) 012013 doi:10.1088/1755-1315/32/1/012013

Effect on $Ca(OH)_2$ pretreatment to enhance biogas production of organic food waste

H Junoh[1]
Industrial Science, Master student, Universiti Tenaga Nasional, Kajang, Selangor.
CH Yip[2]
Senior Lecturer, Universiti Tenaga Nasional, Kajang, Selangor.
P Kumaran[3]
Centre Renewable Energy, Universiti Tenaga Nasional, Kajang, Selangor.

Abstract. This study investigated the effect of calcium hydroxide, $Ca(OH)_2$ pretreatment in optimizing COD solubilisation and methane production through anaerobic digestion process. Two different parameters, chemical concentration (40-190 mEq/L) and pretreatment time (1-6 hours) were used to pretreat food waste. A central composite design and response surface methodology (RSM) was applied in obtaining the optimized condition for COD solubilisation. Result showed COD solubilisation was optimized at 166.98 mEq/L (equivalent to 6.1 g $Ca(OH)_2$/L) for 1 hour. These conditions were applied through biomethane potential test with methane production of 864.19 mL/g $VS_{destructed}$ and an increase of 20.0% as compared to untreated food waste.

1. Introduction

Currently, active studies have been pursued on every possible sustainable waste resources including organic food waste as substrate for anaerobic digestion (AD). Anaerobic digestion constitutes of a basic process which includes hydrolysis, acidogenesis, acetogenesis and methanogenesis. However the most critical and rate limiting step in anaerobic digestion is the hydrolysis process where complex molecule of organic waste need to be broken down into monomers that requires longer time with slow rate of solubilisation. This hydrolysis process requires assistance in reducing the time taken during digestion process called pretreatment.

Pretreatment is required to improve accessibility of soluble organic material as represented by value of the soluble chemical oxygen demand (SCOD) to be an essential source for microbial population during digestion process. SCOD represent the source of energy available to maintain the activity of microbes and its condition throughout the entire process. Therefore, by providing the pretreated substrate with readily available monomers will eventually uplift methane production in shorter retention time. According to Jayashree et al. [1], SCOD is one of the parameters used in assessing particulate material for optimal solubilisation and to identify the quantity of soluble organic content in the subjected substrate. The SCOD value expresses solubilisation, S as described in (1) [2,3].

$$\text{Solubilisation, } S\ (\%) = [\ SCOD_f - SCOD_i\ /\ TCOD\ ] * 100 \quad (1)$$

[1] E-mail: hafizan_junoh@yahoo.com
[2] E-mail: chyip@uniten.edu.my
[3] E-mail: kumaran@uniten.edu.my

where $SCOD_f$ and $SCOD_i$ refer to final and initial value of SCOD respectively. And TCOD represent total chemical oxygen demand.

In optimizing solubilisation in term of soluble COD, higher SCOD indicates the readiness of a substance to produce methane through anaerobic digestion. Torres and Espinosa Lloréns [2] had reported in his study, a chemical pretreatment with 62.0 mEq/L for 6 hours was optimized with 11.5% increase in COD solubilisation and 150 mL CH_4/g VS (172% higher) as compared to untreated substrate. This paper aims to investigate the effect of $Ca(OH)_2$ in enhancing solubilisation of organic food waste as a promising way to provide soluble component to enhance digestion rate and reduce hydraulic retention time. Response surface method is an approach used to determine the maximum response for an optimum condition within the selected range of factors.

2. Methodologies

2.1. Preparation of substrates and inoculum

Food waste (FW) prepared manually represented the main component in Malaysian food waste composition comprising steamed rice, vegetable (cooked cabbage and carrot), bread, fried chicken, fried fish and fried noodle. The indigestible material such as bones were removed manually. The component mixture was further homogenized using a crusher and sieved to produce a homogeneous substrate with <2mm particle size. Sewage sludge was used as a source of inoculum.

2.2. Chemical pretreatment

Chemical pretreatment was carried out using a stainless steel container equipped with stirrer. The substrate was tested for range 40-190 mEq/L for 1-6 hours. Chemical oxygen demand (COD) were tested according to the Standard Methods for the Examination of Water and Wastewater (APHA). Soluble COD was obtained from filtered sample through 0.45 μm filter paper.

2.3. Experimental design and procedure

Central composite design (CCD) was employed for the selected factors to obtain the desired output from the experimental design with Design Expert 6.0.8 software. The level and code of variables considered in this study are shown in Table 1.

Table 1. Level code for Central Composite Design.

Factor	Variable	Coded levels		
		-1	0	1
Chemical Concentration, mEq/L	A	40	115	190
Pretreatment Time, hour	B	1	3.5	6

For two variables (n=2), and three levels (low, medium and high, coded as -1, 0 and +1), the total number of 11 runs were determined by the expression 2^n (2^2=4 factorial points), 2n (2*2=4 axial points), 3 centre points (point of replications) as given in Table 3. The average of triplicated values of each run was taken as response in term of COD solubilisation.

A quadratic polynomial equation was developed to predict the response as a function of independent variables and their interactions. In general, the response for the quadratic polynomials is described in (2):

$$Y = \beta_0 + \sum\beta_i x_i + \sum\beta_{ii} x_i^2 + \sum\sum\beta_{ij} x_{ij} x_j \quad (2)$$

where Y is the response (COD solubilisation), β_0 is the intercept coefficient, β_i is the linear terms, β_{ii} is the squared terms and β_{ij} is the interaction terms.

2.4. Biomethane potential (BMP) tests

The BMP tests were used to analyse the potential methane production for chemically treated food waste as compared to untreated one. Batch assays were performed in 2L glass AD vessel (Schott Duran, Germany) filled with 1.8L of substrate which contain 90% inoculum and 10% of food waste under mesophilic condition (room temperature). All the AD vessel were connected to a 1L Schott bottle containing an alkaline solution 2M (scrubber) before releasing into gas collection set up. The scrubber was used to absorb CO_2 and H_2S in biogas. The headspace of digester was flushed with nitrogen gas for about 2 minutes for anaerobic environment. Control bottles were prepared using the food waste without pretreatment. All samples were adjusted to pH 7.0 with either 1 M HCl or 1 M NaOH solution and were stirred for one hour per three hours interval. Daily methane production was monitored by measuring the volume of water displaced from eudiometer. The biogas composition was analysed by using gas analyser (BINDER, Germany) to detect methane content.

3. Result and discussion

A Central Composite Design (CCD) under RSM was used to optimize COD solubilisation based on the interactive effect of chemical concentration and treatment time. The baseline range followed Torres and Espinosa Llorens [2] that treated organic fraction of municipal solid waste (OFMSW). The preliminary result (data was not shown) indicated solubilisation continued increasing after 100 mEq/L $Ca(OH)_2$ at 6 hours. Therefore, the concentration was increased up to 190 mEq/L for the optimization of $Ca(OH)_2$ pretreatment.

3.1. Chemical pretreatment

Based on the matrix design of coded variables tabulated in Table 2, both experimental and predicted value were shown as representing the value of response (COD solubilisation). All the values were obtained from quadratic model using the software Design Expert. $Ca(OH)_2$ pretreatment was tested for different concentration and treatment time to enhance solubilisation prior to biogas production. The quadratic model given by the Software for thermo-chemical pretreatment is represented in (3):

$$Y = 3.17 + 1.25A - 0.41B - 1.881A^2 + 1.69B^2 - 1.29AB \qquad (3)$$

where Y (yield) is COD solubilisation (%); A is the temperature (°C); *B* is the treatment time (hour). The statistical model was checked by *F*-test and analysis of variance (ANOVA) for the response surface quadratic model as tabulated in Table 3. The Model *F*-value of 169.44 implies the model is significant. There is only a 0.01% chance that Model *F*-value this large could occur due to noise. Value of "Prob > F" less than 0.0500 indicate model terms are significant. The Lack-of-Fit "*F*-value" of 13.22 implies the Lack of fit is insignificant. There is a 7.11% chance that Lack-of-Fit "*F*-value" could occur due to noise. Non-significance of Lack of fit is good as we need the model to be fit. Therefore, model coefficients, namely A, B, A^2, B^2 and AB are significant (Table 3) to affect COD solubilisation. In addition to that, determination of coefficient (R^2) and adjusted R^2 can also be used to check the model effectiveness. R^2 value represents the amount of variance accounted for in the relationship between two or more variables. The closeness of R^2 value to 1 illustrates that the model exactly explains the variability in dependent variable, Fig. 1. Hence, the value larger than 0.5 can be considered a significant relationship [4]. The value of adjusted R^2 (0.9883) indicates that 98.83% of variance can be predicted from the independent variables and only 1.17% of the total variation cannot be explained by the model. The predicted R^2 (0.9404) is in reasonable agreement with adjusted R^2. "Adeq precision" measures the signal to noise ratio and a ratio greater than 4 is desirable. The ratio of 42.487 of the Model (Table 3) indicates an adequate signal. This model can be used to navigate the design space.

For optimization purpose, the function of desirability was applied to obtain predicted solubilisation for the model at its lower and upper limit of each variables ($Ca(OH)_2$ concentration and treatment time) based on contour and surface plot obtained in Fig 1. The possible solutions generated by the software which fulfilled all required conditions will give desirability value close to 1. In this study, 166.98 mEq/L and 1 hour was selected for biomethane potential test. For validation purposes, the predicted solubilisation (6.13%) was close to the experimental value (5.95%) that was performed with only 2.94% error.

Table 2: CCD with experimental and predicted values of COD solubilisation (%)

Chemical Concentration (mEq/L)	Pretreatment Time (hour)	COD solubilisation (%)	
		Experimental	Predicted
40.00	1.00	0.69	0.85
40.00	3.50	0.27	0.04
40.00	6.00	2.54	2.61
115.00	1.00	5.38	5.27
115.00	3.50	3.17	3.17
115.00	3.50	3.23	3.17
115.00	3.50	3.10	3.17
115.00	6.00	4.37	4.45
190.00	1.00	5.99	5.93
190.00	3.50	2.34	2.54
190.00	6.00	2.68	2.53

Table 3. ANOVA for the response surface quadratic polynomial model

Source	Sum of square	*df*	Mean square	*F*-value	Probability (p)>*F*
Model	29.88	5	5.98	169.44	<0.0001
A	9.40	1	9.40	266.50	<0.0001
B	1.02	1	1.02	28.83	0.0030
A^2	8.97	1	8.97	254.35	<0.0001
B^2	7.22	1	7.22	204.68	<0.0001
AB	6.66	1	6.66	188.71	<0.0001
Residual	0.18	5	0.035		
Lack of fit	0.17	3	0.056	13.22	0.0711
Pure error	8.467*E*-003	2	4.233*E*-003		
Cor. total	30.06	10			

R^2=0.9941; adjusted R^2=0.9883; predicted R^2=0.9404 and adeq precision=42.487

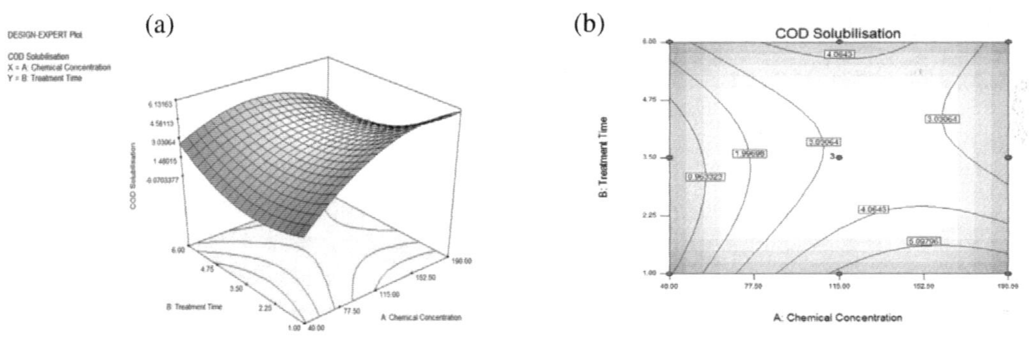

Fig. 1 Three dimensional surface plot (a) for NaOH pretreatment (b) and $Ca(OH)_2$ pretreatment

3.2. BMP tests

The batch anaerobic digestions were tested to investigate the effect of chemical pretreatment, $Ca(OH)_2$ on the production of methane from food waste based on the optimal condition from the pretreatment process. Anaerobic digestion is expressed by volume of methane produce per mass of volatile solid (VS) of a substrate that is destructed throughout the digestion process (mL CH_4/g $VS_{destructed}$) [5]. Cumulative and specific methane production were represented in Fig. 2. The pretreated food waste showed similar pattern to untreated sample in daily basis methane production. In Fig. 2(a), methane production started to increase after day 6 for both treated and untreated food waste as supported by [6] that biogas production can be optimally produced as early as in 8 days as insignificant COD changes in further degradation. Furthermore, there is an increase of 5.4% in cumulative methane production for $Ca(OH)_2$ pretreated food waste which illustrates enhancement of anaerobic digestion process by providing soluble components in pretreatment process. Compared to study by [2], higher $Ca(OH)_2$ concentration was used in this study which resulted in increment of specific methane potential by 20.0% when compared with control digester. Higher production of methane in this study occurred due to higher volatile solids content (84,860 mg/L) as compared to the study by [2] with organic fraction municipal solid waste (OFMSW) as a substrate. Other studies had shown positive effect of $Ca(OH)_2$ pretreatment with increased biogas production for extruded rice straw [7] and feather waste [8].

(a)

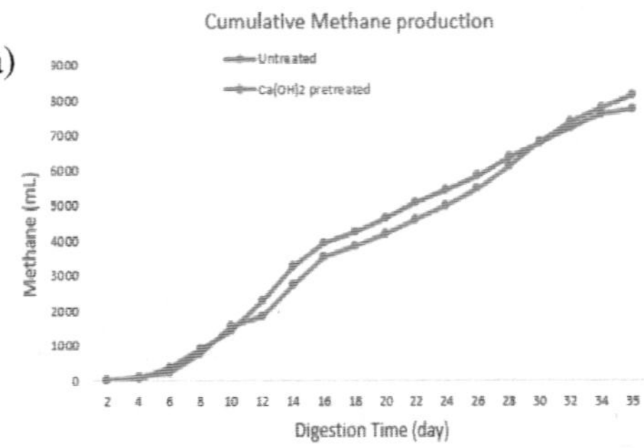

(b)

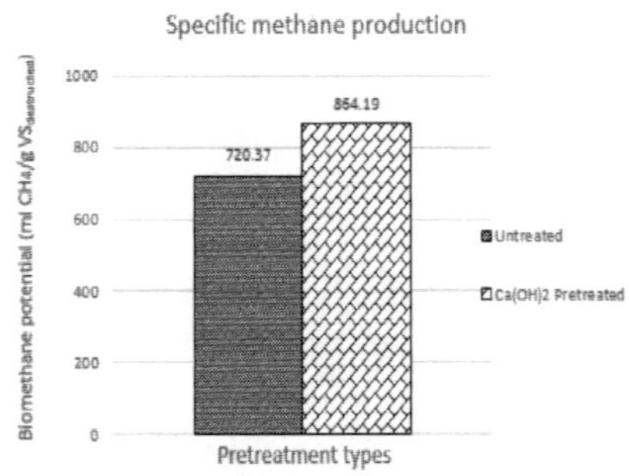

Fig. 2 Methane production (a)cumulative production; (b)specific production throughout 35 days of digestion

4. Conclusion

The aim of this study is to investigate the effect of chemical pretreatment on solubilization in enhancing methane production. Calcium hydroxide showed potential as an alkaline chemical to be used in pretreatment of food waste that can enhance up to 20.0% of specific methane production compared to untreated food waste.

Acknowledgement

Authors wish to acknowledge assistance and encouragement from colleagues especially to Uniten Centre Renewable Energy (CRE), Ministry of Higher Education (11012013ERGS), Akaun Amanah Industri Bekalan Elecktrik (AAIBE) and Hepzhibah David for continuous support and providing laboratory facilities for this project.

Reference

[1] Jayashree C, Janshi G, Yeom IT, Adish Kumar S, & Rajesh Banu J 2014 Effect of low temperature thermo-chemical pretreatment of dairy waste activated sludge on the performance of microbial fuel cell *International Journal of Electrochemical Science* 9 5732-5742.

[2] Torres M L, & Espinosa Lloréns M D C 2008 Effect of alkaline pretreatment on anaerobic digestion of solid waste *Waste Management* 28(11) 2229-2234.

[3] Marin J, Kennedy K J, & Eskicioglu C 2010 Effect of microwave irradiation on anaerobic degradability of model kitchen waste *Waste Management* 30(10) 1772-1779.

[4] Moo K C, 2010 R2 *Encyclopedia of research design* ed N J Salkind (SAGE publications, Inc.) pp 1187-1190.

[5] Wang M, Sun X, Li P, Yin L, Liu D, Zhang Y, Li W, & Zheng G 2014 A novel alternate feeding mode for semi-continuous anaerobic co-digestion of food waste with chicken manure *Bioresource Technology* 164 309-314.

[6] Eskicioglu C, Kennedy K J, & Droste R L 2006 Characterization of soluble organic matter of waste activated sludge before and after thermal pretreatment *Water Research* 40 3725-3736.

[7] Gu Y, Zhang Y, & Zhou X 2015 Effect of $Ca(OH)_2$ pretreatment on extruded rice straw anaerobic digestion *Bioresource Technology* 196 116-122.

[8] Forgács G, Niklasson C, Horváth I S, Taherzadeh M J 2014 Methane production from feather waste pretreated with $Ca(OH)_2$: Process development and economical analysis *Waste Biomass Valorization* 5(1) 65-73.

A comparative study of artificial intelligent-based maximum power point tracking for photovoltaic systems

Ammar Hussain Mutlag[1,2], **Azah Mohamed**[1], **Hussain Shareef**[3]
[1] Department of Electrical, Electronic and Systems Engineering, Faculty of Engineering and Built Environment, Universiti Kebangsaan Malaysia, 43600 Bangi, Selangor, Malaysia
[2] College of Electrical and Electronic Engineering Techniques, Middle Technical University, Baghdad, Iraq.
[3] Department of Electrical Engineering, United Arab Emirates University, 15551 Al-Ain, UAE

ammar_alqiesy@yahoo.com, azah@eng.ukm.my, shareef@uaeu.ac.ae

Abstract. Maximum power point tracking (MPPT) is normally required to improve the performance of photovoltaic (PV) systems. This paper presents artificial intelligent-based maximum power point tracking (AI-MPPT) by considering three artificial intelligent techniques, namely, artificial neural network (ANN), adaptive neuro fuzzy inference system with seven triangular fuzzy sets (7-tri), and adaptive neuro fuzzy inference system with seven gbell fuzzy sets. The AI-MPPT is designed for the 25 SolarTIFSTF-120P6 PV panels, with the capacity of 3 kW peak. A complete PV system is modelled using 300,000 data samples and simulated in the MATLAB/SIMULINK. The AI-MPPT has been tested under real environmental conditions for two days from 8 am to 18 pm. The results showed that the ANN based MPPT gives the most accurate performance and then followed by the 7-tri-based MPPT.

1. Introduction

Photovoltaic (PV) is considered a widely used renewable energy (RE) source because it is free, clean, and environmental friendly [1]. PV system may be installed in the form of stand-alone and grid-connected systems [2]. The drawbacks of a PV system are that it exhibits intermittent power generation under varying weather conditions and the amount of generated power from a solar cell depends on the nonlinear current–voltage (I–V) and power–voltage (P–V) characteristics, which vary with irradiance and temperature. However, there is a unique point on the P–V curve which is the maximum power point (MPP). To increase the performance of PV systems, it is crucial to operate near to the MPP [3].

Many MPPT algorithms have been developed such as the conventional MMPT techniques which include the incremental conductance (IC), perturb and observe (P&O), and constant voltage (CV) techniques [4]. The P&O technique has been widely utilized because of its simple control algorithm and minimal number of input parameters. However, the disadvantage of the P&O technique is that enormous oscillation is exhibited in the region of MPP and such oscillation leads to a certain loss in power. The IC technique can somehow eliminate oscillations in the region of MPP [5] but it requires accurate sensors for measuring either the voltage or current.

[1] To whom any correspondence should be addressed.

Recently, AI based-MPPT has been applied because it is robust and easily adapted to complex systems without the need of a mathematical model [6]. In this paper, a comparative study for three AI-based MPPT is made for a PV system by considering three artificial intelligent techniques, namely, artificial neural network (ANN), adaptive neuro fuzzy inference system with seven triangular fuzzy sets (7-tri), and adaptive neuro fuzzy inference system with seven gbell fuzzy sets (7-gbell).

2. PV modelling

The equivalent electrical circuit of a PV cell shown in Figure 1 can be utilized to obtain the characteristics of a PV cell. From the mathematical model of the circuit, the output of the cell current (I), can be expressed as [7]:

$$I = I_{ph} - I_o\left(e^{\left(\frac{q(V+I.R_s)}{n.K_B.T}\right)} - 1\right) - \frac{V + I + R_s}{R_{sh}} \tag{1}$$

where I is the cell output current, I_{ph} is the light-generated current, I_o is the cell reverse saturation current or dark current, q is the electronic charge (1.6×10^{-19} C), V is the cell output voltage, R_s is the series resistance, n is the ideality factor, K_B is the Boltzmann's constant (1.38×10^{-23} J/K), T is the cell temperature (K), and R_{sh} is the shunt resistance.
The I_{ph} can be calculated as:

$$I_{ph} = \left[I_{sc,n} + \alpha(T - T_n)\right]\frac{G}{G_n} \tag{2}$$

where $I_{sc,n}$ short-circuit current at the nominal condition, α is short-circuit current temperature coefficient, T_n is nominal temperature, G is irradiance, and, G_n is the nominal irradiance.

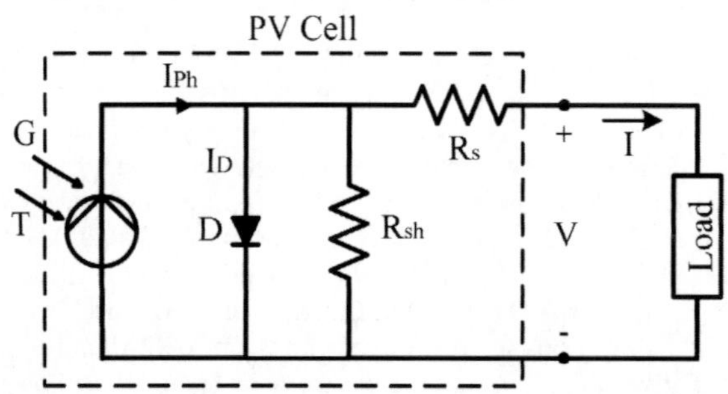

Figure 1. Equivalent electrical circuit of a PV cell

3. Artificial intelligent based-MPPT

Three AI based-MPPT techniques are used in this study, namely, ANN, adaptive neuro fuzzy inference system (ANFIS) with seven triangular fuzzy sets (7-tri), and ANFIS with seven gbell fuzzy sets (7-gbell). ANN is a distributed processing system consisting of neurons which are simple connected elements. The ANN model considered in this study is the multi-layer feedforward model with back propagation (BP) algorithm. The ANN model consists of three layers, namely, the input, hidden, and output layers with 2, 20 and 1 neurons, respectively. The ANFIS considered in this study uses a fuzzy inference system model to transform a given input into a target output.

Since electric power is the product of current and voltage, therefore a power-voltage (P-V) characteristic curve of a solar cell can be obtained for a given radiation level as shown in Figure 2. However, there is one particular point at which the solar cell can deliver maximum power for a given radiation intensity, and this operating point is the maximum power point (MPP). From Equations (1) and (2), the cell output current is shown to be nonlinear and dependent on irradiation and temperature. These equations can be used to calculate reference current (I_{MPP}) which eventually provides MPP by considering the cell output voltage.

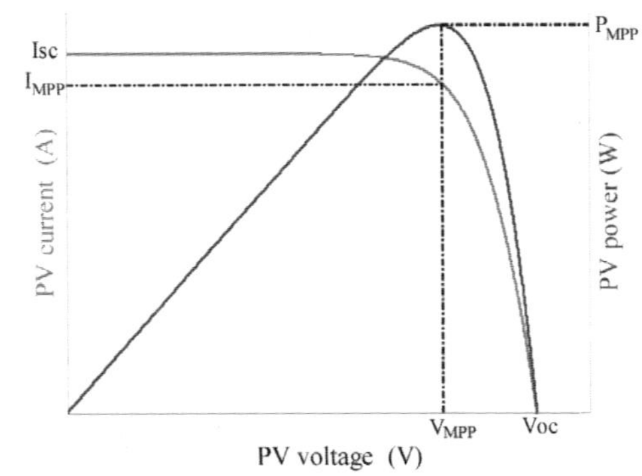

Figure 2. Current-Voltage (I-V) and Power-Voltage (P-V) characteristic curves of a solar cell

In this work, 300,000 input data samples which correspond to G and T have been generated using the following equations (3) and (4).

$$G_i = rand * (G_{max} - G_{min}) + G_{min} \quad (3)$$

$$T_i = rand * (T_{max} - T_{min}) + T_{min} \quad (4)$$

where i from 1 to 300,000. The minimum and maximum limits for the G and T are selected based on the historical data. In Malaysia, G typically varies between 0 to 1200 W/m^2, while the T fluctuates in the range of 20 to 40 °C. Therefore G_{min} , G_{max}, T_{min} , and T_{max} in (3) and (4), are chosen as 0, 1500 W/m^2, 0, and 45 °C, respectively so as to ensure that the generated data covers the typical data range.

4. Results and Discussion

In this study, 25 SolarTIFSTF-120P6 PV modules were employed to supply a 3 kW peak. The modules were arranged in series-connected configuration to produce a DC output voltage of 435 V. To evaluate the performance of the developed AI-based MPPT, field input data, namely, G and T, were collected from Universiti Kebangsaan Malaysia on the 9th and 10th of August 2014, from 8 am to 6 pm at a sampling rate of 30 s. The goal is to compare the performance of the three developed AI-based MPPT under the same strict conditions. Data acquisition was performed with the HOBOwarePro. Figures 3 and 4 show the meteorological conditions (G and T) and the output maximum electrical power registered from the developed AI-based MPPT for the studied cases.

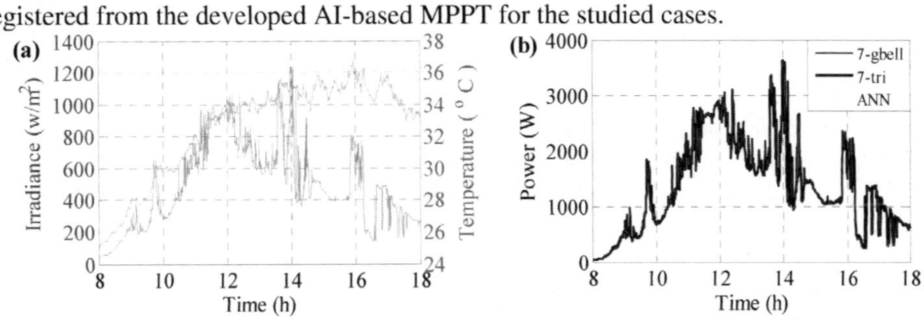

Figure 3. Metrological conditions (a) and extracted electrical power (b) on 9th of August 2014

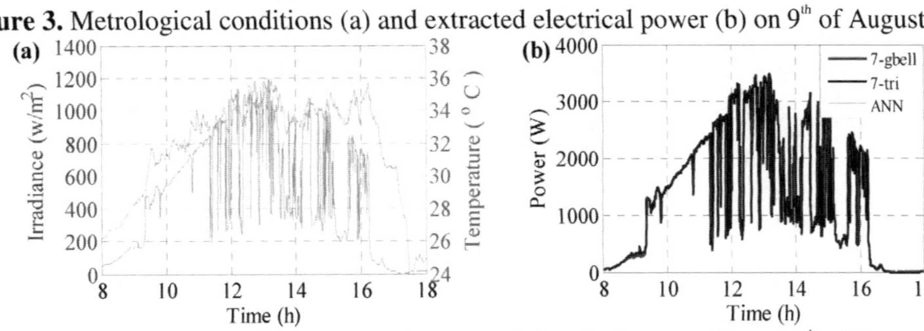

Figure 4. Metrological conditions (a) and extracted electrical power (b) on 10th of August 2014.

To evaluate the performance of the developed AI-based MPP trackers, three indices are used, namely, standard deviation (SD) for the error, mean absolute error (MAE), and correlation (R^2) using the q-q plot. The comparison of results is shown in Table 1. A low SD indicates that the data points tend to be very close to the mean, whereas a high SD indicates that the data points are spread out over a large range of values. The best performance is boldfaced. Table 1 clearly shows that the ANN based MPP tracker performs better than the other two ANFIS-based MPP tracker in terms of SD. For the second index, MAE, the decrease in its value mean the output is moving closer to the MPP. In terms of MAE values, the ANN based MPP tracker performs better than the ANFIS-based MPP tracker as depicted in Table 1. A statistical analysis was conducted to calculate the correlation (R^2) using the q-quantile (q-q) plots for two days. The q-q plot is an exploratory graphical method to ensure the validity of a distributional assumption for a set of data. When the data deviate from the assumed distribution, the points on the q-q plot fall on a straight line. As shown in Table 1, the ANN-based MPPT and 7-tri based MPPT perform better than the 7-gbell based MPPT which achieved R^2=1.

Table1. Performance indicators of the three AI-based MPP trackers.

	Standard deviation (SD)			Mean absolute error (MAE)			Correlation (R^2)		
Experiment data	7-gbell	7-tri	ANN	7-gbell	7-tri	ANN	7-gbell	7-tri	ANN
9th of August	11.6390	1.14178	**0.04199**	10.2166	0.37268	**0.01735**	0.9997	**1**	**1**
10th of August	11.7489	2.72587	**0.08046**	7.49826	1.42912	**0.03517**	0.9998	**1**	**1**

5. Conclusion

Three AI-based MPP trackers using ANN, ANFIS with seven triangular fuzzy sets (7-tri), and ANFIS with seven gbell fuzzy sets for a PV system were developed and the performances were evaluated using SD, MAE and correlation (R^2). Actual field data were used to measure the performance indices. Based on the results, the ANN-based MPP tracker is found to perform better than the two ANFIS-based MPP tracker in which it gives the lowest SD and MAE. A comparative analysis based on R^2 value reveals that the ANN and ANFIS with 7-tri achieved a robust correlation, which is (R^2=1).

Acknowledgment
The authors are grateful to Universiti Kebangsaan Malaysia for supporting this research financially under research project ETP-2013-044.

References

[1] Mutlag A H, Shareef H and Mohamed A. Design and Analysis of a High Performance Three-Phase Voltage Source Inverter Based on Fuzzy Logic Controller. Student Conference on Research and Development (SCOReD) 2013 Proceeding, Putrajaya, Malaysia. 204–208.
[2] Shareef H, Mutlag A H and Mohamed A. A novel approach for fuzzy logic PV inverter controller optimization using lightning search algorithm. *Neurocomputing* 2015 **168** 435–453.
[3] Subiyanto, Mohamed A and Shareef H. Hopfield Neural Network Optimized Fuzzy Logic Controller for Maximum Power Point Tracking in a Photovoltaic System. *Int. J. Photoenergy* 2012 **2012** 1–13.
[4] Yusof Y, Sayuti S H, Latif M A and Wanik M Z C. Modeling and Simulation of Maximum Power Point Tracker for Photovoltaic System. National Power & Energy Conference (PECon) 2004 Proceeding, Kuala Lunipur, Malaysia. 88–93.
[5] Liu F, Duan S, Liu F, Liu B and Kang Y. A Variable Step Size INC MPPT Method for PV Systems. *IEEE Transactions on Industrial Electronics* 2008 **55** 2622–2628.
[6] Houssamo I, Locment F and Sechilariu M. Experimental analysis of impact of MPPT methods on energy efficiency for photovoltaic power systems. *Electrical Power and Energy Systems* 2013 **46** 98–107.
[7] Ghani Z A, Hannan M A and Mohamed A. Simulation model linked PV inverter implementation utilizing dSPACE DS1104 controller. *Energy and Buildings*. 2013 **57** 65–73.

IOP Conf. Series: Earth and Environmental Science **32** (2016) 012015 doi:10.1088/1755-1315/32/1/012015

Wet Load Study of Gross Pollutant Traps; Kemensah River, Malaysia

N. M. Zahari[1], L. M. Sidek[1], H. Basri[1], N. F. Md Said[2], M. S. F. Md Noor[2], M. Jajarmizadeh[2], M. R. Zainal Abidin[2], N. Mohd. Dom[3]

[1] Department of Civil Engineering, College of Engineering, Universiti Tenaga Nasional, Jalan IKRAM-UNITEN 43000 Kajang, Selangor, Malaysia.
[2] Centre for Sustainable Technology and Environment, Universiti Tenaga Nasional, Jalan IKRAM-UNITEN 43000 Kajang, Selangor, Malaysia
[3] Humid Tropics Centre, Department Irrigation and Drainage, Jalan Ledang, Taman Duta, 50480 Kuala Lumpur, Wilayah Persekutuan Kuala Lumpur, Malaysia.

Email: Nazirul@uniten.edu.my

Abstract. Efforts have been made by the government agency to install the Gross Pollutant Trap (GPT) to protect of entering the gross pollutants from direct discharge into the river. Few studies have been performed in related to GPT and there is a need to do research on the performance of GPTs in Malaysia. The performance of GPTs is strongly dependent on installed location. Moreover, contribution of landuse has also potential to evaluate the GPTs performance. The main objective of this paper is to analyse the wet load that has been trapped inside GPTs. In this study, Installation of GPTs are performed based continuous deflective separation (CDS) and downstream defender (DD). Kemensah River is chosen as a case study and GPTs are installed at drainage systems due to analysis the impact of GPTs on reduction of solid waste from discharge to the Kemensah River. The result shows that the GPTs can reduce the gross pollutants before entering the river system in optimum way. Consequently, the area of catchment, dense residential area and rainy season have important role on increasing the gross pollutant.

1.0 Introduction

Urbanization also often relates to deterioration of stormwater quality due to many factors such as uncontrolled pollution, increasing non-point source and point source and chemical waste disposal [1]. Currently, more people live in urban areas than in rural areas, with 54% of the world's population residing in urban areas in 2014. In 1950, 30% of the world's population was urban, and by 2050, 66% of the world's population is projected to be urban [2]. Urbanization can lead to the improvement of socioeconomic life in one area and it can also bring a wide range of challenges to sustain the environment. Thus, management of river improvement in urban areas is an extremely challenging mission.

Gross pollutant is a debris item larger than five millimetres that includes litter, debris and coarse sediments [3]. Gross pollutants are main targeted for the river cleaning application removal. Urban storm water pollutants include gross pollutants, trace metals and nutrients that are associated with sediments, and dissolved pollutants [3]. According to Goonetilleke and Thomas [4] and New South Wales Department of Environment and Conservation [5] several issues that can relate to gross pollutant trap performance i.e. inadequate treatment capacity, incapable to remove very fine and dissolved sediments, collected water in the system may become polluted during a dry period and high maintenance cost.

Efforts have been made by the authority (Malaysia Department of Irrigation and Drainage) to install the Gross Pollutant Trap (GPT) in order to reduce gross pollutants from entering the river system. Globally, few studies have been performed in related to GPT [1, 3, 6]. There is a need to do research on performance of GPTs in Malaysia and attributed impact of using them on entering the pollution to the river system. However, it is important to note that these devices will only function well with periodic and proper maintenance method. The performance of GPTs is strongly dependent upon the specific site criteria including type of land use, hydrological regime and maintenance frequency. The objective of this paper is to analysis the wet load trap after GPTs cleaning at Kemensah River for evaluate the performance of the GPTs.

2.0 Study area

Figure 1. Kemensah River located at Kuala Lumpur (Left) and Location of GPTs at Kemensah River (Right)

The study area lies in north east of Capital city of Malaysia, Kuala Lumpur. River namely as Kemensah is located at the upper stream of Klang River. The upper stream of Klang River flows through a major residential and commercial area. Kemensah River is surrounded by residential area and village at upper stream Kemensah River. National Zoo, one of interesting place to visit in Malaysia is located at downstream of Kemensah river. Figure 1 shows the GPTs location at Kemensah River. Each GPT has an own unique identification numbers to differentiate between GPTS (Figure 1).

3.0 Proprietary of GPT and Cleaning Method

There are two proprietaries of GPT installed in Kemensah River which is continuous deflective separation (CDS) and downstream defender (DD). All GPTs at Kemensah River are installed at the end of drainage and it is believed that it can reduce the solid waste from direct discharge towards into Kemensah River. Eight units of CDS, one unit of DD and the total of GPTs were installed in Kemensah River is nine. Figure 2 and Figure 3 show that cross section of CDS and cross section of DD respectively. Every month, GPTs are cleaned with a manual method which is using a net as medium to collect a wet load. The collection of wet load are weighed using the portable weight equipment and recorded in every cleaning process. The duration of cleaning process is done from August 2014 to December 2014.

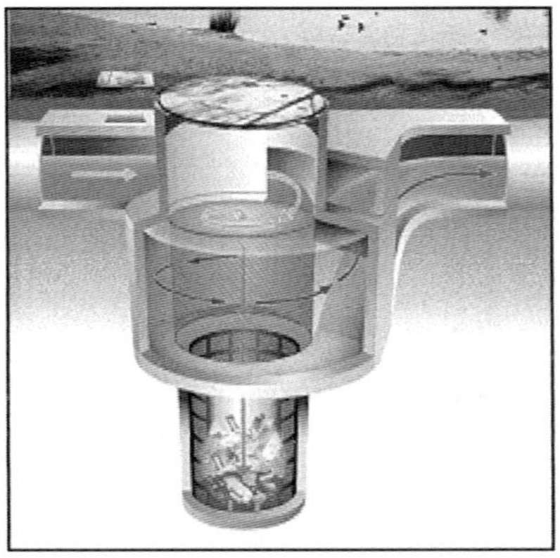

Figure 2. Cross Section of CDS

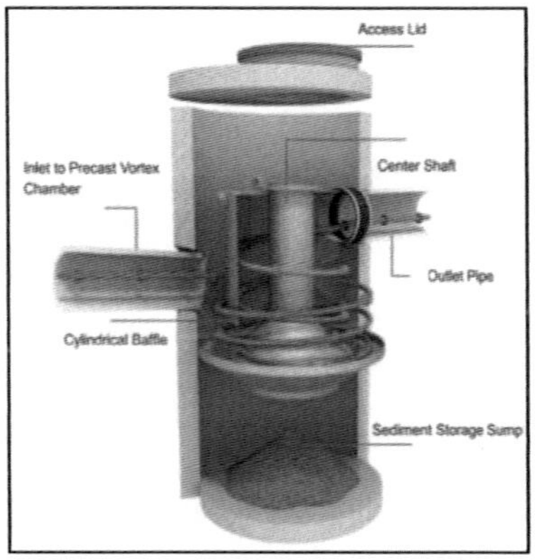

Figure 3. Cross Section of DD

4.0 Results and Discussion

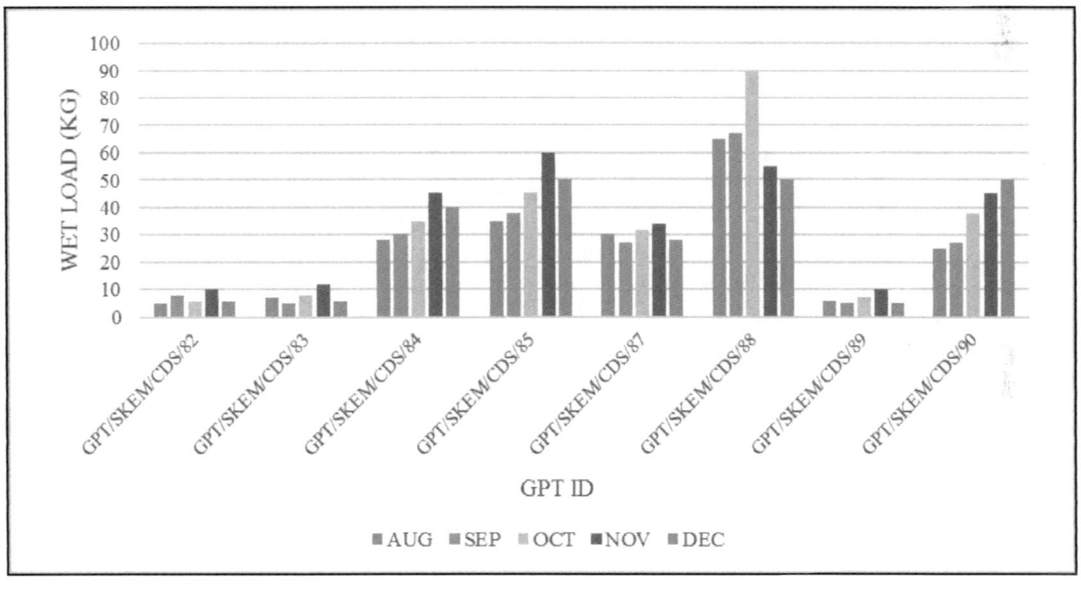

Figure 4. Wet Load for CDS from August 2014 to December 2014

Figure 4 shows the wet load data for 8 units of CDS which are installed at Kemensah River. The highest wet load is recorded 90 kg on October 2014 which is from GPT/SKEM/CDS/88. The reason for the high wet load could be due to the large catchment area, dense residential area and rainy season. The GPT with number 82, 83, 89 show the lowest wet load among other GPTs. It might be related with location of GPT including less dense residential area. The average gross pollutant of wet load trapped in all CDS are 29.38 kg for a period of August to December.

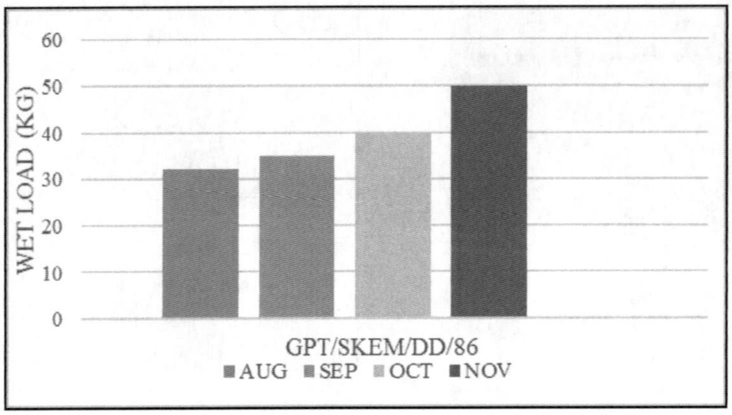

Figure 5. Wet Load for DD from August 2014 to November 2014

Figure 5 shows the wet load data for DD installed at Kemensah River. The highest wet load data was recorded from GPTs maintenance is 50kg on November 2014 at GPT/SKEM/DD/86. Higher wet load indicates on November 2014 could be due to the monsoon season (wet season) in Malaysia.

5.0 Conclusion

This paper explains an analysis of wet load of Gross Pollutant Trap (GPT) which is installed at Kemensah River. This analysis indicates that installed GPT at Kemensah River can reduce gross pollutant from entering the river system. Perhaps, some reasons of higher of wet load could be due to the large catchment area, dense residential area and rainy season. Moreover, authority still continuously develop the GPTs to enhance the environment quality especially the river system in Klang Valley.

Acknowledgements

The authors would like to acknowledge and express appreciation to Humid Tropics Centre, Department of Irrigation and Drainage, Kuala Lumpur for their support and contribution towards preparing the fund for this study.

References

[1] Pouraghniaei M J 2002 *Effects of Urbanization on Quality and Quantity of Water in the Watershed* Natural Resources Research Center of Semnan, Semnan Province, Iran.

[2] Nations U 2014 *Word Urbanization Prospect* Retrieved from http://esa.un.org/unpd/wup/Highlights/WUP2014-Highlights.pdf

[3] Allison R, Chiew F, and Mcmahon T 1997 *Stormwater Gross Pollutants Industry Report 97/11*, Cooperative Research Centre For Catchment Hydrology, Monash University, Australia

[4] Goonetilleke A, and Thomas E C 2003 *Urban Water Quality And The Triple Bottom Line - Can We Reconcile The Irrecocncibles?*, In Search of Sustainability

[5] NSW EPA 2008, New South Wales Environment Protection Authority (2000). *Unpublished Costing Data For Stuctural Stormwater Quality BMPs Derived from NSW Local Goverments*, NSW EPA, Sydney, New South Wales.

[6] Lariyah, M.S.. Mohd Nor M.D., Norazli O., Md. Nasir M.N., Hidayah B., and Zuleika Z. (2011). *Gross Pollutant Analysis in Urban Residential Area for a Tropical Climate Coutry*. 12th International Conference on Urban Drainage., Porto Alegre/Brazil.

IOP Conf. Series: Earth and Environmental Science **32** (2016) 012016 doi:10.1088/1755-1315/32/1/012016

Development of High Efficiency and Low Emission Low Temperature Combustion Diesel Engine with Direct EGR Injection

R J Ho[1], Kumaran P.[2], M Z Yusoff[2]
[1]JKM, Politeknik Kuching Sarawak, 93050 Kuching, Sarawak, Malaysia
[2] UNITEN Putrajaya, Jalan IKRAM-UNITEN, 43000 Kajang, Selangor, Malaysia.

E-mail: horuijin@poliku.edu.my

Abstract. Focus on energy and environmental sustainability policy has put automotive research & development directed to developing high efficiency and low pollutant power train. Diffused flame controlled diesel combustion has reach its limitation and has driven R&D to explore other modes of combustions. Known effective mode of combustion to reduce emission are Low temperature combustion (LTC) and homogeneous charge combustion ignition by suppressing Nitrogen Oxide(NO_x) and Particulate Matter (PM) formation. The key control to meet this requirement are chemical composition and distribution of fuel and gas during a combustion process. Most research to accomplish this goal is done by manipulating injected mass flow rate and varying indirect EGR through intake manifold. This research paper shows viable alternative direct combustion control via co-axial direct EGR injection with fuel injection process. A simulation study with OpenFOAM is conducted by varying EGR injection velocity and direct EGR injector diameter performed with under two conditions with non-combustion and combustion. n-heptane (C_7H_{16}) is used as surrogate fuel together with 57 species 290 semi-detailed chemical kinetic model developed by Chalmers University is used for combustion simulation. Simulation result indicates viability of co-axial EGR injection as a method for low temperature combustion control.

1. Introduction

The main contributors to green house gases are power generation and transportation industries. Diesel engine is considered as one of the most efficient power plant widely used in transportation industries contributed by its high compression ratio and low throttling loses design [1]. Study showed an estimate of 6% to 32% of Nitric Oxide (NO_x) emission is contributed by passenger automobile in overall and NO_x and Particulate Matter(PM) is by-product a conventional diesel combustion engine [2]. Extensive research has been done to reduce NO_x and PM by reducing flame temperature. It has been a challenging task to increase engine thermal efficiencies while suppressing combustion temperature as both NO_x and PM formation has strong relation to combustion temperature [3]. Research this direction are also known as Low Temperature Combustion(LTC), High Temperature Air Combustion (HTAC),Flameless Oxidation (FLOX), Low Nitrix Oxide Injection (LNI), Homogeneous Charged Compressed Ignition(HCCI), and Mild Combustion.

2. Background Studies

Although diesel engine are an efficient powertrain, but it also produces high NO_x and PM emission. Improvement to conventional diffused flame control diesel engine combustion has reached its limitation as described in figure 1. Compressed stroke of a conventional diesel engine compresses rises combustion air above diesel fuel autoignition temperature. Fuel injection takes place slightly before Top Dead Center (TDC) when the injected fuel mixed with hot compressed combustion air forming air-fuel mixture. Autoignition takes place at stoichiometrc air-fuel mixture forming high temperature region known as flame front. Remaining combustion is maintained by premixed turbulent jet surrounded diffused flame [2]. This high flame front temperature forms NO_x and rich fuel ratio entrained behind the flame front forms PM in such combustion process. NO_x formation are mainly contributed from thermal with presence of oxygen and hyrdroxyl radical [4, 5]. PM formation follow different path where it begins with fuel pyrolysis with unburned fuel is exposed to high temperature at low oxygen condition [2].

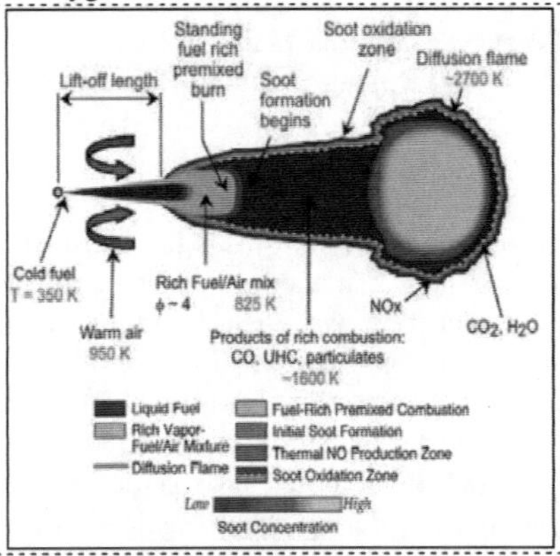

Figure 1. Conventional direct injection diesel spray combustion schematics. [1]

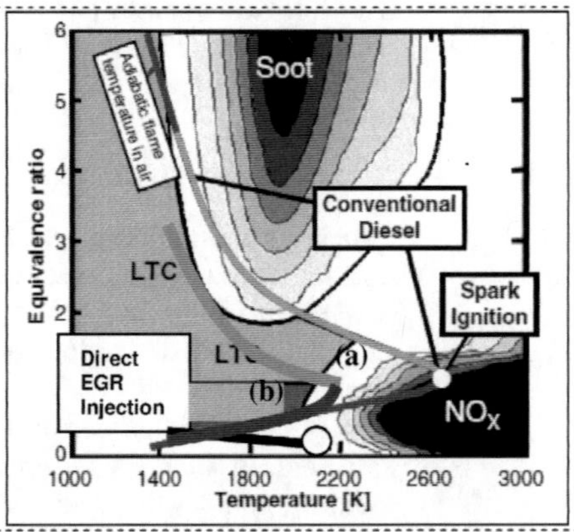

Figure 2. Pollutant formation condition and distribution. [6]

A systematic simulated research approach indicates that highest efficiency were attained from increase of compression ratio, decrease of equivalence ratio and increase of EGR [7]. HCCI and LTC combustion mode are able to meet similar performance characteristics [6]. HCCI suppresses combustion temperature by low equivalence ratio. However, HCCI requires complex premixed fuel-air preparation for mixture homogeneity prior to combustion which limits such application to low load condition and a challenge for high load condition [1]. LTC involves high EGR to keep oxygen concentration low enough to suppress flame front development but high equivalence ratio increases fuel pyrolysis. Hence, rapid mixing of fresh combustion air with fuel vapor cloud decreases equivalence ratio with direct EGR injection at the right injection velocity and timing to prevent soot and NO_x formation as shown in line (b) in figure 2. Objective of this investigation is to identify the prospect of direct EGR injection in developing low emission high efficiency diesel spray combustion.

3. Research Methodology

The simulation platform applied for this research is Open-Sourced Field Operation and Manipulation (OpenFOAM) computational fluid dynamics simulation software with a modified dieselFoam solver to model diesel spray breakup, fuel evaporation and gas dynamics. The simulated model implemented for gas dynamic and thermodynamics for diesel spray combustion model is based on approach used in this research in diesel spray combustion [13]. The primary reference fuel used as surrogates fuel is n-heptane together with CO_2 as EGR is modelled with reduced chemical kinetic model developed by Golovitchev (57 species, 290 reaction) [12] for this parametric research to identify optimal configuration. The boundary conditions are as listed in table 1. Case 1-9 are setup by varying co-axial nozzle diameter and EGR injection velocity and compared with conventional diesel spray combustion, Case 10 & 11. Co-axial injection is intended to improve fuel evaporation and mixing promoting LTC to reduce both NO_x and soot formation at the same time. The results is compared with turbulent flame controlled diesel spray combustion.

Table 1. Boundary Condition and Case Setups

Boundary Conditions	
Air Concentration	21% O2, 79% N_2
Air Temperature	1000K
Injection Configurations	
Pressure	150MPa
Duration	6.8 ms
Reference Fuel	n-Heptane(nC7H16)
Nozzle Diameter	0.1mm
EGR Temperature	950K
EGR Duration	5.5ms

Direct EGR Injection Condition		
Case	EGR Nozzle Diameter (mm)	Injection Velocity (m/s)
1	0.2	100
2	0.2	150
3	0.2	200
4	0.3	100
5	0.3	150
6	0.4	100
7	0.4	150
8	0.6	100
9	0.6	150
10	Reference with Golovitchev Reduced Mechanism	
11	Reference with full n-Heptane Chemkin Model	

4. Result and Discussion

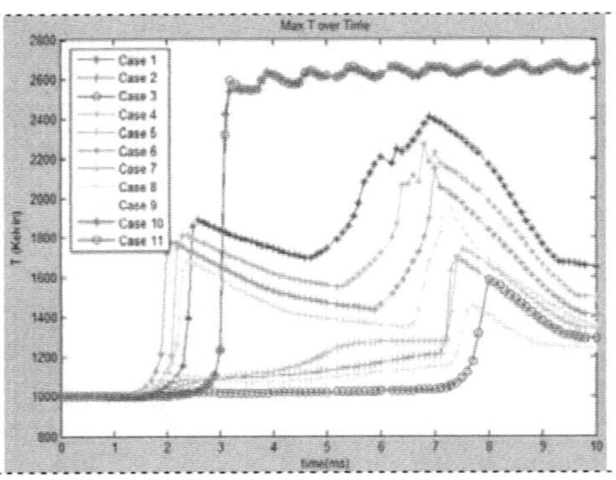

Figure 3. Peak combustion temperature.

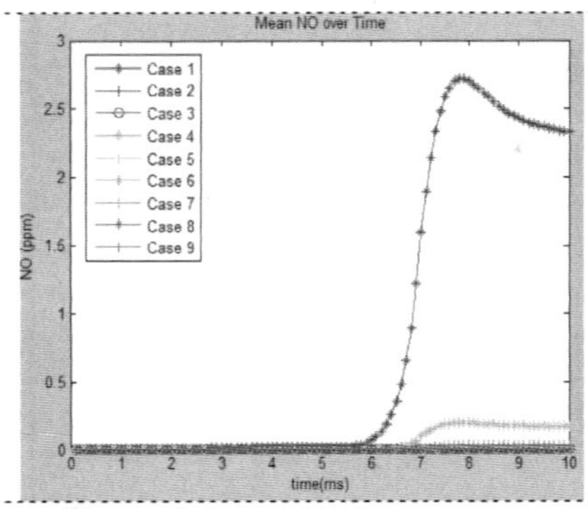

Figure 4. Formed NO comparison.

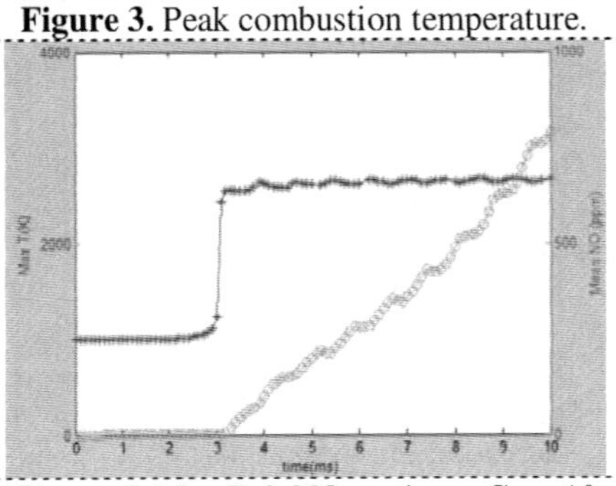

Figure 5. Max T & NO vs time - Case 10

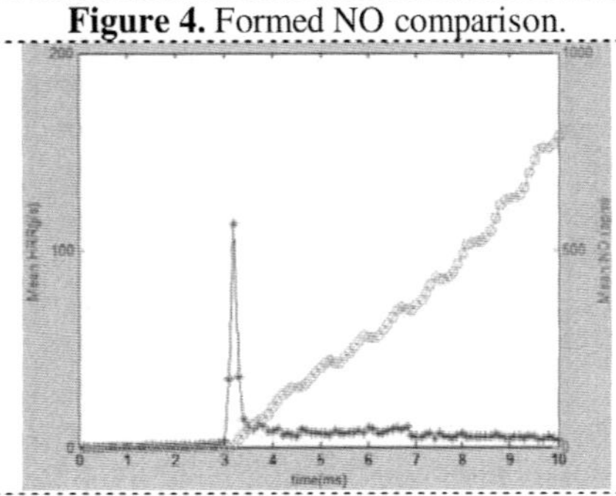

Figure 6. Mean HRR & NO vs time - Case 10

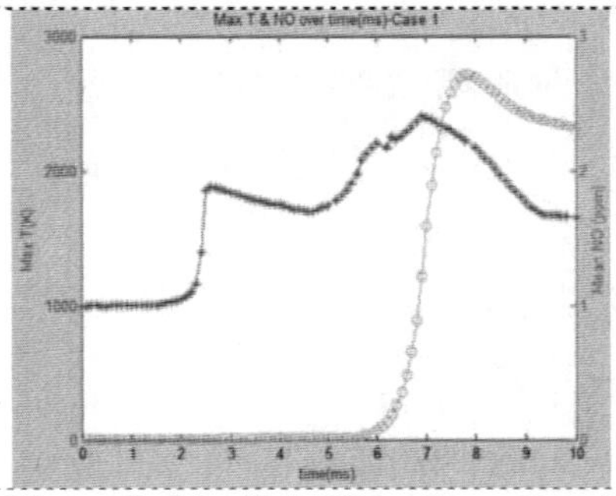

Figure 7. Max T & NO vs time - Case 1

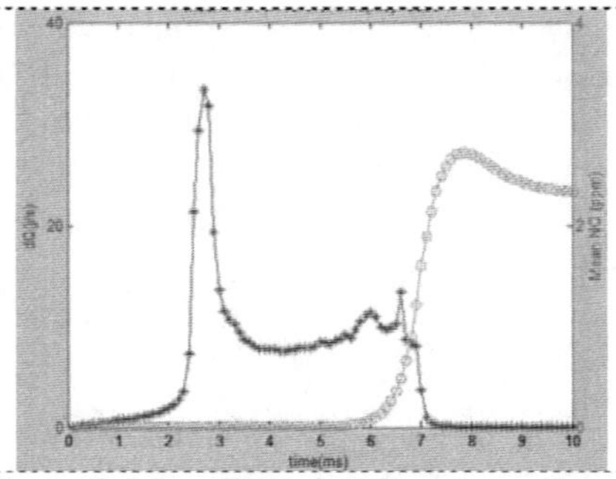

Figure 8. Mean HRR & NO vs time - Case 1

Figure 3-4 are summary for all cases. Comparison between Golovitchev reduced mechanism (Case 10) and full chemkin model (Case 11) in figure 3, shows agreement in modelling n-heptane diesel spray combustion chemistry. All co-axial EGR injection case showed reduced peak combustion temperature and negligible NO formation shown in figure 3 & 4. Co-axial injection suppressed thermal NO formation by reducing N_2 residence time exposed to high combustion temperature and suppresses high combustion temperature flame front by rapid mixing induced by co-axial direct EGR injection. Case 10 in figure 5 indicates conventional combustion maintained elevated peak combustion temperature above thermal NO formation limit produces continuous increase of NO formation after ignition if compared to Case 1 from figure 7 that showed lower peak combustion temperature with negligible NO formation. Case 10 from figure 6 shows high HRR after ignition indicating pre-mixed flame combustion later followed by lower mean HRR turbulent diffused flame mixing controlled combustion. Case 1 from figure 8 showed reduced ignition delay with co-axial injection with prolonged premixed combustion duration with lower premixed combustion mean HRR and followed lower combustion temperature mixing controlled combustion. This could be explained as co-axial injection break up fuel droplets and promotes fuel evaporation reducing ignition delay. High injection velocity sustains longer premixed combustion duration and a reduces the surface volume of mixture at stoichiometric at an instance unlike conventional diesel spray combustion.

Case 1-3 compares different injection velocities between 100m/s(Case 1), 150m/s(Case 2) and 200m/s (Case -3) indicates higher co-axial injection velocity above Case 1 showed reduced or negligible premixed combustion and maintains combustion at low temperature mixing controlled combustion. Case 1, Case 4, Case 6, Case 8 compares different EGR injection volume with larger EGR injection nozzle. Results showed higher EGR injection volume showed reduced or negligible premixed combustion. In both situation of reduced or negligible premixed combustion are unfavourable as it would diminish the benefits of HCCI combustion at start of injection and solely relies on LTC to maintain combustion. Preliminary conclusion indicates low EGR injection volume at lower velocities describe in Case 1 is optimal at this point. Further investigations with lower EGR volume, EGR injection velocities, EGR injection duration are necessary

5. Conclusion

There has been great interest in LTC and HCCI combustion for high efficiency and low pollution emission diesel engine application, however, posed as great challenge to combustion control. Simulated data indicates that direct EGR Injection has the ability to directly control mixing controlled combustion instead of high emission conventional high diffused flame combustion. Direct EGR injection could serve as a mean of combustion control to expand the range of engine operation under low temperature flameless combustion mode with high efficiency and low emission.

References

[1] J. E. Dec, "Advanced compression-ignition engines - Understanding the in-cylinder processes," *Proceedings of the Combustion Institute* , no. 32, pp. 2727-2742, 2009.

[2] D. R. Tree and K. I. Svensson, "Soot processes in compression ignition engines," *Prgress in Energy and Combustion Science* , no. 33, pp. 272-309, 2007.

[3] S. Gan, H. K. Ng and K. M. Pang, "Homogeneous Charge Compression Ignition (HCCI) combustion: Implementation and effects on pollutants in direct injection diesel engines," *Applied Energy,* no. 88, pp. 559-567, 2011.

[4] S. Saravanan, G. Nagarajan, S. Anand and S. Sampath, "Correlation of thermal NOx formation in compression igntion (CI) engine fuelled with diesel and biodiesel," *Energy,* no. 42, pp. 401-410, 2012.

[5] S. C. Hill and L. D. Smoot, "Modeling of nitrogen oxides formation and destruction in combustion system," *Progress in Energy and Combustion Science,* no. 26, pp. 417-418, 2000.

[6] R. Ho, M. Yusoff and K. Palanisamy, "Trend and future of diesel engine: Development of high efficidency and low esmission low temperature combustion diesel engine," in *4th International Conference on Energy and Environment 2013 (ICEE 2013)*, 2013.

[7] J. A. Caton, "The thermodynamic characteristics of high efficiency, internal combustion engines," *Energy Conversion and Management,* no. 58, pp. 84-93, 2012.

[8] J. Wunning and J. Wunning, "Flameless Oxidation to Reduce Thermal NO-Formation," *Prog. Energy & Combustion,* vol. 23, pp. 81-94, 1997.

[9] R. Weber, J. P. Smart and W. v. Kamp, "On the (MILD) Combustion of Gaseous, Liquid, and Solid Fuels in High Temperature Preheated Air," in *Proceedings of the Combustion Institute,* 2005.

[10] C. Tang, Z. Tang, P. Ma, Q. Lin and X. Xing, "Research on the Three Different Kinds of Technologies to Achieve Flameless Combustion and Their Applications," *IEEE,* pp. 978-1-4244-2487-0/09, 2009.

[11] L. M. Pickett, "Low flame temperature limits for mixing-controlled Diesel combustion," *Proceedings of the Combustion Institute,* no. 30, pp. 2727-2735, 2005.

[12] V. Golovichev, "Mechanisms (combustion chemistry):," [Online]. Available: http://www.tfd.chalmers.se/~valeri/MECH.html. [Accessed 15 March 2014].

[13] R. Novella, A. Garcia, J. Pastor and V. Domenech, "The Role of Detailed Chemical Kinetics on CFD Diesel Spray Ignition and Combustion Modelling," *Mathematical and Computer Modelling,* vol. 54, pp. 1706-1719, 2011.

[14] J. Wunning, "Flameless Oxidation," in *6th HiTACG Symposium*, Essen - Germany, 2005.

Impact Resistance Behaviour of Banana Fiber Reinforced Slabs

Zakaria Che Muda[1], Agusril Syamsir [1], Kamal Nasharuddin Mustapha[3], Muhamad Rifdy Samsudin[2], Sivadass Thiruchelvam[4], Fathoni Usman[1], Salmia Beddu [1], Nur Liyana Mohd Kamal[1], Md Ashraful Alam[1], Ahmed H Birima[1], O S Zaroog[4]

[1]Centre of Sustainable Technology and Environment, Universiti Tenaga Nasional, Malaysia
[2]Former Student of Universiti Tenaga Nasional, Malaysia
[3]Centre of Forensic Engineering, Universiti Tenaga Nasional, Malaysia
[4]Centre of Innovation and Design, Universiti Tenaga Nasional, Malaysia

mzakaria@uniten.edu.my

Abstract. This paper investigate the performance of banana fibre reinforced slabs 300mm x 300mm size with varied thickness subjected to low impact projectile test. A self-fabricated drop-weight impact test rig with a steel ball weight of 1.25 kg drop at 1 m height has been used in this research work. The main variables for the study is to find the relationship of the impact resistance against the BF contents and slab thickness. A linear relationship has been established between first and ultimate crack resistance against BF contents and slab thickness by the experiment. The linear relationship has also been established between the service (first) crack and ultimate crack resistance against the BF contents for a constant spacing for various banana fibre reinforced slab thickness. The increment in BF content has more effect on the first crack resistance than the ultimate crack resistance. The linear relationship has also been established between the service (first) crack and ultimate crack resistance against the various slab thickness. Overall 1.5% BF content with slab thickness of 40 mm exhibit better first and ultimate crack resistance up to 16 times and up to 17 times respectively against control slab (without BF)

1. Introduction

The search of sustainable green materials for the construction industry has reached its critical time to mitigate the negative impact of the climatic change. It is critical to study the impact strength characteristics and assess its performance for eco-green construction materials for various potential use in the building industry. Concretes with natural fibers such as Banana fibers (Musa Sepientum) as reinforcement one of the most promising building materials, represent a variety of an extensive class of composite materials presently widely used in various branches of the industry. There is lack of research investigation been carried out on impact resistance of banana fiber reinforced slab. Impact resistance represents the ability of concrete to withstand repeated blows and absorb energy without adverse effect to cracking and spalling. Impact scenario can also be classified into low velocity impact and high velocity impact. The review paper on impact resistance on concrete target has been published by Q.M. Li et al. [1]. Impact resistance of oil palm shells lightweight concrete slab with bamboo fibers has been studied by Z Che Muda et al. The results indicate that 2% volume fraction of bamboo fibers has an optimal performance in first crack resistance and ultimate crack resistance regardless of its fiber length with a potential to be used as an impact resistance composite structures in the future [2]. An investigation on the effect of polyolefin, polyvinyl alcohol and steel fibers on the fracture energy due to low velocity projectile impact on concrete slabs has been carried out by K.C.G. Ong et al. [3].

The objectives of this research are;

i. To establish relationship between impact resistance against banana fibre (BF) content
ii. To establish a relationship between impact resistance against slab thickness.

2. Materials and Test Set-up.

Ordinary Portland cement complying with ASTM Type I cement are used with 2% of super plasticizer is used in the design mix to achieve the desired workability. Banana fibers (BF) are used to reinforce concrete slab is splice and cut into the required length of 20 mm, 30 mm and 40 mm. The basic mix design for the concretes is 0.6 Water/Cement ratios and 2.0 Sand/Cement. There are three different BF/Cement ratios were used in the mix designed; 0.5%, 1.0%, and 1.5% BF/Cement ratios. The compressive strength and density of these mix design at 28 days are as shown in Table 1.

Table 1 Compressive strength and density at 28 days for 0.5%, 1.0% and 1.5% BF content cubes.

BF Content (%)	Density (kg/m^3)	Compressive Strength (N/mm^2)
0.5	1980	28
1.0	1840	27
1.5	1790	26

The above results comply with the category 3-Structural as classified by the American Concrete Institute (ACI) Committee [4] in according to its strength values 16-42 N/mm^2 and density between 1500-2000 kg/m3. The study used a self-fabricated low velocity drop-weight impact test shown in Figure 1 using a steel ball weighing 1.25 kg with drop height of 1000 mm impacting the specimen of size 300mm x 300mm with thickness of 20 mm, 30 mm, and 40 mm mounted on the steel rack frame. The test sample is 1-way simply supported.

(a) (b)

Figure 1. (a) Banana fiber (b) Low-velocity Drop-weight Impact Test Rig

3. Methodology

The potential energy due to the drop body is absorbed as strain energy, generating stresses that causes cracks in the target element. The width, depth, length of the crack developed and its failure mode is associated with the intensity of the energy, the amount of energy absorbed and the properties of concrete. It is assumed that the total computed energy imparted is fully absorbed by the specimens. The relationship of potential energy of a drop-weight projectile and the strain energy dissipated in cracks development is expressed as following formula as proposed by Kankam [5];

$$N*e = R_u* l_c* d_c* w_c \quad (3.1)$$

Where, N = No. of Blows, e = Energy per blow (Joules), l_c = Total length of all cracks, d_c = Maximum crack depth, w_c = Maximum crack width, R_u = Ultimate crack resistance

Another dimensionless factor "impact crack resistance ratio" was also defined:

$$C_r = R_u* f_{cu} \quad (3.\ 2)$$

Where, C_r = Impact crack resistance ratio, f_{cu} = Cube compressive strength

Residual impact strength ratio as defined in Equation 3.3 helps to evaluate the 'post – crack' behavior and use to measure the ductility of the banana fiber composite imparted in the OPS cement matrix.

$$Residual\ impact\ strength\ ratio\ (\ Irs) = \frac{Energy\ absorbed\ at\ ultimate\ failure}{Energy\ absorbed\ at\ initiation\ of\ first\ crack} \quad (3.3)$$

A total of 72 sample slabs of size 300mm x 300mm with 30 mm, 40mm and 50mm thickness were casted and tested with 0.5%, 1.5% and 1.5% BF.

At the first crack and ultimate (failure) crack, the total crack length, the crack width and the crack depth measured by filler gauge with its total numbers of blows recorded.

4. Results and Discussion

4.1 Crack Resistance of Control Sample

Table 2 indicate the service (first) and ultimate crack resistance of control sample without mesh reinforcement.

Table 2 First and Ultimate Crack Resistance of Control Sample without BF Reinforcement

Control Sample No	T20	T30	T40
Thickness of Control Sample	20 mm	30 mm	40 mm
Service (First) Crack Resistance	3.1 N/mm^2	5.35 N/mm^2	7.93 N/mm^2
Ultimate Crack Resistance	4.95 N/mm^2	7.49 N/mm^2	8.05 N/mm^2

The presence of the BF reinforcement for first crack resistance increases up to 16 times and ultimate crack resistance increase up to 17 times against its control sample without BF reinforcement for slab thickness of 40 mm.

4.2 Relationship between crack resistance and fibre content

There is a strong linear correlation between impact resistances against fiber as shown in Figure 2. Generally the first and ultimate crack resistance increases with increasing fiber content. The first crack resistance increase by 1.28 times and 1.57 times for fiber content of 1% and 1.5%, respectively compared to fiber content of 0.5%. The ultimate crack resistance increase by 1.1 times and 1.3 times for fiber content of 1% and 1.5%, respectively compared to fiber content of 0.5%. Comparing the first and ultimate crack resistance for fiber content 1.5%, the crack resistance increase by 1.2%. The slab with fibre content 1.5% show the best performance in resisting impact load for service and ultimate loading. Generally, a good linear correlation between the first and ultimate crack resistance against the bamboo diameter is obtained from the graphs.

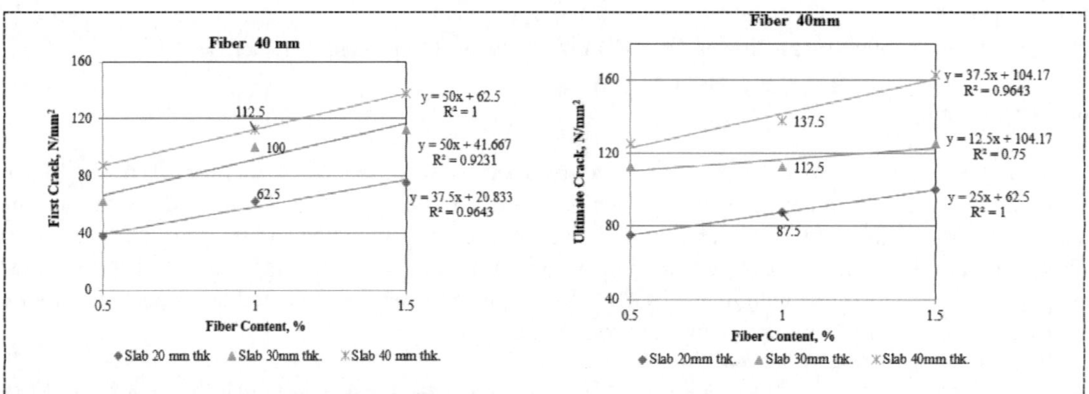

Figure 2 First and ultimate crack resistance against banana fiber content (40 mm length)

4.3 Relationship between Crack Resistance and Slab Thickness

There is a linear relationship between the crack resistance and slab thickness as shown in Figure 3. Increase in crack resistance is more substantial as the thickness increases as compared to the increases in the BF content. For 0.5% fiber, the first crack resistance increase up to 2.33 times and 1.66 times for ultimate crack resistance as the slab thickness increases. For 1% fiber, the first crack resistance increase up to 175% and 157% for ultimate crack resistance. For 1.5% fiber, the first crack resistance increase up to 200% and 137.5% for ultimate crack resistance. The slab with thickness of 40 mm show the best performance in resisting impact load for service and ultimate loading. Generally, a good linear correlation between the first and ultimate crack resistance against the slab thickness is obtained from the graphs.

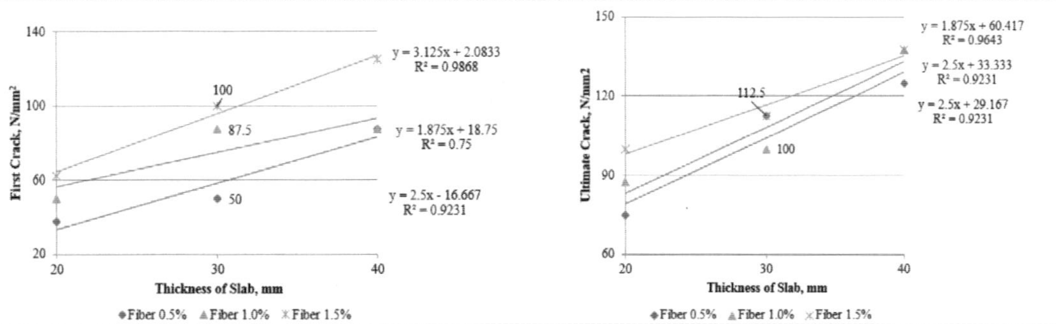

Figure 3 First and ultimate crack resistance against slab thickness

5. Conclusion

The following conclusions can be derived from the experimental results;

- The good linear relationship for the first and ultimate crack resistances against the banana fiber content and the slab thickness are obtained.
- The slab with thickness of 40 mm and fibre content 1.5% show the best performance in resisting impact load for service and ultimate loading.

Acknowledgment

The authors express their gratitude to the technicians and Civil Engineering Department, Universiti Tenaga Nasional, UNITEN, Malaysia who provided the assistance and funding for this study.

References

[1] Q. M. Li, S.R. Reid, H.M. Wen, A. R. Telford, Local impact effects of hard missiles on concrete targets. International Journal of Impact Engineering, Volume 32, August 2005.

[2] Zakaria Che Muda, Salah F A Sharif, Lariyah Bte. Mohd Sidek, Nawfal S Farhan, *Impact Resistance Of Oil Palm Shells Lightweight Concrete Slab With Bamboo Fibers,* International Journal of Science and Engineering Research (IJSER) Volume 4, Issue 1, January 2013.

[3] K.C.G Ong, M. Basheerkhan, P. Paramasivam, Resistance of fibre concrete slabs to low velocity projectile impact, Journal of Cement & Concrete Composite, Volume 21, July 1999.

[4] ACI Committee 544 1996 State-of-the-art report on fiber reinforced concrete ACI Committee 544 report 544.1R-96 Detroit: American Concrete Institute; 1996

[5] CK Kankam . *Impact Resistance of palm kernel fibre-reinforced concrete pavement slab.* J Ferrocement 1999;29(4):279-86. Oct

Introduction of an Emergency Response Plan for flood loading of Sultan Abu Bakar Dam in Malaysia

N. F Md Said [1], L M Sidek [1], H Basri [1], R S Muda[2], and A. Z Abdul Razad[2]

[1]Centre for Sustainable Technology and Environment, College of Engineering, Universiti Tenaga Nasional, Jalan IKRAM-UNITEN 43000 Kajang, Selangor, Malaysia.
[2]TNB Research Sdn. Bhd. No. 1, Lorong Ayer Itam, Kawasan Institusi Penyelidikan, Kajang, 43000, Kajang, Selangor, Malaysia

Email: Farazuien@uniten.edu.my

Abstract. Sultan Abu Bakar Dam Emergency Response Plan (ERP) is designed to assist employees for identifying, monitoring, responding and mitigation dam safety emergencies. This paper is outlined to identification of an organization chart, responsibility for emergency management team and triggering level in Sultan Abu Bakar Dam ERP. ERP is a plan that guides responsibilities for proper operation of Sultan Abu Bakar Dam in respond to emergency incidents affecting the dam. Based on this study four major responsibilities are needed for Abu Bakar Dam owing to protect any probable risk for downstream which they can be Incident Commander, Deputy Incident Commander, On-Scene Commander, Civil Engineer. In conclusion, having organization charts based on ERP studies can be helpful for decreasing the probable risks in any projects such as Abu Bakar Dam and it is a way to identify and suspected and actual dam safety emergencies.

1. Introduction

Essentially, to prevent emergency incidents with colossal number of human from happening in regular life is very difficult. A dam safety emergency is an event which could potentially lead to dam failure and need to be taken care with a massive plan. Events that could result in a dam safety emergency include large floods, earthquakes, explosions, cracks appearing in the embankment, landslide/slippage, unexplained increases in seepage, operational incidents and sabotage. Therefore, an Emergency Response Plan (ERP) is required in order to prevent critical incidents being occurred. ERP can be defined as the detailed contingency program that includes actions that have to be taken in order to control and/or minimize the impact of an emergency situation [1]. ERP is an important method for dealing with different types of accidents, such as fires, explosions, toxic releases, earthquakes, floods, typhoons, and landslides [2]. One of the important parts of ERP scheme is related with definition of key responsibilities due to management of any probable risk events. The objective of this study is developing an ERP as a guideline to set up the communication human resources between correspond people to the key places at emergency time for Sultan Abu Bakar Dam. In this study, ERP of Sultan Abu Bakar has two stages Emergency Response Plan namely Dam Safety Event and Dam Safety Emergency.

2. Application of Sultan Abu Bakar Dam ERP

The development of a comprehensive ERP requires a systematic review of the dam on-site, and the assumption of possible risk scenarios [3]. When the real emergency situation occurred, the staff could not manage any problem properly without previous training and enough knowledge. Therefore, a complete ERP must be effectively developed and distributed to the appropriate human resources to prevent delaying corrective actions [4]. For the first step, Table 1 has been suggested for Sultan Abu Bakar dam owing to identification required human resources for following emergency management team by attribute authority and responsibility. However, all the personnel in the Emergency Management team chart are not involved with dam safety Emergency Response Plan.

The main authorities and related responsibilities required as key persons in Sultan Abu Bakar Dam are summarised in Table 1. In Table 1, the four authorities are the most important persons to conduct Sultan Abu Bakar Dam ERP. Everyone with responsibilities can delegate some or all of their responsibilities to the next most senior person if possible. Furthermore, if the delegated person is unreachable, the next most senior person should be contacted automatically to assume that role. Basically, Incident commander (IC), who is responsible for management of any probable risk, can take final decision and suggests solutions to perform in ERP outline. Next, Deputy Incident Commander (DIC) can be responsible to assist of IC in most of the events. Moreover, DIC can take final decision when IC is absent. On-Scene Commander is a person who is involved with a team to analyse the disaster and preparing the results and consequently the reports for finding the solutions. Also, management of executive parts can be related with On-Scene Commander. Finally, civil engineers can be assigned to fix and management of sub-sections. Usually, civil engineers should take any order by IC to perform any solution for possible risk.

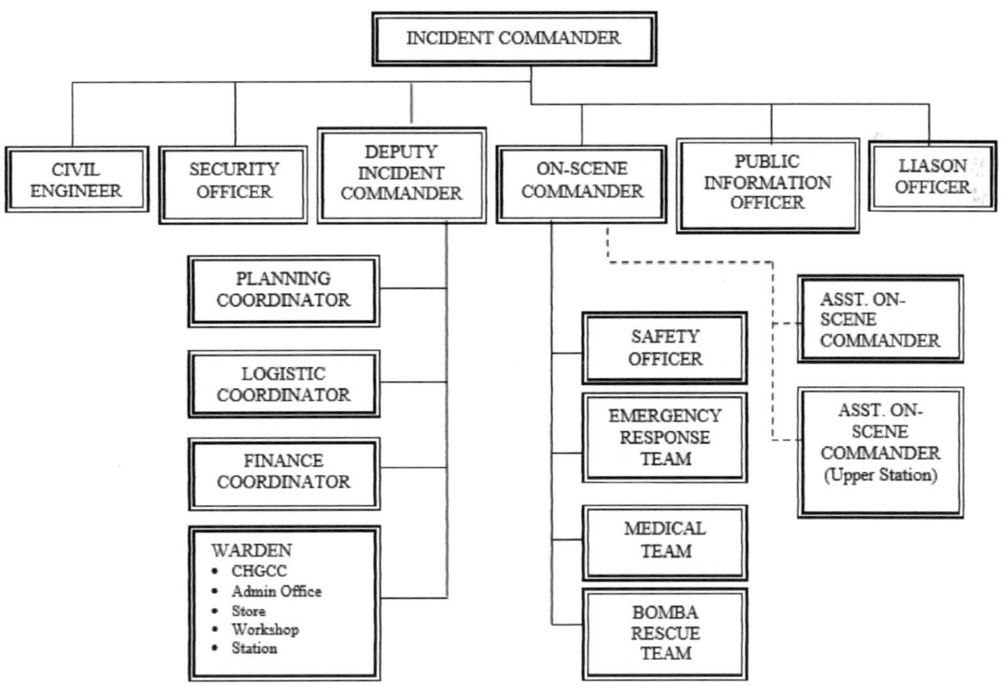

Figure 1. Suggested organization chart for Emergency Management Team

Table 1. Responsibility of Emergency Management Team in Sultan Abu Bakar Dam

Authority	**Responsibilities**
Incident Commander (IC)	• Highest authority for responding to operate directly during emergency events • Managing responsibilities for dam safety emergencies and implementation required procedures • Responsible for the reports of dam safety events and emergencies • Assessing and monitoring of probable risks of Sultan Abu Bakar Dam are well predicted • Establish an Emergency control Centre (ECC) at a safe location • Responsible in maintaining a formal log of the entire event
Deputy Incident Commander	• Next person in charge to take the responsibility of the Incident Commander (IC) if the IC cannot be reached • Assist IC in organizing the ECC • Establish regular meeting to have dynamic and activate mobilize planning, logistic and finance section
On-Scene Commander	• Acting as chief respondent at the beginning of emergency events • Assessment of the scale of incident • Establishment of Incident Command Post (ICP). • Providing updated information pertaining to the development of the event to IC • Management of emergency operations at the scene of the incidents • Managing power generation units during emergencies • Handling and controlling situations with Emergency Response Team (ERT) before the arrival of aid team
Civil Engineer	• Conducting dam inspection and reports on the condition of Sultan Abu Bakar Dam under Incident Commander • Performing possible remedial actions as instructed under the Incident Commander

2.1. Establishment of ECC

During both Dam Safety Event and Dam Safety Emergency, Incident Commander must establish an ECC at a safe and stable location with the most communication channels available – landline telephone, mobile telephone, radio telephone, fax, e-mail, and radio and television facilities. As the conditions during dam emergencies are uncertain, it may be a good option to have a satellite phone or trunk radio that is useful during power breakdown or when the telecommunication signal go off.

2.2. Contents in Dam Safety Event and Dam Safety Emergency

For section in Dam Safety Event and Dam Safety Emergency which including flood loading, earthquake loading and normal loading Dam Safety Event, each of these section has a further three parts:

- **Information and triggers**, which describes potential dam safety incidents and prescribes trigger levels – it is essential for users to read,
- A **flowchart**, which illustrates how the emergency incident will be handled and splits the response into four distinct sections - Figure 2 shows the flowchart interpretation using colour code and box types. 'Event Identification' (yellow), 'Initial Response' (green), 'Dam Safety Emergency' (red) and 'End Event/Emergency' (blue).

- An **action list**, which contains the details behind the various sections of the flowchart – the 'who does what during a dam safety emergency, and when, where and how'.

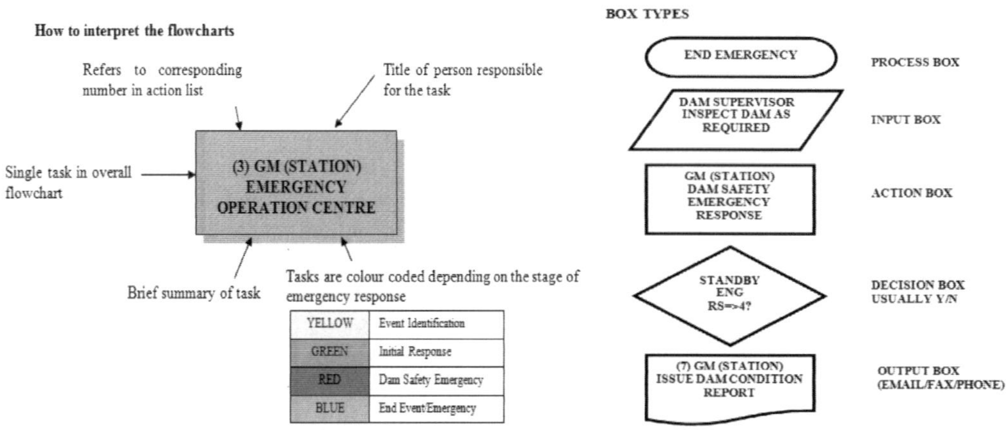

Figure 2. Intepretation of the Flowchart

3. Triggering Level in Sultan Abu Bakar Dam ERP

A dam safety event gives rise to concern that the safety of the dam may be deteriorating such that action must be taken to prevent the development of an emergency condition. Notification is made to appropriate emergency response agencies to enable them to prepare for the implementation of downstream responses such as evacuation of the population at risk. Dam safety event types have been grouped together based on the loading condition under type of occurrence such as flood loading, earthquake loading and normal operation dam safety events. Table 2 describes potential dam safety incidents and prescribes trigger levels. A flood loading dam safety event is initiated when the water level obtained at Dam Safety Level and increased. While such flood is still well within the design capacity of the spillway at Sultan Abu Bakar Dam, it represents a prudent dam safety practice. Considerable flooding downstream of the dam can be expected.

Table 2. Sultan Abu Bakar Dam Safety Triggering Level

Triggering Level	Action	Message Recipients
Dam Safety Event	• Declare Dam Safety Event • Confirm complete evacuation at Bertam Valley • Activate control spilling • Inform state agencies	1. TNB* 2. MKN** Pahang 3. Police Pahang 4. Fire Brigade Pahang
Dam Safety Emergency	• Declare Dam Safety Emergency	1. TNB 2. MKN Pahang
End of Event	• End of Dam Safety Emergency	1. TNB 2. MKN Pahang

*TNB-Tenaga Nasional Berhad

**MKN-Majlis Keselamatan Negara

4. Response to the flood loading dam safety

Once a flood Dam Safety Event is identified, the On-Scene Commander shall alert the Incident Commander by issuing a 'Dam Safety Event' message. This then triggers a process where the dam and its surrounds are immediately inspected and reported on by the Civil Engineer for the duration of the flood to detect signs indicative of potential dam failure. The Incident Commander must continue to monitor the situation from the ECC. If the water level continues to rise above Dam Safety Emergency level and it is predicted that serious structural damage to the dam is suspected or confirmed, the Incident Commander must declare a Dam Safety Emergency. If the water level reaches near of Event level and the procedure pretends that the water level is decreasing, and doesn't cause any serious structural damage to the dam, it is considered that the event is finished.

5. Conclusion

Emergency Response Plan (ERP) is prepared for dam management and supporting personnel in the early detection of the emergency conditions which could endanger the integrity of the dams. ERP provides emergency condition and timely notification to relevant emergency management agencies. The Emergency Management Team need to comply with the assigned obligations regarding mischances after the ERP has been started by the incident commandant. The responsible person in the team should do their responsibilities with the fastest and proper way according to established ERP to prevent the loss of life, environmental degradation and structural damage to the dam. The implementation of ERP through training is important for emergency preparedness to the team in minimising the impact to the society and environment due to dam failure.

Acknowledgements

The authors would like to acknowledge and express our appreciation to Tenaga Nasional Berhad Research (TNBR) for attributed support and contribution towards fund project "Refinement Development of Agencies Emergency Response Plan (TNB ERP) and Development of Agencies ERP for Cameron Highlands – Batang Padang Hydroelectric Scheme under Dam Safety Programme".

References

[1] Lariyah M S, Hidayah B, Sivadass T, Rahsidi S M, Azwin Zailti A R and Zuraidah A 2014 Implementation of Dam Safety Management Program: From Theory to Practice in Malaysia, *Applied Mechanics & Materials* , **567** 583-588.

[2] Tseng J M, Shu C M, Horng, J J, Kuan C M, and Hsu H I 2007 Planning an Emergency Response Center in Southern Taiwan science park , *Trans IChemE Part B,* **85**(B2) 125-132.

[3] Fitzgerald M E 1996 The Emergency Response Plan: Key to Compliance with the Emergency Response Provisions of the Hazardous Waste Operations and Emergency Response Standard, *Applied Occupational and Environmental Hygiene*, **11**(9), 1154-1162.

[4] Tseng J M, Liu M Y, Chang R H, Su J L, and Shu C M 2008 Emergency Response Plan of Chlorine Gas for Process Plants in Taiwan, *Journal of Loss Prevention in the Process Industries,* ***21*** 393-399.

Impact Resistance Performance of Kenaf Fibre Reinforced Concrete

Zakaria Che Muda[1], Nur Liyana Mohd Kamal[1], Agusril Syamsir[1], Chiam Yung Sheng[2], Salmia Beddu[1], Kamal Nasharuddin Mustapha[3], Sivadass Thiruchelvam[4] Fathoni Usman[1], Md Ashraful Alam[1], Ahmed H Birima[1], O S Zaroog[4]

[1]Centre of Sustainable Technology and Environment, Universiti Tenaga Nasional, Malaysia [2]Former Student of Universiti Tenaga Nasional
[3]Centre of Forensic Engineering, Universiti Tenaga Nasional, Malaysia
[4]Centre of Innovation and Design, Universiti Tenaga Nasional, Malaysia

mzakaria@uniten.edu.my

Abstract. This paper investigate the performance of kenaf fibre mesh reinforced concrete (KFMRC) with varied kenaf fibre mesh reinforcement content for the concrete slab of 300mm x 300mm size reinforced with different mesh diameter at constant spacing with varied slab thickness subjected to low impact projectile test. A self-fabricated drop-weight impact test rig with a steel ball weight of 1.236 kg drop at 0.40 m height has been used in this research work. The main variables for the study is to find the relationship of the impact resistance against the amount of mesh reinforcement and slab thickness. A linear relationship has been established between first and ultimate crack resistance against kenaf fiber diameters by the experiment. The linear relationship has also been established between the service (first) crack and ultimate crack resistance against the slab thickness. The threshold (highest) values for service crack and ultimate crack is 47.9 N/mm2 and 130.58 N/mm2 respectively observed and computed for 50 mm slab with 7 mm diameter mesh.

1. Introduction

In the search of sustainable green materials, it is critical to study the impact strength characteristics and assess its performance for eco-green construction materials for various potential uses in the building industry.

There is lack of research investigation been carried out on impact resistance of kenaf fibre mesh reinforced concrete (KFMRC). Impact resistance represents the ability of concrete to withstand repeated blows and absorb energy without adverse effect to cracking and spalling. Impact scenario can also be classified into low velocity impact and high velocity impact. According to Dancygier [1], the nuclear structures are designed to resist impact loading caused by projectile or missiles travelling up to 1000 m/s. The response of reinforced concrete structures under impact loading is different from the static loading especially in the case of high velocity impact of rigid projectile. For impact velocities up to 10 ms^{-1}, the failure modes are generally the same as the static failure, except there is increased tendency for local damage or shear failure to occur. A repeated impact test, a weighted pendulum Charpy-type impact test, a projectile impact test, and explosion-impact test, a constant strain rate test, a split Hopkinson bar test, and an instrumented pendulum impact test could be used to measure the impact resistance of fiber reinforced composites [2]. The impact resistance of material also can be measured by using the criteria such as the energy of fracture of the specimen, repeated impact tests, and the velocity and the size of the spall the specimen subjected to a blast loading surface [3]. However, the drop weight impact test which is recommended by the ACI Committee 544 [4] is the simplest method. The review paper on impact resistance on concrete target has been published by Z Che Muda et al [5]. Impact resistance of oil palm shells lightweight concrete slab with bamboo fibers has been studied by Z Che Muda et al. The results indicate that 2% volume fraction of bamboo fibers has an optimal performance in first crack resistance and ultimate crack resistance regardless of its fiber length with a potential to be used as an impact resistance composite structures in the future [6].

The objectives of this research are;

i. To establish relationship between impacts resistance against content of kenaf fibre mesh reinforcement.
ii. To establish a relationship between impact resistance against slab thickness.

2. Materials and Test Set-up.

The kenaf fiber is obtained from private supplier MZM solution that based in Kedah. The kenaf fiber is graded as grade AA which is capable to produce the highest tensile strength of 900 MPa according to the supplier. The fibres are twisted into the required bundled diameters of 3 mm, 5 mm and 7 mm diameter as shown in Figure 1b and tied to form a mesh at a constant spacing of 50 mm and placed in the mid-depth of the slab

Ordinary Portland cement complying to ASTM Type I cement are used with 2% of super plasticizer is used in the design mix to achieved the desired workability.

The basic mix design for the KFMRC is shown in Table 1.

Table 1. Mix Design for Kenaf fibre mesh reinforced concrete (KFMRC)

Cement (kg/m^3)	Fine Aggregates (kg/m^3)	Course Aggregates (kg/m^3)	Water/cement Ratio	Slump (mm)	Compressive Strength (N/mm2)
403	687	1030	0.6	72	31.7

The study used a self-fabricated low velocity drop-weight impact test set up are as shown in Figure 1b using a steel ball weighing 1.236 kg with drop height of 400 mm impacting the specimen of size 300mm x 300mm with thickness of 20 mm, 30 mm, and 40 mm mounted on the steel rack frame. The test sample is 1-way simply supported.

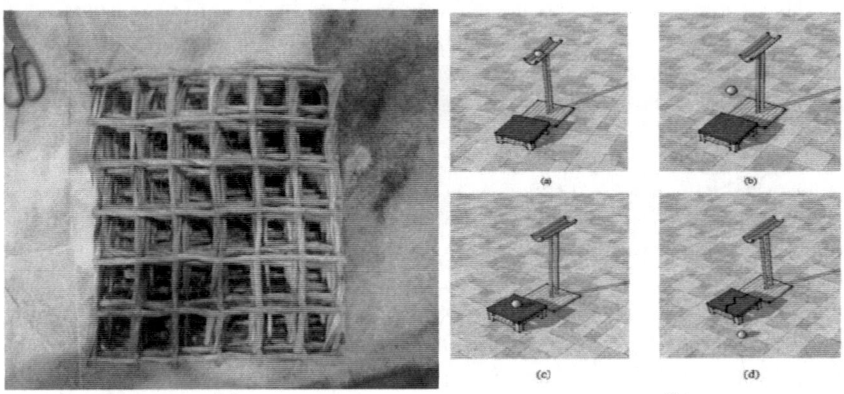

(a) (b)

Figure 1. (a) Kenaf Fibre Mesh (b) Low-velocity Drop-weight Impact Test Set-up

3. Methodology

The potential energy due to the drop body is absorbed as strain energy, generating stresses that causes cracks in the target element. The width, depth, length of the crack developed and its failure mode is associated with the intensity of the energy, the amount of energy absorbed and the properties of concrete. It is assumed that the total computed energy imparted is fully absorbed by the specimens. The relationship of potential energy of a drop-weight projectile and the strain energy dissipated in cracks development is expressed as following formula as proposed by Kankam [7];

$$N*e = R_u * l_c * d_c * w_c \quad (3.1)$$

Where, N = No. of Blows, e = Energy per blow (Joules), l_c = Total length of all cracks, d_c = Maximum crack depth, w_c = Maximum crack width, R_u = Ultimate crack resistance

A total of 27 sample slabs of size 300mm x 300mm with 30 mm,40mm and 50mm thickness were casted with 3 mm, 5mm and 7 mm bundled mesh diameter.

At the first crack and ultimate (failure) crack, the total crack length, the crack width and the crack depth measured by filler gauge with its total numbers of blows recorded.

4. Results and Discussion

4.1 Crack Resistance of Control Sample

Table 2 indicate the service (first) and ultimate crack resistance of control sample without mesh reinforcement.

Table 2. First and Ultimate Crack Resistance of Control Sample without Mesh Reinforcement

Control Sample No	T30	T40	T50
Thickness of Control Sample	30 mm	40 mm	50 mm
Service (First) Crack Resistance	2.15 N/mm^2	2.33 N/mm^2	3.86 N/mm^2
Ultimate Crack Resistance	2.69 N/mm^2	3.92 N/mm^2	5.66 N/mm^2

The presence of the mesh for first crack resistance increases up to 13.91 times and ultimate crack resistance increase up to 23.1 times against its control sample without mesh reinforcement.

4.2 Relationship between Crack Resistance and Kenaf Mesh Diameter

There is a linear correlation between impact resistance against mesh diameter as shown in Figure 2. Generally the first and ultimate crack resistance increases with increasing diameter. The crack resistance increase by up to 5.04 times for first crack and up to 3.47 times for ultimate crack as the kenaf mesh diameter increases for the given thickness. The highest ultimate crack resistance of 130.58 N/mm2 is observed for 50 mm slab thickness with kenaf mesh of 7 mm diameter whilst the lowest is 14.63N/mm2 for 30 mm slab with 3 mm diameter kenaf mesh.

That means that the kenaf mesh has the ability to control cracks and resisted a lot of energy until the crack started to widen and propagate further and lose its bonding and tensile, which then lead to failure.

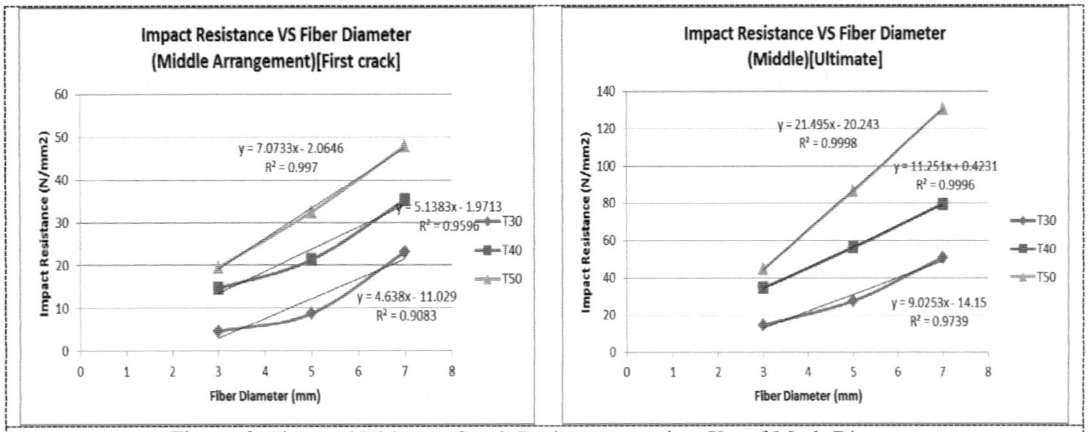

Figure 2. First and Ultimate Crack Resistance against Kenaf Mesh Diameter

4.3 Relationship between Crack Resistance and Slab Thickness

There is a linear relationship between the crack resistance and slab thickness as shown in Figure 3. The crack resistance increase by up to 4.27 times for first crack and up to 3.13 times for ultimate crack as the slab thickness increases for the given mesh diameter. Figure 3 show clearly that the thickness of the slab has an impact on the service and ultimate failure crack. The threshold (highest) values for service crack and ultimate crack is 47.9 N/mm^2 and 130.58 N/mm^2, respectively.

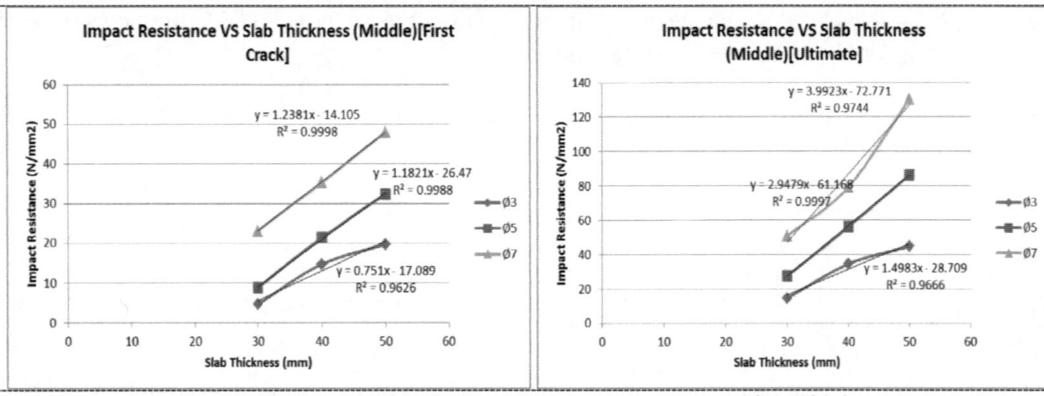

Figure 3. First and Ultimate Crack Resistance against Slab Thickness

5. Conclusion

The following conclusions can be derived from the experimental results;

- The good linear relationship for the first and ultimate crack resistances against the slab thickness and kenaf mesh diameter.
- The presence of the mesh for first crack resistance increases up to 13.91 times and ultimate crack resistance increase up to 23.1 times against its control sample without mesh reinforcement.
- The crack resistance increase by up to 5.04 times for first crack and up to 3.47 times for ultimate crack as the kenaf mesh diameter increases for the given thickness.
- The crack resistance increase by up to 4.27 times for first crack and up to 3.13 times for ultimate crack as the slab thickness increases for the given mesh diameter.
- The threshold (highest) values for service crack and ultimate crack is 47.9 N/mm^2 and 130.58 N/mm^2 respectively observed and computed for the thickest 50 mm slab with its biggest 7 mm diameter mesh.

References

[1] Dancygier, A N 1997 *Effect of reinforcement ratio on the resistance of reinforced concrete to hard projectile impact.* Nuclear Engineering and Design 172, pp. 233-245.

[2] Gopalartnam, V. S., Shah, S. P., and John, R. A modified instrumented Charpy test for cement based composites." Exp. Mech. 1984; 24(2): 102–111.

[3] Balaguru, P. N., and Shah, S. P. Fibre-reinforced cement composites, McGraw-Hill, London.1992.

[4] ACI Committee 544. State-of-the-art report on fiber reinforced concrete. ACI Committee 544 report 544.1R-96. Detroit: American Concrete Institute. 1996.

[5] Zakaria Che Muda, Kong Sih Ying, Salah F A Sharif, Lariyah Bte. Mohd Sidek,Nawfal S. Farhan, *A Review - Local Failure On Concrete Target Due To Projectile Impact.* International Journal of Science and Engineering Research (IJSER) Volume 4, Issue 1, January 2013.

[6] Zakaria Che Muda, Salah F A Sharif, Lariyah Bte. Mohd Sidek, Nawfal S Farhan, *Impact Resistance Of Oil Palm Shells Lightweight Concrete Slab With Bamboo Fibers,* International Journal of Science and Engineering Research (IJSER) Volume 4, Issue 1, January 2013.

[7] CK Kankam. *Impact Resistance of palm kernel fibre-reinforced concrete pavement slab*. J Ferrocement 1999;29(4):279-86. Oct

Impact Resistance Behaviour of Light Weight Rice Husk Concrete with Bamboo Reinforcement

Zakaria Che Muda[1], Salmia Beddu [1], Agusril Syamsir[1], Joshua Sigar Ating[2], Nur Liyana Mohd Kamal[1], Kamal Nasharuddin Mustapha[3], Sivadass Thiruchelvam[4], Fathoni Usman[1], Md Ashraful Alam[1], Ahmed H Birima[1], O S Zaroog[4]

[1]Centre of Sustainable Technology and Environment, Universiti Tenaga Nasional, Malaysia
[2]Former Student of Universiti Tenaga Nasional, Malaysia
[3]Centre of Forensic Engineering, Universiti Tenaga Nasional, Malaysia
[4]Centre of Innovation and Design, Universiti Tenaga Nasional, Malaysia

mzakaria@uniten.edu.my

Abstract. This paper investigate the performance of lightweight rice husk concrete (LWRHC) with varied bamboo reinforcement content for the concrete slab of 300mm x 300mm size reinforced with varied slab thickness subjected to low impact projectile test. A self-fabricated drop-weight impact test rig with a steel ball weight of 1.236 kg drop at 0.65 m height has been used in this research work. The main variables for the study is to find the relationship of the impact resistance against the amount of bamboo reinforcement and slab thickness. A linear relationship has been established between first and ultimate crack resistance against bamboo diameters and slab thickness by the experiment. The linear relationship has also been established between the service (first) crack and ultimate crack resistance against the bamboo reinforcement diameter and slab thickness. 5% RH content exhibit better first and ultimate crack resistance up to 1.80 times and up to 1.72 times respectively against 10% RH content.

1. Introduction

The search of sustainable green materials for the construction industry has reached its critical time to mitigate the negative impact of the climatic change. It is critical to study the impact strength characteristics and assess its performance for eco-green construction materials for various potential use in the building industry.

There is lack of research investigation been carried out on impact resistance of light weight rice husk concrete. Impact resistance represents the ability of concrete to withstand repeated blows and absorb energy without adverse effect to cracking and spalling. Impact scenario can also be classified into low velocity impact and high velocity impact. According to Dancygier [1], the nuclear structures are designed to resist impact loading caused by projectile or missiles travelling up to 1000 m/s. The response of reinforced concrete structures under impact loading is different from the static loading especially in the case of high velocity impact of rigid projectile. For impact velocities up to 10 ms^{-1}, the failure modes are generally the same as the static failure, except there is increased tendency for local damage or shear failure to occur. The review paper on impact resistance on concrete target has been published by Q.M. Li et al. [2]. Impact resistance of oil palm shells lightweight concrete slab with bamboo fibers has been studied by Z Che Muda et al. The results indicate that 2% volume fraction of bamboo fibers has an optimal performance in first crack resistance and ultimate crack resistance regardless of its fiber length with a potential to be used as an impact resistance composite structures in the future [3].

The objectives of this research are;

i. To establish relationship between impact resistance against diameter of bamboo reinforcement.
ii. To establish a relationship between impact resistance against slab thickness.

2. Materials and Test Set-up.

Buloh kuning (*Bambusa vulgaris Schrad)* is used to prepare the bamboo reinforcement material. The bamboo are splice and cut into the required diameters of 2.5 mm , 5 mm and 7.5 mm diameter and tied to form a mesh at a constant spacing of 50 mm.

Rice husk (RH) is taken from the rice production mill in Sarawak. Thus out of the 2.2 Million tons of paddy available in Malaysia (2007 statistics), 121 thousand tons of rice husk ash can be produced. This huge renewable resource should not be wasted and if not used poses disposal problems [4].

Ordinary Portland cement complying to ASTM Type I cement are used with 2% of super plasticizer is used in the design mix to achieved the desired workability.

The basic mix design for the concretes are 0.6 Water/Cement ratio and 2.0 Sand/Cement. There are two different RH/Cement ratios were used in the mix designed; 0.05 RH/Cement ratios and 0.10 RH/Cement ratios. The compressive strength and density of these mix design at 28 days are as shown in Table 1.

TABLE 1 Compressive strength to BS 1881: Part 110:1983 and density at 28 days for 5% and 10% RH content cubes.

RH Content (%)	Density (kg/m^3)	Compressive Strength (N/mm^2)
5	1940	23.6
10	1728	17.9

The above results comply with the category 3-Structural light weight as classified by the American Concrete Institute (ACI) Committee.

The study used a self-fabricated low velocity drop-weight impact test shown in Figure 1 using a steel ball weighing 1.236 kg with drop height of 650mm impacting the specimen of size 300mm x 300mm with thickness of 20 mm, 30 mm, and 40 mm mounted on the steel rack frame. The test sample is 1-way simply supported. A total of 33 test sample slabs of size 300mm x 300mm with 30 mm, 40mm and 50mm thickness were casted and tested which is consist of 6 control samples and 27 samples with bamboo reinforcement.

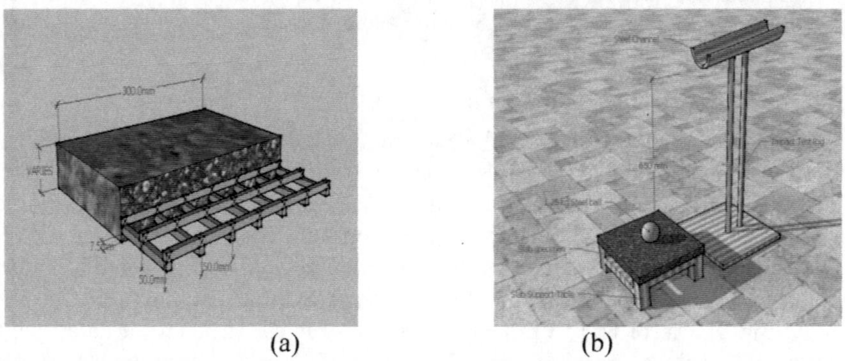

(a) (b)

Figure 1 (a) Bamboo Reinforced RH mortar (b) Low-velocity Drop-weight Impact Test Rig

3. Methodology

The potential energy due to the drop body is absorbed as strain energy, generating stresses that causes cracks in the target element. The width, depth, length of the crack developed and its failure mode is associated with the intensity of the energy, the amount of energy absorbed and the properties of concrete. It is assumed that the total computed energy imparted is fully absorbed by the specimens. The relationship of potential energy of a drop-weight projectile and the strain energy dissipated in cracks development is expressed as following formula as proposed by Kankam [5];

$$N*e = R_u * l_c * d_c * w_c \quad (3.1)$$

Where, N = No. of Blows, e = Energy per blow (Joules), l_c = Total length of all cracks, d_c = Maximum crack depth, w_c = Maximum crack width, R_u = Ultimate crack resistance

At the first crack and ultimate (failure) crack, the total crack length, the crack width and the crack depth measured by filler gauge with its total numbers of blows recorded.

4. Results and Discussion

4.1 Relationship between Crack Resistance and Bamboo diameter

The relationship of impact resistance against bamboo diameter are as shown in Figure 2 and Figure 3. Generally the first and ultimate crack resistance increases with increasing diameter. For 5% RH, with the given thickness, crack resistance increase by up to 1.43 times for first crack and up to 1.24 times for ultimate crack as the bamboo diameter increases. For 10% RH, crack resistance increase by up to 1.53 times for first crack and 1.58 times for ultimate crack as the bamboo diameter increases for the given thickness.

Generally, a good linear correlation between the first and ultimate crack resistance against the bamboo diameter is obtained from the graphs.

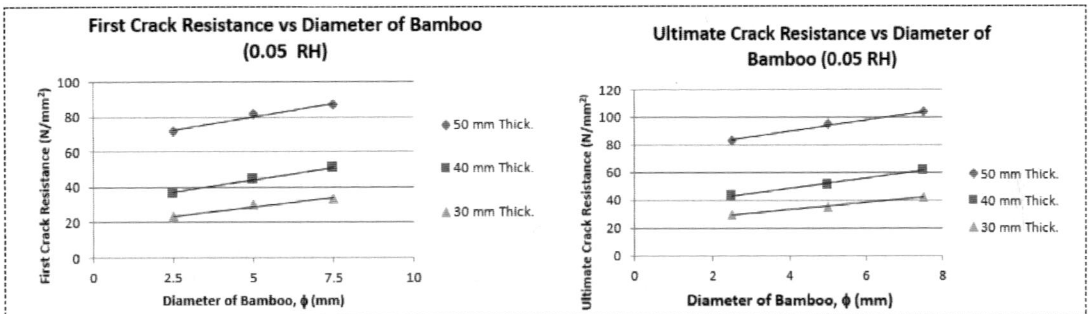

Figure 2 First and Ultimate Crack Resistance against Bamboo Diameter for 5% RH Content

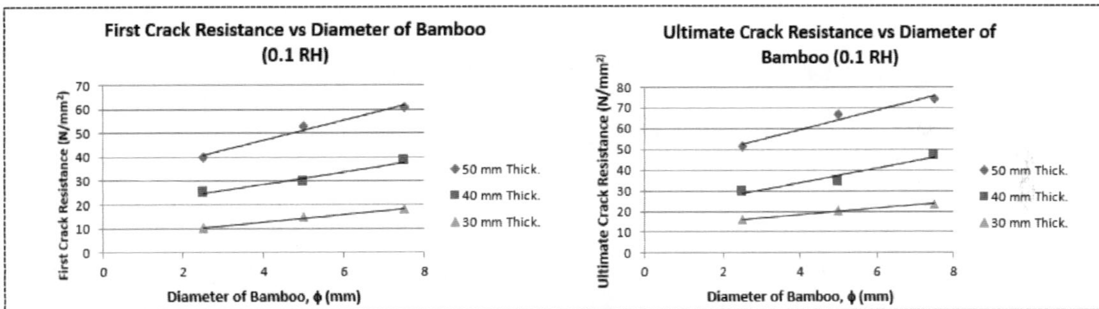

Figure 3 First Crack and Ultimate Crack Resistance against Bamboo Diameter for 10% RH Content

4.2 Relationship between Crack Resistance and Slab Thickness

There is a linear relationship between the crack resistance and slab thickness as shown in Figure 4 and Figure 5. Increase in crack resistance is more substantial as the thickness increases as compared to the increases in the diameter. For 5% RH, with the given bamboo diameter, the first crack resistance increase up to 3.09 times and 2.8 times for ultimate crack resistance as the slab thickness increases. For 10% RH, with the given bamboo diameter, the first crack resistance increase up to 3.9 times and 3.32 times for ultimate crack resistance. Thickness of the slab has more effect on the crack resistance as compare to the diameter of the bamboo reinforcement. Generally, a good linear correlation between the first and ultimate crack resistance against the thickness is obtained from the graphs.

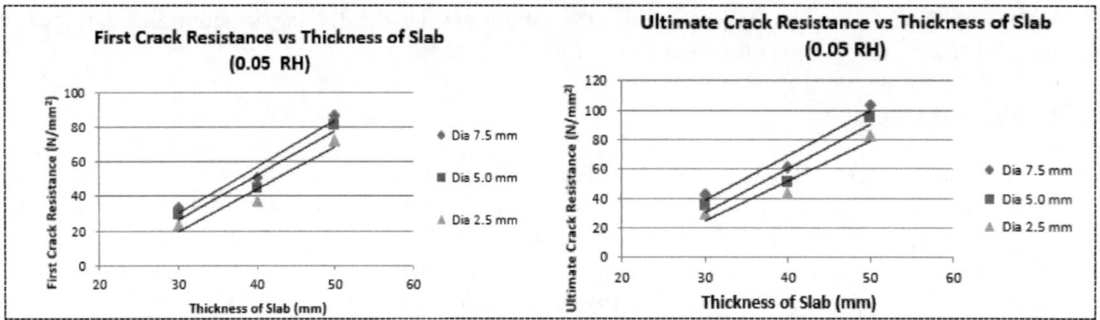

Figure 4 First and Ultimate Crack Resistance against Slab Thickness for 5% RH Content

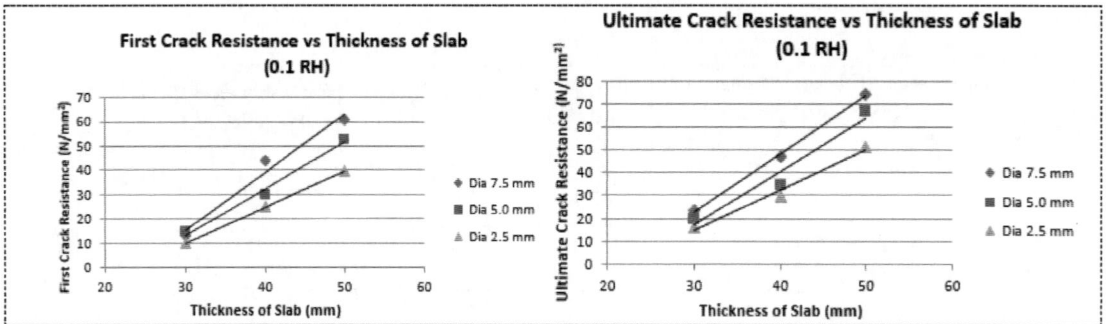

Figure 5 First and Ultimate Crack Resistance against Slab Thickness for 10% RH Content

5. Conclusion

The following conclusions can be derived from the experimental results;

- The good linear relationship for the first and ultimate crack resistances against the slab thickness and bamboo diameter.
- For 5% RH, crack resistance increase by up to 1.43 times for first crack and up to 1.42 times for ultimate crack with the increase in bamboo diameter for 30 mm slab thickness. For 10% RH, crack resistance increase by up to 1.77 times for first crack and 1.58 times for ultimate crack as the bamboo diameter increases for 30 mm slab thickness.
- For 5% RH, the first crack resistance increase by up to 3.09 times and 2.8 times for ultimate crack resistance with the increase in the slab thickness for 2.5 mm bamboo diameter. For 10% RH, with the given bamboo diameter, the first crack resistance increase up to 3.9 times and 3.32 times for ultimate crack resistance

References

[1] Dancygier, A N 1997 *Effect of reinforcement ratio on the resistance of reinforced concrete to hard projectile impact.* Nuclear Engineering and Design 172, pp. 233-245.

[2] Q. M. Li, S.R. Reid, H.M. Wen, A. R. Telford, Local impact effects of hard missiles on concrete targets. International Journal of Impact Engineering, Volume 32, August 2005.

[3] Zakaria Che Muda, Salah F A Sharif, Lariyah Bte. Mohd Sidek, Nawfal S Farhan, *Impact Resistance Of Oil Palm Shells Lightweight Concrete Slab With Bamboo Fibers,* International Journal of Science and Engineering Research (IJSER) Volume 4, Issue 1, January 2013.

[4] Jha, J.N. and K.S. Gill Effect of rice husk ash on lime stabilization. J. Inst. Eng., 2006,87: 33-39.

[5] CK Kankam . *Impact Resistance of palm kernel fibre-reinforced concrete pavement slab.* J Ferrocement 1999;29(4):279-86. Oct

The Semantic Network of Flood Hydrological Data for Kelantan, Malaysia

[1]Aziyati Yusoff, [1]Norashidah Md Din, [2]Salman Yussof and [3]Samee Ullah Khan

[1]College of Engineering, Universiti Tenaga Nasional (UNITEN), Malaysia
[2]College of Information Technology, Universiti Tenaga Nasional (UNITEN), Malaysia
[3]Department of Electrical and Computer Engineering, North Dakota State University (NDSU), United States of America

Email: aziyati@mohr.gov.my, norashidah@uniten.edu.my , salman@uniten.edu.my , samee.khan@ndsu.edu

Abstract. Every year, authorities in Malaysia are putting efforts on disaster management mechanisms including the flood incidence that might hit the east coast of Peninsular Malaysia. This includes the state of Kelantan of which it was reported that flood is just a normal event occurred annually. However, the aftermath was always unmanageable and had left the state to struggle for its own recoveries. Though it was expected that flood occurred every year, among the worst were in 1967, 1974, 1982 and recently in December 2014. This study is proposing a semantic network as an approach to the method of utilising big data analytics in analysing the huge data from the state's flood reading stations. It is expected that by using current computing edge can also facilitate mitigating this particular disaster.

1. Introduction

The study on the patterns of flood incidence had triggered further studies to be done for an effective warning system and an efficient evacuation system. Researchers had proposed numerous approaches on the flood disaster management. The need for precise computation had challenged the experts in Geographical Information System (GIS) to be updated with the latest technology in the world of computing. For this research, we are interested to study on the flood patterns that were reported to occur annually in the state of Kelantan, Malaysia.

In most cases, the governmental agencies had been using renowned computer software for flood data and mapping. Among others, ArcGIS software by Esri Inc. is always preferable for mapping and geospatial measurement Other prominent geospatial mapping tools include Global Precipitation Measurement (GPM) [3], SERVIR Labs [4], MODIS NRT Flood Mapping [5], Global Flood Monitoring System (GFMS) [6], Global Disaster Alert and Coordination System (GDACS) [7], and Dartmouth Flood Observatory [8].

The flood management is under the administration of Department of Irrigation and Drainage (DID), Ministry of Natural Resources and Environment (NRE). According to DID, every year it is estimated that 29,800 kilometers of Malaysian land is prone to flood incidence. In year 2006 and 2007, Malaysia was facing a very bad flood disaster that total loss was estimated as RM1.1 billion nationally and RM776 million for regional state assets excluding the personal loss of the affected communities nationwide [9]. However, the authority body that manage disaster occurrence in this country is the National Disaster Management Agency (APBN), Prime Minister's Department. This agency has the responsibility to coordinate the management of national disaster, and to ensure that all of the national disaster policies and mechanisms are adhered and implemented at all level of disaster management units [10].

As an example, one of the worst hit flood had occurred in December 2014. Kelantan had the highest figures with losses and damages. Report from the officials of Malaysia [8] had been recorded and the profile of the incidence as illustrated in Table 1.

Table 1. Profile of the latest flood incidence in the state of Kelantan (2014)

Facts	**Figures**
Country	Malaysia (States worst-hit: Kelantan, Terengganu, Pahang)
Date	21 to 31 December 2014
Death	At least 14 persons
Injured	8 persons
Affected institutions	1,415 villages, 237 schools, 32 religious sites, 17 government offices
Evacuees	132,000 (21,686 for Kelantan state)
Affected infrastructures	Neighbourhoods turned into islands surrounded by a sea of murky brown water. Streets doubled as rivers. Cars were swept away and toppled over. Strong currents and interrupted power supply. Clean water cannot be delivered.
Affected national production	184,500 hectares (455,000 acres) of palm oil production. Output dropped about 20 per cents (1.4 million metric tons).

2. Semantic network of hydrology data

This research is to study on a feasible method for handling bulk hydrology data that being archived by the state authorities, of which were never being fully utilised for the sake of effective flood management. As a result, an approach of semantic network is proposed for this research. Figure 1 has been developed to represent the classes and relationships that can be abstracted from this study.

The study is proposing that its main class is 'big data analytics'. The sub-classes for this analytics will involve the 'hydrology data', 'GIS data' and data from 'NASA Earth Observatory (NEO)', which is downloadable from its web portal.

As illustrated in Figure 1, the hydrology data has the property of sub-classes of: (i) rainfall data, (ii) water level data, and (iii) reading stations. They have equal relationship of determining the type of data being read and stored in the class of 'hydrology data'. From the classes of 'rainfall data' and 'water level data', statistical inferences were made. The null hypothesis represented here is that the occurrence of flood incidence is determined by these two (2) variables. As a consequence, the study on the two variables are reflected from alternative hypotheses that can be deduced from the data readings and were computed as the analysis in the big data research.

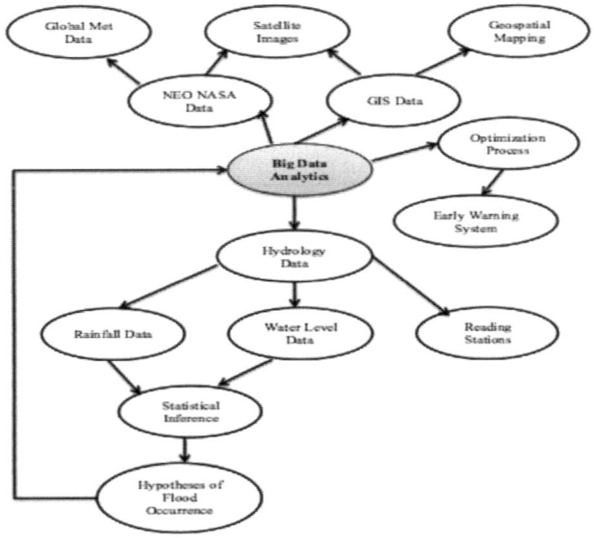

Figure 1. The semantic network of big data analytics for hydrology data.

In addition, the main class of 'big data analytics' should also be able to facilitate relative variables from sub-classes of 'NEO NASA data' and 'GIS data'. Though the type of data readings from the three sub-classes of 'big data analytics' might be in different format, it is expected that the proposed early warning flood system would be able to accommodate each class of data type. It is also expected that, the hydrology data was meant to represent the comma separated values (csv) data, the GIS data was to represent the Geography Markup Language (GML) or GeoRSS feed, and the NEO NASA data was represented by Geography Tag Image File Format (GeoTIFF).

3. The Knowledge Abstraction of Semantic Network

From a semantic network, we are able to define the knowledge representations and predicate calculus abstracted from the relationships and properties of each class and subclasses identified. The biggest contribution for this step is the fact that we are able to do decision making and knowledge management [11].

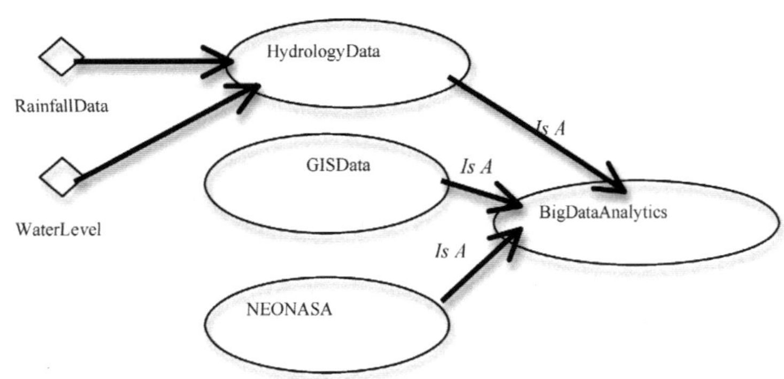

Figure 2. The class and instances relationship

The property of 'hydrology data' which is the subclass of 'big data analytics' is also illustrating on the fact that our limitations of study are contained within these two main classes. Hence other subclasses of the main class of 'big data analytics' hold TRUE to its identified classes until they are defined otherwise. The predicate calculus for this semantic includes: (i) *isA* (*HydrologyData, BigDataAnalytics*), (ii) *isA* (*GISData, BigDataAnalytics*), and (iii) *isA* (*NEONASA, BigDataAnalytics*), so on and so forth. This is as illustrated in Figure 2. From the figure, two instances were given for the class 'HydrologyData' consisting of 'RainfallData' and 'WaterLevel'. Instances can be further extended from other classes too. It is nevertheless, from this semantic that it is expected to design an effective data management and build efficient expert systems for flood management in Kelantan.

4. Conclusion

This research is proposing on using semantic network to facilitate the classes and relationships to the study of big data analytics of flood hydrological data. The study is specifically for the state of Kelantan, Malaysia. An introduction to the state flood history and previous reports were primarily highlighted for identifying the classes and relative parameters. As a result, a semantic network was designed and the class and instances relationship of the predicate calculus were also developed.

5. References

[1] A.S. Zakaria, "The Geomorphology of Kelantan Delta (Malaysia)." Catena 2 (1975): 337-349.
[2] TCPD. 2002. Malaysia Structure Plan, Malaysia Department of Town and Country Planning. Unpublished report. TCPD: Kuala Lumpur.
[3] Global Precipitation Measurement (GPM): Precipitation Measurement Missions, URL: "*http://pmm.nasa.gov/GPM*", Jul 2015
[4] SERVIR: Crest Viewer, "*http://ags.servirlabs.net/crestviewer/*", Jul 2015
[5] NRT Global Flood Mapping, "*http://oas.gsfc.nasa.gov/floodmap/*", Jul 2015
[6] Global Flood Monitoring Services (GFMS), "*http://flood.umd.edu/*", Jul 2015
[7] Global Disaster Alert and Coordination System, "*http://www.gdacs.org*", Jul 2015
[8] Dartmouth Flood Observatory, "*http://floodobservatory.colorado.edu/*", Jul 2015
[9] Pengurusan Bencana Tanggungjawab Bersama. URL: "*http://portalbencana.mkn.gov.my/Portal/Board/Detail?board=141&entity=7524*", 2014
[10] Majlis Keselamatan Negara. "Arahan No. 20 (Semakan Semula): Dasar dan Mekanisme Pengurusan Bencana Negara". Jabatan Perdana Menteri, Malaysia. 2012
[11] E. Turban, R. Sharda, D. Delen, and T. Efraim, "Decision Support And Business Intelligence Systems". Pearson Education India, 2007.

Acknowledgments

The authors wish to express the gratitude to the research team of Early Flood Warning System and Big Data Analytics of College of Engineering, UNITEN under the JICA-SATREPS Project on Research and Development for the reduction geo-hazard damage in Malaysia caused by landslide and flood and also to Big Data research team of North Dakota State University (NDSU). Heartiest appreciation goes to Public Service Department of Malaysia (JPA) as the main sponsor for this research, and Department of Irrigation and Drainage (JPS) Malaysia for providing the flood big data.

IOP Conf. Series: Earth and Environmental Science **32** (2016) 012022 doi:10.1088/1755-1315/32/1/012022

Urban Stormwater Management Model and Tools for Designing Stormwater Management of Green Infrastructure Practices

H Haris [1], M F Chow[1], F Usman[1], L M Sidek[1], Z A Roseli[2], M D Norlida[2]

[1]Centre for Sustainable Technology and Environment, Universiti Tenaga Nasional, Jalan IKRAM-Uniten, 43000 Selangor, Malaysia.
[2]Humid Tropics Centre Kuala Lumpur, No. 2 Jalan Ledang off Jalan Duta, 50480 Kuala Lumpur, Malaysia.

E-mail: harizah@uniten.edy.my; chowmf@uniten.edu.my; lariyah@uniten.edu.my

Abstract. Urbanization is growing rapidly in Malaysia. Rapid urbanization has known to have several negative impacts towards hydrological cycle due to decreasing of pervious area and deterioration of water quality in stormwater runoff. One of the negative impacts of urbanization is the congestion of the stormwater drainage system and this situation leading to flash flood problem and water quality degradation. There are many urban stormwater management softwares available in the market such as Storm Water Drainage System design and analysis program (DRAINS), Urban Drainage and Sewer Model (MOUSE), InfoWorks River Simulation (InfoWork RS), Hydrological Simulation Program-Fortran (HSPF), Distributed Routing Rainfall-Runoff Model (DR3M), Storm Water Management Model (SWMM), XP Storm Water Management Model (XPSWMM), MIKE-SWMM, Quality-Quantity Simulators (QQS), Storage, Treatment, Overflow, Runoff Model (STORM), and Hydrologic Engineering Centre-Hydrologic Modelling System (HEC-HMS) . In this paper, we are going to discuss briefly about several softwares and their functionality, accessibility, characteristics and components in the quantity analysis of the hydrological design software and compare it with MSMA Design Aid and Database. Green Infrastructure (GI) is one of the main topics that has widely been discussed all over the world. Every development in the urban area is related to GI. GI can be defined as green area build in the develop area such as forest, park, wetland or floodway. The role of GI is to improve life standard such as water filtration or flood control. Among the twenty models that have been compared to MSMA SME, ten models were selected to conduct a comprehensive review for this study. These are known to be widely accepted by water resource researchers. These ten tools are further classified into three major categories as models that address the stormwater management ability of GI in terms of quantity and quality, models that have the capability of conducting the economic analysis of GI and models that can address both stormwater management and economic aspects together.

1. Introduction

Urbanization is growing rapidly in Malaysia. Rapid urbanization has known to have several negative impacts towards hydrological cycle due to decreasing of pervious area and deterioration of water quality in stormwater runoff. One of the negative impacts of urbanization is the congestion of the stormwater drainage system and this situation leading to flash flood problem and water quality degradation. In the past, urban waterways have been limited to narrow river corridors with the channels canalised and concrete and other man-made materials forming the bed and banks of the river. At the same time, the waterways located in or near urban area are adversely influenced by solids in stormwater runoffs [1].

In the early 1970, simulation of water quantity has been growing rapidly by US Government Agency (US Environment Protection Agency). By using computer model, the mathematical relationships that represent the behaviour of the system are solved. There are two models in a hydrological model which are stochastic and deterministic model. Stochastic model is when a variable is considered as random variable that has probability distribution and it always produce a different response model. While deterministic model produces identical results when the same input parameter is inserted. From this deterministic model it could also be a stochastic model if the random variable replaced by their mean values. Both models may be classified into empirical or conceptual depending on whether the model is based on physical laws or not [2].

Most of the urban stormwater management tools or green infrastructure tools available in the market come in the software version. To compete with other well known softwares in the market, Department of Irrigation and Drainage (DID) Malaysia has come out with a prototype of MSMA Design Aid and Database software. This MSMA design aid and database software is developed guided by Manual Saliran Mesra Alam (MSMA 2nd Edition).

Green Infrastructure (GI) is one of the main topic that has been widely discussed all over the world. Every development in the urban area is related to GI. GI can be defined as green area build in the develop area such as forest, park, wetland or floodway. The role of GI is to improve life standard such as water filtration or flood control [3]. GI can offer a wide-ranging of ESS such as decreasing urban heat island, improve air quality, adaptation of climate change, improving the community liveability and improve the ethical values of community. Nevertheless, in site or locality scale, GI is normally defined as a key of managing the stormwater runoff by fine the solution how to make the water infiltrate into the ground or fine the way to reuse it in the daily life [4].

2. Urban Stormwater Model

Urban storm water models have been applicable for usage such as operational tools. Nevertheless, they are more frequently used as either planning or design tools. There are two basic modules of an urban storm water model which are [2].

i) Rainfall runoff modelling – The creation of surface and sub-surface runoff from rainfall, wash-off and build-up of pollutants from impervious surface.

ii) Transport modelling - Routing of flows and pollutants through the storm water infrastructure, for example storages, open channel and pipe network.

The illustration of link between these processes is in Figure 8.

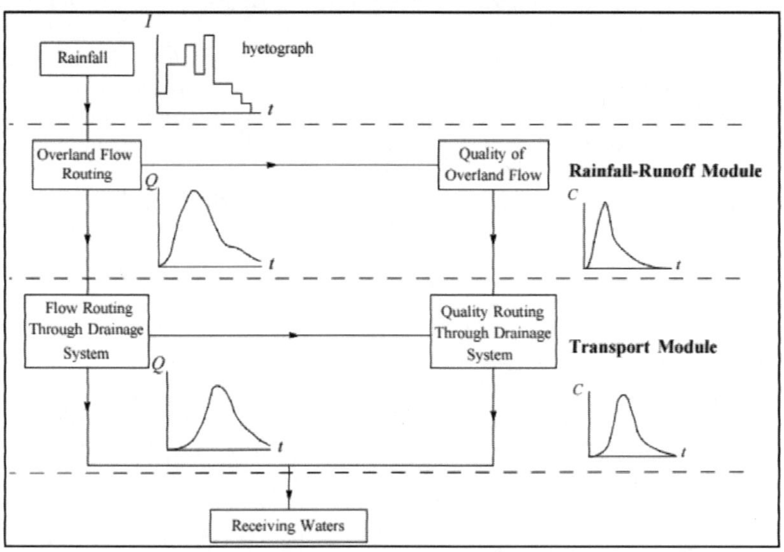

Figure 8: Overview of process incorporated in a storm water model [5].

Rainfall simulation is a well-known type of simulation and has been used for a very long time almost a hundred years. Advances in technology and knowledge about rainfall and interaction between rain and land also depend on the rainfall simulator development. Now, there are different types of rainfall simulator used around the world. Rainfall simulation can also be used to study the influences of a certain management practice on the properties of a particular material, or it can be used to improve our knowledge about of processes, such as erosion, runoff and infiltration [6].

2.1. Hydrological models

In order to develop the hydrological model we need to use hydrological methods. In the hydrological method the Muskingum method, unit hydrograph, nonlinear storage and lumped continuity or storage models need to be considered. The spatial variability is going to be ignored in the problem and the hydrological method is only based on the conservation of mass [2]. Stormwater Management Software is commonly commercially available. Nevertheless, different software has provided different types of design facilities and modelling.

2.1.1. Software name: DRAINS

In January 1998, Watercom Pty Ltd released one multipurpose Window program called Storm Water Drainage System design and analysis program (DRAINS). Watercom Pty Ltd based in New South Wales, Australia. DRAINS is use for designing and analysing urban storm water drainage system and catchments. This software capable to model drainage systems of all sizes, from small to up to 10 km^2 by using sub-catchment with ILSAX hydrology and greater using storage routing model hydrology. DRAINS will simulate and convert the rainfall patterns to storm water runoff hydrographs by working through a number of time steps that befall during the course of a storm event. The simulation runoff will route through channel, networks of pipes and streams. The connections to CAD and GIS program, automatic design procedure for piped drainage systems and an in-built Help system is also include in this software.

In a single package, DRAINS can model the hydrological model using ILSAX, rational method and a storage routing models together with unsteady hydraulic modelling of systems of pipes, open channels and, and in the premium hydraulic model, surface overflow routes. The important functions that are not included in DRAINS are [7]:

(a) Continuous modelling over long periods including wet and dry conditions
(b) Water quality modelling
(c) 2-dimensional unsteady flow modelling
(d) Life cycle cost

2.1.2. Software name: MOUSE

Based in Pennsylvania, USA, Urban Drainage and Sewer Model (MOUSE) is created as a comprehensive modelling system. This software is used for analysis of urban drainage and sewer systems including links to GIS. Other than that, MOUSE also used to simulate spatial variations in flows, water levels, sediment transport and pollution in pipes and open drains. Single or multiple events can be defined by the user and the time period up to several years.

Deterministic mathematical modelling tool is use as mathematical formulation of MOUSE model. With user defined the time interval MOUSE produces output a large number of variables. MOUSE has advantage in the aspect of no limits on the size of model area or number of input elements to be included in the simulation and capable to branched and looped network.

Continuous type of simulation is used in the MOUSE for nature of simulation. Besides that, this software also can predict the hydraulic deficiencies, overflow sites, flood inundation areas, and effect of real-time control. For rainfall-runoff model, the contributing area and hydrological losses are part of model parameter. Topographic maps, drainage plans and aerial photos are an input requirement in MOUSE. The Input data requirement is in spread-sheet format [8].

2.1.3. Software name: InfoWork RS

InfoWorks River Simulation (InfoWork RS) is a hydrodynamic modelling software includes full solution modelling of open channels, floodplains, embankments and hydraulic structures. Using both event based and conceptual hydrological methods, the rainfall-runoff simulation is also available using geographical plan views, sectional view, long sections, spreadsheet and time varying graphical data in full interactive views of data [9].

The underlying data can be accessed from any graphical and geographical view. In addition to presentation of results in geographical plan, tables, cross section views and long section and model also provide animation showing how a flood event progress. Full flood-mapping capability is provided based on a sophisticated flood-interpolation model overlaid onto an imported ground model [9].

InfoWorks RS combine the advanced flow simulation engine, both hydrological and hydraulic models, GIS functionality and database storage within one single environment. The basic system architecture is an "Integrated Network Model" links data storage using a GIS to hydrologic/hydraulic modelling software suite embedded in InfoWorks RS [10].

2.1.4. Software name: HSPF

Hydrological Simulation Program-Fortran (HSPF) simulating the extended periods of time the hydrologic processes on pervious and impervious land surfaces and in streams and well-mixed impoundments. HSPF uses continuous rainfall and other meteorological records to compute stream flow hydrographs. This software is used to simulate one or many pervious or impervious unit areas discharging to one or many river reaches or reservoirs. Any time series for frequency-duration analysis can be done. From 1 minute to 1 day at any time that divides equally into 1 day can be used. HSPF also can simulate any period from a minute to hundreds of years. This software is generally used to evaluate the effects of land-use change, reservoir operations, point or nonpoint source treatment alternatives, or flow diversions [11].

Programs which available separately, will support data in pre-processing and post processing for statistical and graphical analysis of data saved to the Watershed Data Management (WDM) file. The model contains hundreds of process algorithms developed from theory, laboratory experiments, and empirical relations from instrumented watersheds. HSPF simulated sediment routing by particle size, channel routing, reservoir routing, and constituent routing. Meteorological records of precipitation and estimates of potential evapotranspiration are required for watershed simulation. Physical measurements and related parameters are required to describe the land area, channels, and reservoirs [11]

2.1.5. Software name: DR3M

Distributed Routing Rainfall-Runoff Model (DR3M) is one of the storm water modelling. This software is created to simulate the storm runoff. It can simulate the routing storm either in system or pipes or natural channel. By using rainfall as an input, this software model the detail simulation of storm runoff according to user period time selected.

DR3M is commonly is used to simulating the storm runoff for small urban basins. To calculate the infiltration and pervious area rainfall excess the Green-Ampt equation is used. The disadvantage of this software is it does not simulate interflow and base flow of the basin. Daily precipitation, daily evapotranspiration, and short-interval precipitation are required for data requirement. To optimize and calibrate the model short interval discharge is needed [12].

2.1.6. Software name: SWMM

Developed in 1971, The USEPA Storm Water Management Model (SWMM) has been used for over 30 years and widely used as utilized model in detailed hydrological and hydraulic modelling of storm water and watershed of the catchment. This software is capable to simulate the precipitation movement from the ground surface through channel/pipe network. Single event and a long continuous period of event can be simulating by using SWMM. For a long time, SWMM version is a freeware and being maintained by several numbers of individual and organisation Now, SWMM engine has been rewritten and known as a version SWIMM 5, SWMM 5 is maintaining by USEPA.

In this hydrological model concept, every sub-catchment in a basin is treated as non-linear reservoir with a single inflow or rainfall input and will produce the outflow or discharge in term of infiltration, evaporation and surface runoff. The development of a number of SWMM Graphical User Interface (GUI) wrappers such as MIKESWMM, XP-SWMM has been created through a year of revolution [13].

2.1.7. Software name: XP-SWMM

XP Storm Water Management Model (XPSWMM) is an inclusive software package for planning, modelling and managing sustainable drainage systems. This software is used to simulate the stormwater and sanitary sewer flows as well as treatment in typical LID (WSUD) systems. Hydraulically, flows are simulated in 1D channel and pipes and coupled to a 2D surface grid for comprehensive flood modelling and mapping.

Scientists, engineers as well as resource and asset managers use this software to simulate natural rainfall-runoff processes and the performance of engineered systems that manage our water resources. XPSWMM is used to develop link-node and spatially distributed models that are used for the analysis, design and simulation of storm and wastewater systems. XPSWMM also models flow in natural systems including rivers, lakes, and floodplains with groundwater interaction [14].

2.1.8. Software name: MIKE-SWMM

MIKE-SWMM is one of the SWMM Graphical User Interface (GUI) wrappers. In this MIKE-SWMM, the function is similar to SWMM which is to analyse the urban drainage system and sanitary sewers. Combination of hydrology, hydraulics and water quality, in MIKE-SWMM provides complete graphical and user friendly interface for user [15]. In this hydrological model concept, every sub-catchment in a basin is treated as non-linear reservoir with a single inflow or rainfall input and will produce the outflow or discharge in term of infiltration, evaporation and surface runoff [13].

2.1.9. Software name: QQS

QQS stand for Quality-Quantity Simulators. The advantage of this software is by using fine interval such as five minutes time interval this software capable to perform continuous or single event simulation. QQS is used to simulating the flows in channels or pipes using an implicit finite difference approximation of the kinematic wave equation, backwater analysis, storage routing and pipes under pressure. This software is suitable to use in urban area catchment to simulate it urban storm water modelling [5].

2.1.10. Software name: STORM

Storage, Treatment, Overflow, Runoff Model (STORM) is a modelling software that has been developed by US Corps of Engineers. This software is capable to simulate the runoff in the urban or rural catchment area in response to precipitation. STORM not only can be used for single event but it also can be used to simulate continuous model by using hourly time steps. Runoff of every sub-catchment will simply accumulate in one catchment by using hourly precipitation. To calculate hourly runoff unit hydrograph method, coefficient method and soil-complex-cover method can be chosen to use. Runoff is a linear relationship between runoff and the precipitation minus rainfall interception [5].

2.1.11. Software name: HEC HMS

In this study, Hydrologic Engineering Centre-Hydrologic Modelling System (HEC-HMS) the model that designed by US Army Corps of Engineers will be used to model the basic hydrological model [15]. This software is updated program from the HEC-1. The advantage of HEC-HMS compare to HEC-1 which is designed working with MS-DOS program, it designed for more function and run with a windows appearances [17].

The development from HEC-1 to HEC-HMS is one of the useful tool for the hydrologist because it is more user friendly [18]. HEC-HMS can be used to help set up the hydrological model system and also for simulating the rainfall-runoff process of a catchment [19]. Numerous methods are used to simulate surface runoff, by apply several methods in HEC-HMS model different result can be reached [19].

HEC-HMS is designed with four main components. Below is the list of main components in HEC-HMS [21]:

i) An advanced graphical user interface illustrating hydrologic system components with interactive features.
ii) A system for storing and managing data, specifically large, time variable data sets.
iii) An analytical model to calculate overland flow runoff and also channel routing.
iv) A means for displaying and reporting model outputs.

HEC-HMS can apply for both short term and long term simulations [22]

2.1.12. Software name: MSMA Design Aid and Database

MSMA Design Aid and Database is a computer design aid system that includes ecohydrological and ecostormwater design tool and data storage. This tool is used for the analysis of ecohydrological and ecostormwater variables that may indicate the risk for flash flood, polluted river water and mud flood at a designated site of interest, which is defined herein as data-collection on site.

MSMA Design Aid and Database is needed to provide available information to assess stormwater component database and design stormwater components. Design Aid is commonly described as an interactive computer program that integrates models and complex data to help managers make planning-level decisions for complex situations that may be rapidly changing and not well-defined in advance. In many applications, Design Aid provides a standard method to demonstrate the potential effectiveness (and limitations) of various levels of stormwater management and will therefore provide the necessary information for defensible recommendations for MSMA Design Aid and Database components which covers all aspects including concepts, detailed design and performances.

MSMA Design Aid and Database module framework covers six main components; Database, Hydrological component, Design Quantity Control, Design Quality Control, Conveyance component and Checking Tools component.

3. Summary of urban stormwater management software

From the urban stormwater management software that has been briefly mentioned above, the summary of the functionality and accessibility, characteristic and components in the quantity analysis of the hydrological design software that available in the market is compiled as shown in Table 1, Table 2 and Table 3.

Table 1: Functionality and accessibility of the representative models

Urban Models	**Functionality**			**Accessibility**	
	Planning	Operational	Design	Public Domain	Commercial
DRAINS	✓	✓	✓		✓
MOUSE		✓	✓	✓	
Infowork RS		✓	✓		✓
HSPF	✓		✓	✓	
DR_3M	✓		✓	✓	
SWMM	✓		✓	✓	
MIKE_SWMM	✓	✓	✓		✓
QQS	✓		✓		
STORM	✓			✓	
HEC-HMS		✓			✓
MSMA Design Aid and Database	✓	✓	✓		✓

Table 2: Characteristics of representative models

Urban Models	**Model Characteristics**				
	Routing Level			Time modelling scale	
	Simple storage	Hydrologic	Hydraulic	Continuous	Event
DRAINS	✓	✓	✓	✓	✓
MOUSE	✓	✓	✓	✓	✓
Infowork RS		✓	✓	✓	✓
HSPF	✓	✓		✓	✓
DR_3M	✓	✓	✓		✓
SWMM	1	✓	2		✓
XP-SWMM		✓	✓		✓
MIKE-SWMM	✓	✓	✓	✓	✓
QQS	✓	✓	✓	✓	✓
STORM	✓				✓
HEC-HMS		✓		✓	✓
MSMA Design Aid and Database	✓	✓		✓	✓

1, Flow balance only, 2, With EXTRAN module

IOP Conf. Series: Earth and Environmental Science **32** (2016) 012022 doi:10.1088/1755-1315/32/1/012022

Table 3: Components in the quantity analysis in representative models

Urban Models	Model quantity component					
	Pipes	Open channel	Retarding basins	Others	Natural Streams	Rainfall runoff
DRAINS	✓	✓			✓	✓
MOUSE	✓	✓			✓	✓
Infowork RS		✓			✓	✓
HSPF	✓	✓	1		✓	✓
DR_3M	✓	✓	✓		✓	✓
SWMM	✓	✓	✓	4		✓
XP-SWMM	✓	✓			✓	
MIKE-SWMM	✓	✓	✓	2-7	✓	✓
QQS	3	✓	✓	2		✓
STORM						
HEC-HMS		✓			✓	✓
MSMA Design Aid and Database	✓	✓	✓	✓	✓	✓

1, reservoir module; 2, weirs and pumps; 3 pressurised pipes; 4, gutter and pumps; 5, surcharges; 6, bridges; 7, overland flow.

4. GI Practice and the development of Modelling Tools

Among the twenty models that has been compared to MSMA SME, ten models were selected to conduct a comprehensive review for this study. These are known to be widely accepted by water resource researchers. These ten tools are further classified into three major categories which are;

1) Models that address the stormwater management ability of GI in terms of quantity and quality,
2) Models that have the capability of conducting the economic analysis of GI and
3) Models that can address both stormwater management and economic aspects together.

A literature review for each tool is conducted extensively by referring their user guides, design manuals, fact sheets, case studies, journal articles, conference proceedings and book chapters. Table 4 shows tools for green infrastructure modelling [23].

5. Modelling Tools that Address Stormwater Quality and Quantity

GI practices attempt to replicate the pre development scenarios of a site in order to reduce the runoff quantities and improve the runoff quality [24]. Three models which have the ability of predicting the responses of different GI practices on runoff management are discussed within this category. Table 5 shows the list of reviewed Modelling Tools.

Table 4: Tools for Green Infrastructure Modelling [23]

Modelling Tools	**Reference and Case Studies**	**Supported GI Practices Comments**	**Comments**
Environmental Protection Agency (EPA) Green Long Term Control- EZ Template	(Schmitt et al.,	Green Roofs, Vegetated Swales, Bio Retention Basins, Permeable Pavements, Rain Barrels	• Planning tool for combined sewer overflow control. • Can be used in small communities.
Water Environmental Research foundation (WERF) BMP SELECT Model	(Reynolds et al., 2012)	Extended Detention ,Bio retention, Wetlands, Swales, Permeable Pavements	• Examines the effectiveness of alternative scenarios for controlling stormwater pollution. • Water quality parameters that can be simulated are Total Suspended Solids, Total Nitrogen, Total Phosphorus and Total Zinc
Virginia Runoff Reduction Method (VRRM)	(Bork and Franklin, 2010)	Green Roofs, Downspout Disconnection, Permeable Pavements, Grass Channels, Dry Swales, Bio Retention, Infiltration, Extended Detention Ponds, Wet Swales, Constructed Wetlands, Wet Ponds	• Incorporates built-in incentives for environmental site design, such as forest preservation and the reduction of soil disturbance and impervious surfaces.
WERF BMP and LID Whole Life Cycle Cost Modelling Tools	(Reynolds et al., 2012)	Green Roof, Planters, Permeable Pavements, Rain Gardens, Retention Ponds, Swales, Cistern, Bio Retention, Extended Detention Basins	• Planning level cost estimation for GI practices. • Different spreadsheet tools are designed for different practices.

Continue Table 4…

Modelling Tools	Reference and Case Studies	Supported GI Practices Comments	Comments
Centre for Neighborhood Technology(CNT) Green Values National Stormwater Management Calculator	(Jaffe, 2011, Guo and Correa, 2013)	Green Roofs, Planter Boxes, Rain Gardens, Cisterns, Native Vegetation, Vegetation Filter Strips, Amended Soils, Swales, Trees, Permeable Pavements	• Allows the user to select a runoff reduction goal and select the combination of GI practices that provides the optimum runoff reduction in a cost effective way.
CNT Green Values Stormwater Management Calculator	(Kennedy et al., 2008, Wise et al., 2010, Jaffe et al., 2010)	Roof Drains, Rain Gardens, Permeable Pavements, Trees, Porous Pavements, Drainage Swales	• Tool which helps to get an approximation of financial and hydrologic conditions for a user defined site.
Chicago Department of Environment Stormwater Ordinance Compliance Calculator	(Emanuel, 2012)	Green Roofs, Planter Boxes, Rain Gardens, Native Vegetation, Vegetated Filter Strips, Swales, Trees	• Used to evaluate the opportunities of GI with regard to the guidelines of Chicago's stormwater management ordinance.
EPA Stormwater Management Model (SWMM)	(Huber and Singh, 1995, Tsihrintzis and Hamid, 1998, Huber, 2001, Khader and Montalto, 2008, Rossman, 2010)	Bio Retention, Infiltration Trenches, Porous Pavement, Rain Barrels, Vegetative Swales, Green Roofs, Street Planters, Amended Soils	• Planning, analysis and design related to stormwater runoff, combined sewer overflows and drainage systems. • Complex model with variety of features.
Delaware Urban Runoff Management Model (DURMM)	(Lucas, 2004, Lucas, 2005)	Filter Strips, Bio Retention Swales, Bio Retention, Infiltration Swales	• Spreadsheet tool to assist GI design.

Continue Table 4...

Modelling Tools	Reference and Case Studies	Supported GI Practices Comments	Comments
Stormwater Investment Strategy Evaluator (StormWISE) Model	(McGarity, 2006, McGarity, 2010, McGarity, 2011)	Riparian Buffers ,Filter Strips, Wetland/Rain Garden, Bio Retention/Infiltration Pits, Rain Barrel/Cisterns, Land Restoration By Impervious Surface Removal, Permeable Pavements, Green Roofs	• Studies on GI projects based on pollutant load reduction and cost benefits.
Program for Predicting Polluting Particle Passage through Pits, Puddles, & Ponds (P8 Urban Catchment Model)	(Elliott and Trowsdale, 2007, Obeid, 2005)	Detention Tanks, Ponds, Wetlands, Infiltration Trenches, Swales, Buffer Strips	• Model the generation and transportation of pollutants through urban runoff and the effectiveness of GI for improving the water quality.
Long-Term Hydrologic Impact Assessment (L-THIA)	(Tang et al., 2005, Bhaduri, 1998, Bhaduri et al., 2001, Engel et al., 2003)	Bio Retention/Rain Gardens, Grass Swale, Open Wooded Space, Permeable Pavement, Rain Barrel/Cisterns, Green Roof.	• Consists of calculations for Stormwater runoff and pollutant loading.
GI Valuation Tool Kit	(GiVAN, 2010)	Green Cover	• Evaluate the dollar value of environmental and social benefit of GI.
RECARGA	(Dietz, 2007, Atchison et al., 2006)	Bio Retention, Rain Garden, Infiltration	• Performance evaluation

Continue Table 4…

Modelling Tools	Reference and Case Studies	Supported GI Practices Comments	Comments
EPA System for Urban Stormwater Treatment Analysis and Integration (SUSTAIN)	(Lai et al., 2006, Lai et al., 2007, Lai et al., 2009, Lai et al., 2010, Shoemaker et al., 2013)	Bio Retention, Cisterns, Constructed Wetlands, Dry Ponds, Grassed Swales, Green Roofs, Infiltration Basins, Infiltration Trenches, Permeable Pavements, Rain Barrels, Sand Filters (Surface And Non-Surface),Vegetated Filter Strips ,Wet Ponds	• Implementation planning for flow and pollution control. • Selects the most cost effective solution in stormwater quality and quantity management.
Model for Urban Stormwater Improvement Conceptualization (MUSIC)	(Wong et al., 2002, Deletic and Fletcher, 2004, Wong et al., 2006, Dotto et al., 2011)	Bio Retention Systems, Infiltration Systems, Media Filtration Systems, Gross Pollutant Traps, Buffer Strips, Vegetated Swales, Ponds, Sedimentation Basins, Rainwater Tanks, Wetlands, Detention Basins.	• Assists in decision making of GI selection for stormwater management in urban development.
Low Impact Development Rapid Assessment (LIDRA)	(Montalto et al., 2007, Behr and Montalto, 2008, Yu et al., 2010)	Green Cover	• Evaluates the effectiveness of green space in reducing stormwater runoff.
WinSLAMM (Source Loading and Management Model for Windows)	(Pitt and Voorhees, 2002)	Infiltration/Bio filtration Basins, Street Cleaning, Wet Detention Ponds, Grass Swales, Filter Strips, Permeable Pavement	• Evaluates how effective the GI practices in reducing runoff and pollutant loadings. • The cost effectiveness of practices and their sizing requirements also been modelled.

IOP Conf. Series: Earth and Environmental Science **32** (2016) 012022 doi:10.1088/1755-1315/32/1/012022

Continue Table 4…

Modelling Tools	Reference and Case Studies	Supported GI Practices Comments	Comments
MSMA Integrated Stormwater Management Eco Hydrology Design Aid and Database System (MSMA SME Design Aid and Database System)	Eco-Friendly Integrated Green Technology Expert System for Sustainable Green Infrastructure (GENIUS)	Design Rainfall (Empirical Method & IDF Curve Creator), Design Runoff (Rational Method, Rational Method Hydrograph Method, Time Area Method), Rainwater Harvesting (MSMA & SPAH), Detention Pond, On-site Detention, i) Water Quality Volume, Infiltration Facilities, Bioretention System, Water Quality Pond, Wetland, Erosion and Sediment Control, Green Roof Design, Greywater, Lined Drain Design, Composite Drain, Swale, Engineered Channel, Pipe Drains, Road Gutter, Inlet Spacing, Curb Opening Inlet, Porous Pavement, Checking tools (Bridge Afflux & Pollutant Loading)	• Build up with individual database. • Consists of calculations for Stormwater runoff and pollutant loading. • Bioretention System consists of permeable and impermeable type of system; complies with orifice and weir type of overflow structure, and can be editable due to the typical design measurements and coefficient values. • Erosion and Sedimentation Control supports with auto-calculation properties of the settling zone depth and width for sediment basin design, and also specially designed for all states in Peninsular Malaysia. • Green Roof Design complies with the US METHOD design manual standard, and provides editable function of the porosity coefficient values. • Infiltration Facilities consist of the trench and basin infiltration type including with design criteria table references. • Pollutant loading complies with editable function of the Event Mean Concentrations (EMC) values for each pollutant parameter and it also supports more than 50 sub-catchments in one calculation process. • Pipe Drain supports more than 50 units of pipe connection flows in one calculation process

Table 5: Reviewed Modelling Tools [23]

Modelling Tools	**Owner**	**Availability**	**Intended Uses in GI Modelling**	**Reference**
RECARGA	University of Wisconsin-Madison, water resources group	Available to download freely http://dnr.wi.gov/topic/stormwater/standards/recarga.html	To design and understand performances of bio retention, infiltration basins and rain gardens.	(Atchison and Severson, 2004)
Program for Predicting Polluting Particle Passage through Pits, Puddles, & Ponds (P8 Urban Catchment Model)	William W. Walker, Jr., Ph.D. Environmental Engineer Massachusetts	Available to download freely http://www.wwwalker.net/p8/	To predict the generation and transportation of pollutants in urban runoff and design GI to achieve Total Suspended Solids Reduction	(Walker Jr, 1990)
EPA Stormwater Management Model (SWMM)	United States Environmental Protection Agency	Available to download freely http://www.epa.gov/nrmrl/wswrd/wq/ models/swmm/	To plan, design and analysis of the performances of different GI in runoff quality improvement and quantity reduction	(Huber et al., 1988, Rossman, 2010)
Water Environment Research Foundation(WERF) BMP and LID Whole Life Cycle Cost Modelling tools	Water Environment Research Foundation, Alexandria	Freely available to download http://www.werf.org/i/a/Ka/Search/ResearchProfile.aspx?ReportId=SW2R08	To evaluate whole life cycle cost for GI practices	(Water Environment Research Foundation, 2009)
The Green Infrastructure Valuation Toolkit	Natural Economy North West UK	Freely available to download http://www.greeninfrastructurenw.co.uk/html/index.php?page=projects&GreenInfrastructureValuationToolkit=true	To evaluate the environmental and economic benefit of GI in monetary terms	(Natural Economy Northwest, 2010)

Continue Table 5…

Modeling Tools	Owner	Availability	Intended Uses in GI Modelling	Reference
Centre for Neighborhood Technology (CNT) Green Values National Stormwater Management Calculator	Center For Neighborhood Technology, Chicago	Free online assessment tool http://greenvalues.cnt.org/national/calculator.php	To compare GI cost, performance and benefits	(Center for Neighborhood Technology, 2009)
EPA System for Urban Stormwater Treatment and Analysis Integration Model (SUSTAIN)	United States Environmental Protection Agency	Freely available to download http://www.epa.gov/nrmrl/wswrd/wq/models/sustain/	To develop implementation plans for flow and pollution control, evaluate cost effectiveness of GI	(Lai et al., 2007)
Model for Urban Stormwater Improvement Conceptualization (MUSIC)	eWater , Australia	Proprietary software http://www.ewater.com.au/products/ewater-toolkit/urban-tools/music/	To evaluate GI practices in order to achieve stormwater quantity reduction, quality improvement and cost effectiveness.	(Wong et al., 2002)
Low Impact Development Rapid Assessment (LIDRA)	eDesign Dynamics, New York	Open Source Web Based tool http://www.lidratool.org/database/database.aspx	To study runoff cost reductions with GI	(Yu et al., 2010)
WinSLAMM (Source Loading and Management Model for Windows)	PV & Associates USA	Proprietary software http://winslamm.com/winslamm_updates.html	To study the quality of urban runoff and the role of GI in runoff quality improvement	(Pitt and Voorhees, 2004)

Continue Table 5…

Modeling Tools	Owner	Availability	Intended Uses in GI Modeling	Reference
MSMA Integrated Stormwater Management Eco Hydrology Design Aid and Database System (MSMA SME Design Aid and Database System)	HTC & UNITEN	To be described by Co-Inventor (JPS/UNITEN)	• To design and understand performances of bio retention, wetland, green roof and etc. • To evaluate GI practices in order to achieve stormwater quantity reduction and quality improvement • To study the quality of urban runoff and the role of GI in runoff quality improvement • To plan, design and analysis of the performances of different GI in runoff quality improvement and quantity reduction	-

Conclusion

In general, urban stormwater management model is the standard tool used to help engineer to simulate the hydrological process. There are many models with the different applications range from small catchment to a large catchment have been developed. Every software has its own specific characteristics and respective applications. Several models are used to model gauged and ungauged study area. Each model has several disadvantages such as lack of user friendly characteristics, large data requirements and nonappearance of clear declarations of their boundaries. In order to overcome these imperfections, it is required for the models to include rapid advances in remote sensing technologies. New distributed models can be developed for modelling gauged and ungauged basins by applying the new technologies. One of the challenges is regarding the use of large quantity of data and hence new facilities are to be included for the efficient storing, managing and manipulation of the large data. Each software manufacturer should be required to provide a clear statement of their border and issued a corresponding guide and enlighten the dominant physical process. A different way than model evaluation required for accurate prediction. Parameters measures will reflect the causes of errors in modelling. To avoid adverse effect on model calibration the modeller need to have proper knowledge about subsurface flow pathways and hydraulic characteristics. In general, the development of MSMA Design Aid and Database is complete, with all components in place. The MSMA Design Aid and Database have met the two main purposes. First, to assist in solution development of MSMA Design Aid and Database components and integrated stormwater management based on expert data and user input and second is to provide better understanding on the MSMA Design Aid and Database components which cover all aspects including concepts, detailed design and performances

Acknowledgement

The authors would like to express their gratitude to Universiti Tenaga Nasional and Humid Tropics Centre (HTC) for the grant UGNCR 011502.

Reference

[1] Roesner, L.A., Amy Pruden And Elizabeth M.Kidner (2007). Improved Protocol For Classification And Analysis Of Stormwater Borne Solid, Water Environment Research Foundation, Colorado State University.
[2] Zoppou, C., (2001). Review of urban storm water model Environmental Modelling & Software 16(3) 195–231 Available at: http://linkinghub.elsevier.com/retrieve/pii/S1364815200000840.
[3] Foster, J. (2011). The Value of Green Infrastructure.
[4] WISE, S. (2008). Green Infrastructure Rising. Planning, 74, 14-19.
[5] Zoppou, C. (1999). Review of storm water models, canberra,australia
[6] Twahirwa, J. (2010). Evaluation of infiltration, run-off and sediment mobilisation using rainfall simulations in the Riebeek-Kasteel Area, Western Cape-South Africa Available at: http://scholar.google.com/scholar?hl=en&btnG=Search&q=intitle:Evaluation+Of+Infiltration+,+Run-Off+And+Sediment+Mobilisation+Using+Rainfall+Simulations+In+The+Riebeek-Kasteel+Area+,+Western+Cape#0 [Accessed September 29, 2014].
[7] O'Loughlin, G., Stack, B., (2014) DRAIN user manual. Available at: www.watercom.com.au
[8] Group, R. S., (1999). Hydrologic modeling inventory model description Form. Available at: http://www.usbr.gov/hmi
[9] Salarpour, M., Rahman, N. A., and Yusop, Z., (2011). Simulation of flood extent mapping by infoworks rs-case study for tropical catchment. Journal of Software Engineering, 5, 127–135.
[10] Darrien, Y. S. M., (2007) Use of infoworks river simulation (rs) in sungai sarawak kanan modelling The Institution of Engineers Malaysia, 68, 3–4
[11] U S Geological Survey, (2008) Summary of HSPF.
[12] Survey U S G, Summary of DR3M. usgs science for a changing world. Available at: http://water.usgs.gov/cgi-bin/man_wrdapp?dr3m.
[13] Lockie, T., Catchment modelling using SWMM.
[14] XP Software, (2013) Technical Descriprion XP Solution.

[15] Board, T. R., (2004) Effective methods for environmental justice assessment, Washington, D.C. Available at: www.national-academies.org/trb/bookstore.

[16] Ramli, M. L., Harun, S. Hydrologic modelling using HEC-HMS Sungai Padang Terap, Jitra Kedah, 10, 1–7.

[17] Hasan, Z. A., Hamidon, N., Suffian, M., (2009). Integrated river basin management (IRBM) : hydrologic modelling using HEC-HMS for sungai Kurau Basin, Perak, International Conference on Water Resources (ICWR 2009).

[18] Kalita, D. N. (2011). A study of basin response using hec-hms and subzone reports of cwc 13th National Symposium on Hydrology with focal theme on "Inflow Forecasting during Extremes."New Delhi. Available at: http://www.researchgate.net/publication/264387915_A_STUDY_OF_BASIN_RESPONSE_USING_HEC-HMS_AND_SUBZONE_REPORTS_OF_CWC.

[19] Chang, C.,(2009). Application of scs cn method in HEC-HMS in Shihmen watershed - simulation of rainfall- runoff hydrologic model. Florida State University. Available at: http://diginole.lib.fsu.edu/etd Recommended.

[20] Arekhi, S., (2012). Runoff modeling by HEC-HMS model (case study : kan watershed , Iran). International Journal of Agriculture and Crip Sciences, 4(23), 1807–1811.

[21] Bajwa, H. S., Tim, U. S., (2002). Toward immersive virtual environments for gis-based floodplain modeling and visualization Proceedings of 22nd ESRI User Conference San Diego TX USA

[22] Yilma, H., Moges, S. A., (2007). Application of semi-distributed conceptual hydrological model for flow forecasting on upland catchments of blue nile river basin, a case study of gilgel abbay catchment. FWU Water Resources Publications, 6, 200.

[23] Jayasooriya, V., Ng, A. W., (2010). Tools for modelling of Stormwater Management and Economics of Green Infrastructure Practices: A Review.

[24] Davis, A. P., 2005. Green engineering principles promote low-impact development. Environmental science & technology, 39, 338A-344A.

Development of Flood Forecasting Using Statistical Method in Four River Basins in Terengganu, Malaysia

M S F M Noor[1], L M Sidek[1], H Basri[1], M M M Husni[1], A S Jaafar[1], M H Kamaluddin[2], W H A W A Majid[2], A H Mohammad[2] and S Osman[2]

[1]Center for Sustainable Technology and Environment, College of Engineering, Universiti Tenaga Nasional, Jalan IKRAM-UNITEN, 43000 Kajang, Selangor, Malaysia

[2] Water Resources Management and Hydrology Division, Department Irrigation and Drainage Malaysia. KM 7, Jalan Ampang, 68000 Kuala Lumpur. Malaysia.

E-mail: siddiqfarouq@gmail.com

Abstract. One of the critical regions in Malaysia is Terengganu which is located at east coast of Peninsular Malaysia. In Terengganu, flood is experienced regularly because of attributed topography and climate including northeast monsoon. Moreover, rainfall is with high intensity during the November to February in Terengganu as forcing factor to produce of flood. In this study, main objectives are water stage forecasting and deriving the related equations based on least squared method. For this study, it is used two methods which called inclusion of residual (Method A) and non-inclusion residual (Method B) respectively. Result depicts that Method B outperformed to forecast the water stage at selected case studies (Besut, Dungun, Kemaman, Terengganu).

1. Introduction

Among all natural hazards, floods are the most frequent, destructive, and costly disasters in Malaysia. The east coast of Peninsular Malaysia have experienced floods regularly because of their geographical location that exposed to northeast monsoon that bring heavy rainfall and eventually contribute large volume of runoff to the relatively large catchment areas [1]. The need for accurate predictions of floods has been highlighted by the recent occurrences of floods Terengganu. The aim of flood forecasting analysis is to provide in-time useful information for making crucial decisions such as issuing alerts or activating required protection measures and mitigating a potential disaster [2]. The forecasting process during the non-flood period involving field data collection of past water level from the stations at upstream in respective basin is significance and can be used to assist the authorities to forecast the severity of upcoming flood event at downstream; hence reduce the overall impact of severe events. Main reason of flood is involved with the incrementing of water level at upstream areas which is dangerous for settlement near the rivers [2]. Using conventional methods as basic and primary analysis provide a valuable review on behaviour of water stage in vulnerable area [2]. The process of statistical method involves regression process between water level at upstream and downstream. The objective of this study is presentation of forecasted water levels with two methods related with inclusion of residual (Method A) and non-inclusion residual (Method B). Also, related empirical equations are derived for this study based on multiple regression.

2. Methodology

The concept of least square method reduces the sum of the squared vertical distance between the observed and forecasted values. After coefficient is obtained, multiple linear regression analysis is used to generate the equation. It is used as a procedure for fitting an equation to a set of data that might be comprises two or more variables [3]. Next, extracting the water level data in upstream are used to forecast water level in downstream data (e.g. outlet station). The water level data at upstream, which are used for regression, should have same period and validity with water level data at downstream. The data are analysed by two methods; with inclusion of residual (Method A) and non-residual (Method B) based on Department of Irrigation and Drainage of Malaysia (DID) guidelines [4].

3. Research area

Figure 1 shows four river basins in Terengganu that discharge into South China Sea. In this study, multiple linear regression is used to model relationship of explanatory variables (water level at upstream) and the outcome of the response variable (water level at downstream).

Multiple Linear Equation Model, $Y= CX_1^aX_2^b X_3^c$ (1)

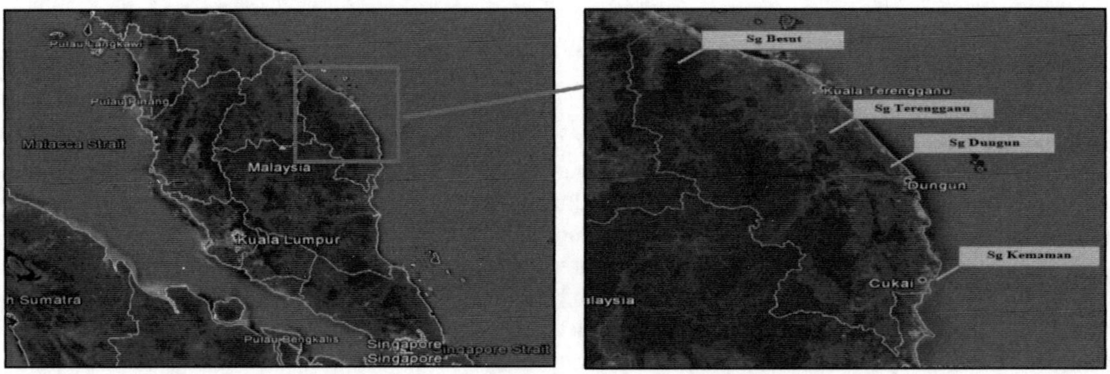

Figure 1: Location of four rivers basin in Terengganu

Data are correspond with maximum water level observed at gauge in upstream area of the river. Table 1, 2, 3, and 4 show the descriptions of the the water level stations in each river. Next, the information regarding operation service of available stations in this study as follows; Besut River (1997-2015), Dungun River (1993-2015), Kemaman River (1997-2015) and Terengganu River (2004-2015). In addition, each river has different number of upstream station available whereby three stations exist at upstream of Kemaman River, two stations for Dungun River, and finally only one station for both Besut and Terengganu River.

Table 1. Descriptions of water level stations in Besut River

Station Name	Station No	Data Transmission
Kg La	5524401	Logger
Jambatan Jertih	5724411	Logger

Table 2. Descriptions of water level stations in Dungun River

Station Name	Station No	Data Transmission
Rumah Pam Paya Kempian	4529401	Logger
Kg Surau	4730401	Logger
Jambatan Jerangau	4832441	Logger

Table 3. Descriptions of water level stations in Kemaman River

Station Name	Station No	Data Transmission
Sg Cherul at Ban Ho	4131453	Logger
Jambatan Air Puteh	4232401	Logger
Rantau	4232452	Logger
Jambatan Tebak	4332401	Logger

Table 4. Descriptions of water level stations in Terengganu River

Station Name	Station No	Data Transmission
Kuala Ping	5129438	Logger
Kg Bukit	5229436	Logger

4. Results and Discussions

In this section, results on forecasting water level are discussed briefly. Method B selected as chosen method because it has provide overestimation on the forecasted water levels at downstream at majority events in comparison with Method A; therefore suitable for safety precautions and warning system. Figure 1 and 2 illustrates the observed and forecasted water levels for all four rivers based on Method A and B. It indicates Method B obtained better results for forecasting water levels compare to Method A. The longer period of data collection increase the data reliability to forecast the flood.

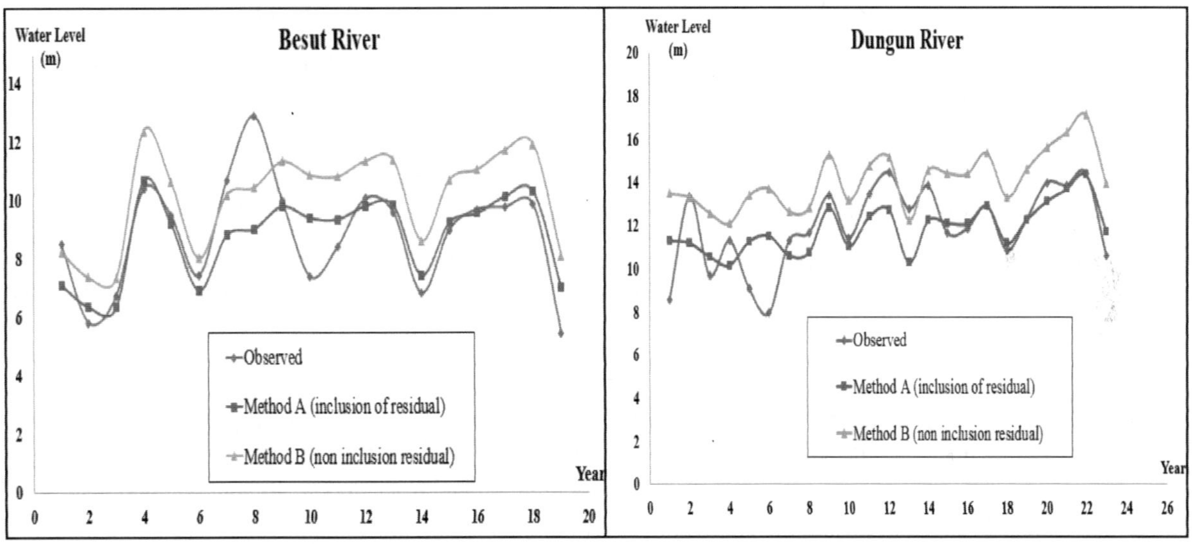

Figure 2. Comparison between Method A, Method B and observed water level for Besut River (left) and Dungun River (right).

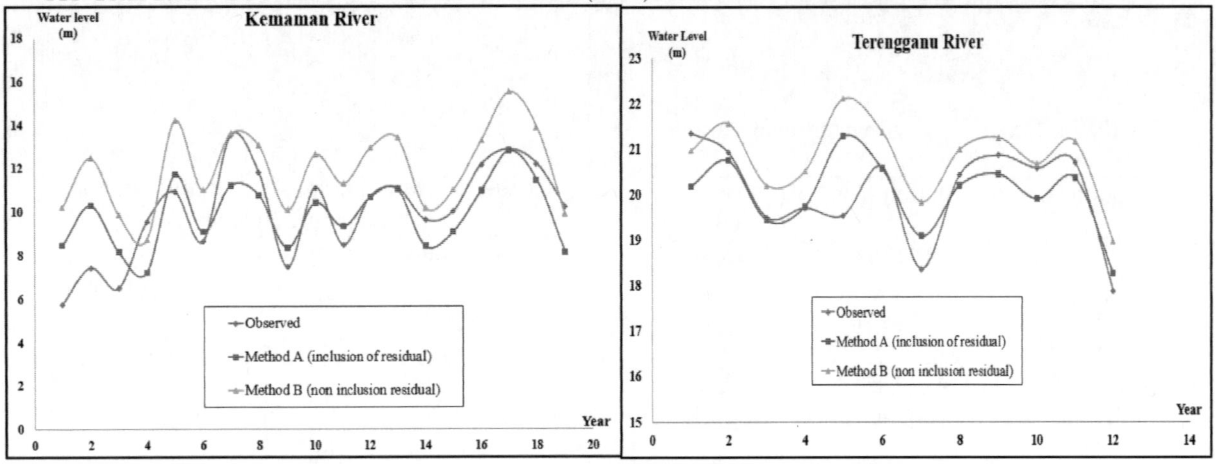

Figure 3. Comparison between Method A, Method B and observed water level for Kemaman River (left) and Terengganu River (right).

Table 5: Equation for both methods for each river basin.

River Basin	Method A (inclusion residual)	Method B (non-inclusion residual)
Besut	$Y_{T+t} = 0.003247X_1^{(1.491411985)}$	$Y_{T+t} = 0.003754X_1^{(1.491411985)}$
Dungun	$Y_{T+t} = (1.2047 \times 10^{-9}) \times X_1^{2.433032} \times X_2^{0.429334}$	$Y_{T+t} = (1.4355 \times 10^{-9}) \times X_1^{2.433032} \times X_2^{0.429334}$
Kemaman	$Y_{T+t} = (2.35846 \times 10^{-6}) \times X_1^{1.195391} \times X_2^{0.197861} \times X_3^{1.035824}$	$Y_{T+t} = (2.85983 \times 10^{-6}) \times X_1^{1.195391} \times X_2^{0.197861} \times X_3^{1.035824}$
Terengganu	$Y_{T+t} = 0.22551 \times X_1^{0.4709529}$	$Y_{T+t} = 0.234246 \times X_1^{0.4709529}$

Y = Forecasted Level at downstream station in time T + t
X_1, X_2, X_3 = Observed level at upstream stations
T = Lag time = 4 hours

5. Conclusion

Flood forecasting is a very challenging task due to high variability and noise in the data for both upstream and downstream stations. In this study, the observed and forecasted data presents non-inclusion of residual (Method B) outperformed the inclusion residual (Method A) for all four river basins. The equation for Method B is more applicable to forecast the water level at downstream area since it capable to provide overestimation of the water level at majority event compare to method A; hence provide sufficient time for safety precautions and relevant to become as early warning system for the citizens. In a nutshell, the Method B suited for flood forecasting process.

Acknowledgements

The authors wish to acknowledge express our gratitude to Water Resources Management and Hydrology Division, Department of Irrigation and Drainage of Malaysia (DID) for collaborate in this study in the project of "Development of Flood Forecasting System Using Simple Model".

References

[1] Barzani G Jumaat H A Mohd Ekhwan H T Sahibin A R Hafizan H J 2007 Coastal Flood Phenomenon in Terengganu, Malaysia: Special Reference to Dungun *Research Journal of Environmental Sciences* **1(3)** 102- 109. ISSN 1819-3412.

[2] Tamoto N Endo J Yoshimoto K Yoshida T Sakakibara T 2008 Forecast-based operation method in minimizing flood damage in urban area. Int. *Conf. Urban Drain*. 11th, Edinburgh, Scotland, UK.

[3] Kitadinis PK Bras R L 1980 Real-time forecasting with a conceptual hydrologic model: 2 applications and results. *Water Resour. Res.* **16,** 1034–1044

[4] Department of Irrigation and Drainage (DID) 2009 LAPORAN Banjir 2008- 2009. Unit Hidrologi Jabatan Pengairan dan Saliran Negeri Terengganu.

Enhancing a Distributed Rainfall Intensity for Flood Analysis within a GIS Framework in an Urban Area (Kajang Flood)

S H M Salleh[1], **L M Sidek**[1],
[1]Universiti Tenaga Nasional, Jalan IKRAM-UNITEN, 43000 Kajang, Selangor, Malaysia.

Email: sitihawa.rho1@gmail.com

Abstract. The town of Kajang has experienced major flood events in previous decades since 1970s. Due to Eastern and Western Monsoon seasons in Malaysia, Kajang has potential to face with number of flood events. One of the critical elements in flood analysis is rainfall distribution. Therefore, flood studies need to have an optimum cognition of rainfall distribution as main input. In this study, HEC GeoHMS model was used in GIS (Geographic Information System) for Sungai Jeluh catchment (Kajang) due to estimation (visualization) of distributed rainfall in Kajang. In comparison with conventional methods, which they produced rainfall in lumped mode (e.g. Thiessen's polygon), HEC GeoHMS visualized and tabulate a full distribution of rainfall for each small part (pixel of map) in a case study. HEC GeoHMS model for Sungai Jeluh has been set up by feeding high resolution of spatial and temporal resolution data (precipitation). Result of this study shows that rainfall by high intensity is distributed near urban area (downtown) in comparison with upstream which involved with less rainfall intensity.

1. Introduction

The United Nations had commissioned a report for the Coordination of Humanitarian Affairs (OCHA) where Malaysia is identified as *'Extreme Flood Hazard'* for the coastal area of Peninsular Malaysia and *'Significant Flood Hazard'* for mountainous area (Titiwangsa Range) which main state of Selangor and Wilayah Persekutuan Kuala Lumpur were reported as *'Projected Increased Extreme Precipitation'* (Jabatan Pengairan dan Saliran Malaysia, 2003). In 2002, the Annual Average Damage (AAD) of whole Malaysia total up to approximately RM 915 million due to flood barely using year 2000 pricing value (Jabatan Pengairan dan Saliran Malaysia, 2003). Usually, flash floods cause the greatest damage and loss of life associated with intense rainfalls in urban areas. Sungai Jeluh is a potential river in Kajang Town which frequently has heavy rainfall and consequently flood. This study presents development of a distributed rainfall system via HEC-RAS Geospatial Hydrologic Modeling Extension (HEC GeoHMS) for Kajang to have contribution of rainfall for each part of catchment in visualization and tabulated form which is suitable for hydrological model as input data.

2. Methodology

2.1 Case study

Sungai Jeluh is one of the main rivers in Kajang town (Figure 1). The sub-watershed drains about 17.65 km^2. Kajang is extensively flooded during heavy rainfall in both eastern and western monsoon of the Peninsular Malaysia. This flood is due to frequent and unknown distribution of rainfall with contribution of other factors such as high urbanization, soil characteristics, and flat plain. Also, insufficient capacity of local drainage system will deteriorate the events. Therefore, there is a need to study the rainfall characteristic in this study.

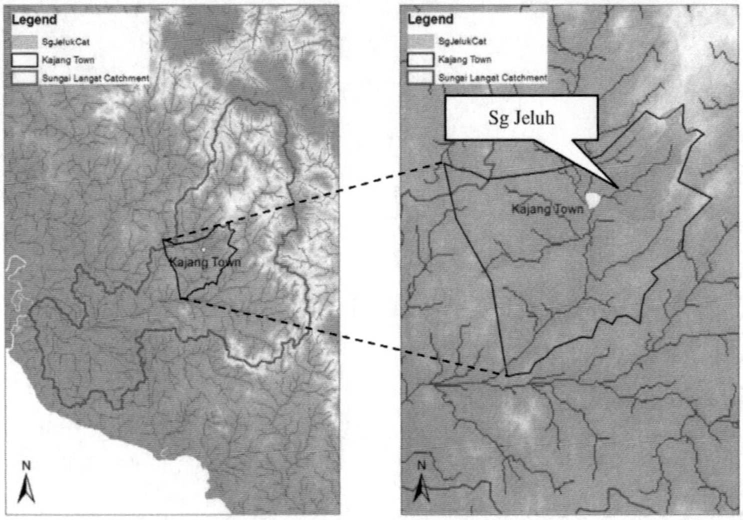

Figure 1: Left: Area of Kajang Town (119 km^2) and Sungai Jeluh (17.65 km^2), right: Watershed delineation of Sungai Jeluh, Malaysia

2.2 HEC GeoHMS

There are various classes of hydrological models that are classified as lumped or distributed (Jajarmizadeh *et al*, 2012). In this research, the physically-based distributed model namely HEC GeoHMS is chosen to derive the fully distribution of rainfall in a distributed system. HEC GeoHMS needs digital terrain data and typical topographical maps as required database. Moreover, HEC GeoHMS provides main catchment, sub- catchments, main river alignment and main urban drainage (Figure 1).

2.3 Preparation of distributed rainfall and land use with HEC GeoHMS

The selection of rainfall duration is based on conservative technique, or namely typical lumped model. From this method, 180 minutes rainfall duration is identified as a critical rainfall storm for Sungai Jeluh catchment. The other issue, which is important for prediction of flood or recognition of flood prone area, is development of land use for urban region besides having distributed rainfall. Especially, HEC GeoHMS integrated with USDA Soil Conservation Services (SCS) curve number technique that needs primary information of land use and rainfall. Therefore, it has prepared a land use map for Kajang area beside derivation of distributed rainfall in first phase. Finally, it is feasible to visualize flood prone area by having distributed rainfall intensity and land use types identification via HEC GeoHMS (Figure 2 and Figure 3).

3. Result and discussion

Figure 3 shows that delineation of land cover of Kajang catchment is involved with residential area (32%) which is potential for flood production. Also, distributed rainfall intensity (Figure 4) shows that intense rainfall (darker blue colour) happened near urban area which is downstream. Visualization shows that the distribution of rainfall is gradually increasing from upstream to downstream in Kajang. Therefore, downstream area is a place with more contribution of high intensity of rainfall which leads to have proper drainage system for mitigation of unmanaged fresh waters on urban region. Distributed rainfall analysis provides an optimal information to conclude exact locations of the area which has a high contribution to intense rainfall. This is due to the distributed rainfall system has more logical meaning by real pattern of rainfall in comparison with lumped concept such as Thiessen's polygon.

IOP Conf. Series: Earth and Environmental Science **32** (2016) 012024 doi:10.1088/1755-1315/32/1/012024

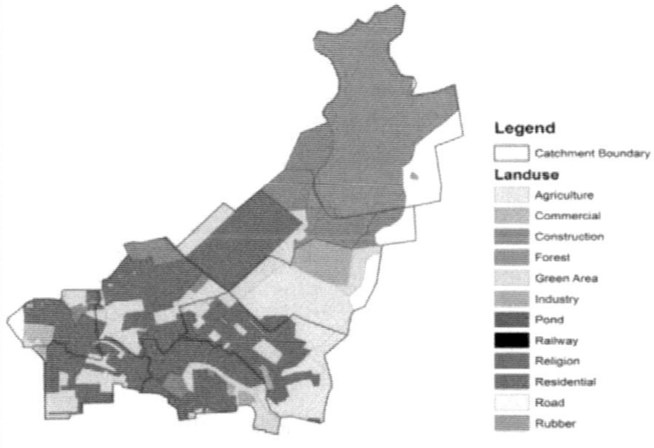

Landuse Type	Area (ha)	%
Forest	375.196	19
Green Area	254.5957	13
Construction	167.028	9
Residential	624.5921	32
Industrial	48.315	2
Agricultural	178.088	9
Commercial	17.5873	1
Pond	0.907	0
Others	279.3361	14
Grand Total	1945.64518	100

Figure 2: Landuse of Sungai Jeluh catchment for curve number map development

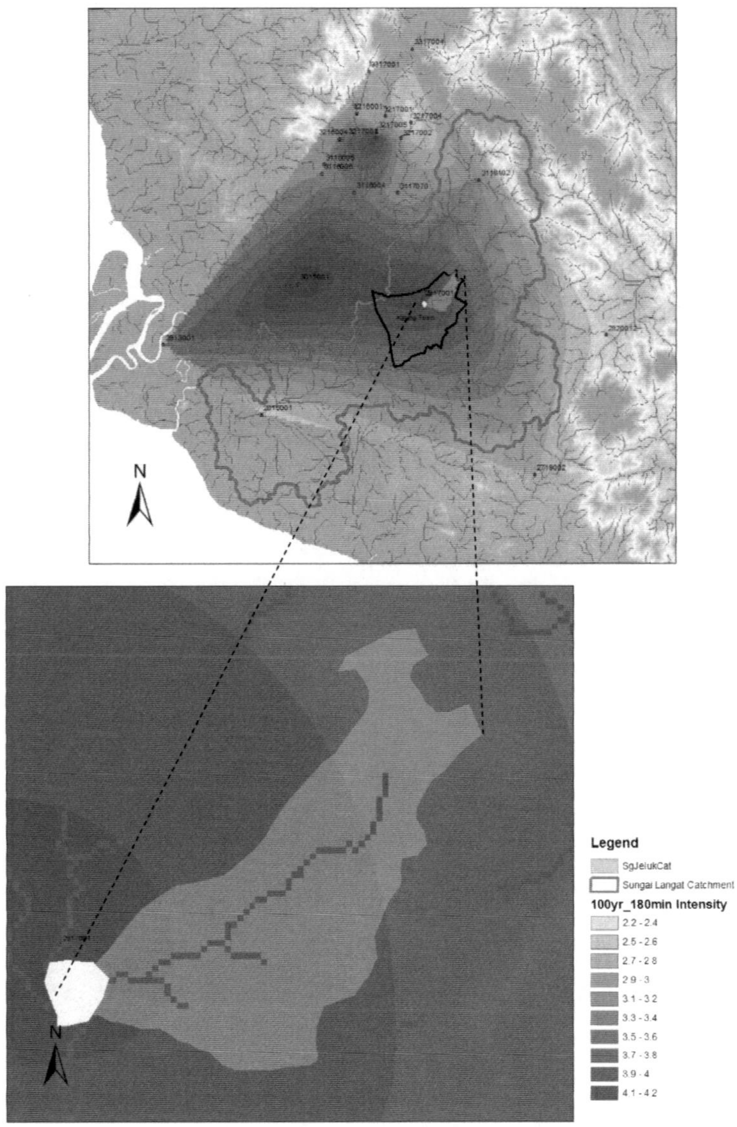

Figure 3: Distributed rainfall intensity of Kajang area

4. Conclusion

One of the critical topics in flood subject is a pattern verification of rainfall in urban regions. In this study, Kajang (Malaysia) has been identified as a flood prone area by development of distributed rainfall system and land use map via HEC GeoHMS. Results show that pattern of rainfall is lower at upstream than downstream base on intensity. It has been suggested to use distributed rainfall for input data in hydrological model to have improved hydrological analyses.

Acknowledgements

The authors wish to acknowledge and express our gratitude to Water Resources Management and Hydrology Division, Department of Irrigation and Drainage of Malaysia (DID) and ZHL-Engineers Sdn. Bhd. for collaborating in this study of "Enhancing Distributed Rainfall Intensity for Flood Analysis within a GIS Framework in an Urban Area (Kajang Flood)" project.

References

[1] Department of Statistics, Malaysia, Key Summary Statistics for Local Authority Areas, Malaysia 2010. Zheng Fang, Enhancing a Distributed Hydrologic Model: Addition of Storm Water Pipe Analysis within a GIS Framework in an Urban Area, Journal of Hydrologic Engineering, 2006.

[2] http://www.careclimatechange.org/publications/global-reports/42-%20humanitarian-implications-of-climate-change=42

[3] Department of Irrigation and Drainage, Malaysia, National Register of River Basins, 2003. Bedient, P. B.;W. C., Huber, 2002, Hydrology and Floodplain Analysis, 3rd Ed., Prentice Hall, Inc., Upper Saddle River, NJ 07458.

[4] Jajarmizadeh, M., Sobri Harun., Salarpour, M. 2012. A Review on Theoretical Consideration and Types of Models in Hydrology. Journal of Environmental Science and Technology. 5(5):249-261.

Effect of Mesh Distribution on Impact Resistance Performance of Kenaf Fibre Reinforced Concrete

Zakaria Che Muda[1], Sivadass Thiruchelvam[4], Agusril Syamsir[1], Chiam Yung Sheng[2], Salmia Beddu [1], Kamal Nasharuddin Mustapha[3], Fathoni Usman[1], Nur Liyana Mohd Kamal[1], Md Ashraful Alam[1], Ahmed H Birima[1], O S Zaroog[4]

[1]Centre of Sustainable Technology and Environment, Universiti Tenaga Nasional, Malaysia, [2]Former Student of Universiti Tenaga Nasional
[3]Centre of Forensic Engineering, Universiti Tenaga Nasional, Malaysia
[4]Centre of Innovation and Design, Universiti Tenaga Nasional, Malaysia

mzakaria@uniten.edu.my

Abstract. This paper investigate the effect of the mesh distribution on the impact performance of kenaf fibre mesh reinforced concrete (KFMRC) for the concrete slab of 300mm x 300mm size reinforced with varied thickness and mesh diameter subjected to low impact projectile test. A self-fabricated drop-weight impact test rig with a steel ball weight of 1.236 kg drop at 0.40 m height has been used in this research work. The objective of this research is to study the effect of the mesh distribution on the impact resistance kenaf fibre mesh concrete for various slab thickness and mesh diameter. 2-layers one Top and one Bottom mesh distribution kenaf mesh is the most efficient in the ability to control crack formation and propagation against impact energy followed by 1-layer Middle mesh distribution and lastly the 1-layer Top mesh distribution is the least effective.

1. Introduction

In the search of sustainable green materials, it is critical to study the impact strength characteristics and assess its performance for eco-green construction materials for various potential uses in the building industry.

There is lack of research investigation been carried out on kenaf fibre mesh reinforced concrete (KFMRC) and the effect of the fibre distribution on the impact resistance. Impact resistance represents the ability of concrete to withstand repeated blows and absorb energy without adverse effect to cracking and spalling. Impact scenario can also be classified into low velocity impact and high velocity impact. According to Dancygier [1], the nuclear structures are designed to resist impact loading caused by projectile or missiles travelling up to 1000 m/s. The response of reinforced concrete structures under impact loading is different from the static loading especially in the case of high velocity impact of rigid projectile. For impact velocities up to 10 ms^{-1}, the failure modes are generally the same as the static failure, except there is increased tendency for local damage or shear failure to occur. A repeated impact test, a weighted pendulum Charpy-type impact test, a projectile impact test, and explosion-impact test, a constant strain rate test, a split Hopkinson bar test, and an instrumented pendulum impact test could be used to measure the impact resistance of fiber reinforced composites [2]. The impact resistance of material also can be measured by using the criteria such as the energy of fracture of the specimen, repeated impact tests, and the velocity and the size of the spall the specimen subjected to a blast loading surface [3]. The tensile properties of kenaf fiber for various condition of chemical fiber surface modifications has been studied by Mahjoub Reza et al. [4]. The review paper on impact resistance on concrete target has been published by Z Che Muda et al [5]. Impact resistance of oil palm shells lightweight concrete slab with bamboo fibers has been studied by Z Che Muda et al. The results indicate that 2% volume fraction of bamboo fibers has an optimal performance in the crack resistance regardless of its fiber length with a potential to be used as an impact resistance composite structures in the future [6].

The objective of this research is to study the effect of the mesh distribution on the impact resistance kenaf fibre mesh concrete.

2. Materials and Test Set-up.

The kenaf fiber is obtained from private supplier MZM solution that based in Kedah. The kenaf fiber is graded as grade AA which is capable to produce the highest tensile strength of 900 MPa according to the supplier. The fibres are twisted into the required bundled diameters of 3 mm, 5 mm and 7 mm diameter as shown in Figure 1a shown a typical mesh arrangement at a constant spacing of 50 mm and placed in either 1-layer at the top only , 1-layer at mid-depth only and 2- layers one at top & one at bottom of the slab.

Ordinary Portland cement complying to ASTM Type I cement are used with 2% of super plasticizer is used in the design mix to achieved the desired workability.

The basic mix design for the KFMRC is shown in Table 1.

TABLE 1 Mix Design for Kenaf fibre mesh reinforced concrete (KFMRC)

Cement (kg/m^3)	Fine Aggregates (kg/m^3)	Course Aggregates (kg/m^3)	Water/cement Ratio	Slump (mm)	Compressive Strength (N/mm2)
403	687	1030	0.6	72	31.7

The study used a self-fabricated low velocity drop-weight impact test set up are as shown in Figure 1b using a steel ball weighing 1.236 kg with drop height of 400 mm impacting the specimen of size 300mm x 300mm with thickness of 20 mm, 30 mm, and 40 mm mounted on the steel rack frame. The test sample is 1-way simply supported.

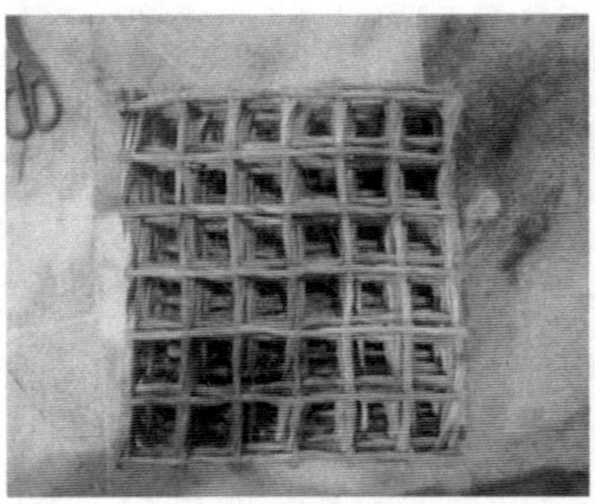

(a)

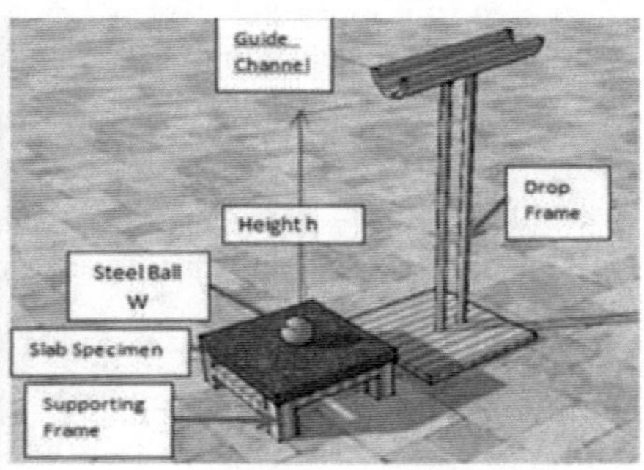

(b)

Figure 1 (a) Typical Kenaf Fibre Mesh arrangement (b) Low-velocity Drop-weight Impact Test Set-up

3. Methodology

The potential energy due to the drop body is absorbed as strain energy, generating stresses that causes cracks in the target element. The width, depth, length of the crack developed and its failure mode is associated with the intensity of the energy, the amount of energy absorbed and the properties of concrete. It is assumed that the total computed energy imparted is fully absorbed by the specimens. The relationship of potential energy of a drop-weight projectile and the strain energy dissipated in cracks development is expressed as following formula as proposed by Kankam [7];

Crack resistance, $Ru = (N*e)/(l_c* d_c* w_c)$ (3.1) Impact energy (Resistance) $= N*e = N*m*g*h$ (3.2)

Where, N = No. of Blows, e = Energy per blow (Joules), lc = Total length of all cracks (mm), d_c = Maximum crack depth (mm), w_c = Maximum crack width (mm), R_u = Crack resistance, m=mass of ball (kg), g = 9.81m/s^2, h = height of impact (mm).

A total of 108 sample slabs of size 300mm x 300mm with 30 mm,40mm and 50mm thickness were casted with 3 mm, 5mm and 7 mm bundled mesh diameter.

The following mesh distribution were investigated to study its impact on service (first) and ultimate (failure) crack resistance.

i. 1-layer top mesh ii. 1-layer mid depth mesh iii. 2 layers one at top and one at bottom

At the first crack and ultimate (failure) crack, the total crack length, the crack width and the crack depth measured by filler gauge with its total numbers of blows recorded.

4. Results and Discussion

4.1 Relationship between Crack Resistance and Kenaf Mesh Distribution

Figure 2, Figure 3 and Figure 4 indicate 2-layers one Top& one Bottom mesh distribution kenaf mesh is the most efficient in the ability to control crack formation and propagation against impact energy followed by 1-layer Middle mesh distribution and lastly the 1-layer Top mesh distribution.

The crack resistance of 50 mm thick slab, 7 mm diameter for 2-layers one Top & one Bottom mesh arrangement increase by up to 1.31 times for first crack and up to 1.89 times for ultimate crack against 1-layer Middle distribution and increase by up to 1.47 times for first crack and up to 2.27 times 1-layer Bottom only mesh distribution. Whilst the 1-layer Middle only mesh distribution arrangement increase by up to 1.12 times for first crack and up to 1.21 times for ultimate crack against 1-layer Top mesh distribution.

The highest service crack of 62.84 N/mm2 and ultimate crack resistance of 246.65 N/mm2 is observed for 2-layers one Top & one Bottom mesh arrangement with 50 mm slab thickness and 7 mm mesh diameter whilst lowest service crack of 4.59 N/mm2 and ultimate crack resistance of 14.63 N/mm2 is observed for 1-layer Top mesh arrangement for 30 mm slab with 3 mm diameter mesh. The first crack resistance value indicate the initiation the first crack and the ultimate crack resistance value indicate the ultimate failure modes with its fracture zones.

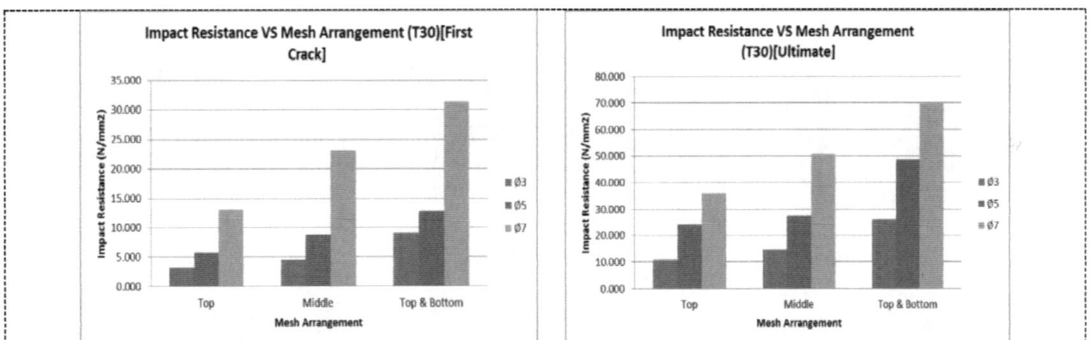

Figure 2 First and Ultimate Crack Resistance of Top, Middle and Top & Bottom Mesh Distribution for 30 mm Thick Slab

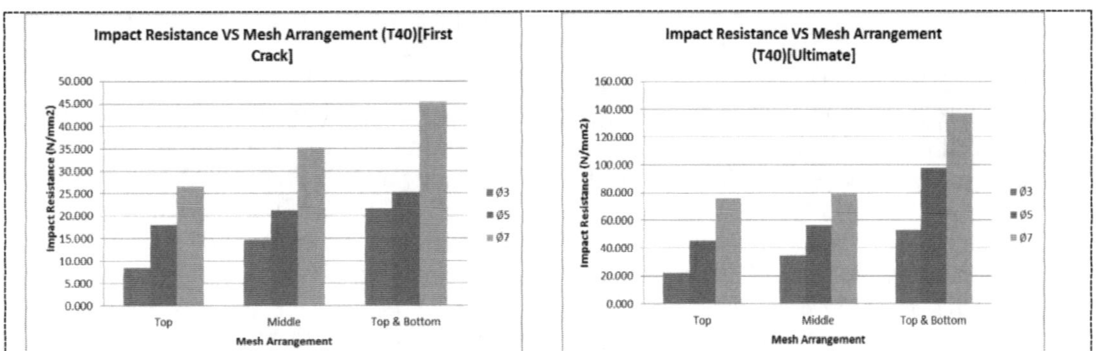

Figure 3 First and Ultimate Crack Resistance of Top, Middle and Top & Bottom Mesh Distribution for 40 mm Thick Slab

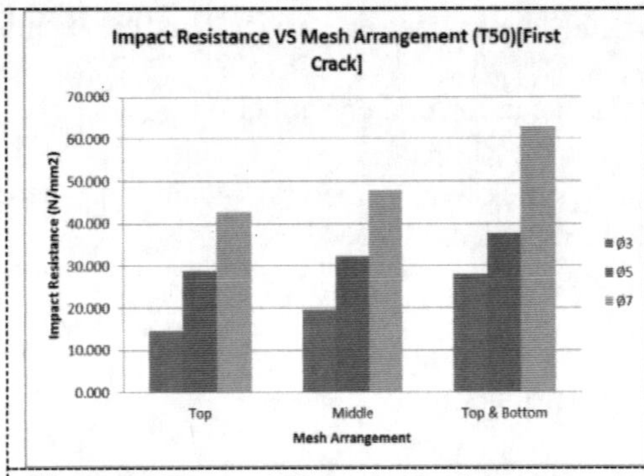

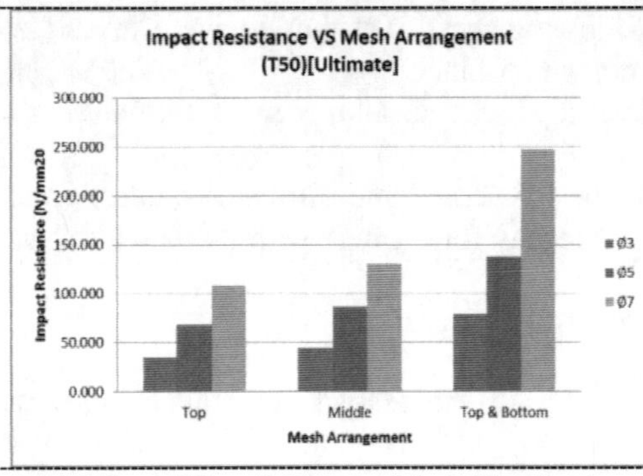

Figure 4 First and Ultimate Crack Resistance of Top, Middle and Top & Bottom Mesh Distribution for 50 mm Thick Slab

2-layers one Top & one Bottom mesh distribution is the most efficient as compare to 1-layer Middle mesh and 1-layer Top distribution in the ability to control cracks and resisted a lot of energy until the crack started to widen and propagate further and lose its bonding and tensile, which then lead to failure.
1-layer Top mesh arrangement with 30 mm thick slab or 3 mm mesh diameter is the least efficient configuration to control service and ultimate crack resistance against impact energy.

5. Conclusion

The following conclusions can be derived from the experimental results;

- 2-layers one Top& one Bottom mesh distribution kenaf mesh is the most efficient distribution followed by 1-layer Middle mesh distribution and lastly the 1-layer Top mesh distribution which is the least effective.
- 2-layers one Top & one Bottom mesh arrangement has a first crack and ultimate crack resistance by 1.47 times and 2.27 times respectively against 1-layer Top mesh Middle distribution for 50 mm thick slab and 7 mm diameter mesh.
- 1-layer Middle mesh distribution has a first crack and ultimate crack resistance by 1.12 times and 1.21 times respectively against 1-layer Top mesh Middle distribution for 50 mm thick slab and 7 mm diameter mesh.
- The threshold (highest) values for service crack is 62.84 N/mm2 and ultimate crack resistance is 246.65 N/mm2 observed for 2-layers one Top & one Bottom mesh distribution

References

[1] Dancygier, A N 1997 *Effect of reinforcement ratio on the resistance of reinforced concrete to hard projectile impact.* Nuclear Engineering and Design 172, pp. 233-245.
[2] Gopalartnam, V. S., Shah, S. P., and John, R. A modified instrumented Charpy test for cement based composites." Exp. Mech. 1984; 24(2): 102–111.
[3] Balaguru, P. N., and Shah, S. P. Fibre-reinforced cement composites, McGraw-Hill, London.1992.
[4] Mahjoub, Reza. Tensile properties of kenaf fiber due to various conditions of chemical fiber surface modifications. Construction and Building Materials 55(2014): 103-113.
[5] Zakaria Che Muda, Kong Sih Ying, Salah F A Sharif, Lariyah Bte. Mohd Sidek,Nawfal S. Farhan, *A Review - Local Failure On Concrete Target Due To Projectile Impact.* International Journal of Science and Engineering Research (IJSER) Volume 4, Issue 1, January 2013.
[6] Zakaria Che Muda, Salah F A Sharif, Lariyah Bte. Mohd Sidek, Nawfal S Farhan, *Impact Resistance Of Oil Palm Shells Lightweight Concrete Slab With Bamboo Fibers,* International Journal of Science and Engineering Research (IJSER) Volume 4, Issue 1, January 2013.
[7] CK Kankam. *Impact Resistance of palm kernel fibre-reinforced concrete pavement slab*. J Ferrocement 1999;29(4):279-86.

Effect of Thickness and Fibre Volume Fraction on Impact Resistance of Steel Fibre Reinforced Concrete (SFRC)

Zakaria Che Muda[1,3], Fathoni Usman[1], Agusril Syamsir[1], Chen Shao Yang[2], Kamal Nasharuddin Mustapha[3], Salmia Beddu [1], Sivadass Thiruchelvam[4] , Nur Liyana Mohd Kamal[1], Md Ashraful Alam[1], Ahmed H Birima[1], Zarina Itam[1], O S Zaroog[4]

[1]Centre of Sustainable Technology and Environment, Universiti Tenaga Nasional, Malaysia [2]Former Student of Universiti Tenaga Nasional
[3]Centre of Forensic Engineering, Universiti Tenaga Nasional, Malaysia
[4]Centre of Innovation and Design, Universiti Tenaga Nasional, Malaysia

mzakaria@uniten.edu.my

Abstract. This paper investigate the effect of the thickness and fibre volume fraction (VF) on the impact performance of steel fibre reinforced concrete (SFRC) for the concrete slab of 300mm x 300mm size reinforced subjected to low impact projectile test. A self-fabricated drop-weight impact test rig with a steel ball weight of 1.236 kg drop at 0.57 m height has been used in this research work. The objective of this research is to study the relationship of impact resistance SFRC against slab thickness and volume fraction. There is a good linear correlation between impact resistances of SFRC against slab thickness. However the impact resistance of SFRC against percentage of volume fraction exhibit a non-linear relationship.

1. Introduction

The impact resistance of a structure is very important to protect the users from being harmed and injured during impact strikes the concrete structure. Important structures such as dams, military defense structure, power plant and so on is very crucial to impact events as if the structure does not have enough impact resistance, it might cause serious lost in property, human life, and economic lost.

Impact resistance represents the ability of concrete to withstand repeated blows and absorb energy without adverse effect to cracking and spalling.

There is lack of research investigation been carried out on the effect of steel fibre distribution on the impact resistance of steel fibre reinforced concrete (SFRC).

Holschemacher et al, it found that increasing of fibre strength will help in increasing the strength, ductility, and post cracking behavior of fibre reinforced concrete because high-strength fibre prevent the fibre from breaking when impact strikes and cracks happened [1]. Xu et al. carried out drop-weight experimental tests on concrete reinforced with 7 types of fibres and study their properties on the impact resistance.[2]. According to A.A. Aliabdo et al, impact resistance of concrete will affected by fibre type and content, shape of steel fibre, coarse aggregate type, and water/cement ratio [3].

The drop weight impact test which is recommended by the ACI Committee 544 [4] is the simplest method. The review paper on impact resistance on concrete target has been published by Z Che Muda et al [5]. Impact resistance of oil palm shells lightweight concrete slab with bamboo fibers has been studied by Z Che Muda et al.[6].

The objective of this research is to study the effect of the steel fibre distribution on the impact resistance SFRC and its mode of failures.

2. Materials and Test Set-up.

The hooked end steel fibre has a fibre length of 50mm with an aspect ratio of 65. The tensile strength of the steel fibre is 1100 $\pm$ 80 MPa.

Ordinary Portland cement complying to ASTM Type I cement are used with 2% of super plasticizer is used in the design mix to achieved the desired workability.

The basic mix design for the Steel Fibre Reinforced Concrete (SFRC) is shown in Table 1.

TABLE 1 Mix Design for Steel Fibre Reinforced Concrete (SFRC)

Cement (kg/m^3)	Fine Aggregates (kg/m^3)	Course Aggregates (kg/m^3)	Water/cement Ratio	Slump (mm)	Compressive Strength (N/mm2)
440	902	889	0.35	72	39

The study used a self-fabricated low velocity drop-weight impact test rig and its set up are as shown in Figure 1 using a steel ball weighing 1.236 kg with drop height of 570 mm impacting the specimen of size 300mm x 300mm with a thickness of 20 mm, 30 mm, and 40 mm with 1% VF, 2% VF and 3% VF mounted on the steel rack frame. The test sample is 1-way simply supported.

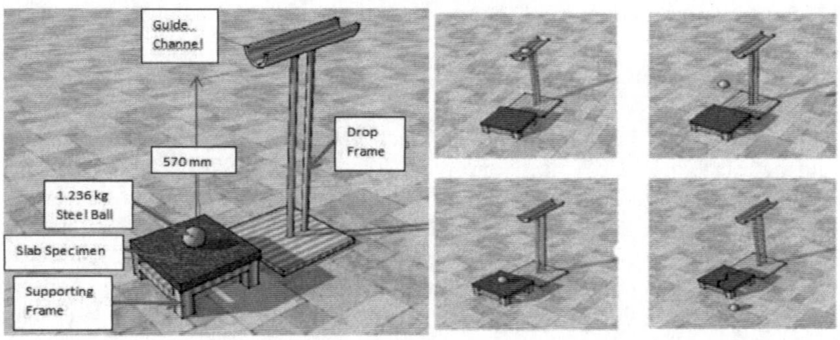

Figure 1 Low-velocity Drop-weight Impact Test Rig and Test Set-up

3. Methodology

The potential energy due to the drop body is absorbed as strain energy, generating stresses that causes cracks in the target element. The width, depth, length of the crack developed and its failure mode is associated with the intensity of the energy, the amount of energy absorbed and the properties of concrete. It is assumed that the total computed energy imparted is fully absorbed by the specimens. The relationship of potential energy of a drop-weight projectile and the strain energy dissipated in cracks development is expressed as following formula as proposed by Kankam [7];

$$N*e = R_u * l_c * d_c * w_c \tag{3.1}$$

Where, N = No. of Blows, e = Energy per blow (Joules), l_c = Total length of all cracks, d_c = Maximum crack depth, w_c = Maximum crack width, R_u = Ultimate crack resistance

A total of 54 sample slabs of size 300mm x 300mm with 30 mm, 40 mm and 50 mm thickness were casted with at control (no fibre), 1%, 2 % and 3% steel fibre volume fraction volume (VF). Each of the combination have 3 samples in order increase its accuracy and an average value is obtained.

The following mesh distribution were investigated to study its impact on service (first) and ultimate (failure) crack resistance.

i. Slab thickness.
ii. Percentage of fibre volume fraction.

At the first crack and ultimate (failure) crack, the total crack length, the crack width and the crack depth measured by filler gauge with its total numbers of blows recorded.

4. Results and Discussion

4.1 Relationship between Crack Resistance and Slab Thickness

There is a good linear correlation with minimum R^2=0.9403 between the crack resistance and slab thickness as shown in Figure 2. Increase in crack resistance is more substantial as the thickness increases as compared to the increases in the fiber volume fraction. The highest value of first crack and ultimate crack resistance obtain is 15.50N/mm^2 and 86.43 N/mm^2 respectively for 50 mm thick slab at 3% VF. The first crack and ultimate crack impact resistance for 50 mm slab are 1.62 times and 1.65 times respectively as against the 30 mm slab at 3%VF steel fibre.

Figure 2(b) shows the ultimate crack resistance without fibre (control slab) is 3.92 N/mm^2 for 50 mm thick slab. The ultimate crack resistance for 50 mm slab thickness with 3% FV increased by 22.05 times against its control sample.

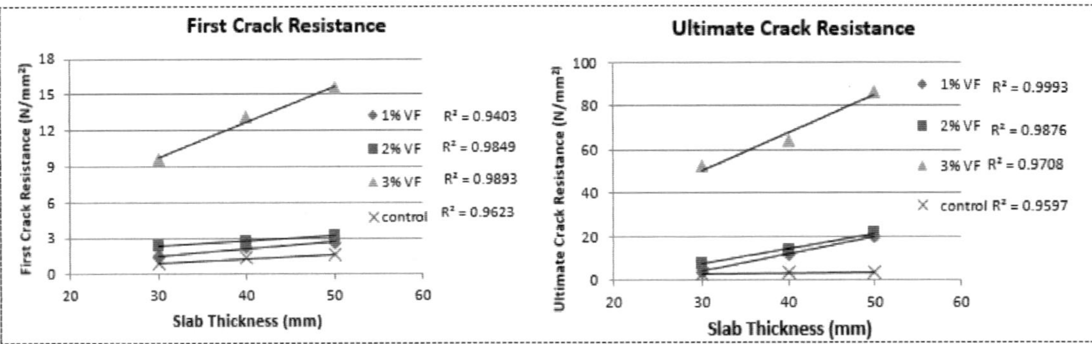

(a) (b)

Figure 2 First and Ultimate Crack Resistance against Slab Thickness

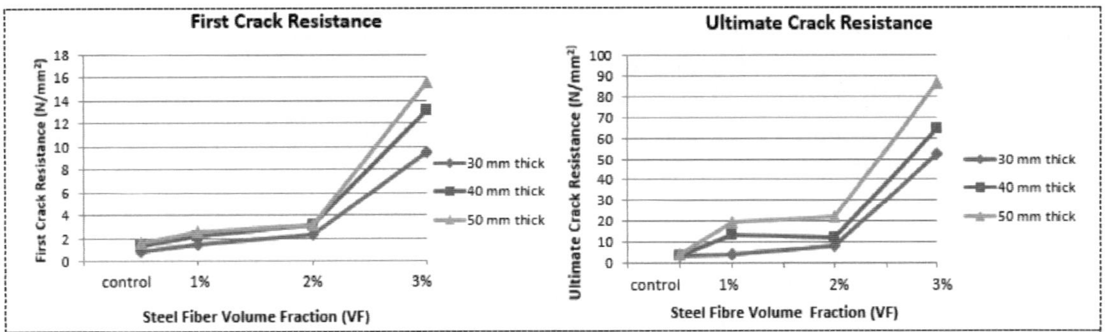

(a) (b)

Figure 3 First and Ultimate Crack Resistance againts Fibres Volume Fraction

4.1 Relationship between Crack Resistance and Fibre Volume Fraction (VF)

There is a non-linear relationship of impact resistance against fibre volume fraction are as shown in Figure 3. Generally the first and ultimate crack resistance increases with increasing fibre VF, however the optimum threshold value is with a 3% VF steel fibre. First crack resistance for 50 mm slab increase by 6.03 times and 4.41 times for ultimate crack at 3% VF as compare with 1% VF steel fibre.

There is not much differences in values of first and ultimate crack resistance between 1%VF and 2%VF steel fibre. This indicate that there is no significant gain of first crack and ultimate crack resistance for fibre less than 3% VF.

4.2 Mode of Failures

Fibres play a very important role in bridging cracks, absorbing energy, transferring loads, and developing micro cracks distribution system. First crack of 0.075mm normally initiate at the bottom surface of the slab due to the limiting tensile strain and propagated upward until neutral axis. When the crack move up and reaches the top part of the concrete, the ultimate crack eventually cause a complete segmental failures. If impact happen continuously, initial crack will subsequently propagate and link up into a larger crack known as a fracture zone. The 50 mm thick slab with 3% VF in Figure 4 (b) has five segmental failures along the fracture zone and more micro-cracking in the other parts of the slab as compare with 2% VF in Figure 4 (a) has four segmental failures along its fracture zone.

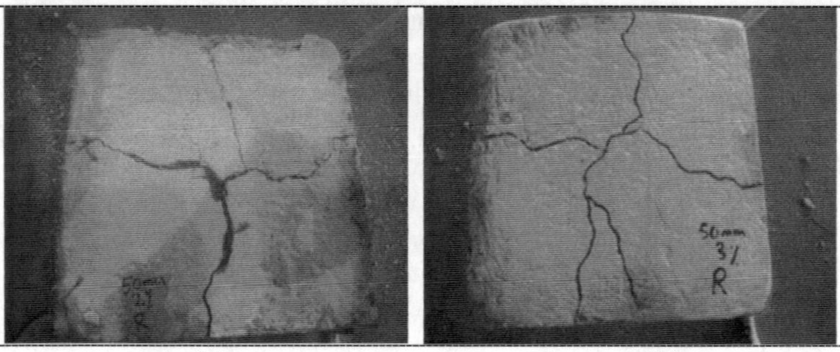

(a) **(b)**

Figure 4 Failure Modes at distal surface of 50 mm slab with (a) 2% VF (b) 3% VF.

5. Conclusion

The following conclusions can be derived from the experimental results;

- There is a good linear correlation with minimum R^2=0.9403 between the crack resistance and slab thickness.
- There is non-linear exponential relationship between the crack resistance and the percentage of VF fibre.
- The first crack and ultimate crack impact resistance at 3%VF steel fibre for 50 mm slab are 1.62 times and 1.65 times respectively as compare with 30 mm slab.
- The ultimate crack resistance for 50 mm slab thickness with 3% VF increased by 22.05 times against its control sample.
- First crack resistance for 50 mm slab increase by 6.03 times and 4.41 times for ultimate crack at 3% VF as compare with 1% VF steel fibre.
- There is no significant gain of first crack and ultimate crack resistance for fibre less than 3% VF.
- The slab with a higher VF has more segmental failures along the fracture zone and micro-cracking in the other parts of the slab as compare with the slab with lower VF fibre.

References

[1] Holschemacher K, Muekker T, Ribakov Y. 'Effect of steel fibres on mechanical properties of high-strength concrete'. Mater Des 2010;31:2604-15.

[2] Xu Z, Hao H, Li HN. 'Experimental study of dynamic compressive properties of fibre reinforced concrete materials with different fibres'. Mater Des 2012;33:42-55.

[3] Ali Abd Elhakam Aliabdo, Abd Elmoaty Mohamed Abd Elmoaty, Mohamed Hamdy. 'Effect of internal short fibres, steel reinforcement, and surface layer on impact and penetration resistance of concrete'. Alexandria Engineering Journal. 2013;52, 407-417.

[4] ACI Committee 544. State-of-the-art report on fiber reinforced concrete. ACI Committee 544 report 544.1R-96. Detroit: American Concrete Institute. 1996.

[5] Zakaria Che Muda, Kong Sih Ying, Salah F A Sharif, Lariyah Bte. Mohd Sidek,Nawfal S. Farhan, A Review - Local Failure On Concrete Target Due To Projectile Impact. International Journal of Science and Engineering Research (IJSER) Volume 4, Issue 1, January 2013.

[6] Zakaria Che Muda, Salah F A Sharif, Lariyah Bte. Mohd Sidek, Nawfal S Farhan, Impact Resistance Of Oil Palm Shells Lightweight Concrete Slab With Bamboo Fibers, International Journal of Science and Engineering Research (IJSER) Volume 4, Issue 1, January 2013.

[7] CK Kankam . Impact Resistance of palm kernel fibre-reinforced concrete pavement slab. J Ferrocement 1999;29(4):279-86. Oct

The Effect of Thickness and Mesh Spacing on the Impact Resistance of Ferrocement Slab

Zakaria Che Muda[1,3], Md Ashraful Alam[1], Agusril Syamsir[1], Sorefan Sulleman[2], Salmia Beddu [1], Kamal Nasharuddin Mustapha[3], Sivadass Thiruchelvam[4] , Firas B. Ismail[5], Fathoni Usman[1] , Nur Liyana Mohd Kamal[1], Ahmed H Birima[1], Zarina Itam[1], O S Zaroog[4]

[1]Centre of Sustainable Technology and Environment, Universiti Tenaga Nasional, Malaysia [2]Former Student of Universiti Tenaga Nasional [3]Centre of Forensic Engineering, Universiti Tenaga Nasional, Malaysia [4]Centre of Innovation and Design, Universiti Tenaga Nasional, Malaysia, [5]Centre of Power Generation, Universiti Tenaga Nasional, Malaysia.

mzakaria@uniten.edu.my

Abstract. This paper investigates the effect of the thickness and mesh spacing on the impact of ferrocement for the concrete slab of 300mm x 300mm size reinforced subjected to low impact projectile test. A self-fabricated drop-weight impact test rig with a steel ball weight of 1.236 kg drop at height of 150 mm, 350mm, and 500mm has been used in this research work. The objective of this research is to study the relationship of impact resistance of ferrocement against slab thickness and mesh reinforcement spacing. There is a good linear correlation between impact resistance of ferrocement against slab thickness and its mesh spacing. The first and ultimate crack impact resistance for 40 mm slab are 2.00 times and 1.84 times respectively against the 20 mm slab with the same mesh spacing. The first and ultimate crack impact resistance for 40 mm slab with 20 mm mesh spacing are 2.24 times and 3.70 times respectively against 50 mm mesh spacing with the same slab thickness. The mesh with higher content of reinforcement provides more contribution to the slab resistance as compare with the thickness.

1. Introduction

In the search of sustainable green materials, it is critical to study the impact strength characteristics and assess its performance for eco-green construction materials for various potential uses in the building industry. Important structures such as vessels, dams, military defense structure, power plant and so on is very crucial to impact events as if the structure does not have enough impact resistance, it might cause serious lost in property, human life, and economic lost.

The fact that ferrocement has a good ability to resist impacts and damage occur in a localized region, Al-Rifaie proposes several applications of this composite material where conventional reinforced concrete can be replaced hence maximizing obvious economic advantages of ferrocement. The positive characteristics of ferrocement upon impact are its resistance against disintegration, damage occurs in localized fashion and damage is easy to repair [1]. The review paper on impact resistance on concrete target has been published by Z Che Muda et al [2]. The combination of mortar and steel mesh results in the composite ferrocement element. Throughout literature, it can be noted that ferrocement has favourable characteristics in impact, as collisions between boats or with rocks are numerous [3]. Ferrocement has high strength, crack resistance, high ductility and energy absorption characteristics, all useful to be used in a dynamic environment and application media [4]. The drop weight impact test which is recommended by the ACI Committee 544 [5] is the simplest method. Impact resistance of oil palm shells lightweight concrete slab reinforced with geogrid has been studied by Z Che Muda et al. indicate an impact resistance improvement for first and ultimate crack up to 5.9 times and 20.1 times against the control sample without geogrid [6].

The objective of this research is to study the impact on service (first) and ultimate (failure) crack resistance against the thickness and mesh spacing of ferrocement slab.

2. Materials and Test Set-up.

Ordinary Portland cement is complying to ASTM Type I cement. The square steel welded mesh as shown in Figure 1(a) has a 1 mm diameter with a spacing of 20 mm, 40 mm and 50 mm and the tensile strength of 450 MPa.

The properties of the fine aggregates used is in the cement mortar matrix was selected according to the ACI Committee 549, (1997). The sieve analysis of the sand shows that the different sand grain sizes are within the limit of a zone 2 of a well graded aggregate sizes with the fineness modulus was found to be 2.3.

The basic mix design for the mortar is as shown in Table 1.

TABLE 1 Mix Design for Mortar

Cement (Part by weight)	Fine Aggregates (Part by Weight)	Water/cement Ratio	Slump Class to BS8500 (Range)	Compressive Strength (N/mm2)
1	2	0.50	180 mm Class S4 (160-210 mm)	24.25

The study used a self-fabricated low velocity drop-weight impact test rig is shown in Figure 1 (b) using a steel ball weighing 1.236 kg with drop height of 570 mm impacting the ferrocement slab of size 300mm x 300mm with a thickness of 20 mm, 30 mm, and 40 mm with ferrocement welded mesh of 1mm diameter @ 20mm, 40mm and 50 mm spacing. The test sample is mounted on the steel rack frame with 1-way simply supported and a test up procedure is given in Figure 1(c).

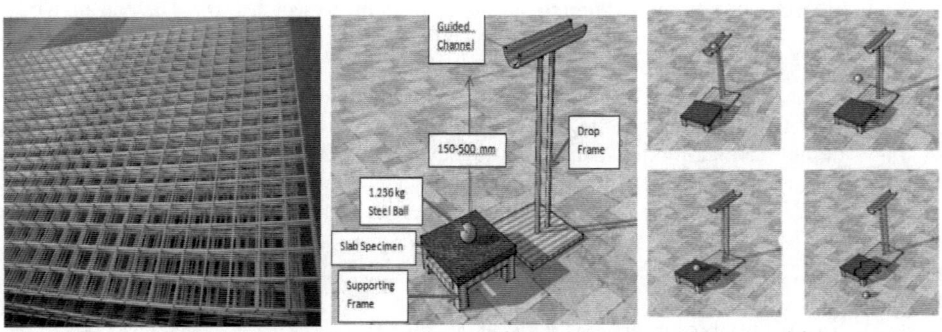

(a) (b) (c)

Figure 1

(a) Square Wire Mesh with 1 mm diameter with a spacing of 20mm, 40mm and 50mm
(b) Low-velocity Drop-weight Impact Test Rig (c) Impact Test Set-up

3. Methodology

The potential energy due to the drop body is absorbed as strain energy, generating stresses that causes cracks in the target element. The width, depth, length of the crack developed and its failure mode is associated with the intensity of the energy, the amount of energy absorbed and the properties of concrete. It is assumed that the total computed energy imparted is fully absorbed by the specimens. The relationship of potential energy of a drop-weight projectile and the strain energy dissipated in cracks development is expressed as following formula as proposed by Kankam [7];

$$N*e = R_u * l_c * d_c * w_c \quad (3.1)$$

Where, N = No. of Blows, e = Energy per blow (Joules), lc = Total length of all cracks, d_c = Maximum crack depth, w_c = Maximum crack width, R_u = Ultimate crack resistance

A total of 63 sample slabs of size 300mm x 300mm with 30 mm, 40 mm and 50 mm thickness were casted with at control (no fibre), 20mm, 40 mm and 50 mm spacing of 1 mm diameter welded mesh. Each set of data have 3 samples in order increase its accuracy and its average value taken. The wire mesh is place at mid-depth of the slab.

A drop height of 150 mm is use for control (no reinforcement) or thin slabs, 350 mm drop is for medium slab thickness and 500 mm drop height is used for thicker slab. The drop was targeted at the centre of the slab and for each drop the bottom surface - the tension zone - was checked for cracks.

At the first crack and ultimate (failure) crack, the total crack length, the crack width and its depth measured with its total numbers of blows are recorded.

4. Results and Discussion

4.1 Crack Resistance for Control Samples

The service (first) and ultimate crack resistance are shown in Table 2. The crack resistance increases with the increase in thickness.

TABLE 2 First and Ultimate Crack Resistance of Control Sample without Mesh Reinforcement

Control Sample No	25M/20T/0R	25M/30T/0R	25M/40T/0R
Thickness of Control Sample	20 mm	30 mm	40 mm
Service Crack Resistance	22.62 N/mm^2	56.08 N/mm^2	91.43 N/mm^2
Ultimate Crack Resistance	30.16 N/mm^2	84.12 N/mm^2	109.72 N/mm^2

4.2 Relationship between Crack Resistance and Slab Thickness

There is a good linear correlation for service and ultimate crack resistance against its thickness as shown Figure 2 with a minimum R^2=0.97644. As the slab thickness increases, the service and ultimate crack resistance increases proportionally.

The highest value of first crack and ultimate crack resistance obtain is 147.36N/mm^2 and 405.23N/mm^2 respectively for 40 mm thick slab with 20 mm mesh spacing. The first crack and ultimate crack impact resistance for 40 mm slab with 20 mm mesh spacing are 2.00 times and 1.84 times respectively against the sample 20 mm slab with the same mesh spacing and 1.61 times and 3.69 times respectively as against its control sample 25M/40T/0R.

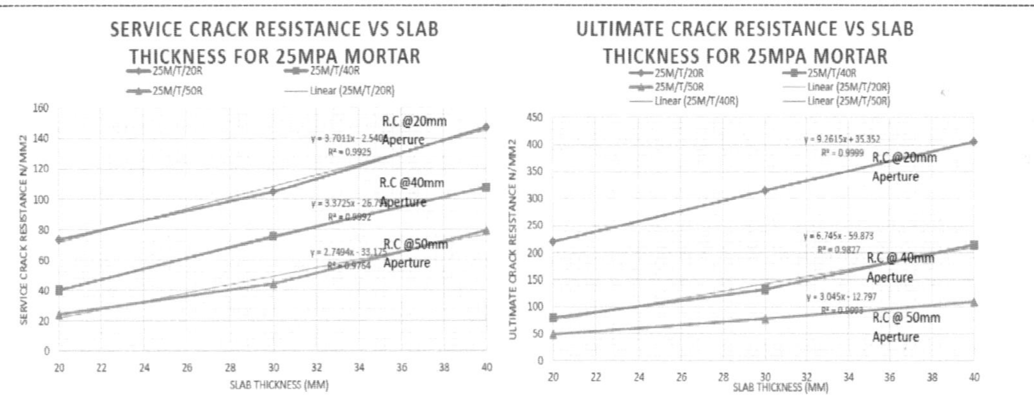

(a) Service (First) Crack (b) Ultimate Crack

Figure 2 Service (First) and Ultimate Crack Resistance Against Slab Thickness

4.3 Relationship between Crack Resistance and Mesh Aperture Spacing

There is a good linear correlation for service and ultimate crack resistance against its mesh spacing (aperture) as shown in Figure 3 with a minimum R^2=0.9574. The highest value of first crack and ultimate crack resistance obtain is 147.36N/mm^2 and 405.23N/mm^2 respectively for 40 mm thick slab with 20 mm mesh spacing. The first crack and ultimate crack impact resistance for 40 mm slab with 20 mm mesh spacing are 2.24 times and 3.70 times respectively against 50 mm mesh spacing the slab with the same thickness.

The first crack and ultimate crack impact resistance for 40 mm slab with 50 mm mesh spacing are 0.72 times and 1 times respectively against its control sample 25M/40T/0R. Thus the large 50 mm mesh spacing does not contribute to any gain in the crack resistance.

The presence of the mesh with the smallest mesh of 20 mm spacing provide more contribution to the impact resistance against first and ultimate crack than the slab thickness by 1.61 times and 3.69 times respectively. Increase in the thickness is less significant for the crack resistance as compared to the mesh spacing.

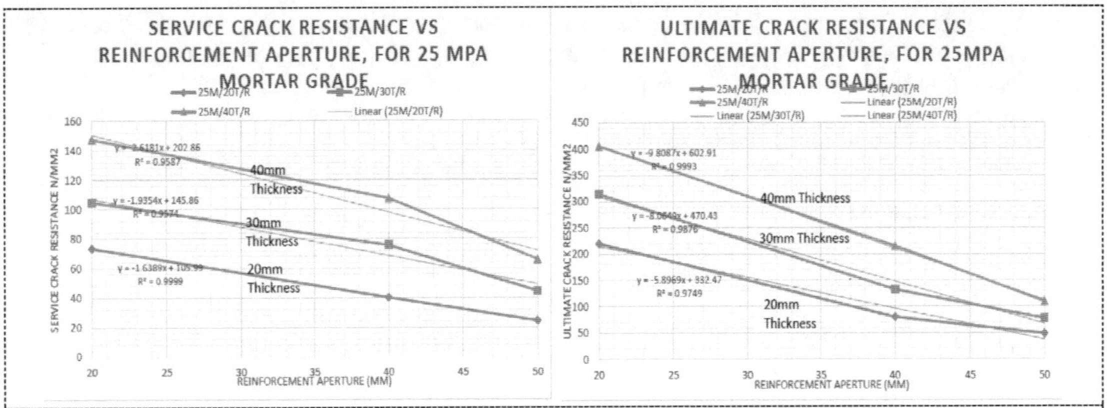

(a) Service (First) Crack (b) Ultimate Crack

Figure 3 First and Ultimate Crack Resistance Against Mesh Aperture Spacing

5. Conclusion

The following conclusions can be derived from the experimental results;

- There is a good linear correlation with minimum R^2= 0.97644 between the crack resistance and slab thickness.
- There is a good linear correlation for service and ultimate crack resistance against its mesh spacing (aperture) with a minimum R^2=0.9574
- The first crack and ultimate crack impact resistance for 40 mm slab with 20 mm mesh spacing are 2.00 times and 1.84 times respectively against the 20 mm slab with the same mesh spacing.
- The first crack and ultimate crack impact resistance for 40 mm slab with 20 mm mesh spacing are 2.24 times and 3.70 times respectively against 50 mm mesh spacing the slab with the same thickness.
- There is no significant gain of first crack and ultimate crack resistance for large 50 mm mesh spacing.
- The mesh with the smallest spacing of 20 mm provides more contribution to the impact resistance than the slab thickness.

References

[1] Al-Rifaie, W.N. (2006). Ferrocement Wall: Penetration Testing, Proceedings of Eight International Symposium and Workshop on Ferrocement and Thin Reinforced cement Composites. 06-08 February, Bangkok Thailand, IFS, 177-185.

[2] Zakaria Che Muda, Kong Sih Ying, Salah F A Sharif, Lariyah Bte. Mohd Sidek,Nawfal S. Farhan, A Review - Local Failure On Concrete Target Due To Projectile Impact. International Journal of Science and Engineering Research (IJSER) Volume 4, Issue 1, January 2013.

[3] Iorns, M. E., and Watson, L. L., Jr. (1977), Ferrocement Boats Reinforced with Expanded Metal. Journal of Ferrocement (Bangkok), V. 7, No. 1, July 1977, pp. 9-16

[4] Yousry B.I. Shaheen, Noha Mohamed Soliman and Doha El Metwally Kandil (2013). Influence of Reinforced Ferrocement Concrete Plates under Impact Load. International Journal of Current Engineering and Technology, 21 October 2013, Vol.3, No.4

[5] ACI Committee 544. State-of-the-art report on fiber reinforced concrete. ACI Committee 544 report 544.1R-96. Detroit: American Concrete Institute. 1996.

[6] Z C Muda, G.Malik, S Beddu,M.A Alam K N Mustapha, A H Birima, OS Zarroq L M Sidek and M ARashid, Impact Resistance of Sustainable Construction Material Using Lightweight Oil Pal Shells Reinforced Geogrid Concrete Slab, IOP Conf. Series : Earth and Environmental Sciences 16(2013) 012062 doi:10.1088/1755-1315/16/1/012062

[7] CK Kankam . Impact Resistance of palm kernel fibre-reinforced concrete pavement slab. J Ferrocement 1999;29(4):279-86. Oct.

Effect of Steel Fibres Distribution on Impact Resistance Performance of Steel Fibre Reinforced Concrete (SFRC)

Zakaria Che Muda[1], Nur Liyana Mohd Kamal[1], Agusril Syamsir[1], Chen Shao Yang[2], Salmia Beddu[1], Kamal Nasharuddin Mustapha[3], Sivadass Thiruchelvam[4], Fathoni Usman[1], Zarina Itam[1] , Md Ashraful Alam[1], Ahmed H Birima[1], O S Zaroog[4]

[1]Centre of Sustainable Technology and Environment, Universiti Tenaga Nasional, Malaysia [2]Former Student of Universiti Tenaga Nasional
[3]Centre of Forensic Engineering, Universiti Tenaga Nasional, Malaysia
[4]Centre of Innovation and Design, Universiti Tenaga Nasional, Malaysia

mzakaria@uniten.edu.my

Abstract. This paper investigate the effect of the mesh distribution on the impact performance of steel fibre reinforced concrete (SFRC) for the concrete slab of 300mm x 300mm size reinforced with varied thickness and fraction volume subjected to low impact projectile test. A self-fabricated drop-weight impact test rig with a steel ball weight of 1.236 kg drop at 0.57 m height has been used in this research work. The objective of this research is to study the effect of the mesh distribution on the impact resistance SFRC for various slab thickness and fraction volume. Random fibre distribution is the more effective than the top and bottom fibre distribution in terms of absorption of impact energy, crack resistance, the ability to control crack formation and propagation against impact energy.

1. Introduction

The impact resistance of a structure is very important to protect the users from being harmed and injured during impact strikes the concrete structure. Important structures such as dams, military defense structure, power plant and so on is very crucial to impact events as if the structure does not have enough impact resistance, it might cause serious lost in property, human life, and economic lost.

Impact resistance represents the ability of concrete to withstand repeated blows and absorb energy without adverse effect to cracking and spalling.

There is lack of research investigation been carried out on the effect of steel fibre distribution on the impact resistance of steel fibre reinforced concrete (SFRC).

Holschemacher et al, it found that increasing of fibre strength will help in increasing the strength, ductility, and post cracking behavior of fibre reinforced concrete because high-strength fibre prevent the fibre from breaking when impact strikes and cracks happened [1]. Xu et al. carried out drop-weight experimental tests on concrete reinforced with 7 types of fibres and study their properties on the impact resistance.[2]. According to A.A. Aliabdo et al, impact resistance of concrete will affected by fibre type and content, shape of steel fibre, coarse aggregate type, and water/cement ratio [3].

The drop weight impact test which is recommended by the ACI Committee 544 [4] is the simplest method. The review paper on impact resistance on concrete target has been published by Z Che Muda et al [5]. Impact resistance of oil palm shells lightweight concrete slab with bamboo fibers has been studied by Z Che Muda et al.[6].

The objective of this research is to study the effect of the steel fibre distribution on the impact resistance SFRC and its mode of failures.

IOP Conf. Series: Earth and Environmental Science **32** (2016) 012028 doi:10.1088/1755-1315/32/1/012028

2. Materials and Test Set-up.

The hooked end steel fibre has a fibre length of 50mm with an aspect ratio of 65. The tensile strength of the steel fibre is 1100 $\pm$ 80 MPa.

Ordinary Portland cement complying to ASTM Type I cement are used with 2% of super plasticizer is used in the design mix to achieved the desired workability.

The basic mix design for the Steel Fibre Reinforced Concrete (SFRC) is shown in Table 1.

TABLE 1 Mix Design for Steel Fibre Reinforced Concrete (SFRC)

Cement (kg/m^3)	Fine Aggregates (kg/m^3)	Course Aggregates (kg/m^3)	Water/cement Ratio	Slump (mm)	Compressive Strength (N/mm2)
440	902	889	0.35	72	39

The study used a self-fabricated low velocity drop-weight impact test rig and its set up are as shown in Figure 1 using a steel ball weighing 1.236 kg with drop height of 570 mm impacting the specimen of size 300mm x 300mm with a thickness of 20 mm, 30 mm, and 40 mm with 1% VF, 2% VF and 3% VF mounted on the steel rack frame. The test sample is 1-way simply supported.

Figure 1 Low-velocity Drop-weight Impact Test Rig and Test Set-up

3. Methodology

The potential energy due to the drop body is absorbed as strain energy, generating stresses that causes cracks in the target element. The width, depth, length of the crack developed and its failure mode is associated with the intensity of the energy, the amount of energy absorbed and the properties of concrete. It is assumed that the total computed energy imparted is fully absorbed by the specimens. The relationship of potential energy of a drop-weight projectile and the strain energy dissipated in cracks development is expressed as following formula as proposed by Kankam [7];

$$N*e = R_u * l_c * d_c * w_c \tag{3.1}$$

Where, N = No. of Blows, e = Energy per blow (Joules), lc = Total length of all cracks, d_c = Maximum crack depth, wc = Maximum crack width, Ru = Ultimate crack resistance

A total of 54 sample slabs of size 300mm x 300mm with 30 mm, 40 mm and 50 mm thickness were casted with at control (no fibre), 1%, 2 % and 3% steel fibre fraction volume (FV). Each of the combination have 3 samples in order increase its accuracy and an average value is obtained.

The following mesh distribution were investigated to study its impact on service (first) and ultimate (failure) crack resistance.

i. Random fibres distribution.
ii. Top and bottom fibres distribution.

The random fibres distribution are uniformly distributed throughout the depth of the slab. The top and bottom fibres distribution has the fibres distributed within 10mm depth at the top and bottom of the slab.

At the first crack and ultimate (failure) crack, the total crack length, the crack width and the crack depth measured by filler gauge with its total numbers of blows recorded.

4. Results and Discussion

4.1 Relationship between Crack Resistance and Steel Fibres Distribution

In Figure 2 and Figure 3 indicate that random fibre distribution has a better crack resistance values than the top and bottom fibre distribution as the volume fraction increases. Random fibre is generally more efficient arrangement in its ability to control crack formation and propagation against impact energy as compared with the top & bottom fibre distribution.

The fibre distribution arrangement is insignificant against crack resistance when the FV is below 3%. The significant fibre distribution arrangement between random and top & bottom can only be seen at 3% FV for the first and ultimate crack resistance.

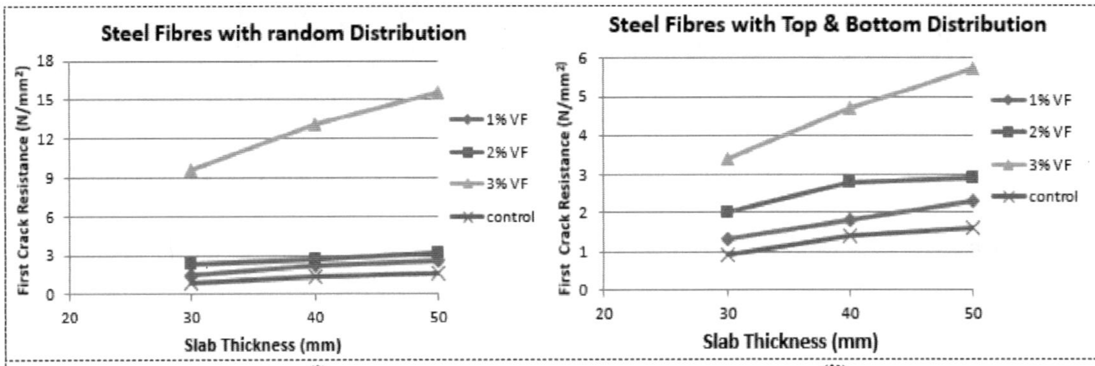

(i) (ii)

Figure 2 First Crack Resistance i. Random Fibres Distribution and ii. Top & Bottom Fibres Distribution

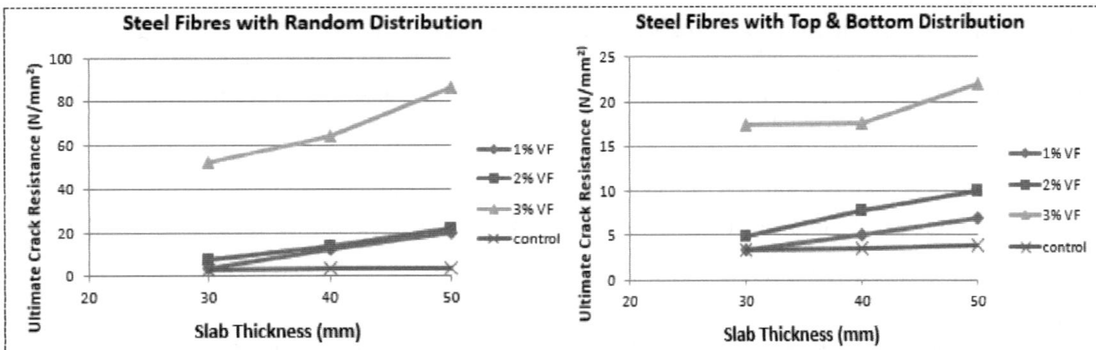

Figure 3 Ultimate Crack Resistance of i. Random Fibres Distribution and ii. Top & Bottom Fibres

The highest values of the first crack for random distribution is 15.5 N/mm^2 and the top & bottom fibre distribution is 5.7 N/mm^2. The first crack resistance for random distribution increases by 2.7 times as compare with the top & bottom fibre distribution.

The random distribution with 3% FV steel fibres has the better ability of transferring stresses across a cracked section uniformly, bridging and provide more resistance to the cracks formation and propagation, transferring impact loads, absorbing more impact energy and developing more micro cracks distribution in the slab. In top and bottom fibre distribution, the crack bridging is absence due to the missing fibres in the mid-depth of the slab, thus providing lower impact energy and resistance for the first and ultimate cracks.

4.2 Mode of Failures

Fibres play a very important role in bridging cracks, absorbing energy, transferring loads, and developing micro cracks distribution system. First crack of 0.075mm normally initiate at the bottom surface of the slab due to the limiting tensile strain and propagated upward until neutral axis. When the crack move up and reaches the top part of the concrete, the ultimate crack eventually cause a complete segmental failures.

If impact happen continuously, initial crack will subsequently propagate and link up into a larger crack known as a fracture zone. The random distribution in Figure 4 (a) has five segmental failures along the fracture zone with more micro-cracking in the other parts of the slab as compare with the top and bottom distribution in Figure 4 (b) has only three segmental failures along the fracture zone for ultimate crack of 3% FV fibers and 50 mm thick slab.

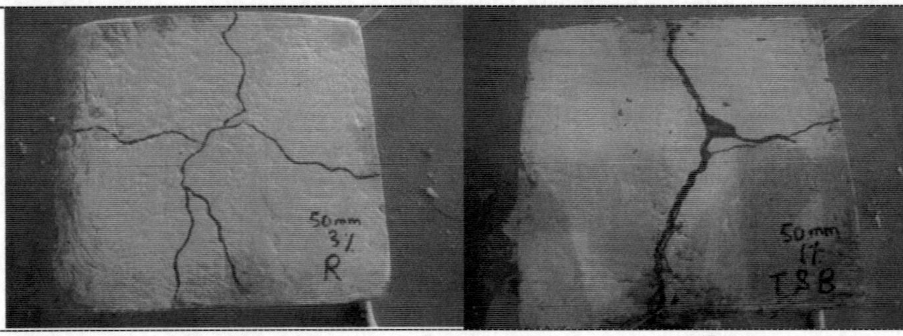

(a) **(b)**

Figure 4 Ultimate Crack Resistance Failure Modes for fibers with 3% FV and 50 mm thick slab. (a) Random Fibres Distribution (b) Top & Bottom Fibres Distribution.

5. Conclusion

The following conclusions can be derived from the experimental results;

- Random fibre is generally more efficient arrangement in its the ability to control crack formation and propagation against impact energy as compared with the top & bottom fibre distribution.
- The fibre distribution arrangement is insignificant against crack resistance when the fraction volume is below 3%.
- The first crack resistance for random distribution increases up to 2.7 times as compare with the top & bottom fibre distribution.
- In top and bottom fibre distribution, the crack bridging is absence due to the missing fibres in the mid-depth of the slab, thus providing lower impact energy and resistance for the first and ultimate cracks.
- The random distribution with higher fraction volume has more segmental failures along the fracture zone with more micro-cracking in the other parts of the slab as compare with the lower fraction volume.

References

[1] Holschemacher K, Muekker T, Ribakov Y. 'Effect of steel fibres on mechanical properties of high-strength concrete'. Mater Des 2010;31:2604-15.

[2] Xu Z, Hao H, Li HN. 'Experimental study of dynamic compressive properties of fibre reinforced concrete materials with different fibres'. Mater Des 2012;33:42-55.

[3] Ali Abd Elhakam Aliabdo, Abd Elmoaty Mohamed Abd Elmoaty, Mohamed Hamdy. 'Effect of internal short fibres, steel reinforcement, and surface layer on impact and penetration resistance of concrete'. Alexandria Engineering Journal. 2013;52, 407-417.

[4] ACI Committee 544. State-of-the-art report on fiber reinforced concrete. ACI Committee 544 report 544.1R-96. Detroit: American Concrete Institute. 1996.

[5] Zakaria Che Muda, Kong Sih Ying, Salah F A Sharif, Lariyah Bte. Mohd Sidek,Nawfal S. Farhan, A Review - Local Failure On Concrete Target Due To Projectile Impact. International Journal of Science and Engineering Research (IJSER) Volume 4, Issue 1, January 2013.

[6] Zakaria Che Muda, Salah F A Sharif, Lariyah Bte. Mohd Sidek, Nawfal S Farhan, Impact Resistance Of Oil Palm Shells Lightweight Concrete Slab With Bamboo Fibers, International Journal of Science and Engineering Research (IJSER) Volume 4, Issue 1, January 2013.

[7] CK Kankam . Impact Resistance of palm kernel fibre-reinforced concrete pavement slab. J Ferrocement 1999;29(4):279-86. Oct

IOP Conf. Series: Earth and Environmental Science **32** (2016) 012029 doi:10.1088/1755-1315/32/1/012029

The Effect of Mortar Grade and Thickness on the Impact Resistance of Ferrocement Slab

Zakaria Che Muda[1,3], Agusril Syamsir[1], Kamal Nasharuddin Mustapha[3], Sorefan Sulleman[2], Salmia Beddu [1], Sivadass Thiruchelvam[4], Firas B. Ismail[5], Fathoni Usman[1], Nur Liyana Mohd Kamal[1], Md Ashraful Alam[1], Ahmed H Birima[1], Zarina Itam[1], O S Zaroog[4]

[1]Centre of Sustainable Technology and Environment, Universiti Tenaga Nasional, Malaysia [2]Former Student of Universiti Tenaga Nasional [3]Centre of Forensic Engineering, Universiti Tenaga Nasional, Malaysia [4]Centre of Innovation and Design, Universiti Tenaga Nasional, Malaysia, [5]Centre of Power Generation, Universiti Tenaga Nasional, Malaysia.

mzakaria@uniten.edu.my

Abstract. This paper investigate the effect of the thickness and mesh spacing on the impact of ferrocement for the concrete slab of 300mm x 300mm size reinforced subjected to low impact projectile test. A self-fabricated drop-weight impact test rig with a steel ball weight of 1.236 kg drop at height of 150 mm, 350mm, and 500mm has been used in this research work. The objective of this research is to study the relationship of impact resistance of ferrocement against the mortar grade and slab thickness. There is a good linear correlation between impact resistance of ferrocement against the mortar grade and the thickness of ferrocement slab. The first and ultimate crack impact resistance of mortar grade 43 (for 40 mm thick slab with mesh reinforcement) are 1.60 times and 1.53 times respectively against the mortar grade 17 slab (of same thickness with mesh reinforcement). The first and ultimate crack impact resistance for 40 mm thick slab (mortar grade 43 with mesh reinforcement) are 3.55 times and 4.49 times respectively against the 20 mm thick slab (of same mortar grade with mesh reinforcement).

1. Introduction

In the search of sustainable green materials, it is critical to study the impact strength characteristics of such construction materials for various potential uses in the building industry. Important structures such as vessels, dams, military defense structure and power plant is very crucial to impact loads to prevent the serious lost in property, human life, and economic lost.

Ferrocement has a good ability to resist impacts and damage occur in a localized region, where it could be easily replaced hence maximizing obvious economic advantages of ferrocement. The positive characteristics of ferrocement upon impact are its resistance against disintegration, damage occurs in localized fashion and damage is easy to repair [1]. The review paper on impact resistance on concrete target has been published by Z Che Muda et al [2]. The combination of mortar and steel mesh results in the composite ferrocement element. Throughout literature, it can be noted that ferrocement has favourable characteristics in impact, as collisions between boats or with rocks are numerous [3]. Ferrocement has high strength, crack resistance, high ductility and energy absorption characteristics, all useful to be used in a dynamic environment and application media [4]. The drop weight impact test which is recommended by the ACI Committee 544 [5] is the simplest method. Impact resistance of oil palm shells lightweight concrete slab reinforced with geogrid has been studied by Z Che Muda et al. indicate an impact resistance improvement for first and ultimate crack up to 5.9 times and 20.1 times against the control sample without geogrid [6].

The objective of this research is to study the impact on first and ultimate crack resistance against the mortar grade and the thickness of ferrocement slab.

2. Materials and Test Set-up.

Ordinary Portland cement used in this work is complying to ASTM Type I cement. The square steel welded mesh as shown in Figure 1(a) has a 1 mm diameter with a spacing of 20 mm and the tensile strength of 450 MPa.

The properties of the fine aggregates used is in the cement mortar matrix was selected according to the ACI Committee 549, (1997). The sieve analysis of the sand shows that the different sand grain sizes are within the limit of a zone 2 of a well graded aggregate sizes with the fineness modulus was found to be 2.3.

The basic mix design for the mortar is as shown in Table 1.

Table 1. Mix Design for Mortar

Cement (Part by weight)	Fine Aggregates (Part by Weight)	Water/cement Ratio	Slump Class to BS8500 (Range)	Compressive Strength (N/mm2)
1	3	0.6	Class S3 (160-210 mm)	16.6
1	2	0.50	Class S3 (160-210 mm)	24.3
1	1	0.35	Class S4 (100-150 mm)	43.1

The study used a self-fabricated low velocity drop-weight impact test rig is shown in Figure 1 (b) using a steel ball weighing 1.236 kg with drop height of 150 mm, 350mm and 500mm impacting the ferrocement slab of size 300mm x 300mm with a thickness of 20 mm, 30 mm, and 40 mm with ferrocement welded mesh of 1mm diameter @ 20mm. The test sample is mounted on the steel rack frame with 1-way simply supported and a test up procedure is given in Figure 1(c).

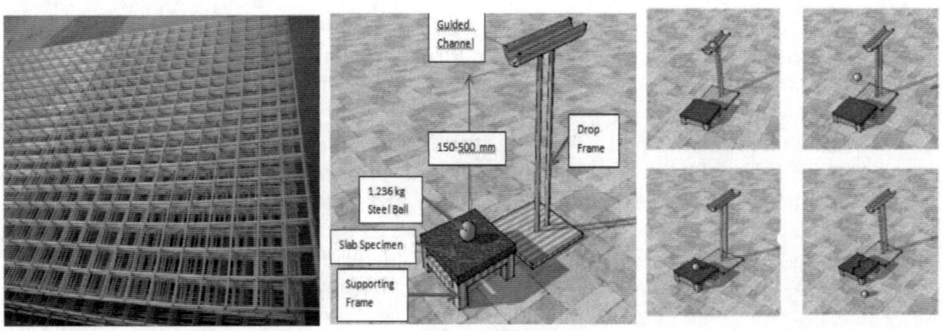

(a) (b) (c)

Figure 1.
(a) Square Wire Mesh with 1 mm diameter with a spacing of 20mm, 40mm and 50mm
(b) Low-velocity Drop-weight Impact Test Rig (c) Impact Test Set-up

3. Methodology

The potential energy due to the drop body is absorbed as strain energy, generating stresses that causes cracks in the target element. The width, depth, length of the crack developed and its failure mode is associated with the intensity of the energy, the amount of energy absorbed and the properties of concrete. It is assumed that the total computed energy imparted is fully absorbed by the specimens. The relationship of potential energy of a drop-weight projectile and the strain energy dissipated in cracks development is expressed as following formula as proposed by Kankam [7];

$$N*e = R_u * l_c * d_c * w_c \qquad (3.1)$$

Where, N = No. of Blows, e = Energy per blow (Joules), l_c = Total length of all cracks, d_c = Maximum crack depth, w_c = Maximum crack width, R_u = Ultimate crack resistance

A total of nine (9) sets for each slab of size 300mm x 300mm of mortar grade 16.6, 24.3 and 43.1 having 30 mm, 40 mm and 50 mm thickness with mesh reinforcement and nine (9) sets of control (no mesh reinforcement) for each grade and thickness were casted for 28 day curing strength. Each set of data have 3 samples in order increase its accuracy and its average value taken. The 1 mm diameter at 20 mm spacing wire mesh reinforcement is place at mid-depth of the slab.

A drop height of 150 mm is used for control (no reinforcement) or thin slabs, 350 mm drop is for medium slab thickness and 500 mm drop height is used for thicker slab. The drop was targeted at the centre of the slab and for each drop the bottom surface - the tension zone - was checked for cracks.

At the first and ultimate crack, the total crack length, the crack width and its depth are measured with its total numbers of blows recorded.

4. Results and Discussion

4.1 Crack Resistance for Control Samples

The first and ultimate crack resistance for the control sample are shown in Table 2. The crack resistance increases with the increase in mortar grade and thickness. The first and ultimate crack impact resistance of mortar grade 43 control sample are up to 4.93 times and 5.82 times respectively against mortar grade 17 control sample of the same thickness.

Table 2. First and Ultimate Crack Resistance of Control Sample (without Mesh Reinforcement)

Mortar Grade N/mm^2	16.6			24.3			43.1		
Thickness of Control Sample (mm)	20	30	40	20	30	40	20	30	40
First Crack Resistance N/mm^2	7.31	21.01	31.2	22.62	56.08	91.43	36.08	67.02	101.95
Ultimate Crack Resistance N/mm^2	10.84	22.15	34.81	24.10	45.06	68.28	63.09	100.91	150.25

4.2 Relationship between Crack Resistance and Mortar Grade

There is a good linear correlation for service and ultimate crack resistance against its thickness as shown Figure 2 with a minimum R^2=0.9886. As the mortar grade increase, the service and ultimate crack resistance increases proportionally. The ultimate crack resistance values are converging as the mortar grade increases for the different thickness of slab.

The highest value of first and ultimate crack resistance obtain is 192.77 N/mm^2 and 539.75 N/mm^2 respectively for the grade 40 mortar with 40 mm thick slab. The first and ultimate crack impact resistance of mortar grade 43 of 40 mm thick slab with mesh reinforcement are 1.60 times and 1.53 times respectively against the mortar grade 17 slab of same thickness with mesh reinforcement. The gain in the service crack resistance gain is very significant as the mortar grades increases for each slab thickness, however the ultimate crack resistance values are converging as the mortar grade reach 40 regardless of its thickness. The ultimate resistance values indicate no significant gain at higher mortar grade of 43 with mesh reinforcement.

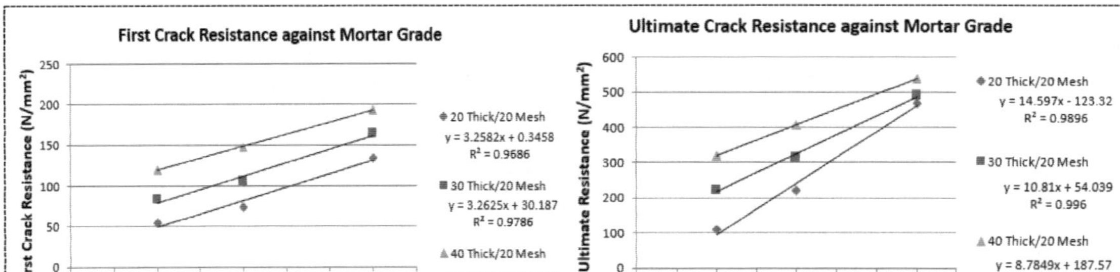

(a) First Crack (b) Ultimate Crack

Figure 2 First and Ultimate Crack Resistance Against Mortar Grade

4.3 Relationship between Crack Resistance and Slab Thickness

There is a good linear correlation for service and ultimate crack resistance against its thickness as shown Figure 3 with a minimum R^2=0.9495. The highest value of first crack and ultimate crack resistance obtain is 192.77 N/mm^2 and 539.75 N/mm^2 respectively for 40 mm thick slab with 20 mm mesh spacing. The first crack and ultimate crack impact resistance for 40 mm slab (mortar grade 43) with 20 mm mesh spacing are 3.55 times and 4.49 times respectively against the sample 20 mm slab (mortar grade 43) with the same mesh spacing. The first crack and ultimate crack impact resistance for 40 mm thick control sample for mortar grade 40 are 2.83 times and 2.38 times respectively against the control sample 20 mm thick slab of the same grade.

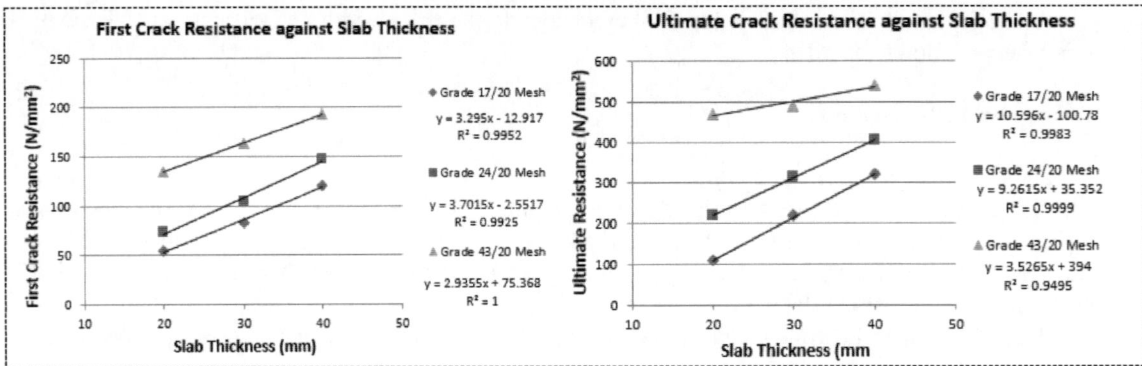

(a) First Crack (b) Ultimate Crack

Figure 3 First and Ultimate Crack Resistance Against Slab Thickness

5. Conclusion

The following conclusions can be derived from the experimental results;

- There is a good linear correlation with minimum R^2= 0.9886 between the crack resistance and mortar grade of the slab.
- There is a good linear correlation for service and ultimate crack resistance against the slab thickness with a minimum R^2=0.9495.
- The highest value of first crack and ultimate crack resistance obtain is 192.77 N/mm^2 and 539.75 N/mm^2 respectively for 40 mm thick slab with 20 mm mesh reinforcement spacing.
- The first and ultimate crack impact resistance of mortar grade 43 for control sample are up to 4.93 times and 5.82 times respectively against the mortar grade 17 control sample of the same thickness.
- The first and ultimate crack impact resistance of mortar grade 43 (for 40 mm thick slab with mesh reinforcement) are 1.60 times and 1.53 times respectively against the mortar grade 17 slab (of same thickness with mesh reinforcement).
- The first and ultimate crack impact resistance for 40 mm thick slab (mortar grade 43 with mesh reinforcement) are 3.55 times and 4.49 times respectively against the 20 mm thick slab (of same mortar grade with mesh reinforcement).

References

[1] Al-Rifaie, W.N. (2006). Ferrocement Wall: Penetration Testing, Proceedings of Eight International Symposium and Workshop on Ferrocement and Thin Reinforced cement Composites. 06-08 February, Bangkok Thailand, IFS, 177-185.

[2] Zakaria Che Muda, Kong Sih Ying, Salah F A Sharif, Lariyah Bte. Mohd Sidek,Nawfal S. Farhan, A Review - Local Failure On Concrete Target Due To Projectile Impact. International Journal of Science and Engineering Research (IJSER) Volume 4, Issue 1, January 2013.

[3] Iorns, M. E., and Watson, L. L., Jr. (1977), Ferrocement Boats Reinforced with Expanded Metal. Journal of Ferrocement (Bangkok), V. 7, No. 1, July 1977, pp. 9-16

[4] Yousry B.I. Shaheen, Noha Mohamed Soliman and Doha El Metwally Kandil (2013). Influence of Reinforced Ferrocement Concrete Plates under Impact Load. International Journal of Current Engineering and Technology, 21 October 2013, Vol.3, No.4

[5] ACI Committee 544. State-of-the-art report on fiber reinforced concrete. ACI Committee 544 report 544.1R-96. Detroit: American Concrete Institute. 1996.

[6] Z C Muda, G.Malik, S Beddu,M.A Alam K N Mustapha, A H Birima, OS Zarroq L M Sidek and M ARashid, Impact Resistance of Sustainable Construction Material Using Lightweight Oil Pal Shells Reinforced Geogrid Concrete Slab, IOP Conf. Series : Earth and Environmental Sciences 16(2013) 012062 doi:10.1088/1755-1315/16/1/012062

[7] CK Kankam . Impact Resistance of palm kernel fibre-reinforced concrete pavement slab. J Ferrocement 1999;29(4):279-86. Oct.

Impact of Sedimentation hazard at Jor Reservoir, Batang Padang Hydroelectric Scheme in Malaysia

Jansen Luis[1], Lariyah Mohd Sidek[1] and Milad Jajarmizadeh[1]

[1]Center for Sustainable Technology and Environment, Universiti Tenaga Nasional, Jalan IKRAM-UNITEN, 43000 Kajang, Selangor, Malaysia

E-mail: [1]jansenla@tnb.com.my

Abstract. Sedimentation in reservoir can be treated as a hazard because it affects the overall safety of the dam. It is a growing concern for reservoir operators throughout the world as it impacts the operability of the hydropower plant and its function as flood control. The objective of the study is to carry out reservoir bathymetric survey to determine the storage volume available at Jor reservoir. The paper intends to discuss the results of two successive surveys carried out in year 2007 and 2010 and comparison with historical data in1968 owing to analyse of sedimentation trend. The result showed that the total storage loss is approximately 43% with an estimated deposited sediment volume of 1.4 million m^3 in year 2010. The sedimentation rate is estimated at 3.3% for the years surveyed which is greater than the world average of 0.93%. The findings from the survey are used to develop a revised elevation-storage curve which could be used by the operator and engineers to carry out future power generation planning and flood study predictions. The findings are also expected to be used to determine the optimum method for sediment management and hydro-mechanical protection.

1. Introduction

Sedimentation in reservoir becomes a hazard to energy infrastructure such as hydropower reservoir and dams when it starts to impact both the functionality of reservoir and economic operability of the hydro power plants which includes dam safety constraints and limiting flood controls capabilities. The understanding on the contributing factors of sediments and ultimately their management can therefore extend the useful life and enhance functions of such infrastructures. Sedimentation processes within the reservoir are complex and dynamic in nature. When sediment deposits affect the reservoir useful life, it is necessary to change the reservoir operation or adopt any other corrective measure [1]. In most of the energy production country by large reservoirs, the sedimentation issues are not regarded as very important for production at short and medium term [2]. However, the average useful life of existing reservoirs in all countries of the world decreased from 100 to 22 years [3]. The annual cost for promoting the removal sediments is estimated in US$ 6 billion. The annual average volume loss in reservoirs due to sediment deposition was of 0.93% varying from one country to another, as well as from one region to another. The estimated loss of storage capacity in reservoirs in the United States due to sedimentation accounts for an annual monetary loss of US$ 100 million [4]. The sedimentation rates for 28 countries are shown in Figure 1 [5]. Meanwhile many other researchers have determined the annual sedimentation rates for the respective countries as shown in Table 1. World sedimentation overview found that mainly the drought affected countries will be confronted with reservoir sedimentation problems. It also predicted that semi-arid regions are also generally more susceptible for reservoir sedimentation problems as they often have relatively larger reservoir capacities and hence higher trapping efficiencies (TE) as seen in China and Tanzania with relative high sedimentation rates at 2.9% and 3.27% respectively. Many other countries in semi-arid regions such as Morocco, Pakistan, Sudan and Tunisia, have sedimentation rates above the world average. Malaysia is generally grouped within the Southeast Asia region where the sedimentation rate is observed at 0.3% [6]. The average annual reservoir sedimentation rate shown in table 2 is 0.003% which is lower than the estimates for Southeast Asia [7]. In this study, Jor reservoir is selected at Batang Padang Hydroelectric Scheme to evaluation of contribution of sedimentation. The two successive surveys performed in 2007 and 2010 and the result is discussed with historical data in 1968.

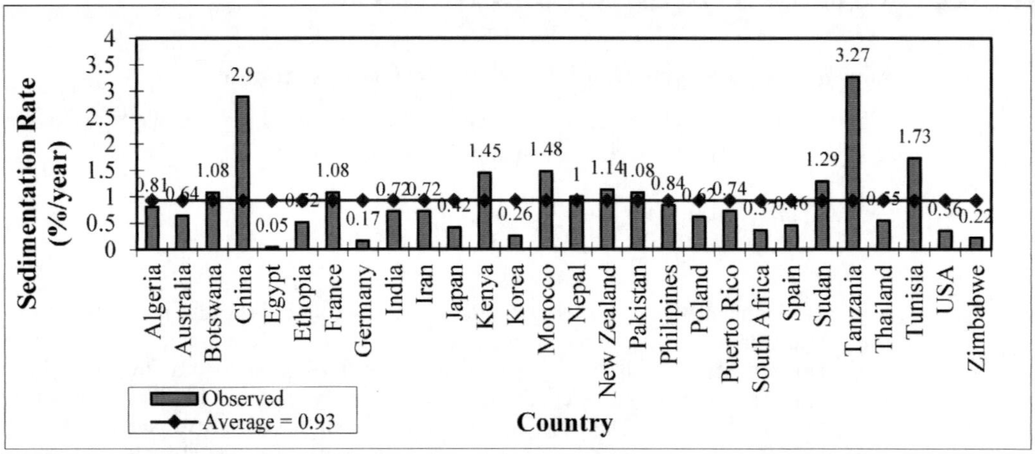

Figure 1. Reservoir sedimentation rates in various other countries [5]

Country	Annual Sedimentation Rate (%)	Source
China	2.30	Hu, 1995
India	0.46	White, 2001
Japan	0.15	NISA, 2001; ANRE, 1984-2000
Southeast Asia	0.30	White, 2001
South Africa	0.34	White, 2001
Turkey	1.50	White, 2001
UK	0.10	White, 2001
USA	0.22	Morris & Fan, 1987
World	1.00	White, 2001

Table 1. Reservoir sedimentation rates by various researchers

Table 2. Reservoir sedimentation rates in Malaysia

Name	Years in Operation	Storage (m^3)	Deposited Sediment Volume (m^3)	Annual Sedimentation Rate (%)
Pedas Reservoir (PWD)	1932 Surveyed 1988 (56 years)	49×10^3 13×10^3	36,000	0.05
Klang Gates Reservoir (PWD/JBA/State)	1959 Surveyed 1976 (17 years)	32×10^6 32.2×10^6	740,100	0.001
Bukit Merah Reservoir (DID)	1906 Surveyed 1965 – 1986 (21 years)	93×10^6 85.4×10^6	7.6×10^6	0.004

2. Material and methods

2.1. Case study

Jor reservoir was constructed to assure flow regulation for the downstream hydroelectric power station, as well as to provide storage for flood control [8]. The reservoir is confined by two earthfill dams namely Jor Dam and Jor Saddle Dam which forms part of the Batang Padang Hydroelectric

Scheme. The flow into Jor reservoir is contributed by the average natural inflows of Sg. Jor, Sg. Batang Padang and Sg. Sekam of 10 m^3/s and the tailrace discharge from Jor power station of 21 m^3/s [9]. Water from Jor Reservoir is fed through the Menglang intake together with other side stream intakes (Lengkok, Bot, Chenes and Tidong) to Sultan Idris II Power Station also known as Woh Power Station. The power station has a rated head of 406.8 m and has a total installed capacity of 150 MW. Jor reservoir covers a total area of 0.32 km^2 and has an original total storage volume of 3.17 million m^3 with approximately 2.33 million m^3 as live storage for hydropower power generation and 0.84 million m^3 as dead storage for sediment accumulations below the minimum operating level 480.06 m. The Batang Padang River was diverted through the diversion tunnel and spillway tunnel on October 1965 and the diversion tunnel was closed on July 1967 when the impounding of the reservoir commenced. The Batang Padang scheme was completed in 1968. Since commencing no detail study was conducted to determine the sedimentation rate and storage capacity. In order to estimate the reservoir sedimentation at Jor reservoir a bathymetric survey up to 3 m above the full supply level of the reservoir and combined with topographic survey to a boundary of 20 m from the reservoir banks was conducted.

3. Result and discussion

The two successive reservoir surveys conducted in 2007 and 2010 was compared to estimate the reservoir sedimentation and storage loss. It is estimated that in year 2007, the gross storage has reduced to 2.12 million m^3 (reduced by 33%) with a live storage at 1.95 million m^3 (reduced by 17%) and dead storage at 0.18 million m^3 (reduced by 79%). While the 2010 surveys show that both live and dead storage amounts have reduced further from year 2007 and the residual live and dead storage volumes at 2010 were about 1.77 (reduced by 24%) and 0.05 million m^3 (reduced by 94%) respectively. Overall the total storage of the reservoir surveyed in 2010 has further decreased by 43%. Table 3 shows the storage allocations between the two successive surveys. The elevation versus storage curve in figure 3 shows a leftward movement of the curve between the years surveyed which indicates a continued loss in storage. It is estimated that the annual sedimentation rate is at an alarming 3.3% which is greater than the world average of 1%.

Table 3. Storage allocation for Jor Reservoir

Storage Allocation	Original (1968) (m^3)	2007 Survey (m^3)	2007 Storage (%)	2010 Survey (m^3)	2010 Storage (%)
Dead Storage	840,000	178,418	21	49,941	6
Live Storage	2,330,000	1,945,306	83	1,769,971	76
Total Storage	3,170,000	2,123,724	67	1,819,912	57

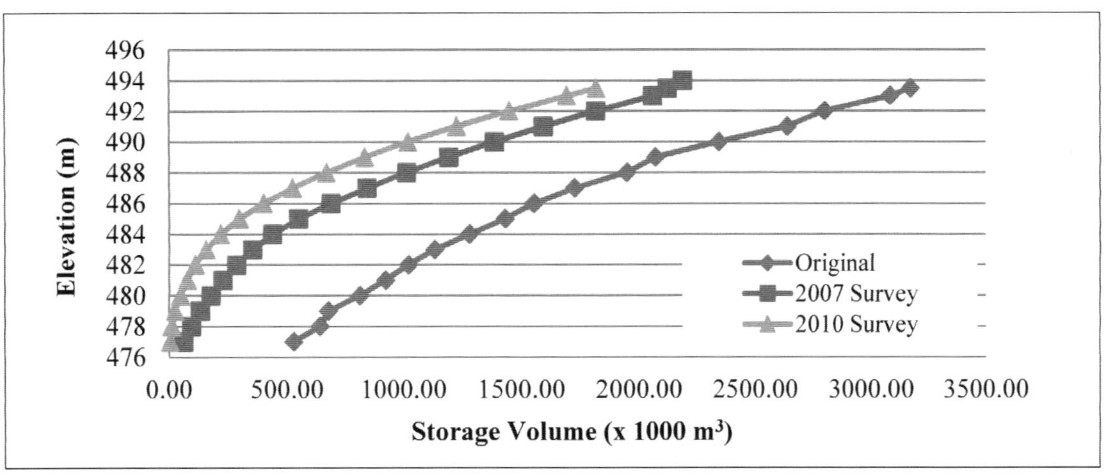

Figure 3. Elevation storage curve for Jor reservoir showing continued loss in storage

3.1 Impact of sedimentation hazard on reservoir

Sediment management in reservoir is a challenging area in the field of civil engineering especially in Malaysia where the legal framework is still not clear. The continuous deposition of sediments in the reservoir may pose a hazard to the overall safety of the dam. The significant impact contributed by sedimentation includes disruptions to the hydropower generation and flood control. Among the reported hydro-mechanical problems includes frequent choking of strainers or filters with sediments, choking and puncturing of cooler tubes, damage to guide vane bushes and their cap seals, damage to seals of intake valve and main inlet valve, seating/sealing problems in hydro-mechanical gates, intake as well as draft tubes and restriction of flows into intake gates and debris accumulation. The capability of the reservoir to contain the probable maximum floods may also be affected should the reservoir continues to loss it storage to sedimentation.

4. Conclusion

The study has shown that Jor reservoir has lost almost half its storage capacity to sedimentation. The data analysed in the study is used to develop a revised elevation-storage curve which is crucial for dam operators to plan for the future power generation and flood control. The revised data is also expected to be used for future dam safety study and flood studies. Since the sediments are deposited vastly throughout the reservoir, the data also enables the engineers to accurately determine the storage volume available at various elevations. Sedimentation in reservoir contributes to serious hazard to the overall safety of the dam especially in area of mountainous terrain. Periodic reservoir survey has shown that the deposited sediment volume has increased many folds since construction contributed mainly by the uncontrolled activities upstream. The operator of the dam is working tirelessly to address the issue by carrying out continued reservoir survey and immediate mitigation such as dredging and plant live extension works.

References

[1] ICOLD. 1989. *Sedimentation control of reservoirs.* International Committee of Large Dams.
[2] Almeida S B and Carvalho N O 1993 Efeitos do assoreamento de reservatórios na geração de energia elétrica: análise da UHE Mascarenhas, ES. *X Brazilian Symposium on Water Resources and I Symposium on Water Resources of the South Cone. Gramado, RS.*
[3] Mahmood K 1987 *World Bank Technical Paper.* Reservoir sedimentation: Impact, extent, and mitigation. No. 71.
[4] Julien P Y 1995 *Erosion and Sedimentation.* Cambridge University Press.
[5] Basson G 2008 Reservoir sedimentation – An overview of global sedimentation rates, sediment yield and sediment deposition prediction. *International Workshop on Erosion, Transport and Deposition of Sediment Berne, ICOLD* .
[6] White W R 2001 *Evacuation of Sediments from Reservoirs.* London, UK: Thomas Telford
[7] Long T Y 1992 An Overview of Reservoir Sedimentation in Malaysia. *Seminar on Methods for Preservation of Useful Reservoir Storage on Heavily Sediment Laden - Rivers.*
[8] LLN/TNB 1961 *Batang Padang Hydroelectric Scheme, Preliminary Report on Batang Padang Scheme.*
[9] Preece Cardew and Rider 1962 *Report on Batang Padang Hydro-electric Scheme*, Vol 2.

Conceptualization of a Collaborative Decision Making for Flood Disaster Management

Siti Nur Aishah Zubir[1], Sivadass Thiruchelvam[1], Kamal Nasharuddin Mustapha[2], Zakaria Che Muda[1], Azrul Ghazali[1], Hazlinda Hakimie[1], Normy Norfiza Abdul Razak[1], Abdul Aziz Mat Isa[1], Hasril Hasini[3], Khairul Salleh Mohamed Sahari[4], Norhayati Mat Husin[5], Mohd Ezanee Rusli[6], Rahsidi Sabri Muda[7], Lariyah Mohd Sidek[8], Hidayah Basri[8], Izawati Tukiman[9],

[1]Centre for Innovation and Design, Universiti Tenaga Nasional, Malaysia
[2]Centre for Forensic Engineering, Universiti Tenaga Nasional, Malaysia
[3]Centre for Fluid Dynamic, Universiti Tenaga Nasional, Malaysia
[4]Centre for Advance Mechatronic and Robotics,Universiti Tenaga Nasional, Malaysia
[5]Centre of Strategic Governance & Policies, Universiti Tenaga Nasional, Malaysia
[6]Centre for Information and Network Security, Universiti Tenaga Nasional Malaysia,
[7]TNB Research Sdn. Bhd, Malaysia [8]Centre for Sustainable Technology and Environment, Universiti Tenaga Nasional, Malaysia [9]International Islamic Univesity, Malaysia

nuraishah@uniten.edu.my

Abstract. Flooding is the utmost major natural hazard in Malaysia in terms of populations affected, frequency, area extent, flood duration and social economic damage. The recent flood devastation towards the end of 2014 witnessed almost 250,000 people being displaced from eight states in Peninsular Malaysia. The affected victims required evacuation within a short period of time to the designated evacuation centres. An effective and efficient flood disaster management would assure non-futile efforts for life-saving. Effective flood disaster management requires collective and cooperative emergency teamwork from various government agencies. Intergovernmental collaborations among government agencies at different levels have become part of flood disaster management due to the need for sharing resources and coordinating efforts. Collaborative decision making during disaster is an integral element in providing prompt and effective response for evacuating the victims.

1. Introduction

Major disaster such as flooding necessitates a coordinated diverse disciplinary (multi-agency) response to guarantee the required relief in the form of food, water, medical supplies and transportation arrives at the right place at the right time (Geale, 2012). Achieving optimal coordination and interoperability between agencies are heavily reliant on dynamic and distributive decision making (Smith & Dowell, 2000). Notwithstanding the pivotal role of collaborative decision making between agencies during multi-agency emergency responses, at present little work has been conducted to determine the factors and processes influencing decision makers during moments of high uncertainty, time pressure and high stakes.

The appropriate factors for emergency responses meant for evacuation of victims, the weight of importance the factors are considered, and how each factor is defined among the decision makers in the lead agencies during the recent flood catastrophe. By having a validated set of factors and their corresponding definitional dimensions, the lead agencies could apprehend the nature and characteristic of the real-time emergency decision making processes compared to the elements outlined in the current Standard Operating Procedures (SOPs). This will further enhance and reinforce the disaster planning and preparedness initiatives to be undertaken in facing any future calamities.

Developing a systematic process and procedure for decision making during disaster especially for evacuation of victims can provide the key stakeholders the information required to maximize the effort taken in search and rescue operations during flood disaster. With a clear perspective of what might transpire during the actual event, a list of proven and tested on the ground factors, definitional dimensions, and weight of importance, the key stakeholders could better anticipate and approach certain scenarios which were not comprehended before.

2. Current State of Decision Making During Disaster

Disasters are serious conditions or situations that disrupt the capacity of public and nations to effectively protect their populations and infrastructure, to reduce both human and property loss, and to rapidly recover ((Shaluf, 2007); (Altay & Green, 2006)). Disasters can be classified into three types: (i) natural; (ii) man-made; and (iii) hybrid. Natural disaster such as flooding has become a recurring phenomenal in most parts of the world. Flooding is the utmost major natural hazard in Malaysia in terms of populations affected, frequency, area extent, flood duration and social economic damage. The recent flood devastation towards the end of 2014 witnessed almost 250,000 people being displaced from eight states in Peninsular Malaysia. Having 189 river basins throughout Malaysia, including Sabah and Sarawak, the rivers and their corridors of flood plains fulfil a variety of functions both for human use and for the natural ecosystem (Sukereman, Sulaiman, & Hussin, 2013). This increases the possibility of threat to entire corridor areas. Therefore, an effective and efficient response to disaster should be in place by adopting a well-structured disaster planning strategies.

Disasters like flood will continue hitting our communities, businesses, and economies. How bad it could be? In his recent announcement addressing changes in the global economic landscape, Honourable Prime Minister, Dato' Seri Najib Razak has endorsed almost RM 2.9 billion for rebuilding infrastructure damaged by the recent flood as part of the overarching strategy to strengthen the economic resilience of Malaysia ("Special address on current developments and government's financial position, 20 January 2015, Putrajaya.," 2015). It is in everyone's interest to understand how we can manage disaster effectively and efficiently. Better management of disaster operations will improve readiness, increase response speed, and ease of recovery. Hence, decision making during disaster could be considered as of high importance. The decision making process during a disaster response contrasts considerably from conventional decision making (Dai, Wang, & Yang, 1994). During the state of a disaster, important attributes of the faced problem are uncertain in terms of its nature, scale and time. We must acknowledge that there is very little time for making a decision but information might not be available. This is further worsening in some situations as there might be abundance of information which might not be reliable.

In the event of a flood disaster outbreak, evacuation is considered as a way to prepare people when at risk from an impending hazard ((Taylor & Freeman, 2010); (Murray-Tuite, 2007)). Evacuation is a process with a very long history as in the early fifth century B.C., the Greek Historian Herodotus described Egyptian evacuation in facing seasonal flooding of Nile River (Perry, 1975). The purpose of evacuation encompasses the action of alerting, warning, preparing, moving and as necessary temporarily holding people, animals, personal belongings and supplies from an actual danger to a place which is relatively safer ((*Evacuation and shelter guidance: Non-statutory guidance to complement Emergency preparedness and Emergency response and recovery*, 2013); (*Evacuation Planning*, 2005)). An evacuation should only be carried out if the benefit of leaving an area significantly outweighs the risk of sheltering in place as it is imperative to realize that this process could be traumatic to vulnerable people and it could hamper business and the local economy (("The effect of flooding on mental health," 2011):(Skertchly & Skertchly, 2000);(Ketteridge & Fordham, 1998)). Collaborative decision making by intergovernmental agencies could address on how to reduce the impact of evacuation to the victims

3. Collaborative Decision Making During Disaster

Multi-agency collaboration could be viewed as joint activity by two or more agencies to increase the delivery value for public by working together rather than individually (Bardach, 1998). For flood disaster in Malaysia, among the lead agencies are the District Office, Fire and Rescue Department, Royal Malaysian

Police, Social Welfare Department, Drainage and Irrigation Department, Health Department and Public Works Department. National Security Council (Majlis Keselamatan Negara) in the Prime Minister's Department is responsible as the coordinator for all these agencies. The responsibilities and functions of the above mentioned agencies are dictated in the National Security Council (NSC) Directive 20 (Malaysia, 1997). In disaster management, collaboration encompasses key elements such as coordination, communication formation of network, partnerships, and interoperability (Naim Kapucu, Arslan, & Demiroz, 2010). The underlying link for all these elements is effective and efficient decision making (Cosgrave, 1996), as poor decisions would lead to poor disaster management which can have very serious consequences. Major disasters such as flooding necessitates a coordinated diverse disciplinary (multi-agency) response to guarantee the required relief in the form of food, water, medical supplies and transportation arrives at the right place at the right time (Geale, 2012). (Katuk, Ku-Mahamud, Norwawi, & Deris, 2009) contended that collaborative decision making during disaster management is of great importance. Effective management of such disaster operations requires collaborative and streamlined effort from various emergency agencies (Subramaniam, Ali, & Shamsudin, 2010). The combination of efforts with efficient use of resources needs collaboration among all stakeholders. Achieving optimal coordination and interoperability between agencies are heavily reliant on dynamic and distributive decision making (Smith & Dowell, 2000).

In the development and assessment of policy for disaster mitigation and response, human behavioural factors are often the least well quantified, understood, and modelled (Carlson et al., 2014). Although comprehensive disaster management follows an all-hazard approach, generalizing policies and plans for all kinds of emergencies, it is interesting to note that we may find different optimal approaches for different specific incidents ((Altay & Green, 2006);(Cosgrave, 1996)). Therefore, we could reach an initial understanding that evacuation process for flood related disaster would be unique compared to other evacuations. As different government agencies have different operational approaches, during disaster, collaborative efforts could be of a vital issue ((Chen, Sharman, Rao, & Upadhyaya, 2008); (McEntire, 1999);(Frishammar, 2003)). Multi-agency working has been recognized as much more difficult to achieve as single-agency strategies due to the different aims and agendas of partners, communications challenges due to organizational differences, more complex accountability issues, inequalities of power and struggles for dominance, and legacy issues (Ranade & W., 1998). Smith & Dowell, (2000)reinforced this notion mentioning each agency diligently pursued its own duty but there was lack of liaison between them. Perception of collaboration during disasters may vary significantly due to differences in organizational goals, objectives, and cultures (Naim Kapucu et al., 2010). However, (Quarantelli) construed that informing other agencies of one's operations could be perceived in two ways: (i) as normal information exercise; and (ii) mutual agreement on which agency is going to perform effectively. Drawing on research into decision making, effective and efficient disaster management operations require the ability of decision making under pressure which is subject to a number of pitfalls: (i) unique situation; (ii) data deficit; (iii) emotional denial; (iv) gambling on probabilities; and (v) positive reinforcement ((Shapiro, 2010); (*The politics of crisis management*, 2005)). Thus, successful disaster operations such as evacuations must be able to leverage the existence of all lead agencies on the ground via collaborative decision making.

4. The Way Forward: Successful Joint Operation During Disaster

Several types of decision making are required in disaster management such as analytical, naturalistic, procedurally based, creative, and distributive decision making (Sinclair, Doyle, Johnston, & Paton, 2012). It is evident that effective and efficient measures during disaster depend on sharing and using information between multi-agencies to make critical decisions under uncertainty (N. Kapucu, 2006). Decision making process during this state is more likely to be a challenging issue due to involvement of great uncertainty, sudden and unexpected events, risk of possible mass casualty, great time pressure and urgency, severe resource shortage, large scale impact and damage, and the disruption of infrastructure support necessary for coordination (Chen et al., 2008). This is further complicated by factors such multi-authority, personal involvement, and conflict of interest (Chen et al., 2008).

These shortcomings provide the basis and motivation of this undertaken study, which represents the initial steps in development of a comprehensive, predictive framework that incorporates human factors in decision making for disaster mitigation and response. One can argue that collaborations are not appropriate in situations where rapid decision making process are needed like disasters. Nevertheless, the culture of working together established through social relations can substantively increase the speed of decision making (Ansell & Gash, 2007). Grounding on Resource Dependence Theory (RDT), this paper will fathom the exchanges of information between organizations. Resources are a driving force in the relationships among organizations, and resource dependence is theorized to shape the nature of relationships among organizations, having both positive and negative connotations (Zakour & Gillespie, 2013). Developing a systematic process and procedure for collaborative decision making during disaster especially for evacuation of victims can provide the key stakeholders the information required to maximize the effort taken in search and rescue operations during flood disaster.

Each year, natural disaster results in significant loss of life, the destruction of homes and public infrastructure, and economic hardship. Disasters strike both rich and poor indiscriminately. Vulnerable population such as the physically challenged, elderly, non-native speaking, children, chronically ill, mentally ill, the impoverished and geographically or culturally isolated, often suffer worse effects from disasters than others. Disaster planning seeks to minimize the consequences resulting from those events. Developing a decision making model is resource intensive and requires the planning and regular participation of key members of a management team and regular practice exercises. A well-developed model includes pre-disaster and post disaster actions and evacuation processes that minimize the short term and long term impacts of natural disaster (Vicki, 2007).

5. Conclusion

The collaborative decision making practice in the Malaysian context of disaster management especially in the area of evacuation for flood disaster encourage broader community of government agencies to improve their decision making process as part of Disaster Risk Reduction (DRR) strategy to face any future calamities. Recommendations will be offered to guide and assist decision makers to achieve successful evacuation decision making without leaving much impact on the flood victims. It contributes mainly in the development of an intergovernmental collaborative decision making model that (i) acts as a reference to better understand the dynamics of decision making during a disaster, and (ii) act as a basis to develop future emergency planning and preparedness exercise for lead agencies.

References

Altay, N., & Green, W. G. (2006). OR/MS research in disaster operations management. *European Journal of Operational Research, 175*(1), 475-493.

Ansell, C., & Gash, A. (2007). Collaborative Governance in Theory and Practice. *Journal of Public Administration Research and Theory, 18*(4), 543-571.

Bardach, E. (1998). *Getting agencies to work together: the practice and theory of managerial craftsmanship.* Washington, D.C: Brookings Institution Press.

Carlson, J. M., Alderson, D. L., Stromberg, S. P., Bassett, D. S., Craparo, E. M., Guiterrez-Villarreal, F., & Otani, T. (2014). Measuring and Modeling Behavioral Decision Dynamics in Collective Evacuation. *PLoS ONE, 9*(2), e87380.

Chen, R., Sharman, R., Rao, H. R., & Upadhyaya, S. J. (2008). Coordination in emergency response management. *Communications of the ACM, 51*(5), 66-73.

Cosgrave, J. (1996). Decision making in emergencies. *Disaster Prevention and Management: An International Journal, 5*(4), 28-35.

Dai, J., Wang, S., & Yang, X. (1994). Computerized support systems for emergency decision making. *Annals of Operations Research, 51*(7), 313-325.

The effect of flooding on mental health. (2011).

. *Evacuation and shelter guidance: Non-statutory guidance to complement Emergency preparedness and Emergency response and recovery*. (2013). Crown.

. *Evacuation Planning*. (2005). Emergency Management Australia (EMA).

Frishammar, J. (2003). Information use in strategic decision making. *Management Decision, 41*(4), 318-326.

Geale, S. K. (2012). The ethics of disaster management. *Disaster Prevention and Management, 21*(4), 445-462.

Kapucu, N. (2006). Interagency Communication Networks During Emergencies: Boundary Spanners in Multiagency Coordination. *The American Review of Public Administration, 36*(2), 207-225.

Kapucu, N., Arslan, T., & Demiroz, F. (2010). Collaborative emergency management and national emergency management network. *Disaster Prevention and Management, 19*(4), 452-468.

Katuk, N., Ku-Mahamud, K. R., Norwawi, N., & Deris, S. (2009). Web-based support system for flood response operation in Malaysia. *Disaster Prevention and Management, 18*(3), 327-337.

Ketteridge, A.-M., & Fordham, M. (1998). Flood evacuation in two communities in Scotland: Lessons from European Res. *International Journal of Mass Emergencies and Disasters, 16*(2), 119-143.

Malaysia, N. S. C. o. (1997). *Directive No. 20: Policy mechanism of national disaster management and relief.* .

McEntire, D. A. (1999). Issues in disaster relief: progress, perpetual problems and prospective solutions. *Disaster Prevention and Management, 8*(5), 351-361.

Murray-Tuite, P. (2007). Perspectives for Network Management in Response to Unplanned Disruptions. *Journal of Urban Planning and Development, 133*(1), 9-17. Perry, R. W. (1975). Evacuation decision making in natural disaster. *Mass Emergencies, 4*, 25-38.

The politics of crisis management: public leadership under pressure. (2005). Cambridge, UK ; New York: Cambridge University Press.

Quarantelli, E. L. (May 27-28). *Research based criteria for evaluating disaster and managing.* Paper presented at the Internatial Seminar on "Chernobyl and Beyond: Humanitarian Assistance to Victims of Technological Disasters" Moscow, Russia.

Ranade, & W. (1998). Making sense of multi-agency groups.

Shaluf, I. M. (2007). Disaster types. *Disaster Prevention and Management, 16*(5), 704-717.

Shapiro, S. (2010). Decision making under pressure. . *THE FUTURIST*, 42-44.

Sinclair, H., Doyle, E. E. H., Johnston, D. M., & Paton, D. (2012). Decision-making training in local government emergency management. *International Journal of Emergency Services, 1*(2), 159-174.

Skertchly, A., & Skertchly, K. (2000). Human aspects of the Katherine-Daly flood disaster. *Australian Journal of Emergency Management, 15*(1), 47-52.

Smith, W., & Dowell, J. (2000). A case study of co-ordinative decision-making in disaster management. *Ergonomics, 43*(8), 1153-1166.

Special address on current developments and government's financial position, 20 January 2015, Putrajaya. (2015, 20 January 2015).

Subramaniam, C., Ali, H., & Shamsudin, F. M. (2010). Understanding the antecedents of emergency response: a proposed framework. *Disaster Prevention and Management, 19*(5), 571-581.

Sukereman, A. S. b., Sulaiman, R. b., & Hussin, K. b. (2013). The issues towards sustainability of water resource management and the current governance practice in managing water crisis in Malaysia. *Proceeding of the International Conference on Social Science Research, ICSSR*, 301-319.

Taylor, M. A. P., & Freeman, S. K. (2010). A review of planning and operational models used for emergency evacuation situations in Australia. *Procedia Engineering, 3*, 3-14.

Vicki, E. H. (2007). *Critical Components of Natural Disaster Planning.* Argosy University.

Zakour, M. J., & Gillespie, D. F. (2013). *Community Disaster Vulnerability: Theory, Research, and Practice*. New York: Springer.

Formulation of an Integrated Community Based Disaster Management for Hydroelectric facilities: The Malaysia Case

Norshamirra Hijazzi[1] Sivadass Thiruchelvam[1] Rahsidi Sabri Muda[2] Kamal Nasharuddin Mustapha[1] Zakaria Che Muda[1] Azrul Ghazali[1] Ahmad Kamal Kadir[1] Hazlinda Hakimie[1] Khairul Salleh Mohamed Sahari[1] Hasril Hasini[1] Lariyah Mohd Sidek[1] Zarina Itam[1] Mohd Fadhli Mohamad[2] Azwin Zailti Abdul Razad[2]

[1]Centre of Innovation and Design, Universiti Tenaga Nasional, Malaysia
[2] Environmental Unit, TNB Research, Malaysia

E-mail: shamirra@uniten.edu.my

Abstract. Dams, however significant their contributions are to the society, are not immune to failures and diminishing lifespan not unlike other structural elements in our infrastructure. Despite continuing efforts on design, construction, operation, and maintenance of dams to improve the safety of the dams, the possibility of unforeseen events of dam failures is still possible. Seeing that dams are usually integrated into close approximities with the community, dam failures may consequent in tremendous loss of lives and properties. The aims of formulation of Integrated Community Based Disaster Management (ICBDM) is to simulate evacuation modelling and emergency planning in order to minimize loss of life and property damages in the event of a dam-related disaster. To achieve the aim above, five main pillars have been identified for the formulation of ICBDM. A series of well-defined program inclusive of hydrological 2-D modelling, life safety modelling, community based EWS and CBTAP will be conducted. Finally, multiple parties' engagement is to be carried out in the form of table top exercise to measure the readiness of emergency plans and response capabilities of key players during the state of a crisis.

1. Introduction

Dams have provided tremendous benefit to the development of civilization, and continue to cater for our improving standard of living. The earliest known dams were primarily used for irrigation and water supply and thereafter expanded to be used for other significant purposes such as flood control, hydropower and recreation. Throughout history, dams have apparently performed well overall. However, dams have periodically failed or suffered incidents throughout history as they are not expected to last forever. This has necessitated for dams to have regular inspections as they hold a potential risk of failure due to many technical and safety problems. Dam failures are regarded as one of the major "small probability, large consequence" events. The floods resulting from dam failures can lead to devastating disasters with tremendous loss of life and property, especially in densely populated areas. Loss of human life is generally accepted as the most important consequence; therefore it often dominates dam safety decisions (McClelland & Bowles, 2002).

In Malaysia, there are almost 51 recorded dams (60% of the dams are of earthfill type) under different ownerships (Othman, 2006). To-date we have not had the experience of any dam-related failures. However, the occurrence of a deadly tragedy on the 23rd October 2013 at Bertam Valley, Cameron Highlands has been treated as a wakeup call for all parties that prior to any dam failure (see Figure 1), there could be other possible scenarios which could claim lives in the vicinity of any dams (e.g. normal riverine flooding).

Figure 1: Aftermath of Bertam Valley incident

It should be realized that there is a dynamic interaction between people, vehicles, buildings and flood wave. Therefore, flooding from dam related disasters usually cause huge loss of lives and destructions of properties and environment. For example, the 1963 failure of Vajont dam in Italy caused 2600 deaths, the 1976 failure of Teton dam in America caused hundred deaths and economic loss about 1 billion dollars, and the 1993 failure of Gouhou dam in China caused 300 deaths (You, Li, Min, & Xiaolei, 2012). The statistical analysis of 534 dam failures from 43 countries before 1974 indicated that earth-rock dam failures accounted for the largest proportion of all failures and included 49% caused by overtopping, 28% seepage in dam body and 29% seepage in foundation (ICOLD, 1998). Dam failures may cause tremendous loss of life and property, and hence professionals experienced in the field of dam engineering have paid continuing efforts on design, construction, operation, and maintenance of dams to improve the safety of dams. However, many existing dams still pose increasing hazards to the downstream areas due to structural deterioration, inadequate design, faulty construction, and poor operation and maintenance. These dams are referred to as distressed dams. Looking from the stand-point of Malaysia, 20% of its dams are in the range which problems often begin to manifest (Othman, 2006). Othman (2006) also highlighted that it is also worth noting that the safety review of the oldest dam, Bukit Merah Dam (1906) was conducted in 1998 and the review for Sungai Perak Hydroelectric Scheme comprising the Temengor (1978) and Chenderoh (1930) dams was carried out in May 2003. Tenaga Nasional Berhad (TNB) as the major key player in the Malaysia Electricity Supply Industry has conducted its latest review on Kenyir Hydroelectric Scheme in 2006.

2. The Current State of Emergency Preparedness in Malaysia

The Chief Government Security Office (CGSO) Malaysia which is an important arm under Prime Minister's Office of Malaysia, has asserted that dams are considered as one of the 15 national strategic targets (sasaran penting negara) (CGSO, 2013). Therefore, it is essential to strengthen national preparedness, timely response, and rapid recovery of this critical infrastructure in the event of an attack, natural disaster, or other emergency. Potential risks associated with the failure or disruption of these assets could be considerable and potentially result in significant destruction, including loss of life (LOL), massive property damage, and severe long-term consequences. Therefore, consistent consequence estimation approaches must be achieved to identify those assets within this critical infrastructure sector whose failure or disruption could potentially lead to the most severe impacts. At a minimum, consequence assessment should focus on the most fundamental impact; those corresponding to human consequences. The exposure of people to the physical effects of a dam failure can result in loss of life. Consistent estimation of consequence variables constitutes a problem of paramount importance. Common definitions, scenarios, assumptions, and metrics are needed to ensure human risk assessment

IOP Conf. Series: Earth and Environmental Science **32** (2016) 012032 doi:10.1088/1755-1315/32/1/012032

efforts can effectively contribute to a shared understanding of the dam risk profile among key stakeholders associated to dam operations in Malaysia. With this information in-hand, authorities could best plan the evacuation related procedures and logistics.

The failure of Banqiao and Shimantan Dam in Henan in China in 1975 was evaluated as the world's most catastrophic dam failures, the number of deaths directly caused by the dam-break flood was 26,000; 5.4 million people were impacted (Graham, 1999). Loss of life due to dam break failure is highly influenced by three factors: (1) population occupying the flood plain downstream the dam; (2) the amount of warning time given to the people; and (3) the severity and magnitude of the flooding (Graham, 1999). In addition, nearly 100 quantitative or categorical variables can be considered to affect the number of flood fatalities due to a dam break (McClelland & Bowles, 2002). Loss of life resulting from dam failure, as measured by fatality rates or total number of fatalities, has generally been higher for populations near the dam than for those located further downstream. The recent incident at Sultan Abu Bakar (SAB) Dam has proven that loss of life could also be contributed by dam associated flash floods (riverine flooding) which are obviously not rare compared to dam failure flooding. Dam failures and associated flash floods can result in high fatality rates, especially when flooding overwhelms an unsuspecting group of people. However, dam failures that produce slowly rising floods tend to result in lower fatality rates. The need for mitigation of human risks due to dam related disasters raises questions about the proper estimation of potential loss of lives. We need to engage with the people who live in areas below dams and who would be affected by dam-related flooding. Dam owner must be able to attain the ability of making decisions during the time of crisis in the aspects of warning and assisting evacuation in relation to dam related flooding. The time available to evacuate is an important mediating factor in the dam-failure context, the techniques to be used to educate and warn community is a major component to be addressed.

Evacuation is the process between the start of leaving the risky area and the arrival at a safe place. Evacuation is not solely about ordering people to move from one place to another. It is imperative to persuade people to move, which relates to warning processes, actual movement process and its management. A vital element in-lieu to this effort would be effective utilization of time. Careful consideration of the time variable is necessary to ensure that the evacuation of all those at risk can actually be affected. Notwithstanding, prior community engagement is fundamental to the success of any warning and evacuation strategy. This is reflected by the nature of warning and evacuating people were never simple and would be of least difficult if the people have a sound prior understanding of the threat which requires them to evacuate. Hence, community must be educated about the threat, about its management by dam owners and emergency responders, and about the actions which people can take to respond in their own interests. Apart from that, it remains vital that the efforts of dam owners and emergency responders are properly integrated so that the two sides of the management of potential dam related disasters are fully incorporated.

Prior to this, there have been several studies being undertaken by TNB on dam break of its major hydroelectric schemes. For instance, in 2003, Sungai Perak hydroelectric scheme has been undertaken whereby the emergency action plan (EAP) was derived using USBR format. However, beginning from 2008, there has been a change in adopting HTC format for formulating EAPs. In 2008, Kenyir hydroelectric scheme has been studied and followed by integration of HTC and USBR format for ERP formulation specifically for Sg. Perak and Kenyir facilities in 2011. This chronology continued with dam break study for Cameron Highlands hydroelectric scheme in 2012. The recent study was conducted in the first quarter of 2014 for Pergau Dam. The nature of these studies was more focused towards the types of dam breaches and characteristics of flooding, such as flood arrival time (FAT), depth and flood velocity. However, loss of lives, establishment of safe haven, and engagement with the public was not looked into with a greater emphasis. The perspective towards these elements has changed after the Bertam Valley incident.

IOP Conf. Series: Earth and Environmental Science **32** (2016) 012032 doi:10.1088/1755-1315/32/1/012032

3. The Way Forward: Enhancing Community Ownership Towards Disaster Management

Five essential criteria have been identified for formulation of ICBDM. The formulation of ICBDM stands upon five main pillars which support the overarching aim of this study (see figure 2). A 2-D flood modelling will be conducted in the initial phase of this project to determine the impact of dam releases on riverine flooding as an input for the Life Safety Simulator (LSS) under various failure modes. Detailed knowledge on the characteristics of the flooding is then used as an input parameter to conduct a numerical analysis for loss of life using the Life Safety Simulator (LSS), which should be able to model the casualties that may occur due to the flooding. This Life Safety and Evacuation Model is capable of allowing dynamic interaction between people, vehicles, and buildings with the flood wave. Based on the findings from the analysis, a Community Based Early Warning System (CBEWS) can be constructed. From there, a feasibility study will be conducted on the early warning system in terms of efficiency in generating and disseminating timely and meaningful warning information to community members, communities, and organizations so that necessary preparedness measures and action could be taken in sufficient time to reduce the possibility of harms or losses. With the information gathered from the feasibility study, a Community Based Training and Awareness Program (CBTAP), which is a community awareness program, will be designed to educate people about the threat and about its management by dam owners and emergency responders, and about the actions which people could take to respond in their own interests. Finally, a Multi-Stakeholders Engagement Program (MSEP) will be carried out which will include multiple parties' engagement in the form of table-top exercise to measure the readiness of emergency plans and response capabilities of key players during the state of a crisis. Once all these procedures have been implemented, all the stakeholders, including the public, should be well educated and prepared with the dam disaster evacuation response system.

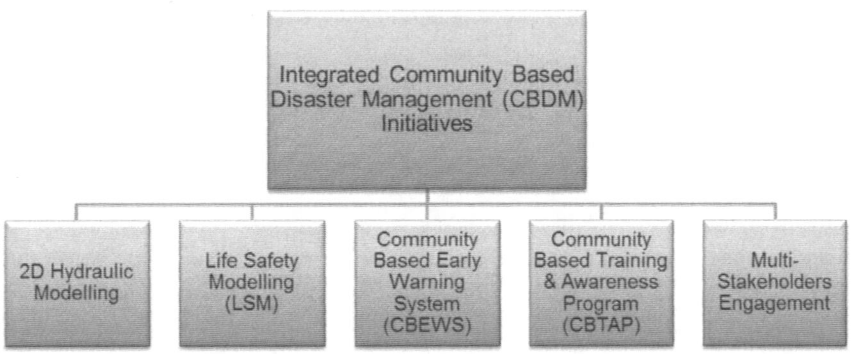

Figure 2: Proposed Integrated Community Based Disaster Management (ICBDM) Scopes

4. Conclusion

In line with current global awareness of water security, the aspect of dam safety has drawn increasing attention from the public as it constitutes the element of a country's national security. This is because the aftermath floods resulting from breaching of dams can lead to devastating disasters with tremendous loss of life and property, especially in densely populated areas. Previous disasters exhibited that the repercussions of dam failures are directly related to the evacuation time available should failures occur. It is therefore very important to study dam failures and their corresponding breaching characteristics, and if possible, to estimate the possible fatalities (loss of life).

References

[1] McClelland, D., & Bowles, D. (2002). Estimating life loss for dam safety risk assessment–a review and new approach. *Institute for Water Resources, US Army Corps of Engineers, Alexandria, VA.*

[2] Othman, Z. A. (2006). Overview of dam safety in Malaysia.

[3] You, L., Li, C., Min, X., & Xiaolei, T. (2012). Review of dam-break research of earth-rock dam combining with dam safety management. *Procedia Engineering, 28*, 382-388.

[4] ICOLD. (1998). Dam-break flood analysis. *Bulletin 111*.

[5] CGSO, Chief Goverment Security Officer (2013). Background. Retrieved 30 January, 2013, from http://www.cgso.gov.my/~cgso/portal/index.php/en/faq.html

[6] Graham, W. J. (1999). *A procedure for estimating loss of life caused by dam failure*: US Department of the Interior, Bureau of Reclamation, Dam Safety Office.

Encapsulation of a Decision-Making Model to Optimize Supplier Selection via Structural Equation Modeling (SEM)

Ruzanna Sahul Hameed[1], Sivadass Thiruchelvam[1], Kamal Nasharuddin Mustapha[2], Zakaria Che Muda[3], Norhayati Mat Husin[4], Mohd Ezanee Rusli[5], Lee Choon Yong[1], Azrul Ghazali[1], Zarina Itam[1], Hazlinda Hakimie[1], Salmia Beddu[1], Nur Liyana Mohd Kamal[1]

[1]Centre of Innovation and Design, Universiti Tenaga Nasional, Malaysia
[2]Centre of Forensic Engineering, Universiti Tenaga Nasional, Malaysia
[3]Centre of Sustainable Technology and Environment, Universiti Tenaga Nasional, Malaysia
[4]Centre of Strategic Governance & Policies, Universiti Tenaga Nasional, Malaysia
[5]Centre of Information & Network Security, Universiti Tenaga Nasional, Malaysia

Ruzanna@uniten.edu.my

Abstract. This paper proposes a conceptual framework to compare criteria/factor that influence the supplier selection. A mixed methods approach comprising qualitative and quantitative survey will be used. The study intend to identify and define the metrics that key stakeholders at Public Works Department (PWD) believed should be used for supplier. The outcomes would foresee the possible initiatives to bring procurement in PWD to a strategic level. The results will provide a deeper understanding of drivers for supplier's selection in the construction industry. The obtained output will benefit many parties involved in the supplier selection decision-making. The findings provides useful information and greater understanding of the perceptions that PWD executives hold regarding supplier selection and the extent to which these perceptions are consistent with findings from prior studies. The findings from this paper can be utilized as input for policy makers to outline any changes in the current procurement code of practice in order to enhance the degree of transparency and integrity in decision-making.

1. Introduction

The procurement and purchasing function in an organisation is viewed as a critical component of supply chain management in construction industry [1]. Hence, an appropriate supplier selection can significantly lessen the purchase cost of the companies and consequently enhance the enterprise competitiveness [2].

However, there is scarce on supplier selection criteria in the Malaysian construction industry. Different criteria need to be considered for the supplier selection process. Degraeve and Roodhooft [3] declared that main focus of supplier selection were given to the price consideration. Research done by Suzari [4], indicates that factors such as supplier relationship management and supplier delivery performance emerged as the two most critical factors in the supplier selection. Their research also shows other tangible and intangible factors should also be given due consideration when making selection decisions, depending on the business context and organisational requirements. The interrelationship among the criteria of supplier selection should be given strong consideration while selecting the supplier. This is due to the multiple supplier selection criteria being interrelated to each other which could affect the supplier selection decision. According to Frödell [5], achieving efficient contractor-supplier relations has been defined as cutting costs, and reducing lead times, through reciprocal involvement by contractor, and supplier in the interface-related value creating processes.

The construction industry activities are vital to the achievement of national socio-economic development goals of providing infrastructure, shelter and employment. It is a risky business and can be classified as the most fraudulent industry worldwide, providing the perfect environment for ethical dilemmas, with its low-price mentality and fierce competition [6]. The scenario in the Malaysian construction industry is no exception, where graft and malpractices are numerously reported in the media. This lack of control would subsequently inhibit growth in the construction industries and directly restrict the effectiveness of the procurement process. Therefore, great attention are paid to develop a procurement decision-making and few studies have been conducted to optimize supplier selection in Public Work Department (PWD). However, most studies do not investigate how the critical criteria vary across members and factors that influence supplier selection. This paper aims to examine the factors that influence supplier selection, analyze the correlation exist among critical criteria and their respective significance to the supplier selection.

2. Supplier Selection Practice In The Malaysian Construction Industry

In Malaysia, construction industry has begun since independence in 1957. The increased of construction output from 1965 to 2003 reflects the industry's important role in providing the infrastructure to satisfy development needs and facilitate investment in other economic sectors [7]. Malaysia has achieved much progress in this direction but there is more to be done. On 28 August 2007, the Malaysian Construction Industry Master Plan (CIMP) was approved by the Cabinet Committee for Investment and Infrastructure which was chaired by the Deputy Prime Minister of Malaysia [8]. This plan outlines the 10-year strategic roadmap for the Malaysian Construction Industry to develop into a world-class, innovative and knowledgeable global solution provider.

However, there have been significant changes in the technical and economic conditions prevailing in the construction industry globally in the last decades. A number of supply chain management (SCM) initiatives had been launched by the construction industry to improve internal and external efficiency, reducing waste and adding value across the entire supply chain and trying to remove their adversarial inter-organizational purchaser-supplier relationships and fragmented business process [9]. Nevertheless, it have not made the breakthrough in construction industry yet, as the attempts to replicate the benefits obtained by supply chains in other industries still testify a scarce of effectiveness and a partial and slow implementation. SCM application has particularly found obstacles in construction sector as a consequence of its particular context of "temporary multiple organization" [10] and due to the difficulties in managing networks of a large number of different companies, supplying materials, multiple services and components [9].

A systematic approach for selection of the most appropriate system is now needed [11]. The procurement and purchasing function in an organisation is viewed as a critical component of supply chain management. It is an important part of supply chain management such that successful organisations have developed comprehensive procurement strategies which are aligned to the companies' business objectives with the purpose of helping the companies boost their business performance in the long term [4]. The role of procurement has extended far beyond merely sourcing products and services to meet internal organizational needs [12].

The new era of globalization and liberalization had caused developing country throughout the world move towards developed and industrialized nation. Thus, causing all kind of industry including construction sector to respond to the changes in demand such as broadening of customers' and suppliers' markets which have become more competitive [13]. One of the most important stages in the course of any supply chain which determines the success of any organization is supplier selection. It is a multi-criteria decision making process involving various criteria which may be quantitative as well as qualitative. The criteria for the supplier selection are determined by the buyer/retailer according to his requirement. The criteria may differ according to the situation [14]. Supplier selection is a critical issue in a supply chain partnership. Such partnerships involve a worldwide network of suppliers, factories, warehouses, distribution centers, and retailers through which raw materials are acquired, transformed, and delivered to customers [15]. Thus, it is important to maximize the purchasing function's contribution to achieving the goals of the organization.

3. Formulation of a Structural Equation Model for Supplier Selection

Supplier selection attained the state of highest significance for companies in the current scenario because of increasing competition. Improper selection of suppliers will have a poor impact on the overall performance of the manufacturer. Thus, it is vital to maximize the supplier selection to achieving the goals of the organization. By data gathering and analysis, the organization could identify which criterion is the most important and compare criteria that influence the supplier selection. According to Suzari (2013), procurement strategies developed by the purchasing department must be integrated with overall business strategies which support the goals and objectives of the organization. Continuous evaluation to develop alternative supply options and competitive suppliers as well as monitoring market trends on critical products and services help organizations to achieve an effective procurement process.

Structural equation modeling (SEM) is a multivariate method that allows assessment of both direct and indirect effects of each variable on the other variables [16]. The use of SEM is justified to avoid excessive multi-collinearity that can lead to bias and unstable findings. Its aim to explore the causal relationship between objects, and to express it through model of causal relationship, path diagram and so on. Figure 1 shows the conceptual model of SEM. There are 5 steps of SEM application which are assumption, identification, estimation, evaluation and modification [17].

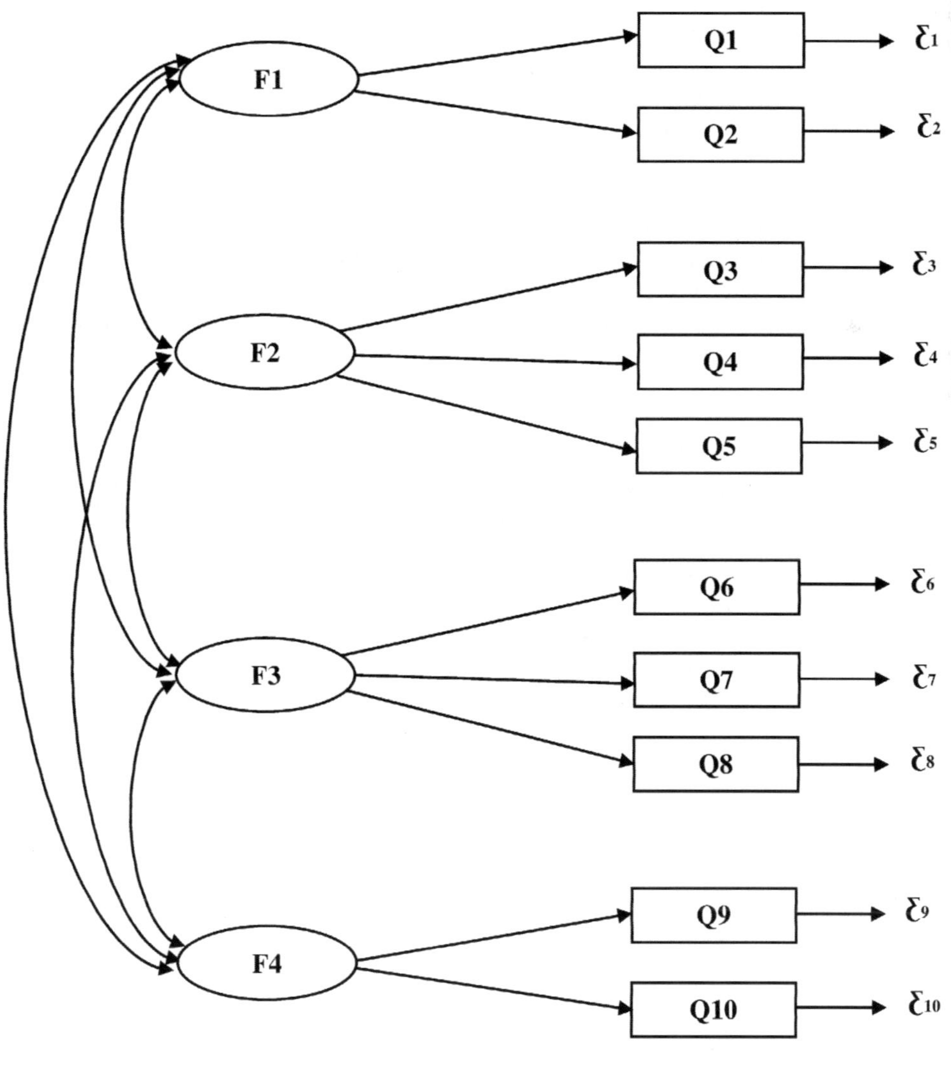

Figure 1: Conceptual model of SEM

4. Conclusion

Supplier selection is an integral part of the supply chain network, and choosing the right suppliers to work with will have an impact on the companies' business operations and performance. The supplier selection criteria will continue to change based on an expanded definition of excellence to include traditional aspects of performance in addition to non-traditional. Effective supplier evaluation and purchasing processes are critical success factors for many organizations. The supplier selection decision is affected by the multiple supplier selection criteria because they are interrelated to each other. Therefore, inter relationship among the criteria of supplier selection should be given strong consideration while selecting the supplier.

5. References

[1] Sarkis, J. and Talluri, S. (2002), "A Model for Strategic Supplier Selection", *Journal of Supply Chain Management*, vol. 38, no. 1, pp. 18-28.

[2] Ghodsypour, S. ., & O'Brien, C. (2001). The total cost of logistics in supplier selection, under conditions of multiple sourcing, multiple criteria and capacity constraint. *International Journal of Production Economics*, *73*(1), 15–27

[3] Degraeve, Z. and F. Roodhooft, *Effectively Selecting Suppliers Using Total Cost of Ownership.* The Journal of Supply Chain Management, 1999. 35(1): p. 5-10.

[4] Suzari, A.R., *Supplier Selection in the Malaysian Telecommunications Industry.* 2013, Brunel University London: Uxbridge, United Kingdom.

[5] Frödell, M., *Criteria for achieving efficient contractor-supplier relations.* Engineering, Construction and Architectural Management, 2011. 18(4): p. 381-393.

[6] Adnan, H., Hashim, N., Mohd, N., Yusuwan, & Ahmad, N. (2012). Ethical Issues in the Construction Industry: Contractor's Perspective. *Procedia - Social and Behavioral Sciences*, *35*, 719–727.

[7] Abdulllah, P. M. D. F., Chiet, C. V., Anuar, K., & Shen, T. T. (2004). An Overview On The Growth and Development Of The Malaysian Construction Industry

[8] Karib, M. S. A. (December 2008). A Report on The Proposal for A Malaysian Construction INdustry Payment and Adjudication Act. Kuala Lumpur.

[9] Aloini, D., Dulmin, R., Mininno, V., & Ponticelli, S. (2012). Supply chain management: a review of implementation risks in the construction industry. *Business Process Management Journal*, *18*(5), 735–761.

[10] Cheng, J. C. P., Law, K. H., Bjornsson, H., Jones, A., & Sriram, R. (2010). A service oriented framework for construction supply chain integration. *Automation in Construction*, *19*(2), 245–260.

[11] Alhazmi, T., & McCaffer, R. (2000). Project Procurement System Selection Model. *Journal of Construction Engineering and Management*, *126*(3), 176–184.

[12] Cousins, P. D., & Lawson, B. (2007). Sourcing strategy, supplier relationships and firm performance: an empirical investigation of UK organizations. British Journal of Management, 18(2), 123-137.

[13] Micheli, G. J. (2008). A decision-maker-centred supplier selection approach for critical supplies. Management Decision, 46(6), 918–932.

[14] Punniyamoorthy, M., Mathiyalagan, P., & Parthiban, P. (2011). A strategic model using structural equation modeling and fuzzy logic in supplier selection. Expert Systems with Applications, 38(1), 458–474.

[15] Khalfan, M. M. A., & Maqsood, T. (2012). Supply chain capital in construction industry: coining the term. *International Journal of Managing Projects in Business*, *5*(2), 300–310.

[16] Chinda, T., & Mohamed, S. (2008). Structural equation model of construction safety culture. *Engineering, Construction and Architectural Management*, *15*(2), 114–131.

[17] Tang, L., Shen, G. Q., Skitmore, M., & Wang, H. (2014). Procurement-Related Critical Factors for Briefing in Public-Private Partnership Projects: Case of Hong Kong. *Journal of Management in Engineering*, 04014096.

A Conceptual Framework for Procurement Decision Making Model to Optimize Supplier Selection: The Case of Malaysian Construction Industry

Ngam Min Chuan[1], Sivadass Thiruchelvam[1], Kamal Nasharuddin Mustapha[2], Zakaria Che Muda[1], Norhayati Mat Husin[1], Lee Choon Yong[1], Azrul Ghazali[1], Mohd Ezanee Rusli[1], Zarina Binti Itam[1], Salmia Beddu[1], Nur Liyana Mohd Kamal[1]

[1]Centre of Innovation and Design, Universiti Tenaga Nasional, Malaysia
[2]Centre of Forensic Engineering, Universiti Tenaga Nasional, Malaysia

Ngam@uniten.edu.my

Abstract. This paper intends to fathom the current state of procurement system in Malaysia specifically in the construction industry in the aspect of supplier selection. This paper propose a comprehensive study on the supplier selection metrics for infrastructure building, weight the importance of each metrics assigned and to find the relationship between the metrics among initiators, decision makers, buyers and users. With the metrics hierarchy of criteria importance, a supplier selection process can be defined, repeated and audited with lesser complications or difficulties. This will help the field of procurement to improve as this research is able to develop and redefine policies and procedures that have been set in supplier selection. Developing this systematic process will enable optimization of supplier selection and thus increasing the value for every stakeholders as the process of selection is greatly simplified. With a new redefined policy and procedure, it does not only increase the company's effectiveness and profit, but also make it available for the company to reach greater heights in the advancement of procurement in Malaysia.

Keywords: Supplier selection, procurement, Malaysian construction industry, metrics

1. Introduction

Supplier selection has always been a major strategic decision for clients of many industry. Supplier selection can be defined as a multi-criteria decision making problem which takes into account both qualitative and quantitative factors. With the advancement of procurement especially in the construction industry, a supplier selection must be able to take on the complicated modern day procurement practices systematically and with outmost transparency. This paper focuses on developing a procurement decision making model to optimize supplier selection in the Malaysian construction industry. This paper will provide the fundamental framework supplier selection metrics for Malaysia's infrastructure building and determining which metrics contributes most in creating a successful procurement in construction projects. To obtain this metrics, studies have to be done on the end user's perspective on the definition and viability of metrics. Degraeve and Roodhooft [1] research states that the main factor supplier selection were given to the price consideration. However, not many studies are being done on determining and weighing other criteria of supplier selection in Malaysian construction industry.

With that said, this conceptual paper proposed a framework to evaluate criteria that influence supplier selection by identifying the importance of each criteria. For this paper, a mix method approach which comprises of qualitative and quantitative methods is used. Next, the scope is towards identifying and defining criteria that stakeholders in Public Works Department (PWD) believe should be use in supplier selection. The product from this paper is foreseen to bring PWD to an advancement in procurement and to the next strategic supplier selection level. The outcome of this research also provides a more in depth understanding on supplier selection drivers in the construction industry. This will not only benefit PWD but every other construction agencies and supplier selection decision making groups in the future. Finally, the value of this research can be put into good use as an initiative to increase awareness in transparency of decision making among stakeholders and policymakers therefore greatly increases integrity in procurement practices.

2. Review on Current State of Procurement Decision Making for Construction Industry

The construction industry is an important role in generating wealth and improving the quality of life for Malaysians. The construction industry makes up an important part of the Malaysian economy. The construction industry in Malaysia has scarcely conducted studies to determine the appropriate supplier selection metrics. However, the actual fundamental constructs of supplier selection in construction projects is rarely addressed. Day and Barksdale [2] states that most published studies offers only few insights into the purchasing process. This is because many if not all researchers did not study who comprises the decision-making process. Hence, the purpose of this research is to study the supplier selection metrics for construction projects and the importance of each metric that has been assigned in the Public Works Department (PWD).

According to Lu and Swaminathan [3], supply chain management (SCM) is a critical aspect in conducting any business. In supply chain management, the emphasis is on how well a group of companies perform in order to create value for the end user [4]. Since the nineties, there has been an increase in interest towards supply chain management system research and theories. Segerstedt and Olofsson [5] says that this is to understand and characterize the deficiencies and to propose solutions to improve the coordination of the often many subcontractors and supplier in the construction supply chain.

In recent years, both private and public sector's organizations have placed more attention on cost reduction and flexibility, concentrating on core competences and outsourcing non-core activities. This was the initial step for construction companies towards becoming more competitive in the ever growing market. A study done by Ancarani and Capaldo [6] states that the relevance of these activities for the organization's performance requires an accurate analysis of what it is outsourced. Project procurement provides an excellent environment for ethical issues with its low-price state of mind and competition. Procurement process has many openings that could contribute to illegal activities or unethical behaviour especially in the construction industry. In 2006 alone, 17.3% of 417 Malaysian government contract projects were considered unhealthy due to the poor performance by the contractors. Therefore, it is important to govern the project procurement, especially during the entire procurement stage to ensure the accountability and transparency of the decision made in awarding the right contract to the best contractor [7].

To explain the supplier selection process further, a pool of suppliers is chosen for procurement according to a predefined set of criteria. Based on current academic literature, Xu [8] have reviewed 78 related articles appeared in the period from 2000 to 2008. They noted that quality is the most popular criterion in which 68 papers or 87.18% or researchers agrees in the supplier selection process. The second most popular criterion is delivery that is with 64 papers or 82.05% agreeing. Surprisingly, price or cost falls in third place with 63 papers or 80.77% which initially thought would be the main criteria in supplier selection. After much more studies, they discover that there was a trend in the supplier evaluation and selection problem to use the multi-criteria decision making approaches. The procurement decision process is concerned with the problem of lot-sizing and other inventory related issues. Finally, in the sourcing evaluation process, overall efficiency and effectiveness of procurement process is assessed [9].

3. The Rising Trend of Procurement Decision Making

Puente [10] defines supplier selection as the process by which firms identify, evaluate and contract with supplier. The selection of the most suitable supplier for a procurement process has a markedly strategic aspect for a company. Supplier selection metrics have evolved over the years from the traditional set comprising of price and quality. With industrialization and globalization phenomena kicking in at fast pace, there have been an emergence of new metrics for the selection for an optimal supplier. This tracks back to the empirical evidence provided by Genovese [11] showing the addition of innovation, flexibility and even environmental responsibility plays an important role to attain a sustainable supply chain. Studies in manufacturing industries have also shown the rise of new metrics to suit the need of current turbulent business environment [12, 13].

The recent upsurge of awareness in corporate governance related matters in Malaysia and other developing countries is a manifestation of its rising importance for strong economic performance of organizations and nations. In reference to Dalton [14], corporate governance refers to a framework of rules and practices according to which an organization is directed and controlled. Therefore, it functions to establish and enforce policies deemed necessary for effective operation of the organization. Autio [15] says that corporate governance can be seen as a mechanism to monitor decision-making teams and advise them in ways that align their interests with stakeholders, improving the quality of the organization's strategies decision-making. Neill [16] assert that corporate governance guides decision-makers to be responsible with the sole intention to maximize the stakeholders' wealth. Effort must be made from within the organization to provide cost savings which can offset any perceived detrimental effects.

By reviewing and improving its internal business processes, PWD could embark on greater cost saving. Scheer and Nüttgens [17] describes a business process as a procedure relevant for adding value to an organization. According to Tan [18], Carr and Pearson [19], the purchasing function is an important strategy of the business process which can support the overall strategy of any firm. Pagell [20] says that this follows the selection of suppliers making it an important activity in purchasing. Verma [21] added that this can be the deciding factor in ensuring superior business performance. Last but not least, Xu and Lin [22] believe that any savings made from supplier selection will benefit the long term financial health of the organisation.

4. Adoption of Dual Construct for Supplier Selection: A Way Forward

In order to achieve a more advance supplier selection method in the future, an orthogonal perspective or also known as dimensions is needed to guide the identification of metrics. The relevant metrics for supplier selection in the construction industry can be viewed in two perspectives. Based on the study done by Thiruchelvan [23], two main constructs namely supplier competitiveness and supplier attractiveness are formed to distinct all the criteria. However, the author thinks that there are criteria which are not included that servers an importance towards the construction industry. Therefore, two new construct consisting of supplier effectiveness and supplier allurement are proposed to cater for the construction industry. Supplier effectiveness shows the main attributes an organization seeks from a supplier and are relatively similar across the market. The main criteria consist of supplier quality, price, delivery and support services. Supplier allurement are attributes that only appears as high priority to certain organizations. This construct includes supplier flexibility, financial performance, customer focus, safety awareness, government policies and others.

5. Conclusion

Based on past papers, it clearly states that the Public Work Department needs to bring procurement to the next level. Instead of focusing merely on the aspect of cost, other criteria in supplier selection plays important roles and carries weight in overall selection. With an investigation into the supplier selection metric, other criteria will be taken into consideration depending on their degree of importance based on stakeholder's views. Results pertaining to this will be used to create awareness and give a better perception on defining the dimension of each metric in supplier selection. This will ultimately bring growth to the organization profit, effectiveness and transparency in the area of procurement. Lastly, this study will cause a significant change in the area of procurement as many other organization in the construction industry begin to practice new optimized supplier selection methods.

6. References

[1] Degraeve, Z. and F. Roodhooft, *Effectively Selecting Suppliers Using Total Cost of Ownership.* The Journal of Supply Chain Management, 1999. 35(1): p. 5-10.

[2] Day, E. and H.C. Barksdale, *Organizational Purchasing of Professional Services: The Process of Selecting Providers.* Journal of Business & Industrial Marketing, 1994. 9(3): p. 44-51.

[3] Lu, L.X. and J.M. Swaminathan, *Supply Chain Management*, in *International Encyclopedia of the Social & Behavioral Sciences*. 2015, Elsevier. p. 709-713.

[4] Forslund, H. and P. Jonsson, *Selection, implementation and use of ERP systems for supply chain performance management.* Industrial Management & Data Systems, 2010. 110(8): p. 1159-1175.

[5] Segerstedt, A. and T. Olofsson, *Supply chains in the construction industry.* Supply Chain Management: An International Journal, 2010. 15(5): p. 347-353.

[6] Ancarani, A. and G. Capaldo, *Supporting decision-making process in facilities management services procurement: A methodological approach.* Journal of Purchasing and Supply Management, 2005. 11(5-6): p. 232-241.

[7] Aliza, A.H., K. Stephen, and T. Bambang, *The Importance of Project Governance Framework in Project Procurement Planning.* Procedia Engineering, 2011. 14: p. 1929-1937.

[8] Ho, W., X. Xu, and P.K. Dey, *Multi-criteria decision making approaches for supplier evaluation and selection: A literature review.* European Journal of Operational Research, 2010. 202(1): p. 16-24.

[9] Choudhary, D. and R. Shankar, *Joint decision of procurement lot-size, supplier selection, and carrier selection.* Journal of Purchasing and Supply Management, 2013. 19(1): p. 16-26.

[10] García, N., et al., *Supplier selection model for commodities procurement. Optimised assessment using a fuzzy decision support system.* Applied Soft Computing, 2013. 13(4): p. 1939-1951.

[11] Genovese, A., et al., *Greener supplier selection: state of the art and some empirical evidence.* International Journal of Production Research, 2013. 51(10): p. 2868-2886.

[12] Park, J., et al., *An integrative framework for supplier relationship management.* Industrial Management & Data Systems, 2010. 110(4): p. 495-515.

[13] Huang, S.H. and H. Keskar, *Comprehensive and configurable metrics for supplier selection.* International Journal of Production Economics, 2007. 105(2): p. 510-523.

[14] Daily, C.M., D.R. Dalton, and A.A. Cannella, *Corporate Governance: Decades of Dialogue and Data.* The Academy of Management Review, 2003. 28(3).

[15] Mustakallio, M., E. Autio, and S.A. Zahra, *Relational and Contractual Governance in Family Firms: Effects on Strategic Decision Making.* Family Business Review, 2002. 15(3): p. 205-222.

[16] Stovall, O.S., J.D. Neill, and D. Perkin, *Corporate Governance, Internal Decision Making, and the Invisible Hand.* Journal of Business Ethics, 2004. 51(2): p. 221-227.

[17] Scheer, A.-W. and M. Nüttgens, *ARIS Architecture and Reference Models for Business Process Management*, in *Business Process Management*, W. van der Aalst, J. Desel, and A. Oberweis, Editors. 2000, Springer Berlin Heidelberg: Berlin, Heidelberg. p. 376-389.

[18] Tan, K.C., *A framework of supply chain management literature.* European Journal of Purchasing & Supply Management, 2001. 7(1): p. 39-48.

[19] Carr, A.S. and J.N. Pearson, *The impact of purchasing and supplier involvement on strategic purchasing and its impact on firm's performance.* International Journal of Operations & Production Management, 2002. 22(9): p. 1032-1053.

[20] Krause, D.R., M. Pagell, and S. Curkovic, *Toward a measure of competitive priorities for purchasing.* Journal of Operations Management, 2001. 19(4): p. 497-512.

[21] van der Rhee, B., R. Verma, and G. Plaschka, *Understanding trade-offs in the supplier selection process: The role of flexibility, delivery, and value-added services/support.* International Journal of Production Economics, 2009. 120(1): p. 30-41.

[22] Xu, X. and J. Lin, *Strategic Supplier Network for Supplier Selection.* Journal of Computers, 2010. 5(6).

[23] Thiruchelvan, S., *Strategic Sourcing Supplier Selection for Capital Equipment*, in *Construction Management, School of Engineering*. 2012, Auckland University of Technology: Auckland, New Zealand.

Nuclear Power Acceptance Among University Staffs and Students

G Hayder[1] and M S Ab Rahim[2]
[1] Department of Civil Engineering, College of Engineering, Universiti Tenaga Nasional, Malaysia
[2] College of Graduate Studies, Universiti Tenaga Nasional, Malaysia
E-mail: gasim@uniten.edu.my

Abstract. The need to consider alternative energy sources becomes very real. Nuclear has been identified as an alternative electricity source. However, media reports seem to indicate that there is a resistance among peoples with regards to harnessing nuclear for energy. This study was conducted to assess the acceptance level of university staff and students towards nuclear energy by asking them to answer a questionnaire. The questionnaire was constructed in a way to gauge their background knowledge on the energy situation of the country, the risks involved with regards to nuclear energy and also what aspects need to be improved in order to have a safe integration of nuclear energy into the national energy mix. The overall result of the questionnaire indicated high level of support for nuclear energy. The main areas of concerns however, were waste management, control and governance and also nuclear accidents. These should be identified as fields that require extra attention. However, the positive result obtained from this survey should not be construed as overall strong support in general. There might be different outcomes if the survey was conducted on to the general population as compared to the university students and staff that were involved in this research.

1. Introduction

China and India are the leading countries in Asia with regards to nuclear electricity and they anticipates an increase in generation capacity [1]. As the energy security of Malaysia heads towards an uncomfortable situation over the next 5 years, it is imperative that policy makers look to diversify the energy mix. Currently, more than 60 percent of Malaysia's electricity is generated from Gas and the remainders come from coal and hydroelectric sources. Considering that almost all of the coal used for electricity generation is imported, a serious question with regards to energy security lingers [2, 3].

It has been projected by the Economic Planning Unit at the Prime Minister's department, that Peninsular Malaysia will face an energy shortage situation post 2019. At this point onwards, it is projected that the demand for energy will outpace the ability to supply energy [4]. Malaysia began its nuclear journey since 1972 and the Atomic Energy Licensing Board (AELB) was established in 1985 [5, 6]. The environmental benefits over the lifecycle of a nuclear power plant in term of the Green House Gas emissions is 100 times lower than that of Gas and Coal power plants [7]. It is also 10 times lower when compared to Solar PV and Biomass power plants [8]. There are some challenges with regards to nuclear safety, while there are initiates from several universities to offer undergraduate and postgraduate nuclear related courses for capacity building [9].

[1] To whom any correspondence should be addressed.

Published under licence by IOP Publishing Ltd

The objective of this report is to look into the possibility of deploying nuclear energy in Malaysia and to gauge the acceptance level of respondents towards nuclear power plants.

2. Methodology

The One of the of the objectives of this study is to gain an understanding of the attitude of students and staffs of Universiti Tenaga Nasional towards nuclear energy and its utilisation in electricity generation. For this purpose, a questionnaire of 8 sections, which comprised of 16 questions was constructed and electronically distributed to the respondents. This questionnaire was adapted from an earlier published survey by [10] to suit the Malaysian locality.

The first section of this questionnaire looks into the demographics of the respondents. The questions asked in this section are on the gender and the age of the respondents. The purpose of this section is to establish if the response provided by the respondents might correlate to either category.

In the second section, the questionnaire asks respondents to choose their preferred choice for electricity generation. This is followed by questions first on their general support for the usage of nuclear power for electricity generation and then a specific question on their support for nuclear energy in Malaysia. The questionnaire then proceeds to investigate on the perceived risks by the respondents on nuclear energy for electricity generation in the fifth section. Among the risks that were presented to the respondents included nuclear waste management, general maintenance and management risks, nuclear accidents, attacks and natural disasters. This section is included to understand the main factor that the respondent considers as critical aspects to be managed in order for nuclear energy to be utilised for electricity at minimum risk. The sixth section of the questionnaire touched on climate change. It was designed to drive the thought process of the respondent to consider the possibility of supporting nuclear energy for electricity if they considered or believed that it could assist in reducing the impacts of climate change. The preceding section focussed on energy security and whether nuclear energy could be utilised to improve the energy security situation of the country. Lastly, the survey ends by asking the respondent again on their support for nuclear energy for electricity. However, in this section the respondent is asked first on their support for nuclear energy and then followed by a specific question on whether they support nuclear energy to be implemented in Malaysia for electricity generation. Out of the total emails sent containing the link for the electronic questionnaire, there were 142 respondents. The outcome of the questionnaire is then compiled and presented into chart form to be analysed. The result of the questionnaire is presented in the following section.

3. Results and discussion

The demographics of the respondents is summarised in Table 1. There were 142 respondents, with a majority of 70% male respondents. The respondents were mainly aged between 18 - 25 years old of age with 62% of the respondents being students.

Table 1. Demographics of Questionnaire Respondents

Gender/Age/ Occupation	Quantity, N	Percentage, %
Male	99	69.7%
Female	43	30.3%
18 - 25	85	59.86%
26 - 35	29	20.42%
36 - 45	20	14.08%
46 - 55	6	4.23%
56 and above	2	1.41%
Students	88	62%
Staff	54	38%

The respondents were asked to indicate their preference of energy sources for electricity generation in the first section of the questionnaire. Across the option of fossil fuels for electricity generation, natural gas was the most favored for of energy source at 9%, comparing to 1% for diesel and coal each. Across all options, renewable energy was the most preferred choice for fuel source. Solar electricity was chosen by 40% of the respondents, followed by 22% choosing hydroelectricity at 22%. Nuclear energy was chosen by 17% of the respondents.

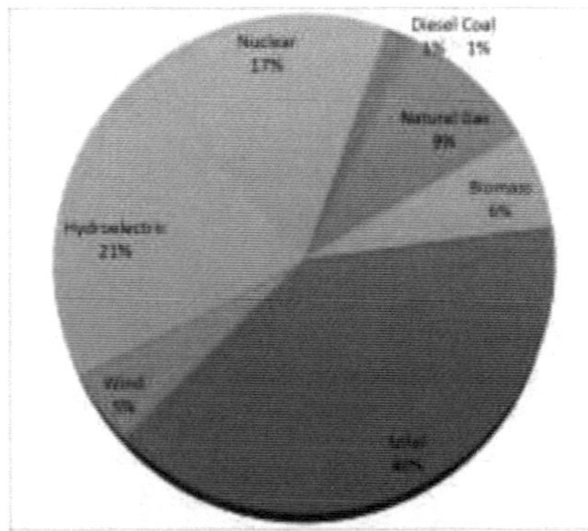

Figure 1. Preferred Source of Energy for Electricity Generation

A question was asked to the respondents on their awareness of any other ASEAN member nation having nuclear electricity as part of their energy mix. This was to gauge their regional awareness of nuclear development. The outcome was somewhat balanced with 47% of respondents claiming to be aware of some ASEAN member countries already having nuclear electricity in their energy mix. The remaining 52% were unaware of this. Currently, no ASEAN country has any operational nuclear power plants yet. However, Vietnam is the most advanced stage when it comes to building a nuclear power plant [9]. Respondents were asked of their opinion and general support for nuclear energy to be used in electricity generation. In principal, 44 percent of respondents expressed support for nuclear energy while 22 percent were opposing the idea. The remaining 26 percent were unsure whether to support or oppose nuclear power for electricity. Another 7 percent had no opinion on the matter.
Interestingly, when the questions was modified a little to ask whether the respondents would support nuclear energy for electricity in Malaysia, the amount of support dropped from 44 percent to 41 percent while the opposition increased from 26 percent to 31 percent. The number of people who were unsure remained at 26 percent.

Table 2. Support for Nuclear Power for Electricity Generation

Choose the option that most closely indicate your level of support for nuclear power plants for electricity generation	Quantity, N	Percentage, %
a) In general, I support the use of nuclear power for electricity generation	63	44.37%
b) In general, I oppose the use of nuclear power for electricity generation.	32	22.54%
c) I am not sure whether I support or oppose the use of nuclear power for electricity generation.	37	26.06%
d) I have no opinion on the matter	10	7.04%
How do you view the future of Nuclear Power Plants in Malaysia		
a) We should build Nuclear Power Plants in Malaysia	59	41.55%
b) We should not build Nuclear Power Plants in Malaysia	45	31.69%
c) I am not sure	38	26.76%
d) I have no opinion on the matter	0	0.00%

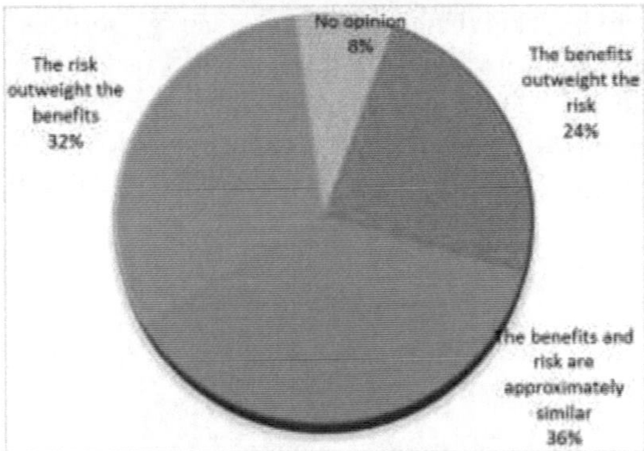

Figure 2. Perceived Risk and Benefits of Nuclear Power for Energy

3.1. Prediction Results

Respondents were asked on the perceived risk and benefits of implementing nuclear power for energy generation. There were three options presented for this question. First, the optimistic option where one sees that the benefits of nuclear power outweighs the risk. This option was selected by 24 percent of respondents. Most of the respondents, 36 percent were more balanced to say that the benefits and risk of implementing nuclear power are approximately the same. The pessimistic view that the risk of nuclear power is higher than the benefits was selected by 32 percent of respondents. A remainder of 8 percent had no opinion on the matter.

A list of potential risks was presented to the respondents. This was to find out, what risk aspect that they consider the most critical to be addressed. The options provided were waste mismanagement risk, quality risks due to construction and/or maintenance, risk of being attacked, governance and regulatory enforcement risk, natural disaster and nuclear accidents.

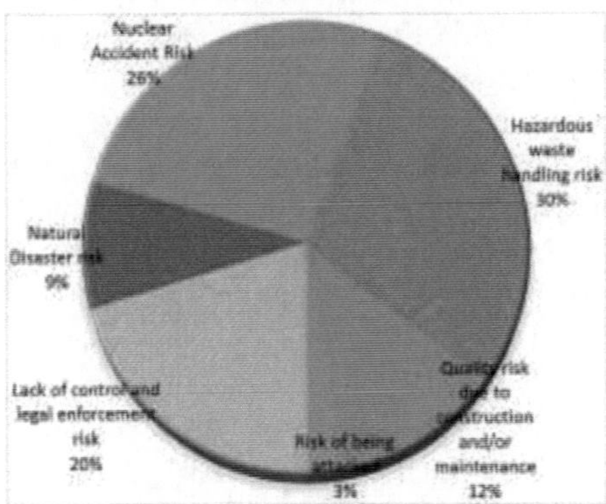

Figure 3. Perceived Risk Associated to Operating a Nuclear Power Plant in Malaysia

Waste handling appears to be the risk aspect that most respondents were concerned of. This option was selected by 30 percent of the respondents. Respondents were secondly concerned over nuclear accidents with 26 percent selecting this option. Next, 20 percent of respondents were concerned with regards to control, regulation and legal enforcement of operating standards. They were least concerned over potential attacks on the power plants with only 3 percent selecting the option.

3.2. Conditional Support for Nuclear Power

The questionnaire was designed to investigate if there were any conditions that might influence the support level of respondents for nuclear power. For this purpose, respondents were asked on several questions with regards to climate change and also energy security. First, they were asked on their opinion on the seriousness of climate change.

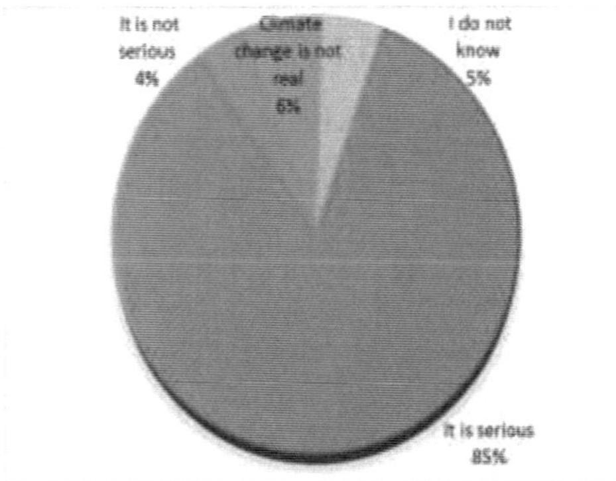

Figure 4. Respondent Opinion on Climate Change

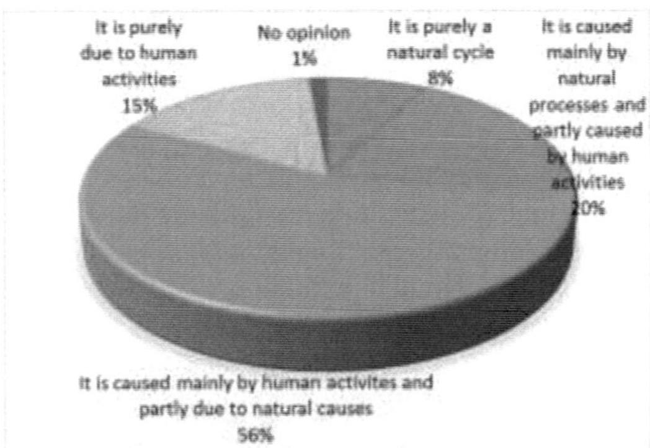

Figure 5. Respondent Opinion on Causes of Climate Change

A strong majority of 85 percent of respondents agreed that climate change is a serious issue. The remainders were almost equally distributed in saying it is not serious, it is not real and have no opinion on the matter. It is believed by 56 percent of the respondent that human activities are the main contributor of climate change. Only 8 percent of respondents believe it to be a natural phenomenon.
Immediately after these two questions, the respondents were asked again on their support for nuclear power for electricity. This time, the focus was to ask the respondents if they would support nuclear power for electricity if it could reduce the effect of climate change. The support level jumped by 21 percent from the initial feedback. This indicates that the respondents were supportive of reducing the impact of climate change.

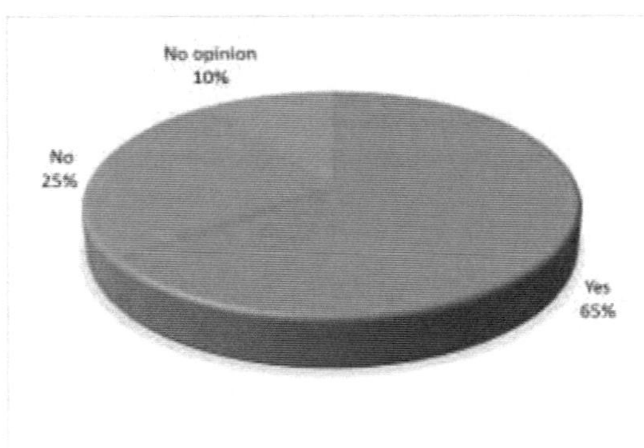

Figure 6. Respondent Support for Nuclear Power Plants if it can reduce the Impact Climate Change

The respondents were then asked on their opinion on whether Malaysia will face energy security issues in the next 50 years. Further, they were asked whether they would support nuclear energy if it could resolve the national energy security concern for the coming 50 years.

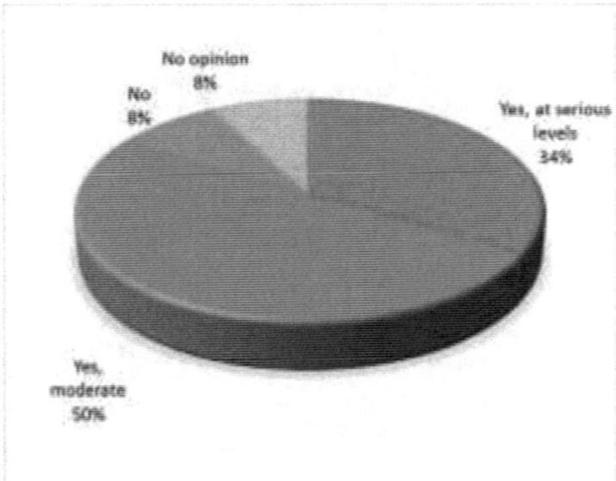

Figure 7. Respondent Opinion on Causes of Malaysia Energy Security

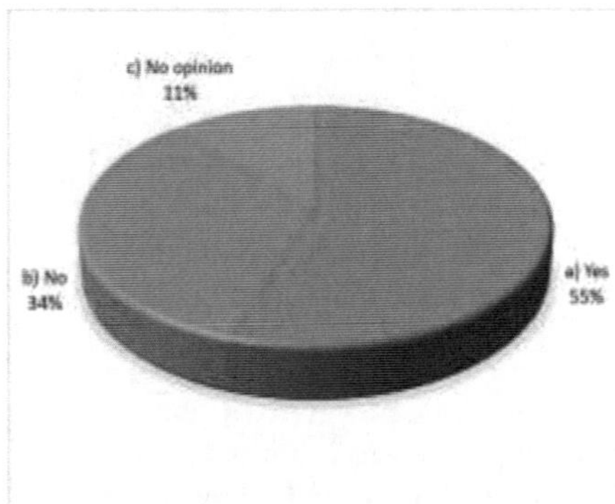

Figure 8. Respondent Support for Nuclear Power Plants if it can Increase Energy Security

This time the support level for nuclear power plants jumped by 15%. At this juncture, one could summarise that if the respondents are made aware of the positive implications of nuclear energy, there would be increased support for the implementation of the technology.

Towards the end of the questionnaire, the respondents were asked whether in principal they support nuclear energy. For this question, 69 percent of respondents supported this idea. At the beginning phase of the questionnaire, only 44 percent of respondents supported nuclear power plants for electricity generation.

When the respondents were asked again on their opinion whether nuclear power plants should be built in Malaysia, 54 percent of respondents had positive opinion compared to only 41 percent in the initial stage.

4. Conclusion

From the literature review, it could be construed that Malaysia will face serious energy security issues in the coming 5 years. As such, it is impediment that the nation diversifies its fuel mix for energy generation. Nuclear could be considered as a viable option as it is more secure, cleaner and it is also a mature technology in a global sense.

Table 3. Support for Nuclear Power for Electricity Generation

In principal do you support nuclear power plants for electricity generation in peaceful and politically stable foreign countries	**Quantity, N**	**Percentage, %**
a) Yes	98	69.01%
b) No	34	23.94%
c) No opinion	10	7.04%
In principal do you support nuclear power plants for electricity generation in Malaysia		
a) Yes	78	54.93%
b) No	48	33.80%
c) No opinion	16	11.27%

The benefits of using nuclear power plants should be viewed in terms of economy, society and the environment. For the economy, it will lower energy prices and increase energy security. Hence development will occur at a rapid pace, job creation will be seen and investments will increase.

The job creation benefits the society. Additionally, the opportunity for higher education will be increased in order to meet the demand of the new sector. As nuclear is much cleaner than gas and coal power sources, the environment will benefit as well. There will be reduction of greenhouse gas emission, particulates and smog.

However, the challenge is to ensure a proper framework mechanism to manage nuclear power and nuclear waste is put in place. The public is greatly concerned on the nuclear waste management system.

A survey was conducted to gauge the general level of acceptance for nuclear power in Malaysia. 142 respondents were engaged. The backgrounds of the respondents were mainly engineering and science.

From the survey, the level of support for nuclear is quite high. In general, 41 percent of the respondents supported in principal the use of nuclear power plants in Malaysia. However, when asked whether they would support nuclear power plants in Malaysia if it reduced the impacts of climate change, the support level increased to 65 percent. When this question was asked again, but in the aspect of energy security, the support level was at 55 percent.

The respondents are mainly concerned on nuclear waste management, lack of control and enforcement and also nuclear accidents. As such, these are the areas that must be focused on and be addressed.

In conclusion, among the learned group, there is enough understanding on the benefits and challenges of nuclear energy. As such, these groups knows enough to support nuclear energy. The government should engage the public and educate them in order to get more public support.

Proper framework and mechanisms should be established in order to ensure proper management of nuclear power when it becomes available. For the long run, in view of energy security issues and greenhouse gas reduction, the government should embark on a nuclear power journey.

5. Acknowledgement

We would like to acknowledge the Universiti Tenaga Nasional staff and student for their feedback and time.

6. References

[1] Rashad, SM. Future of nuclear energy worldwide strategies and policies. In: Proceedings of the national radio science conference, NRSC 2008. Tanta; 18–20 March 2008. p. 1–17.

[2] Oh, T. H., Pang, S. Y., & Chua, S. C. (2010). Energy policy and alternative energy in Malaysia: issues and challenges for sustainable growth. Renewable and Sustainable Energy Reviews, 14(4), 1241-1252.

[3] Abu, M.P. (2011) PROGRESS OF NUCLEAR POWER PROGRAMME IN MALAYSIA NPRO Dialogue Forum on Nuclear Energy Innovations: Common User Consideration for Small and Medium-sized Nuclear PR, Vienna, 10 - 14 October 2011 https://www.iaea.org/INPRO/3rd_Dialogue_Forum/16.INPRO-Abu-Malaysia.pdf

[4] Ibrahim, J.K. (2014) DEVELOPING NUCLEAR ENERGY FOR POWER GENERATION IN MALAYSIA'S ECONOMIC TRANSFORMATION PROGRAMME (ETP), International Construction Week, Kuala Lumpur. https://www.cidb.gov.my/cidbv4/images/pdf/ecobuild/room3/(Room3%20-%2018.09.2014)%20Developing%20of%20Nuclear%20Energy%20for%20Power%20Generation%20Under%20Malaysia's%20ETP.pdf

[5] Ali, R., Daut, I., & Taib, S. (2012). A review on existing and future energy sources for electrical power generation in Malaysia. Renewable and Sustainable Energy Reviews, 16(6), 4047-4055.

[6] Basri, N. A., Ramli, A. T., & Aliyu, A. S. (2015). Malaysia energy strategy towards sustainability: a panoramic overview of the benefits and challenges.Renewable and Sustainable Energy Reviews, 42, 1094-1105.

[7] D. Weisser, A guide to life-cycle greenhouse gas (GHG) emissions from electric supply technologies, Energy 32 (2007) 1543-1559.

[8] van der Zwaan, Bob. "The role of nuclear power in mitigating emissions from electricity generation." Energy Strategy Reviews 1.4 (2013): 296-301.

[9] Caballero-Anthony, M., Cook, A. D., Trajano, J. C. I., & Sembiring, M. The Sustainability of Nuclear Energy in Southeast Asia: Opportunities and Challenges.

[10] Poortinga, W., Pidgeon, N.F., Capstick, S. and Aoyagi, M. (2014) Public Attitudes to Nuclear Power and Climate Change in Britain Two Years after the Fukushima Accident - Synthesis Report (UKERC: London).

Solar Resource Mapping of AL Duqm Industrial Area

Yousuf AL-Hatmi[1,2], C. S. Tan[1,3], Sultan al Al-Yahyai[4], Abdullah Al-Badi[5]

[1] Department of Electrical and Power Engineering, Universiti Tenaga Nasional, Malaysia
[2] PO Box 290, PC 314, Muludah, Oman, Email: alhatmi2012@gmail.com
[3] Institute of Energy Policy and Research, Universiti Tenaga Nasional, Malaysia, E-mail: chingsin@uniten.edu.my
[4] National Meteorological Service, PO Box 1, PC 111, Muscat, Public Authority for Civil Aviation (PACA), Oman, Email: s.alyahyai@met.gov.om
[5] Department of Computer and Electrical Engineering, PO Box 42, PC 123, Muscat-123, Sultan Qaboos University, Oman, Email: albadi@squ.edu.om

Abstract: The Sultanate of Oman has one of the highest percentages of potential solar radiation in the world, which puts it in an outstanding position to horde electricity generation plants through photovoltaic solar systems. Oman consists of 11 governorates, and each governorate is divided into many *Willyiah*, which are made up of small villages. Willyiah of Al Duqm is located in Al Wusta Governorate (19°39'42"N 57°42'17"E) in central-eastern Oman. This paper proposes the use of a monthly average hourly diffuse illuminance from meteorological data and from a Geographic Information System (GIS) to identify the best plots suitable for installing photovoltaic solar farms in the Al Duqm Industrial Area. Solar data collected in this study shows that the highest average annual solar radiation data recorded is 5,764 Wh/m^2 in the Al Duqm area. Solar radiation data from 2000 to 2012 has been collected for further analysis. The Annual Energy Production (AEP) simulated using ArcGIS ranges between 6,028 Wh/m^2 and 3,668 Wh/m^2. This shows a very close estimation between simulation results and the data collected from the meteorological department in Oman.

Keywords: Solar, Resource Mapping, ArcGIS, Duqm, Oman

1. Introduction

Oman government has started to build a Special Economic Zone (SEZ) in the Al Duqm region, which is predicted to need around 1 GWh of electricity in 2020. The Al Duqm SEZ comprises a seaport industrial area, a new town, a fishing harbour, a tourist zone and a logistics centre, which are supported by a multimodal transport system connecting to the Arabian Gulf countries, Middle East, East Africa and South East Asia. However, these rapid developments may face the challenge of providing the sufficient electricity required in this SEZ that is environmentally clean. On other hand, Oman Vision 2020 seeks to reduce Oman's dependence on oil, diversify the economy and create new employment opportunities for all Oman citizens. Vision 2020 stresses the promotion of technology transfer and the increased use of natural and renewable energy resources, with due regard to the social and natural environments [1].

The demand for electricity energy has been continuously increasing in Oman. As projected in the Oman Power and Water Procurement (OPWP) 7-year statement (2012–2018) for long-term planning [2], it is estimated that the electricity demand will increase from 19 TWh to 31.2 TWh between 2011 and 2018, with an increase in the annual average growth rate of 8%.

Among the renewable energy resources, solar energy is becoming one of the most promising forms of power generation in Oman. Many studies have been conducted to determine PV locations in Oman considering their geographical locations. Charabi and Gastli [3,4,5] studied solar power prospects for Al Duqm in the Sultanate of Oman based on the master development plan and GIS solar radiation map. This solar mapping will help the government policy leaders to be more proactive in the development of solar energy and to create a viable solar energy market in the region. The solar mapping was conducted using ArcMap tools embedded in ArcGIS software. They found a very high potential of solar energy and solar electricity generation in Al Duqm throughout the year because of the high ratio of 'sky clearness' and the geographical location, which is located within a country with a very high potential of solar electricity generation that can easily rise up to 1 GW.

The main gaps of the above research include a lack of consideration of populated areas, roads, rivers and sensitive areas in the maps as well as future planning and developments of the country. Moreover, few studies combined solar energy with other energy sources or implementation of solar energy (pilot projects). Filling these gaps will give a more accurate prediction of the actual situation of the area and should be considered when studying the Al Duqm regions. Moreover, implementing pilot projects allows the researchers to obtain complete pictures of the surrounding effects of the renewable energy in that location. Furthermore, spatial analysis gives clear pictures of resource potential in the area of study.

2. Methodology

In this paper, two approaches will be applied to analyse the solar radiation in the area of Al Duqm. First, meteorological data derived from the monitoring station at an elevation of 2 m above mean sea level (MSL) on the coordinates of latitude 19037' N and longitude 57038'E are collected. The data obtained are assumed to represent a circumference of up to 150 km around this point, relying on single data points, and will be used to validate the data in the second approach.

The second approach will complement the first approach by providing the grid points results based on a numerical model using ArcGIS. The ArcGIS system provides solar radiation modelling analysis tools that calculate insolation across a landscape or for specific locations, based on methods from the hemispherical view shed algorithm developed by Rich et al. [6]. The solar radiation analysis tool, which is a part of the spatial analyst extension, enables users to map and analyse the effects of the sun over a geographic area for specific periods. The tool takes into consideration the atmospheric effects, site latitude elevation, steepness (slope) and compass direction (aspect), daily and seasonal shifts of the sun angle, and effects of shadows cast by surrounding topography. However, reflected radiation is not considered in the calculations of the total radiation; therefore, the total radiation is the sum of direct and diffuse radiation.

3. Resource Assessment

3.1 Observation method

The solar radiation data of the Al Duqm area were collected from the meteorology department over a 13-year period from 2000 to 2012. The data were checked, evaluated and treated to the final arrangement for use in this study. The sunshine period from November until the end of March was 11 hours and from April until the end of October was 12 hours. The annual average solar radiation of 13 years is 5,764.394 Wh/m^2.

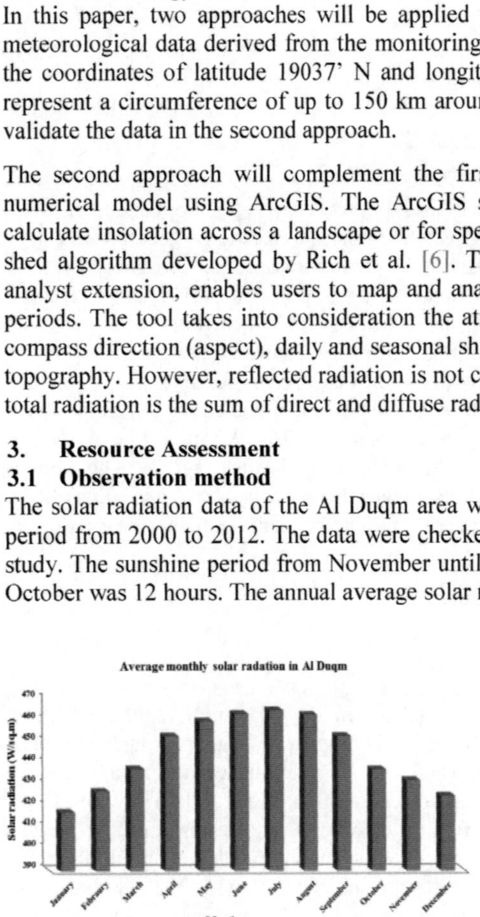

Figure 1: Hourly average solar radiation 2000-2012

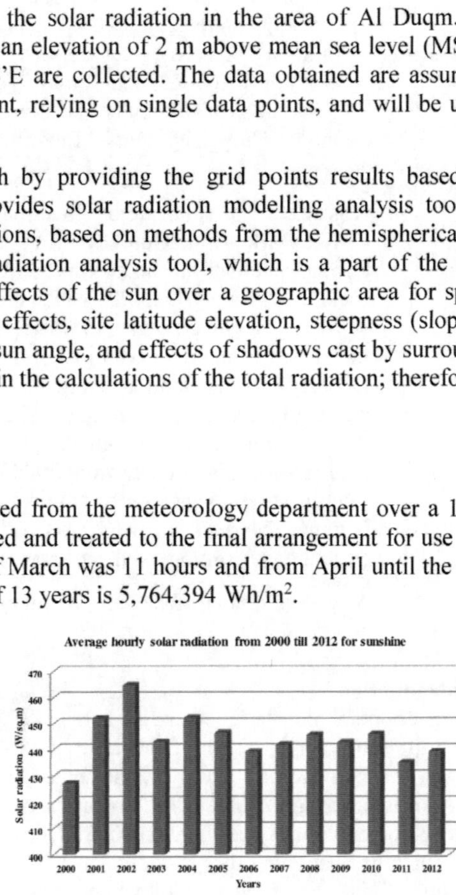

Figure 2: Monthly average solar radiation 2000-2012

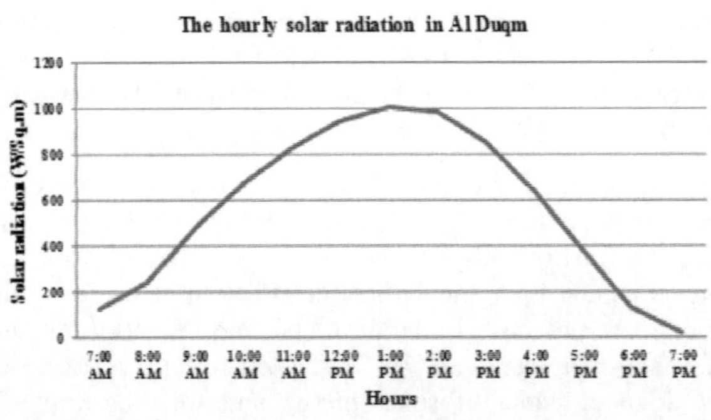

Figure 3: Hourly average solar radiation profile over 24 hours

Figure 1 shows the average hourly solar radiation from 2000 to 2012 is between 427 W/m^2 and 465 W/m^2, proving mostly stable over the 13-year period; thus, future solar radiation is expected to fall into an equivalent range. Figure 2 shows the monthly average solar radiation from 2000 to 2012. The peak solar radiation occurred in July (464.86 W/m2) and the lowest solar radiation occurred in January (417.45 W/m2). Diurnal variation of irradiation is important for the daily operation of any solar based electricity generation system. The hourly average solar radiation for a 24-hour profile is shows in Figure 3, where a peak irradiation of 1,005.56 W/m^2 occurred at 13:00 hrs.

3.2 ArcGIS method

Total monthly and annual global radiation is calculated using the DEM over the land of Duqm. Notice that the highest solar radiation per day (6,028 Wh/m^2) is occured during July (summer solstice) and the lowest (3,668 Wh/m^2) is occured in December (winter solstice); the solar potential is high throughout the year. Figure 5 shows the annual global solar radiation (GSR) over Duqm and the border of the SEZ of Duqm SEZAD.

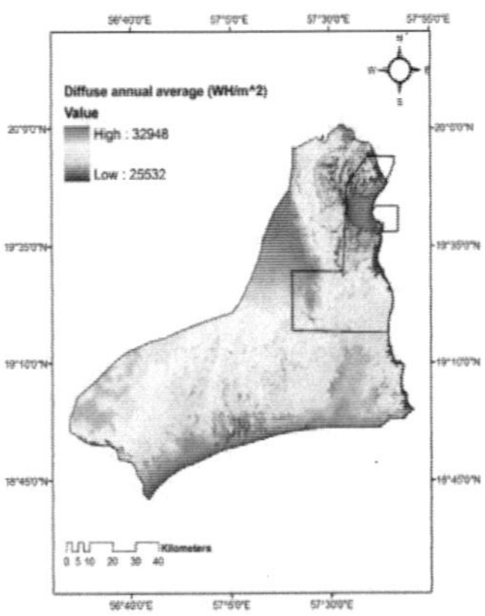

Figure 4: Global annual average solar radiation

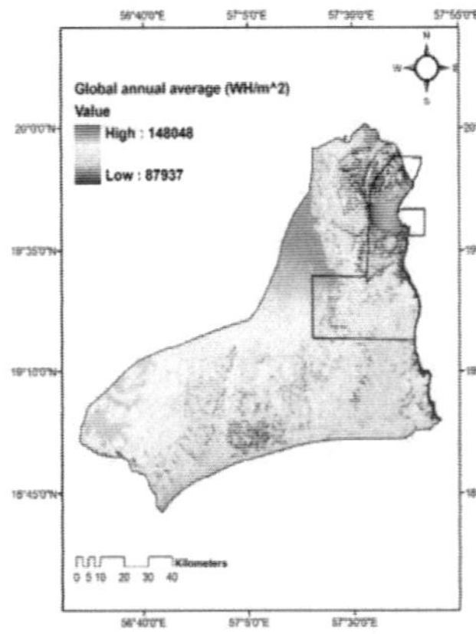

Figure 5: Direct annual average solar radiation

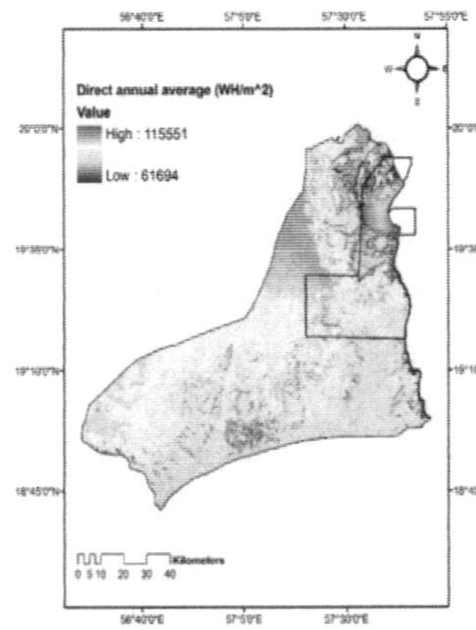

Figure 6: Diffuse annual average solar radiation

As it can be seen that the inland areas of Duqm receive significantly more solar radiation than the coastal areas. On the other hand, the west side of SEZAD has the highest solar radiation values. In this study, the main components of global radiation are direct and diffuse radiations, as shown in Figure 4, 5 and 6, respectively.

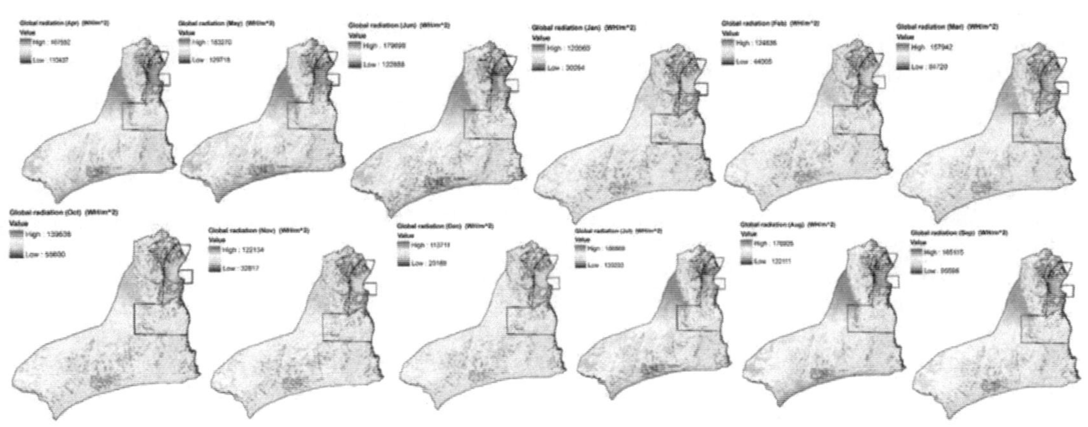

Figure 7: Simulated global monthly solar radiation using ArcGIS model

Direct radiation is intercepted unimpeded in a direct line from the sun, while diffuse radiation is scattered by atmospheric constituents, such as clouds and dust. About 75% of the GSR is direct radiation. In addition to measuring the annual solar radiation, it is very useful to analyse the monthly solar radiation for any resource assessment. Figure 7 shows the monthly GSR. The amount of solar radiation over Duqm increases gradually at the beginning of the year until it reaches its maximum during July and then starts decreasing until reaching its minimum in December, as shown in Figure 8

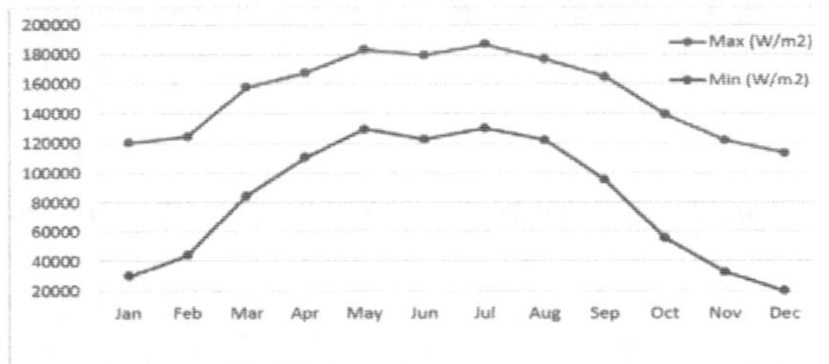

Figure 8: Average monthly variability of maximum and minimum GSR

4. Analysis of Results

Solar energy is a continuing and prevalent source of energy that is perfect for remote electrification in the recent circumstance. Previous studies have confirmed that acceptable production costs of solar electricity occur where radiation levels exceed about 1700 kWh/m²-yr. However, for utilisation of solar energy, it is necessary to know by how much the received solar radiation is weakened by the atmosphere. For clear sky conditions, 9% of direct solar radiation is exhausted by atmospheric scattering, 6% by surface reflection and 3% by other gases, smoke and dust. Moreover, the relative air mass, which is defined as the path length relative to that at the zenith at sea level, is an important parameter to consider when studying solar energy in such a selected location, and it increases as the angle between the source and the zenith increases.

In this study, the observation and ArcGIS methods were used to assess the energy resources in the Al Duqm industrial area in Oman. This study found that the annual average solar radiation over 13 years was 5,764.394 Wh/m^2. No significant difference was found between the results of the two methods. That is, the ArcGIS method recorded the highest daily solar radiation of 6,028 Wh/m^2 in July and the lowest of 3,668 Wh/m^2 in December, while the observation method found the highest daily solar radiation of 6,043 Wh/m^2 in July and the lowest of 5426 Wh/m^2 in December. However, Charabi and Gastli [5] reported a yearly solar radiation of 2357 Wh/m^2 for the Al Duqm area, which is low compared with the results of this study.

5. Conclusion

This paper shows the availability of solar energy resources for the Al Duqm industrial area in the Sultanate of Oman based on observations and analysis of meteorological weather data and the ArcGIS solar radiation map. The observation data set was used to validate the ArcGIS solar radiation map. The outcomes of the results show very high potential of solar energy and solar electricity generation in the Al Duqm over a year. The solar radiation varies between 6,028 Wh/m2/day and 3,668 Wh/m2/day at 1:00 pm and 6:00 pm, respectively. The annual average of solar radiation in the Al Duqm area is 5,764 Wh/m2/day with an average annual duration of sunshine of 11.5 hours.

References

[1] The Duqm Special Economic Zone Authority (SEZAD) (2011). Sultanate of Oman http://www.duqm.gov.om.
[2] OPWP's 7-Year Statement (2012-2018). Oman Power and Water Procurement Co. Issue 6, March 2012. http://www.omanpwp.com/PDF/Final%207YS%202012-2018.pdf
[3] Charabi Y, Gastli A. PV site suitable analysis using GIS-based spatial fuzzy multi-criteria evaluation. Renew Energy 2011;36:2554–61.
[4] Charabi Y, Gastli A. Solar electricity prospects in Oman using GIS-based solar radiation maps. Renew and Sustainable Energy Reviews 2010;14:790–7.
[5] Charabi and Gastli A. GIS assessment of large CSP plant in Duqm, Oman. Renew and Sustainable Energy Reviews 2010;14:835–41.
[6] Rich PM, Dubayah R, Hetrick WA, Saving SC. Using viewshed models to calculate intercepted solar radiation: applications in ecology. American Society for Photo-grammetry and Remote Sens Technical Papers 1994;524–29.

IOP Conf. Series: Earth and Environmental Science **32** (2016) 012037 doi:10.1088/1755-1315/32/1/012037

Effect of H_2 and CO contents in syngas during combustion using Micro Gas Turbine

N F Othman[1] and M H Boosroh
TNB Research Sdn Bhd,
No.1, Lorong Ayer Itam, Kawasan Institusi Penyelidikan, 43000, Kajang, Selangor, Malaysia.
E-mail: fadzilah@tnbr.com.my

Abstract. Synthetic gas or syngas is produced from the gasification process. Its main compositions are hydrogen, H_2; carbon monoxide, CO; methane, CH_4; carbon dioxide, CO_2 and nitrogen, N_2. Syngas is a substitute for the depleting natural gas (80-90%.vol. CH_4). Natural gas is combusted in gas turbine in gas-fired power plant to produce electricity. However, combustion of syngas using gas turbine is expected to show different behavior compared to natural gas combustion. This is because of H_2 and CO contents in syngas have higher adiabatic flame temperature than CH_4. In this study, different quality of syngas with different contents of H_2 (0.6-0.8 %.vol.) and CO (1-3 %.vol.) were combusted using 30kW Micro Gas Turbine (MGT). Performances of different syngas quality were studied using NO_x, CO, CO_2 emissions and combustion efficiency parameters. NO_x and CO are the main pollutants from the combustion process. NO_x emissions were the highest for syngas with H_2 contents of 0.8 %.vol. and CO contents of 3 %.vol. CO emissions were in the range of 220-310 ppm for all the tested syngas. While, CO_2 emissions were in the range of 0.96-1.06 % for all the tested syngas. Combustion efficiencies were reduced for syngas with CO contents of 1 %.vol. and H_2 contents of 0.6-0.8 %.vol. This is most probably due to the dilution effect of N_2 in syngas.

1.0 Introduction

The Integrated Gasification Combined Cycle (IGCC) produces electricity from a solid or liquid fuel. The fuel is first converted to syngas which is a mixture of hydrogen (H_2), carbon monoxide (CO) and methane (CH_4). The syngas is then converted to electricity in a combined cycle power block which consists of a gas turbine process and a steam turbine process with heat recovery steam generator (HRSG) (Figure 1).

The move towards the reduction of greenhouse gas emissions with the use of syngas as a fuel for the production of electricity is predicted to grow sharply in the near future as a result of:

- global warming and ocean surface acidification, which imply the need to control and reduce greenhouse gas emission,
- the low estimated reserves of oil, and
- rising cost as well as demand for natural gas and the need to reduce the dependence on fossil fuel imports from unstable areas.

[1] To whom any correspondence should be addressed.

Syngas has lower heating value than natural gas. When applying low heating value syngas, instead of natural gas, as fuel for an industrial gas turbine engine, some gas turbine performance problems are likely to occur. In this project, syngas is going to be tested using Micro Gas Turbine (MGT), in order to study the combustion performance of the syngas.

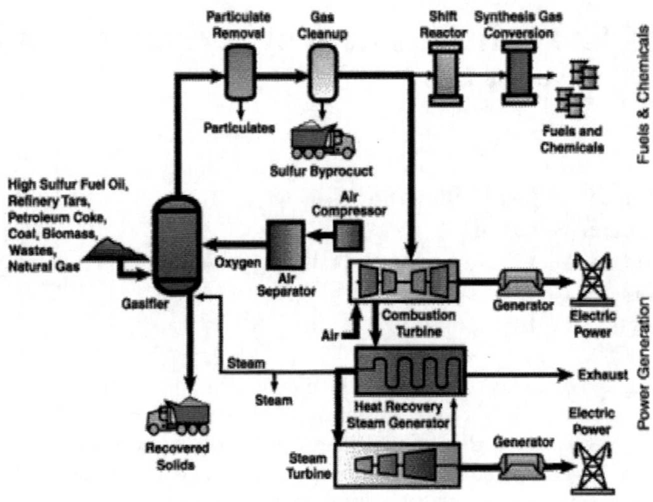

Figure 1. Integrated Gasification Combined Cycle (IGCC) (http://www.gasification.org/).

2.0 Materials and Method

Simulated syngas with different H_2 and CO contents and fixed content of CH_4 (60%.vol) were used as syngas sources (Table 1). Nitrogen, N_2 is the balance for all the syngas samples. Different types of syngas were combusted using Micro Gas Turbine (MGT) 30kWth at TNB Research Sdn Bhd, as shown in Figure 2 [1]. The highest contents of H_2 and CO in syngas that can be accepted for this type of MGT are 1 %.vol and 5 %.vol. Higher contents of H_2 and CO in syngas will contribute to longer flame length and is expected to damage the combustor [2]. Bench Top Gas Analyser, Siemen was used for measuring gas emissions (eg. NO_x, CO, CO_2) during combustion tests, as shown in Figure 3.

Table 1. Syngas with different $H_2$and CO contents

H_2 (%.vol) (in syngas)	CO (%.vol) (in syngas)	CO (%.vol) (in syngas)	CO (%.vol) (in syngas)
0.6	1	2	3
0.7	1	2	3
0.8	1	2	3

Combustion performances of different syngas quality were evaluated using NO_x, CO, CO_2 emissions and combustion efficiency parameters [3]. Combustion efficiency was calculated using Equation 1, as below,

$$\eta_c = 100\% - \frac{\text{Total Heat Losses} \times 100}{\text{Fuel Heating Value}} = \frac{\text{Total Heat Produced} \times 100}{\text{Fuel Heating Value}} \quad (1)$$

3.0 Results and Discussion

Performance parameters of CO, NO_x, CO_2 and combustion efficiency were studied against different contents of H_2 and CO in syngas. It was found that CO emissions from the combustion of syngas using MGT were slightly increased with the increasing of H_2 and CO contents in syngas (Figure 4). CO emissions were increasing as the adiabatic flame temperature increased due to the increasing of the H_2 and CO contents in syngas. High CO emissions also can be related to the incomplete combustion at lower flame temperature from the dilution effect of N_2 in syngas.

Figure 2. Micro Gas Turbine, Capstone 30 kWth.

Figure 3. Bench Top Gas Analyser, Siemen.

Table 2 shows that adiabatic flame temperature for H_2 and CO are higher than CH_4 [4]. This distinction contributes to low emissions from natural gas, which is mostly methane, compared to syngas containing H_2 and CO.

Table 2. Comparison of adiabatic flame temperatures.

Species	**Formula**	**Adiabatic Flame Temperature (K)**
Methane	CH_4	2223
Carbon Monoxide	CO	2381
Hydrogen	H_2	2370

Figure 5 shows that NO_x emissions were the highest for H_2 contents of 0.8 %.vol. and CO contents of 1-3 %.vol. This data trending had shown that higher contents of H_2 in syngas had prominent contribution for higher NO_x emissions compared to the increased of CO content in syngas. However, NO_x emissions were maintained below 50 ppm with the Lean Premix Combustion (LPM) design of MGT [4]. High level of NO_x emissions is contributing to the unhealthy levels of ozone in the air.

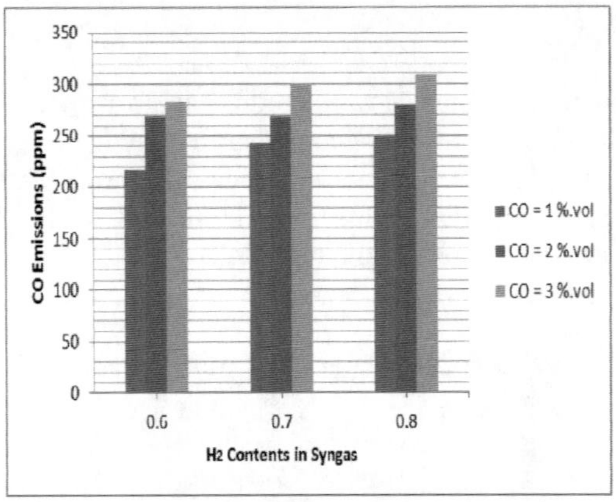

Figure 4. CO emissions from different contents of H_2 and CO in syngas.

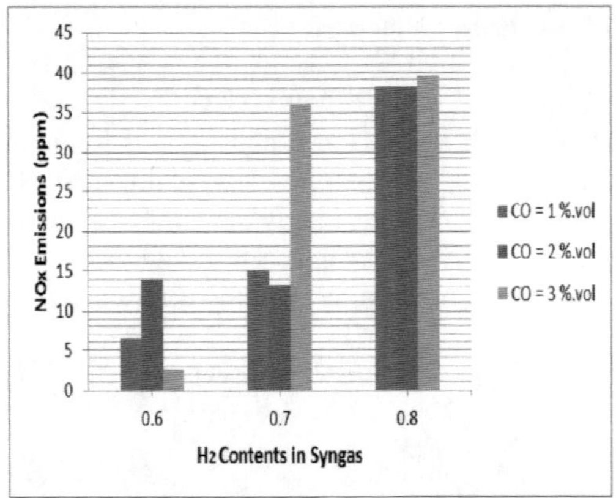

Figure 5. NOx emissions from different contents of H_2 and CO in syngas.

CO_2 emissions from the syngas combustion with different contents of H_2 and CO in syngas had shown steady increased (Figure 6). Especially for the syngas with CO content of 1 %.vol. and H_2 contents of O.6-0.8 %.vol. However, overall CO_2 emission is only within the range of 0.9-1.06 %.

Combustion efficiency for syngas with CO content of 1 %.vol were decreasing as H_2 contents in syngas were increased from 0.6 %.vol. to 0.8 %.vol. The highest combustion efficiency was at CO contents of 1%.vol. and H_2 contents of 0.6 %.vol. The combustion efficiencies for other syngas were slightly changed (Figure 7). Again, the incomplete combustion due to the dilution effect of N_2 in syngas was expected to reduce the combustion efficiency. This finding is aligned with Sadig et al [5]. Overall, combustion efficiency with the presence of H_2 and CO is syngas is higher (>77%) than in 60% CH_4 (~65%) from previous study.

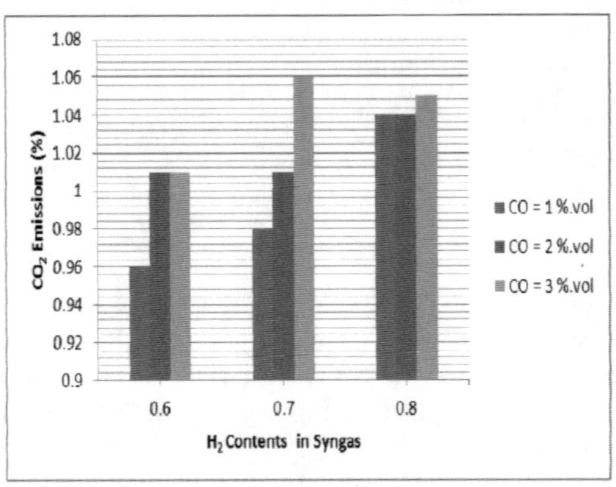

Figure 6. CO_2 emissions from different contents of H_2 and CO in syngas.

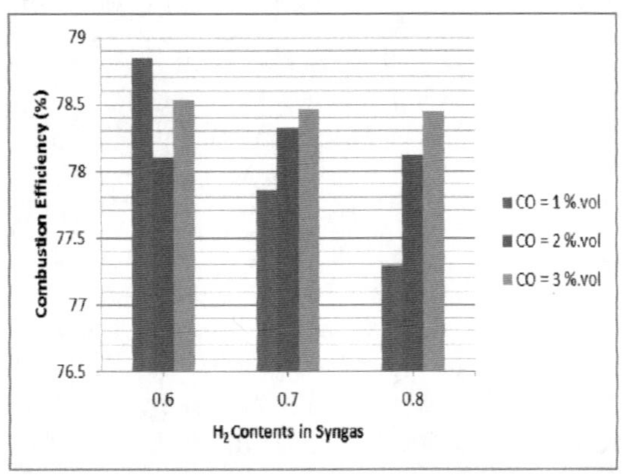

Figure 7. Combustion efficiency from different contents of H_2 and CO in syngas.

4.0 Conclusions

H_2 and CO contents in syngas contributed to higher flame temperature in combustor. It was found that high emissions of CO and NO_x occurred at higher H_2 and CO contents in syngas. However, combustion efficiency can be increased with the presence of H_2 and CO contents in syngas. Combustion efficiency is the highest at H_2 content of 0.6 %.vol. and CO contents of 1%.vol. in syngas.

5.0 References

[1] Technical Reference. Capstone MicroTurbine Fuel Requirements 2009
[2] Verstraete D and Bowkett C 2015 Impact of heat transfer on the performance of micro gas turbines *Applied Energy*. **138** 445-9.
[3] Delattin F, Di Lorenzo G, Rizzo S, Bram S and De Ruyck J 2010 Combustion of syngas in a pressurized microturbine-like combustor: Experimental results *Applied Energy*. **87** 1441-52
[4] Bender W R, Lean pre-mixed combustion 2015 *www.netl.doe.gov/File%20Library/Research/Coal/.../3-2-1.pdf*
[5] Sadig H, Sulaiman S A, Said M A, Yusup S 2014 Performance and emissions of a micro-gas turbine fueled with LPG/producer gas in a dual fuel mode *akademiabaru.com/wvcarmea/docu1052.pdf*

Acknowledgement

This research was jointly supported by Seeding Fund and AAIBE Fund from TNB Research Sdn Bhd and KETTHA.

Numerical analysis of Al coating using different particle shape in LPCS

S N A Yusof[1], A Manap[1] and N F Afandi[1]
[1]Mechanical Department, College of Engineering, Universiti Tenaga Nasional, Kajang, Malaysia.

E-mail: sitinuruakmal.yusof@yahoo.com; Abreeza@uniten.edu.my

Abstract. Cold spray (CS) is a unique spraying process where the spray materials are not melted in a spray gun. Instead, the particles are kinetically deposited on the substrate at low temperature using compressed gas. This study investigates the deposition behaviour of different particle shape of Al coating using low pressure cold sprayed (LPCS) through smoothed particle hydrodynamics (SPH) simulations, which are achieved by modelling the multiple particle impacts on Al substrate. The impact of Al particle on the Al substrate is analysed by evaluating the shape of deformation, porosity between particle, and effect of stress on the substrate. The results show that the irregular particle shapes (horizontal and vertical ellipse) tend to detach the bonded particle from the substrate and thus increase the potential risk of high tensile stress. That is really harmful to the coating quality, which never happens for spherical particle. Deposition using irregular particle exhibits tensile stress at the depth coating, whereas spherical particle exhibits compressive stress. Compressive stress is generally ensure a longer component life due to their positive effect on the fatigue life and wear resistance application.

1. Introduction

Cold spray is the process that produces coating with a very low oxygen and porosity content [1]. Powder particles are accelerated to a velocity ranging from 200 m/s to 1200 m/s through a de Laval type nozzle by a supersonic compressed gas jet at a temperature below the melting point of the powder material [2]. The velocity is approximately 200 – 1200 m/s depending on the particle size and shape, type of process gas and conditions such as pressure and temperature [2]. Bonding in cold spray occurs when the powder particles impact on to the substrate above the critical velocity.

Al is widely known as a non-ferrous metal with high corrosion resistance [3-4]. Lightweight metal combined with the high strength is the reason Al is the major material for aircraft construction for the past sixty years [5]. The metal is ductile and strong which led to the widespread use in transportation and as an effort to reduce fuel consumption which in turn reduces emission and consequent global impact [6].

According to Lee et al, [7] Al coating deposited using LPCS has higher hardness because of the peening effect. Previous study by Yusof et al., [8] proved that LPCS is suitable for the deposition of light materials such as Al. However, it is noticed that the existing works regarding the deposition behaviour of Al particle using the optimum velocity of LPCS mainly focused on the spherical powders, very few studies concerned about the irregular powders. Irregular particle, including horizontal and vertical ellipse are more easily accelerated, obtaining higher impact velocity. But then, they are easier to be produced with shorter time and lower cost in comparison with spherical powder. Therefore, it is of great importance to pay more attention on the different particle shape. Investigating the particle and substrate deformation via experiment is difficult because of the short duration of the deposition process [1]. Based on these results, numerical simulation can best elucidate the LPCS deposition mechanisms [3].

In this study, the deposition mechanism of Al particle using different particle shape were investigated by SPH simulations of multiple particle impact of Al particle on Al substrate. An analysis was conducted by evaluating deformation behaviour, porosity between particle, and residual stress on the substrate to understand the deposition mechanism of LPCS process.

2. Numerical Method

The impact of LPCS particle was modelled by SPH, a mesh free adaptive Lagrangian particle method. This numerical technique uses smoothed particles, instead of a mesh, as interpolation points to represent materials at discrete locations [1, 9]. Al was selected as the particle and substrate material for irregular multiple particle impacts. For the different particle shape simulations, LPCS simulation was performed in this study with initial particle velocity LPCS was at 400 m/s. Ogawa et al. [3] reported that the critical velocity of Al under LPCS is approximately 300 m/s. Based on this finding, any value above this velocity is capable of depositing on the substrate. The particle size was set to 25 μm and the initial temperature for the particle and substrate was assumed to be at room temperature (300 K). The parameters of the Johnson Cook model and the state of equation for Al were obtained from literature [3, 10].

3. Result and discussion

The effect of different particle shape on the impact behaviour was simulated for Al/Al for spherical and irregular particle shape (horizontal and vertical ellipse) using LPCS at 400 m/s. The shape deformation behaviour and temperature contour of different particle shape is given in Figure 1 to Figure 3. The spherical particle shape at the bottom (label 1) in Figure 1 changed to lens shape and deeply penetrated substrate after the impacting process. It is clear that the metal jets are formed at the edge of each particle. The powder particle at the top (label 2) are slight deformed. The powder particle at the bottom (label 1) using horizontal ellipse which is shown in Figure 2 was deformed with a slight deformation and no deformation occurred on the substrate. The results from magnified view show that the deformation of particles after deposition. Besides, it can be seen that the top particle (label 2) of horizontal ellipse almost has no change, the shape still remain which may increase the risk of detachment of bonded particle from the substrate and thus may resulted in large pores forming between these particles. These results are due to the low plastic deformation dissipated in the particles during deposited process. For the vertical ellipse is given in Figure 3 shows that the particle at the bottom (label 1) experience deformation, thus the particle at the bottom completely flatten but there was very slight deformation occurring on the substrate. Unfortunately, there is almost no change occurred for the top particles (label 2) of vertical ellipse. There are no changes for the top particle (label 2) due to the low plastic deformation dissipated in the particles which may increases the pores formation and resulting to the worse bonding between particle and particle. Based on this reason, the optimum velocity of irregular particle need to be higher than that of spherical particle in order to promote better bonding. A study performed by Yin et al., [11] also reported that the optimum velocity of irregular particle shape need to be higher than spherical particle shape.

Figure 4 shows the flattening ratio of the particle deformation for different particles shape. It is obviously seen that the flattening ratio of the vertical ellipse has the highest value, achieving 2.4, followed by horizontal ellipse, achieving to 1.52, then followed by spherical particles, amounting to 1.14. The gap of the flattening ratio between particles at the bottom (label 1) and at the top (label 2) show larger differences for vertical and horizontal. The larger differences of the flattening ratio between label 1 and label 2 will results in higher pores form between particles. This is mainly attributed to the different size of radius particle. Spherical particles shape has a same radius in axis – x and y. But, for the horizontal ellipse, axis – x has a major radius and a minor radius in axis – y. Otherwise, vertical ellipse has a major radius in axis – y and minus radius in axis – x.

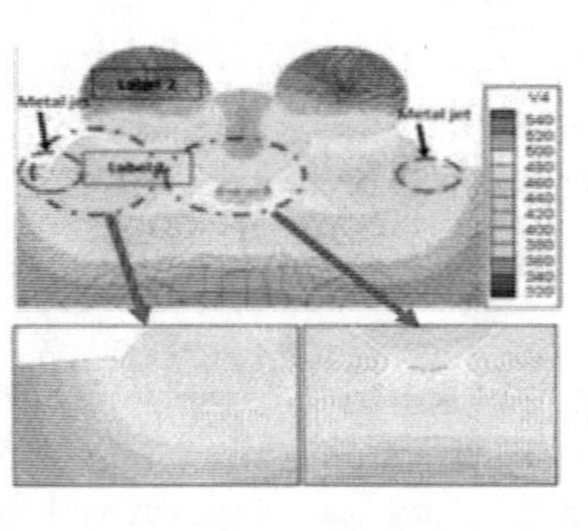

Figure 1. Deformation pattern for spherical particle shape.

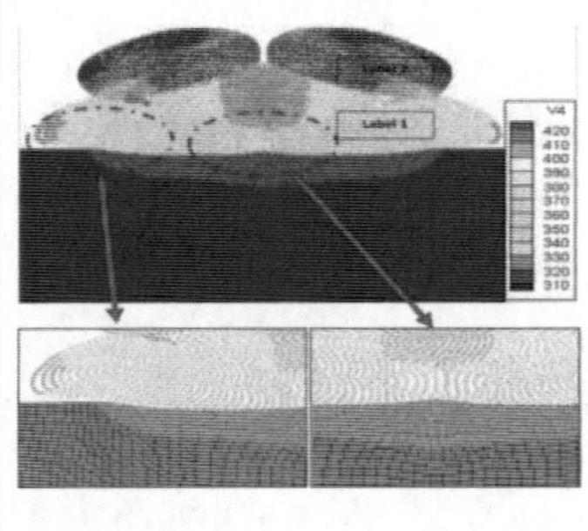

Figure 2. Deformation pattern for horizontal particle shape.

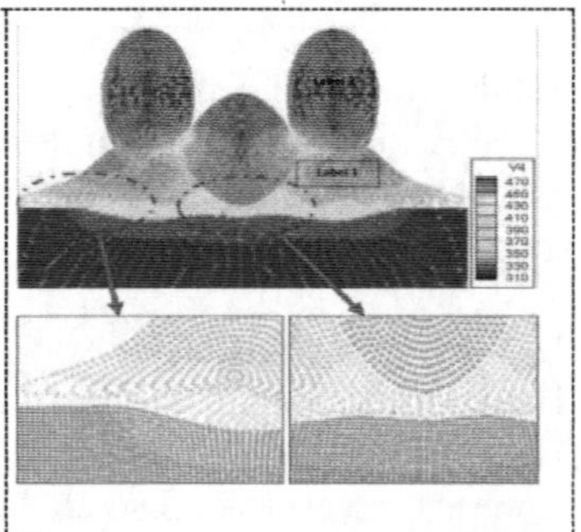

Figure 3. Deformation pattern for vertical particle shape.

However, it is significant that the temperature in the region of the most severely deformed particle near the substrate is the highest mainly due to the subsequent impact from top particles (label 2). Time history of the maximum temperature of different particle shape is given in Figure 5. Clearly, the simulation results demonstrate that for horizontal case, particles at the bottom (label 1) flatten due to the subsequent impact from another particles nearly absorb all the kinetic energy which contribute for the particles flattened on the substrate. For the vertical case, the particles at the bottom (label 1) has the highest value of flattening ratio, but the particles did not experience intensive deformation on the substrate due to the all total energy dissipated in the particles rather than in the substrate. There is a good agreement between the results obtained by this simulation analysis with Yin et al., [11]. A study performed by Yin et al., [11] using irregular particle reported that coating deposited with irregular particle shape shows higher coating porosity and worse bonding than coating deposited with spherical particle shape. A by Ghelichi et al., [12] also claimed a similar result by using different aluminium powders. Both studies provide strong evidence to the numerical simulation in this study.

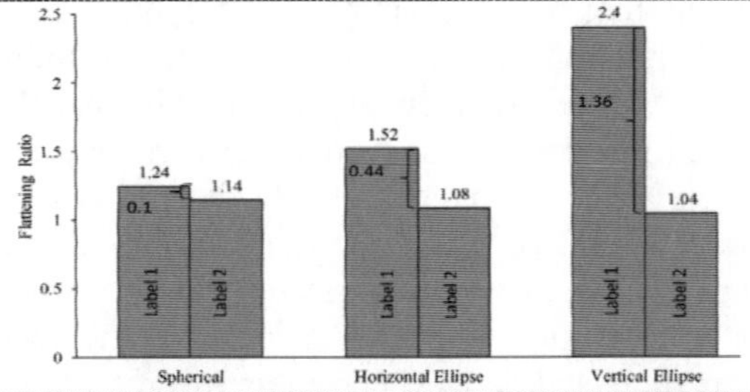

Figure 4. Flattening ratio of different particle shape.

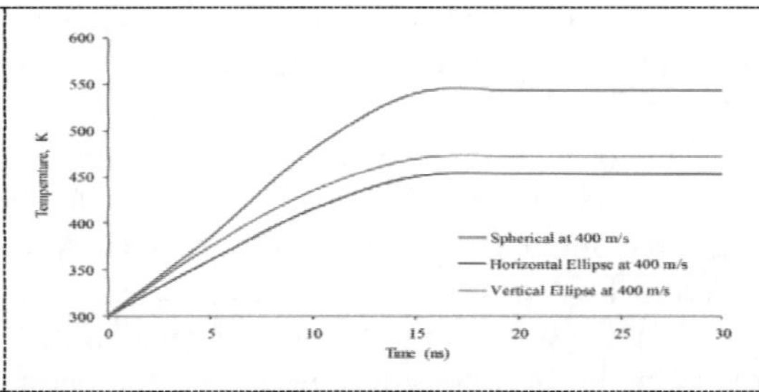

Figure 5. Time history of the maximum temperature of different particle shape.

The through-thickness residual stress profile for the spherical and different particles shape, irregular (horizontal and vertical particles shape) are shown in Figure 6. At the depth of 10 μm, the stress for horizontal which is shown in triangle shape starts compressive above – 400 MPa. Whereas, for vertical ellipse starts with the compressive stress, 325 MPa and begin to reach the tensile stress at the depth of 60 μm, 100 MPa. The dissimilarity of the results is mainly attributed to the different value of the total energy and temperature. These result compare fairly well with Shayegan et al., [13]. A study performed by Shayegan et al., [13] shows that when particles shape (irregular) pass the spherical particles shape, compressive residual stress on the surface increases and tensile stress spreads to the depth and become tensile.

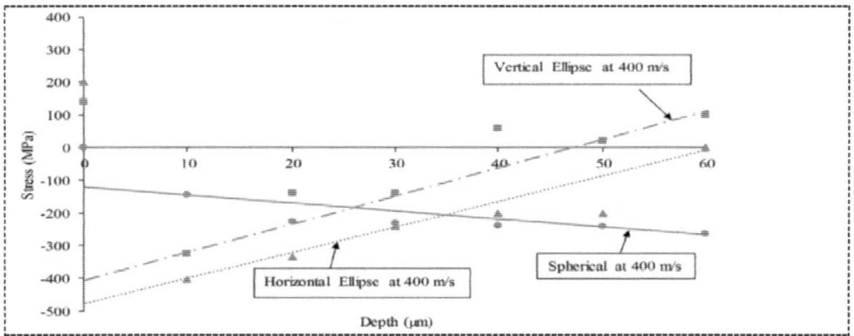

Figure 6. The comparison of the through-thickness residual stress measured in the different particles shape.

4. Conclusion

In this study, numerical analysis was used to investigate the Al coating using different particles shape in LPCS. The optimum velocity of irregular particle need to be higher than that of Al spherical particle in order to promote better bonding. These results are due to the low plastic deformation dissipated in the particles during deposited process. Deposition using irregular particle exhibits tensile stress at the depth coating, whereas spherical particle exhibits compressive stress. Compressive stress is generally ensure a longer component life due to their positive effect on the fatigue life and wear resistance application.

Acknowledgment
This study was supported by the Fundamental Research Grant Scheme (FRGS20130108) and UNITEN Internal Grant (J510050514), which is funded by the Ministry of Higher Education.

Reference

[1] Manap A, Nooririnah O, Misran H, Okabe T and Ogawa K 2014 *Surf. Eng.* **30** 5
[2] Schmidt T, Gartner F, Stoltenhoff H, Assadi T and Kreye H 2006 *Acta Mater.* **54** 729.
[3] Ogawa K, Ito K, Ichimura K, Ichikawa Y, Ohno S and N. Onda 2008 J. *Therm Spray Technol.* 17 728-735.
[4] Soria A. 2007 Prospective Study of the World Aluminium Industry.
[5] Cobden R. and Banbury A 1994 Aluminium: Physical Properties, Characteristics and Alloys. Training in Aluminium Application Technologies.
[6] Ghassemieh E 2011 New Trends and Developments in Automotive Industry 365-394.
[7] Lee H, Shin H, Lee S and Ko K 2008 *Mater. Lett.* **62** 1579-1581.
[8] Yusof S N A, Manap A, Misran H and Othman S Z 2014 Adv. Mater. Res. **974** 147-151.
[9] Randles P W and Libersky L D 1996 Comput. Methods in Appl. Mech. Eng. **139** 375-408.
[10] Batra R C and Kim C H 1991 *Int. J. Eng. Sci.* 29 949-960.
[11] Yin S, He P, Liao H and Wang X 2014 J. *Therm Spray Technol.* **23** (6) 984–990.
[12] Ghelichi R., Bagherifard S., MacDonald D., Fernandez-Pariente I., Jodoin B., and Guagliano M. 2014. *Appl. Surf. Sci.* **288** 26–33.
[13] Shayegan G, Mahmoudi H, Ghelichi R, Villafuerte J, Wang J, Guagliano M and Jahed H 2014 *Mater. Des.* **60** 72-84.

IOP Conf. Series: Earth and Environmental Science **32** (2016) 012039 doi:10.1088/1755-1315/32/1/012039

The Production of Biodiesel and Bio-kerosene from Coconut Oil Using Microwave Assisted Reaction

SAIFUDDIN N., SITI FAZLILI A., KUMARAN P., PEI-JUAN., PRIATHASHINI P.
Centre for Renewable Energy,
Universiti Tenaga Nasional,
Jalan IKRAM-UNITEN, 43000 Kajang, Selangor, Malaysia
Email: saifuddin@uniten.edu.my

Abstract. Biofuels including biodiesel, an alternative fuel, is renewable, environmentally friendly, non-toxic and low emissions. The raw material used in this work was coconut oil, which contained saturated fatty acids about 90% with high percentage of medium chain (C_8-C_{12}), especially lauric acid and myristic acid. The purpose of this research was to study the effect of power and NaOH catalyst in transesterification assisted by microwave for production of biofuels (biodiesel and bio-kerosene) derived from coconut oil. The reaction was performed with oil and methanol using mole ratio of 1:6, catalyst concentration of 0.6% with microwave power at 100W, 180W, 300W, 450W, 600W, and 850W. The reaction time was set at of 3, 5, 7, 10 and 15 min. The results showed that microwave could accelerate the transesterification process to produce biodiesel and bio-kerosene using NaOH catalyst. The highest yield of biodiesel was 97.17 %, or 99.05 % conversion at 5 min and 100W microwave power. Meanwhile, the bio-kerosene obtained was 65% after distillation.

1. Introduction

The use of petroleum still dominates the energy consumption in Malaysia. Policies were introduced to support the development of renewable energy so as to substitute a part of fossil based fuels with biofuels. Methyl (or ethyl) ester is obtained during the process of transesterification of plant oils or animal fats with the presence of methanol or ethanol and some basic catalyst. Biodiesel (most common type of methyl ester) is a nontoxic, biodegradable, clean burning fuel for diesel engines made from renewable fats and oils [1]. Aviation fuels have a number of requirements: they must stay liquid at low temperatures, and have high energy content by volume. Fuels such as diesel, biodiesel or bioethanol do not fit these requirements well. This kind of alternative jet fuel is known as bio-kerosene. Bio-kerosene is another type of methyl ester.

Coconut oils (*Cocos nucifera*) consist of various kinds of triglyceride fatty acids, which saturated fatty acid accounted 90% and are rich of medium chains (C_8-C_{12}), especially lauric acid and myristic acid [2]. Coconut oil as a raw material has advantages compared other vegetable oils, it contain around 65% medium chains of fatty acids. These medium chain fatty acids are good feed stock to obtain other fuels such as bio-kerosene, or jet fuel [2].

The development of biodiesel applying microwave assisted reaction to accelerate the transesterification reaction has been reviewed by many authors [3-5]. Microwave provides many advantages as follows: fast startup time, faster heating, efficient energy and the lower process cost, easy to monitor, precise, selective heating and the better quality of product [6]. The technology of bio-kerosene manufacturing is similar to the technology of biodiesel production. This fuel would have similar properties to petroleum-derived kerosene (the fuel currently used for aviation). However, application of bio-kerosene in aviation needs further detailed investigation and optimization of its production technology. This research was aimed to study the development of the process of the preparation of biodiesel and bio-kerosene from coconut oil by utilizing the microwaves as an energy source and to see the impact of the microwave exposure on the yield of the products.

2. Methods and Experimental Setup

2.1 Materials and Equipment

Food grade coconut oil was procured from a local supermarket. Anhydrous methanol, sodium hydroxide (NaOH) and other chemicals used were all of analytical reagent grade. There were from MERCK-Fluka and obtained from a local chemical supplier.

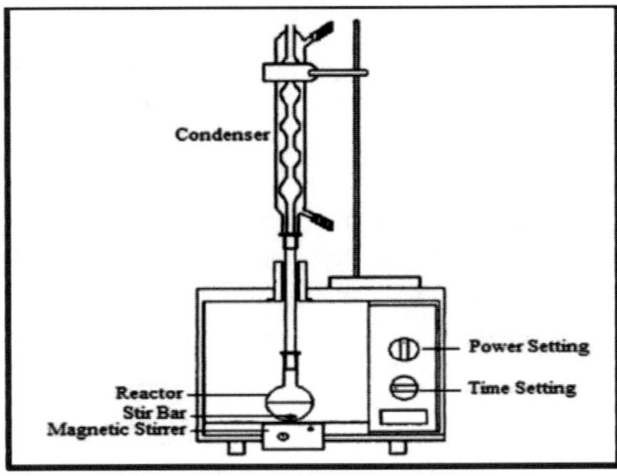

Figure 1. The experiment setup for transesterification process assisted by microwave irradiation

The microwave irradiation was used to carry out the reaction to produce biodiesel. The batch reactor equipment used in this study is shown in Figure 1. The reactor was a flat bottom flask made from Pyrex glass equipped with magnetic stirrer. The microwave irradiation was carried out using a domestic microwave oven (Samsung, CE2877N, S. Korea) with an operating frequency of 2450 MHz. The microwave oven provided microwave radiation at variable power levels of 100 W, 180 W, 300 W, 450 W, 600 W and 850 W and time setting from 0-30 min. The separation of the biofuels into biodiesel and bio-kerosene was carried out by employing a vacuum distillation using a rotary evaporator (Yamato, Japan-YMT255116).

2.2 Transesterification Reaction Using Microwave Heating

Coconut oil was introduced into the reactor and mixture of methanol and sodium hydroxide (NaOH) were added. The molar ratio oil to the methanol was 1:6 (Note: 1 mol oil is estimated as 950 g/mol and 1 mol methanol is 40 mL) and alkaline catalyst used for reaction was 0.6%.wt of the oil. Two hundred grams of coconut oil (0.21 mol) was fed to the reactor and mixed with 50.4 mL methanol (1.26 mol) per run. The solution was reacted in a 500 ml reaction flask attached to a reflux set apparatus. The power was set at 100 W and 3 min. The reaction mixture was stirred at 200 rpm. The process was repeated for different time setting (5, 7, 10 and 15 min) while keeping the power at 100W. Further experiments were carried out at 5 min but with different power setting. After completion, the product was left at room temperature for separation. Separation took only 2 min. The upper layer was methyl ester and the bottom layer was glycerol. The methyl ester was washed using warm water for three times and the product was dried at 100^oC for 1 h in conventional oven to get rid of the water. The methyl ester obtained was followed by vacuum distillation using a rotary evaporator to separate bio-kerosene from biodiesel at 185^oC and 70 mmHg.

3. Results and Discussion

3.1 Effect of Reaction Time on Product Viscosity and Yield

Coconut oil and palm kernel oil are two important tropical plant oils that can be mass produced and satisfy medium chain requirement and have dominant fatty acid as lauric acid, which is the important point for production of bio-kerosene. If the feedstock is proper, then no cracking step is necessary, results in using the simple production technology, reducing investment and production costs for aviation biofuel. The composition of fatty acid of coconut oil shows that the highest contents were 51.02 % of lauric acid and 18.94 % of myristic acid, which were reported in previous studies [2, 6, 7].

Experiment was carried out with 0.6wt % NaOH for various microwave powers, in which input power influenced to the amplitude of wave. If the electric field rises, (greater power) the amplitude of the microwave increases. The rotational speed of the polar molecules has a linear relationship to the amplitude of the microwave.

Results show (figure not shown) that there is an effect of the microwave irradiation time on the viscosity of methyl ester (biofuel) produced. It was found that the viscosity decreased drastically from 28.3 at the initial time of the reaction (taken as control) to 3.05 mm^2/s at 3 min and was relatively constant for the rest of the time until 15 min. It was indicative that the reaction had completed in 3 min. It was interesting to note that the viscosities of products were relatively similar for all the other powers used.

The results (figure not shown) for the yield of product obtained with respect to reaction time conducted at 100W is as follows: The yields of product increased to 91.20% at 3 min and then to 97.17% at 5 min. After 5 min, the yields slightly decreased and then were relatively constant. This indicated that the longer time did not significantly influence the yield of methyl ester (mixture of biodiesel and bio-kerosene). It was characteristic of microwave assisted reaction in biodiesel preparation, that less time is needed for the reaction to achieve high yields of around 97% [3, 8].

Generally, from the results (figure not shown) for the effect of microwave power on yield of biodiesel (conducted at 0.6 wt% NaOH for 5 min. and molar ratio of oil to methanol of 1: 6), there is a decrease in biodiesel yield as the power is increased, which was similar to other works [9]. The increase of microwave power caused the damage of organic substances, such as triglycerides [3]. The energy required for the microwave method was 10 times lower than that of the conventional method. The results showed that the use of microwave in transesterification reaction with NaOH was better conducted within a low power 100 W. It was also found that preparation methyl ester (biodiesel and/or bio-kerosene) by microwave assisted reaction takes only 5 min as compared to conventional technique which takes about 1 h [8]. The highest yield of methyl ester obtained was 97.17% at 100W for molar ratio of oil to methanol 1:6 and 0.6 wt% NaOH.

The preparation of methyl ester conducted above 60°C using NaOH and conventional heating was not effective because of the rivalry between transesterification and saponification reactions as previously noted by some researchers [11]. Therefore, the process of transesterification reaction using 0.6 wt% NaOH in microwaves, carried out with shorter time (5 min) was not only rapid and economical but also gave higher yield of methyl ester. It is anticipated that at low power level of microwave irradiation non-thermal effects of microwave play a role. At low power level the reactive side of the catalyst may undergo conformational changes to favor efficient substrate-catalyst interaction which will enhance the reaction rate [11].

3.2 Separation of Biodiesel and Bio-kerosene

The methyl ester prepared was then followed with the separation process to obtain biokerosene. The separation was conducted by heating at 185°C and 70 mmHg using a rotary evaporator. The time taken was 45 min. The bio-kerosene was evaporated and condensed to be the distillate; meanwhile the portion of the methyl ester that remained which did not escaped because of higher boiling point was the biodiesel. The biodiesel product, which was not evaporate was golden yellow, and the distillate (bio-kerosene) was more transparent. The percentage of bio-kerosene obtained from the methyl ester after distillation was 65% while the remaining was biodiesel.

3.3 Biodiesel Properties of Coconut Biodiesel

The quality of the product is determined by several parameters, such as cetane number, kinematic viscosity, density, and others. Table 1 describes the comparisons of biodiesel properties obtained with those of biodiesel standard value (EN14214). It was found that the product of biodiesel met the European standard (EN) and ASTM standards for biodiesel.

Table 1. Comparison of biodiesel product properties with the biodiesel standard properties values (based on European Standard –**EN 14214**)

Physical Properties	EN14214	Sample from this work
Density at 15°C	860-900 kg/m^3	868 kg/m^3
Viscosity at 40°C	3.5-5.0 mm^2/s	3.91 mm^2/s
Acid number	< 0.50 mg KOH/g	0.07 mg KOH/g
Flash point	120°C (min)	130°C

Table 2. Comparison of bio-kerosene properties produced after distillation with the bio-kerosene standard properties values (based on **ASTM-D1655**).

Physical Properties	ASTM 1566	Sample from this work
Density at 15°C	775-840 kg/m^3	768 kg/m^3
Viscosity at -20°C	8.0 mm^2/s (max)	6.72 mm^2/s
Freezing point	Max -47.0°C	10.2°C
Flash point	38°C (min)	45°C
Lower heating value	42.8 MJ/kg (min.)	42.9 MJ/kg

For the case of bio-kerosene it must comply with the ASTM D1655 standard for some critical parameters such as freezing point, density, flash point, viscosity, lower heating value.

As can be seen in table 2, the bio-kerosene does not meet the criteria for freezing point as well as the flash point. The requirement for freezing point, which is −47°C, is perhaps the most stringent and the hardest to meet. Blending the bio-kerosene with the mineral kerosene may help to improve the parameters. The other way to improve the freezing point is by adding certain amount of aromatic compounds. Typically propylbenzene at 10-20% vol. can be added. No attempt was made in this study to see the effect of additives on the freezing point of the biokerosene.

4. Conclusion

The biofuels, biodiesel and bio-kerosene were successfully prepared using the microwave, which decreased the reaction time. It was found that the greater the power used, the lower was the yield of biodiesel. The highest yield was 97.17 %, at power 100 W for 5 min using 0.6 wt% NaOH. The methyl ester containing mixture of biodiesel and bio-kerosene was separated applying vacuum distillation at 185°C. The distillate (bio-kerosene fraction) obtained was 65%. In general, the biodiesel met with standard quality required, however the bio-kerosene was not up the requirement for jet–fuel quality especially in terms of the freezing point. The quality could be improved by blending with petroleum based kerosene or by addition of aromatics.

References

[1] Oliveira, A. N., Costa, L. R. S., Pires, L. H. O., Nascimento, L. A. S., Angelica, R. S., Costa, C. E. F., Zamian, J. R., & Filho, G. N. R. (2013). Microwave Assisted Preparation of a New Esterification Catalyst from Wasted Flint Kaolin. *Fuel*, *103*, 626-631.

[2] Chowdhury K., Banu L. A., Khan S., and Laif A. (2007). *Studies on the Fatty acid Composition of Edible Oil*. Bangladesh J. Sci .Ind. Res. 42 (3), pp. 311 - 316.

[3] Saifuddin, N., & Chua, K. H. (2004). Production of Ethyl Ester from Used Frying Oil: Optimization of Transesterification Process Using Microwave Irradiation. *Malaysian J. Chem*, *6*, 77-82.

[4] El Sherbiny, S., Refaat, A. A., & El Sheltawy, A. T. (2010). Production of Biodiesel using Microwave Technique. *Journal of Advance Research, 1*, 309-314.

[5] Tippayawong, N., & Sittisun, P. (2012). Continuous Flow Transesterification of Crude Jatropha Oil with Microwave Irradiation. *Scientia IranicaB, 19*(5), 1324-1328.

[6] Jiputti, J., Kitiyanan, B., Rangsunnivit, P., Bunyakiat, K., Attnatho, L., & Jenvanipanjakul. P. (2006). Transesterification of Kernel Oil and Crude Coconut Oil by Different Solid Catalysts. *Chemical Engineering Journal, 116*, 61-66.

[7] Llamas A, Martínez MJG, Lal AMA, Canoira L, Lapuerta M. (2012). Bio-kerosene from coconut and palm kernel oils: production and properties of their blends with fossil kerosene, Fuel 102 (2012) 483–490.

[8] Chen, K. S., Lin, Y. C., Hsu, K. H., & Wang, H. K. (2012). Improving Biodiesel Yield from Waste Cooking Oil by Using Sodium Methoxide and a Microwave Heating System. *Energy*, *38,* 151-158.

[9] Galema, A. S. (1997). Microwave Chemistry. *Chemical Society Reviews, 26*, 233-238.

[10] Pathil, P. D, Gude, V. G, Camacho, L. M., & Deng, S. (2010). Microwave Assisted Catalytic Transesterification of Camela Sativa Oil. *Energy Fuels*, *24*, 1298-1304.

[11] Saifuddin N., Saltanat A., Refal H., (2014). Enhancing the Removal of Phenolic Compounds from Palm Oil Mill Effluent by Enzymatic Pre-treatment and Microwave-Assisted Extraction. Chemical Science Transactions 3(3), 1083-1093.

The Feasibility of Palm Kernel Shell as a Replacement for Coarse Aggregate in Lightweight Concrete

Zarina Itam[1*], Salmia Beddu[1], Nur Liyana Mohd Kamal[1], Md Ashraful Alam[1], Usama Issa Ayash[1]
[1]Department of Civil Engineering, Universiti Tenaga Nasional, Malaysia

E-mail: iZarina@uniten.edu.my

Abstract. Implementing sustainable materials into the construction industry is fast becoming a trend nowadays. Palm Kernel Shell is a by-product of Malaysia's palm oil industry, generating waste as much as 4 million tons per annum. As a means of producing a sustainable, environmental-friendly, and affordable alternative in the lightweight concrete industry, the exploration of the potential of Palm Kernel Shell to be used as an aggregate replacement was conducted which may give a positive impact to the Malaysian construction industry as well as worldwide concrete usage. This research investigates the feasibility of PKS as an aggregate replacement in lightweight concrete in terms of compressive strength, slump test, water absorption, and density. Results indicate that by using PKS for aggregate replacement, it increases the water absorption but decreases the concrete workability and strength. Results however, fall into the range acceptable for lightweight aggregates, hence it can be concluded that there is potential to use PKS as aggregate replacement for lightweight concrete.

1. Introduction

Palm Kernel Shell (PKS) is a waste material obtained during the crushing of palm nuts in the palm oil mills for palm oil extraction. In South East Asia, it is one of the most quantitative waste materials produced. Malaysia produces approximately 4 million tons of PKS annually [1]. Hence, utilizing PKS would impose lower construction costs compared to other waste materials like rubber crump, plastic waste, and others. With proper mix design, PKS can be utilized to develop normal strength concrete, which ranges from 20 to 30MPa [2]. Research has been conducted on PKS as lightweight aggregate to produce lightweight concrete since 1984 [3] which brought immense changes in the concrete industry. It was discovered by Imam et. al (2014) [4] that palm nut shell can be used as construction material in low cost buildings since it has attained a compressive strength of $18 N/mm^2$. This can indirectly facilitate in waste reduction. This research focuses on the contribution of PKS in the improvement of concrete performance in terms of workability, water absorption, density, and compressive strength.

2. Materials Characterization

2.1. Palm kernel shell (PKS)

The PKS used in this research was collected from a palm oil factory in Banting, Selangor. The PKS is air-dried naturally or in the oven, and then soaked for few days under ambient temperature to obtain Saturated Surface Dried (SSD) aggregates. The PKS grading size used was between 10.0 to 12.5mm to replace coarse aggregates at 25, 50, 75, and 100% of the total coarse aggregate value.

Table 1. Properties of aggregates and PKS.

Properties	PKS as coarse aggregate	Coarse aggregate	Fine aggregate
Specific gravity	1.21	2.72	2.6
Water absorption (%)	25.64	0.7	1.1
Bulk density (kg/m^3)	572	1445	------
Fineness modulus	6.24	6.24	1.32
Maximum size (mm)	12.5	20	5
Aggregate impact value (%)	6.65	12.23	------

2.2. Aggregates
Two types of aggregates; coarse aggregate and fine aggregate were used in this research. For the coarse aggregate, crushed granite rocks with a maximum size of 20mm according to the BS 812-103.2 1989 [5] was used. This coarse aggregate was prepared to be in Saturated Surface Dry (SSD) condition. For the fine aggregate, local natural sand with a maximum size of 5mm was used. The physical properties of PKS and aggregates are shown in Table 1.

2.3. Ordinary Portland cement (OPC)
In accordance to BS EN 197-1 2000 [6], OPC Type 1 was used in this research. OPC Type 1 with the specific gravity of 3.14g/cm^3 was preferred because the observation of the concrete properties can be done in normal hydration process.

2.4. Silica fume
Silica fume was obtained from SIKA-Kimia Materials in dry densified form Grade 920 conforming to the mandatory requirements of ASTM C1240 [7]. In this research, silica fume was used as cement replacement material with 15% replacement for all mixes.

2.5. Concrete mix design
The mix design was prepared based on the DOE method. The target 28 days strength was set to 30 to 40MPa. Five mixes were cast with fixed cement, silica fume, fine aggregate, and water content.

Table 2. Concrete mix proportion for PKS replacement with coarse aggregate.

Sample number	Percentage	Cement kg/m^3	w/c	Silica fume kg/m^3	Fine aggregate kg/m^3	Coarse aggregate kg/m^3	PKS kg/m^3	Design slump (mm)
CM	0%	382.5	0.45	67.5	540	1460	0	40-120
M25	25%	382.5	0.45	67.5	540	1095	365	40-120
M50	50%	382.5	0.45	67.5	540	730	730	40-120
M75	75%	382.5	0.45	67.5	540	365	1095	40-120
M100	100%	382.5	0.45	67.5	540	0	1460	40-120

3. Experimental set up
The specimens were cast according to BS1880 [8]. Slump test was conducted to determine the workability of concrete. Concrete density was determined according to BS EN 12390-7 [9]. For normal concrete, the density is approximately 2,400kg/m^3. Modification of the concrete mix will result in concrete with higher or lower density, depending on the amount and density of the aggregate, cement concentration, and the maximum size of aggregate used. Lightweight concrete has a density of between 160 and 1920kg/m^3. Water absorption test was conducted according to the British Standard [10] to determine the susceptibility

of an unsaturated concrete to water penetration. Compressive strength tests on the concrete was carried out conforming to BS 1881: Part 102: 1983 [11]. Concrete cubes of 100x100x100mm were then water-cured and its compressive strength monitored at ages 7, 14, and 28 days.

4. Results and discussion

4.1. Properties of fresh concrete

4.1.1. Slump test

Results show that the slump for CM and M25 are 120 mm and 100 mm respectively, reflecting high workability concrete. The value of slump for M50 is 50 mm, reflecting medium workability. The value of slump for M75 and M100 are 45 mm and 40 mm respectively, reflecting low workability. It can be concluded that the workability of concrete reduces linearly as the amount of PKS percentage increases. This can be that since normal aggregate is denser than PKS aggregate, and the replacement is by weight, the specific surface increases as the PKS content increases. This implies that more cement paste is required for the lubrication of the aggregate, hence reducing the entire fluidity of the mix, thereby reducing the height of the slump. Similar findings have been obtained by Daneshmand and Saadatin (2011) [12], agreeing with the fact that the workability of concrete decreases with the increment of PKS replacement percentage, causing the reduction of height of slump.

4.2. Properties of hardened concrete

4.2.1. Water absorption

Table 3. Results of water absorption test.

Sample number	Wet mass (g)	Dry mass (g)	Water absorption (%)
CM	2198 g	2068 g	6.28
M25	1960 g	1798 g	9.01
M50	1849 g	1693 g	9.21
M75	1790 g	1627 g	10.01
M100	1672 g	1460 g	14.52

Table 3 shows the wet mass for all concrete mixes. Three samples from each mix designs were oven-dried for 24 hours at a temperature of 115°C. The dry mass was then measured. Complying to the British Standard [14], the water absorption of concrete using PKS replacement should fall between the range of 9 to 33%, of which all mixes containing PKS (M25 to M75) abide to.

4.2.2. Compressive strength

Results show that compressive strength of concrete tends to decrease with the increment of PKS. CM gave the highest strength of 33.01 MPa at 28 days of curing compared to other mixes. Mixes M25, M50, M75, and M100 resulted in compressive strengths of 30.2MPa, 25.7MPa, 20.1MPa, and 10.9MPa respectively. This indicates that the amount of PKS affects concrete strength. The reduction of strength could be due to PKS having lower strength when compared to normal aggregate. This is reflected from the impact value for normal aggregate which is double that of PKS (see Table 1). Another factor that may cause the decrease could be attributed to the highly irregular shapes of the PKS, which could prevent full compaction with normal coarse aggregate, thereby affecting the concrete strength. This theory is supported by Daneshmand and Saadatin (2011) [16]. The bonding between PKS and cement paste will also decrease with the inclusion of higher percentage of PKS due to the smooth surface of PKS as mentioned by Mannan [17].

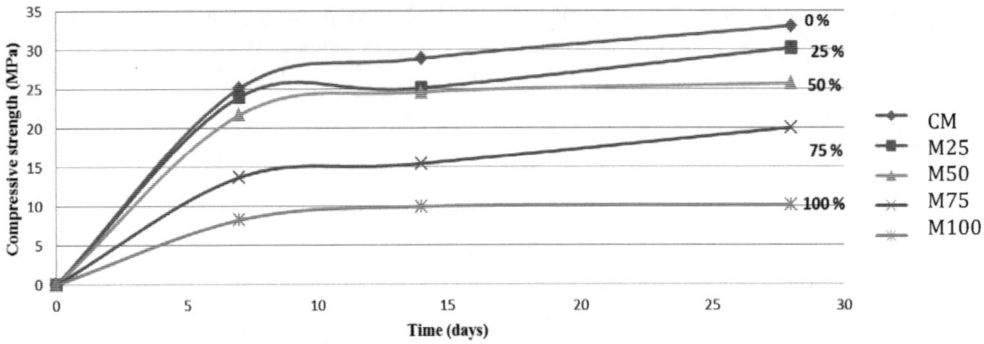

Figure 1. Comparison of compressive strength (MPa) at 7, 14, and 28 days of curing.

4.2.3. Density test

Similar to the compressive strength analysis, it was found that concrete density decreases with the increment of PKS. Starting from CM with the density of 2000kg/m^3, the value decreases linearly with the increment of PKS. Mixes M25 to M100 can be considered as lightweight concrete judging from their densities which are 1960kg/m^3, 1848kg/m^3, 1790kg/m^3, and 1672kg/m^3 respectively.

5. Conclusion

The following conclusions can be derived from the experimental results:

- The workability of fresh concrete mix decreases with the inclusion of PKS.
- Water absorption for concrete containing PKS increases but is still within standard ranges.
- The inclusion of PKS reduces the concrete strength, but the replacement of 50% PKS gives acceptable values as it falls within the range of requirement for structural LWC components.

References

[1] Okpala DC. Palm Kernel Shell as a Lightweight Aggregate in Concrete. Build Environ 1990; 25(4):291–6.

[2] Shafigh P, Jumaat MZ, Mahmud H. Oil Palm Shell as a Lightweight Aggregate for Production High Strength Lightweight Concrete. Constr Build Mater2011; 25:1848–53.

[3] Alengaram, U. Johnson, Baig Abdullah Al Muhit, and Mohd Zamin bin Jumaat. "Utilization of oil palm kernel shell as lightweight aggregate in concrete–a review." Construction and Building Materials 38 (2013): 161-172.

[4] Imam, Hamisu Bara'U., and Nura Usman. "Compressive Strength of Concrete Using Palm Oil Nut Shell as Light Weight Aggregate." Journal of Civil Engineering and Environmental Technology (JCEET) (2014): 15.

[5] BS 812-103.2 1989, Testing aggregates —. Part 103: Methods for determination of particle size distribution

[6] BS EN 197-1, Cement. - Part 1 Composition, specification and conformity criteria for common cements

[7] ASTM-C1240. Standard specification for use of silica fume as mineral admixture in hydraulic cement concrete, mortar and grout",, in Annual Book of ASTM Standards American Society for Testing and Materials. Philadelphia, USA, 2004.

[8] British Standard (BS) 1881:1983 Testing Concrete

[9] BS EN 12390-7:2009. Testing hardened concrete Density of hardened concrete

[10] BS 1881-122:2011 Testing concrete. Method for determination of water absorption

[11] BS EN 12390-4:2000 - Testing hardened concrete. Compressive strength

[12] Daneshmand, Saman, and Omidreza Saadatian. "Influence Of Oil Palm Shell On Workability And Compressive Strength Of High Strength Concrete." Annals of the Faculty of Engineering Hunedoara-International Journal of Engineering 9.2 (2011).

IOP Conf. Series: Earth and Environmental Science **32** (2016) 012041 doi:10.1088/1755-1315/32/1/012041

The drivers to adopt renewable energy among residential users.

Abdul Rahman Zahari[1] and Elinda Esa[2]
[1]Department of Marketing & Entrepreneur Development, Universiti Tenaga Nasional, Sultan Haji Ahmad Shah Campus, 26700 Muadzam Shah, Pahang, Malaysia
[2]Department of Accounting, Universiti Tenaga Nasional, Sultan Haji Ahmad Shah Campus, 26700 Muadzam Shah, Pahang, Malaysia

E-mail: Rahman@uniten.edu.my; Elinda@uniten.edu.my

Abstract. This study aims to examine the drivers to adopt renewable energy (RE) among residential users in Malaysia. Based on the theoretical framework of a consumer's decision making process, an empirical study of the adoption of RE was conducted. A total of 501 residential users were used in this study. This study proved that perceived utility of new technology, perceived utility of new service, and perceived benefit of new technology are the drivers to adopt RE among residential users. These factors are knowing crucial to RE suppliers and producers because it will generates more demand from the residential users and the percentage of energy mix from RE sources can be increase.

1. Introduction

Malaysia population is growing at an average of 1.2% every year and this growth has led to the increase of energy consumption at 2.3%. The demand of energy consumption between 2007 and 2012 is increased by 25% in Peninsular Malaysia. Moreover, in 2012, 36.8% of the energy demand came from transportation segment, 29.8% consumed by industrial sector, and 16% used by non-energy sector i.e. residential and commercial [1]. Past trends state that petroleum has generated a major source of energy mostly in the transportation segment. According to [2], the uses of crude oil, petroleum, natural gas, coal, and hydropower to supply the energy across Malaysia have declined between 2007 and 2012. The main reason behind the slight shortage used of fossil fuel sources is due to the continuous effort by the government to replace sources of energy with renewable sources. Moreover, high dependent use of fossil fuels have caused in increasing the global warming. Due to that, the alternative energy solution which is renewable energy (RE) should be capitalized and commercialized to the users. The slow growth of using RE is not only recorded in developing countries but also reported in developed countries like UK which to date is recorded t only 5% of UK population is using RE. They expect the number will growth to 15% in 2020, however the progress is slow.

The government commitment towards mitigating the climate change is proved through the establishment of Sustainable Energy Development Authority of Malaysia (SEDA Malaysia) in 2011. This statutory body is formed in the conjunction with the implementation of Sustainable Energy Development Authority Act 2011 (Act 726) and RE Act 2011 (Act 725). As at 2012, renewable sources of energy supply such as biomass, biogas, and solar was only at 0.4% [1]. The figure indicated that the policy makers, RE providers and suppliers must combined efforts to create awareness among residential users to adopt the RE. Publics' interest in RE becoming strong and growing because minimizing use of fossil fuels and tendency to use RE will ensures the better global climate change [3;4]. Furthermore, as the demand for RE in the emerging markets is growing significantly, it is important to understand the residential users' attitude towards the intentions to use the RE.

Over the last two decades (1990-2010), a numerous of studies have been executed to identify the factors influencing attitude of the potential users' towards RE in different western countries. Then, those factors can be classified into three major clusters namely social, political or regulatory and technical [5]. However, lack of studies tries to identify the driver factors to the intentions to adopt the RE. Thus, this study aim to study the perceptions of residential users towards the driver factors associated with the adoption of RE. The results of this study later could help the policy makers, RE providers and suppliers to better understand the factors that have contributes to the adoption of RE. The remainder of paper is structured as follows. This study first review the relevant literature. The following section details the method applied. The third part is to discuss the findings and the last part is the conclusion.

2. Literature Review

2.1 Research Model and adoption to use

With regard to adoption behavior related to the new technology, Theory of Reasoned Action (TRA), and Technology Acceptance Model (TAM) have been the most fundamental and influential theories discussed by researchers. These models study about human behaviors' are preceded by subjective norms, individual attitudes, perceived usefulness, and perceived ease-of-use of technology. Therefore, to discuss the residential adoption of RE, it is important to identify residential perceptions that influence intentions to adopt RE. Several studies have discussed adoption behavior with regards to technology such as telephone, social media, search engines, and RE using TRA or TAM [6;7;8;9;5]. The positive beliefs toward innovative technology in products or services perceived by consumers and will drive them to form an adoption intention to use the technology. Most of the time, to fulfill the adoption intention, consumers will proceed to form the purchase intention to buy the product or service. The construct of adoption intention is defined as the likelihood that a person will intend to use an innovative product [10].

2.2 Perceived utility of new technology, perceived utility of RE, and perceived benefit of new technology

Perceived usefulness, perceived ease-of-use, and perceived enjoyment are the crucial beliefs that influence user adoption intention [11;12]. Subsequently, the TAM theory is well adapted to this study. Another essential point is the understanding of the attributes of an innovation. This is because individual beliefs related to these attributes are significant predictors of future adoption and use [13]. According to [10], in the research fields of marketing and advertising, some studies have proposed the concept of perceived utility (tangible or intangible) to discuss the related identified perceptions of consumer acceptance. For that reason, the construct of perceived utility has been adopted in this study. The perceived utility of a new technology refers to the consumer's overall evaluation of the utility based on the perceptions of what is obtained from the technology.

According to [14], the consumers' are focused on the importance of a technology's as the extrinsic value and against the intrinsic value. Moreover, as the extension of Roger's diffusion model, Moore has indicated that the differences between the needs and behaviors of each adopter may stall the transition between stages. Besides, [14] observed that although forerunner adopters will tend to respond to the intrinsic value of a new technology, but most consumers will seek the extrinsic benefits embedded in the service [15]. Due to that, the researchers must focus to the extrinsic benefits with respect to service adoption.

When evaluating a new product or service for possible adoption, consumers weigh the benefit against the cost of adoption, and if the perceived benefit overweighs the perceived cost, consumers are more likely to adopt [16]. As suggested by Construal level theory (CLT), the costs (e.g. monetary cost, learning effort, time commitment) of adopting a new product or service are construed as low-level, whereas the benefits of adopting a new product or service (e.g., that which enables one to do new things) are construed as high-level. Moreover, when considering the possible adoption of a new product or service, feasibility considerations are important only if the adoption is desirable, but desirability remains important whether feasibility is high or low. Desirability of a new product or service would thus result from the perceived benefits of adopting the new product or service, whereas feasibility would be derived from the perceived costs of adopting the new product or service. Thus, the following hypotheses are proposed:

H1: Users' perceived utility of RE technology will have a positive influence on their intention to adopt a RE.

H2: Users' perceived utility of RE services will have a positive influence on their intention to adopt a RE.

H3: Users' perceived benefits of RE will have a positive influence on their intention to adopt a RE.

From the Figure A1 (refer appendix A), the conceptual model is incorporating driver factors as the observed factors for intentions to adopt RE. In this model, drivers of RE are treated as the independent variable and adoption intentions of RE as the dependent variable.

3. Research Methodology

This study was able to obtain 501 respondents from residential users living in Klang Valley. Klang Valley is selected because it is the largest metropolitan areas that support the largest heterogeneous individuals as well as homogenous group of people in Malaysia. Besides, Klang Valley is fortified with modern facilities, compared to the other states in Malaysia [17]. Also, self-administer questionnaire using convenience sampling will be applied for this study. The subjects in this study involved adults over 18 years of age with consumption capability and behavior [18]. A pre-test of the questionnaire was performed with 20 subjects. All comments and suggestions received led to several minor modifications and then, the questionnaire is subsequently been revised. The process of distribution and collection of the questionnaires was carried out over a period of three months. Analysis of the socio-demographic data from the all valid questionnaires also indicated that the seven personal information items were adequately representative of this study sampling requirements. Table A1 (appendix A) presents the sample demographics.

The survey questionnaire is consist two parts. The first part indicates the drivers to adopt RE. The second is the subject's demographic information. Data will be obtained using a six point Likert-type scale, ranging from 'strongly agree' to 'strongly disagree'. A total of 21 items will be used to justify the first part of the questionnaire. All items are adapted from [10] and [19]. Furthermore, the items used is based on three earlier theories developed by [20;10;21]. The data then has been analysed using Partial Least Square (PLS). The analyses were conducted in two stages. First, the measurement model was tested to ensure that the constructs had sufficient psychometric validity. Then, it was followed by an assessment of the structural model in which the hypotheses were tested. A bootstrap resampling procedure also conducted to estimate the coefficients.

4. Findings and discussions

After data collection, measures were subjected to a refinement process to assess their construct reliability, convergent validity, and discriminant validity. The Cronbach alpha values ranged from 0.841 to 0.920 for four constructs that exceed the 0.7 threshold [22]. The values indicates that the high internal consistency of measure reliability and in line with [23]. Likewise, the composite reliabilities for all measures were high, ranging from 0.904 to 0.949. As shown in Table A2 (refer appendix A), twelve items of construct loaded is higher than 0.8, ranging from 0.804 to 0.941. Additionally, the R square value is reported at 0.564. It indicates that 56.4% variations of intentions to adopt RE is explained by the model using perceived utility of new technology, perceived utility of a new service, and perceived benefit of new technology as predictors. The remainder of 43.6% remains unexplained and may be due to the other predictors which are related to intentions to adopt RE.

Next, the item total correlations for these construct (see Table A3 in the appendix A) was examined. The correlation pattern indicated that every item had a stronger correlation with its construct than another construct. Moreover, the average variance extracted (AVE) also represented the value higher than 0.5 which ranged from 0.760 to 0.862, were above the recommended threshold of 0.5 [24]. In general, the results show that all the four constructs namely relative perceived utility of a new service, perceived benefit of new technology, and perceived utility of new technology are all valid measures of their respective constructs. Moreover, the square root of each AVE (shown on the diagonal in Table A3) is greater than the related inter-construct correlations in the construct correlation matrix, indicating adequate discriminant validity for all of the reflective constructs [25]. Therefore, the discriminant validity was confirmed.

To validate the proposed hypotheses and the structural model, the path coefficient between two latent variables is assessed. The results of structural model is used the bootstrap procedure with 5000 times of resampling. Based on previous studies, the path coefficient value needs to be at least 0.1 to account for a certain impact within the model [26;27]. Assessment of the path coefficient (see Table A4 in the appendix A) shows that all proposed hypotheses are supported. From the analysis, supported hypotheses are significant at the level of 0.01, have expected sign directions (i.e., positive) and consist of a path coefficient value (β) ranging from 0.208 to 0.471. Based on the analysis, it indicates that the intentions to adopt RE are influenced by perceived utility of new technology (β=0.471, t=28.985, p<0.001), perceived benefit of new technology (β=0.208, t=16.740, p<0.001), and perceived utility of a new service (β=0.238, t=15.546, p<0.001). These results have indicated the similar findings from previous studies [10;9;28;29].

5. Conclusion

The policy makers, RE technology providers and services could have a chance to recognize and identify the factors that drives the residential users to adopt the RE. Analyzing and understanding those factors is very essential because it will affect the decision to adopt the RE among the residential users. The results have indicated factors i.e. perceived utility of new technology, perceived utility of a new service, and perceived benefit of new technology are significantly influence the residential users to adopt the RE. By knowing these factors it may increase the level of RE usage among the residential users. Thus the support from the RE producers are much needed in order to increase the use of RE in Malaysia.

Since this study was limited to residential users in Klang Valley, therefore in future, the researcher can further expand the geographical areas in Malaysia. Identifying others related factors of motivating and inhibiting the residential users to adopt the RE also needed. By doing so, the policy makers, producers and services of RE technology could have benefits with the future study. Future studies could focus on the specific type of RE sources rather than the whole scope of RE sources. It would also be interesting to compare the adoption of RE sources in Malaysia.

References

[1] Energy Commission Malaysia 2014. *Towards a world-class energy sector: Spark of efficiency ensuring national energy security* vol 1 (Kuala Lumpur: Percetakan Skyline Sdn. Bhd.) p 36-9

[2] U.S. Energy Information Administration 2014 (*Malaysia Primary Energy Consumption Statistics*) retrieved from: http://www.eia.gov/beta/international/analysis_includes/countries_long/Malaysia/malaysia.pdf

[3] Clare H, Ready R, Eshleman J and Yoo J 2012 *Pennsylvanians' attitudes toward renewable energy* (Harrisburg PA: The Center for Rural Pennsylvania) p 1-23

[4] Moriarty P and Honnery D 2009 *Energy Policy* **377** 2469–74

[5] Alam S S and Rashid M 2012 *Energy Research J.* **3(2)** 37-44

[6] Bruner G C II and Kuma A 2005 *J. of Bus. Research* **58** 553–8

[7] Lu J, Liu C, Yu C-S and Wang K 2008 *Information & Management* **45(1)** 52–64

[8] Wu J-H and Wang S-C 2005 *Information & Management* **42(5)** 719–29

[9] Zahari A R, Esa E, Baharudin M H and Amat Paijan M I 2013 *Terengganu International Management and Business J.* **3(2)** 34-46

[10] Teng W, Lu H-P, and Yu H 2009 *Telecommunications Policy* **33** 628-41

[11] Davis F D 1989 *MIS Quarterly* **13** 319–40

[12] Davis F D, Bagozzi R P and Warshaw P R 1992 *J. of Appl. Social Psychology* **22** 1111–32

[13] Rogers E M 1995. *Diffusion of innovations* (vol 4) ed (New York, NY: The Free Press)

[14] Moore G and McKenna R 1999 *Crossing the chasm: Marketing and selling high-tech products to mainstream customers* (New York: Harper Business)

[15] Chakravorti B 2003 *Harvard Business Review* **82(2)** 58–67.

[16] Wang Q, Dacko S, and Gad M 2008 *Advances in Consumer Research* **35** 416-22

[17] Siwar C and Kasim M Y 1997 *International J. of Social Economy* **24(12)** 1524-35

[18] Yuwa H W 2008 *The future and me: Power of the youth market in Asia* (New York: John Wiley & Sons)

[19] Hoeffler S 2003 *J.of Marketing Research* **40** 406-20

[20] Ajzen I and Fishbein M 1980. *Understanding attitudes and predicting social behavior* (Englewood Cliffs, NJ: Prentice-Hall)

[21] Ajzen I 1991*Organizational Behavior and Human Decision Processes* **50(2)** 179-211

[22] Nunnally J C 1978 *Psychometric Theory* (New York: McGrew-Hill)

[23] Nunnally J C and Bernstein I H 1994 *Psychometric Theory* (New York: McGraw-Hill)

[24] Barclay D, Thompson R and Higgins C 1995 Technology Studies **(2:2)** 285-309

[25] Fornell C and Larcker D F 1981 *J. of Marketing Research* **34(2)** 161-88

[26] Hair J F, Ringle C M and Sarstedt M 2011 *J. of Marketing Theory and Practice* **19(2)** 139-151

[27] Wetzels M, Odekerken-Schroder G and van Oppen C 2009 *MIS Quarterly* **33(1)** 177-195

[28] Ahmad A, Rashid M, Alam S S and Omar N A 2014 *J. Pengurusan* **41** 123-131

[29] Alam S S, Nik Hashim N N, Rashid M, Omar N A, Ahsan N and Ismail M D 2014 *Renewable Energy* **68** 255-263

Appendix A

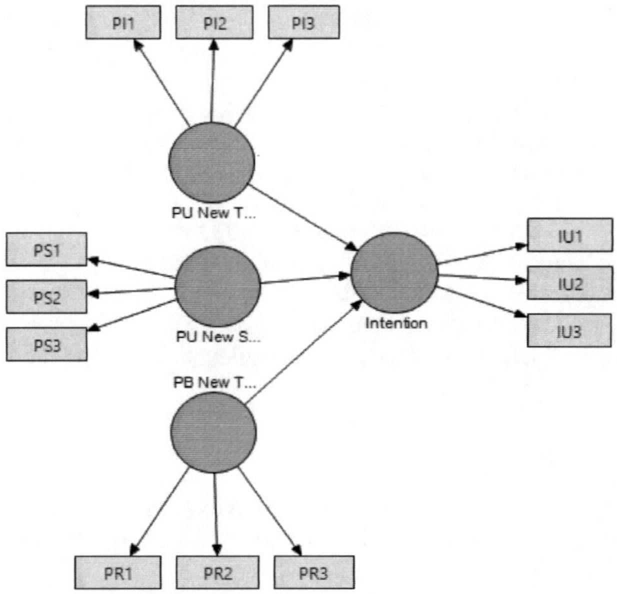

Figure A1. Conceptual framework

Table A1. Demographic

Description	Frequency (N = 501)	%	Description	Frequency (N = 501)	%
Gender			*Ethnicity*		
Male	319	63.7	Malay	353	70.5
Female	182	36.3	Chinese	95	19.0
			Indian	37	7.4
Age			Others	16	3.1
30 years old and below	90	18.0			
31 to 40 years old	241	48.0	*Level of education*		
41 to 50 years old	87	17.4	SPM or STPM	58	11.6
51 years old and above	83	16.6	Diploma	125	25.0
			Bachelor's Degree	232	46.2
Job			Master's or PhD	82	16.4
Professional	145	28.9	Others	4	0.8
Top Management	43	8.6			
Middle Management	139	27.7	*Income*		
Supervisory	33	6.6	Less than RM 2,500	84	16.8
Administrative or Clerical	77	15.4	RM 2,501 to RM 5,000	191	38.0
Technical	41	8.2	RM 5,001 to RM 7,500	103	20.6
Others	23	4.6	RM 7,501 or more	123	24.6

Table A2. Results of measurement model

Construct	Item	Loadings	AVE	Composite Reliability	Cronbach's Alpha
Intentions to adopt RE	IU1	0.904	0.760	0.904	0.841
	IU2	0.903			
	IU3	0.804			
Perceived benefit of new technology	PI1	0.926	0.815	0.929	0.888
	PI2	0.941			
	PI3	0.919			
Perceived utility of a new service	PR1	0.920	0.814	0.929	0.886
	PR2	0.927			
	PR3	0.859			
Perceived utility of new technology	PS1	0.911	0.862	0.949	0.920
	PS2	0.893			
	PS3	0.902			

Note: R square = 0.564

Table A3. Discriminant validity of constructs

	Intentions to adopt RE	Perceived benefit of new technology	Perceived utility of a new service	Perceived utility of new technology
Intentions to adopt RE	**0.872**			
Perceived benefit of new technology	0.414	**0.903**		
Perceived utility of a new service	0.600	0.152	**0.902**	
Perceived utility of new technology	0.713	0.359	0.703	**0.928**

Note: Values in the diagonal (bolded) represent the square root of the AVE while the off-diagonals represent the correlations.

Table A4. Hypothesis testing

Hypothesis	Relationship	Std. Beta	Std. Error	T value	Decision
H1	Perceived utility of new technology -> Intentions to adopt RE	0.471	0.016	28.685	Supported*
H2	Perceived utility of a new service -> Intentions to adopt RE	0.238	0.015	15.546	Supported*
H3	Perceived benefit of new technology -> Intentions to adopt RE	0.208	0.012	16.740	Supported*

* = $p<0.001$

Evaluating the Effect of Syngas Composition on Micro gas turbine Performance

N. Fareeza[1], E.S. Tan[1], Kumaran P. [1], T.M. Indra[1], N. Fadzilah[2], K. Yoshikawa[3]
[1] Department of Mechanical Engineering, Universiti Tenaga Nasional
[2] Tenaga Nasional Berhad Research.
[3] Department of Environmental Science and Technology, Tokyo Institute of Technology

E-mail: eesann@uniten.edu.my

Abstract. Syngas in nature can be derived from many sources for instance biomass, natural gas and any hydrocarbon feedstock. Therefore it can replace our reliance on the fossil fuel in future, as they are finite. This research covers the analysis on the combustion efficiency on eight syngas composition that have various hydrogen and carbon moxide (H_2/CO) ratio and the comparisons on the nitric oxide (NO_x) and carbon monoxide (CO) emissions. Each syngas was tested from 0kW load to 3kW load and all the data was taken upon the micro gas turbine (MGT) reaching stable and steady state conditions. It was found that high amount of hydrogen content in syngas lead to an increase in the combustion efficiency and emits more NO_x emissions. CO emissions were released when the combustion efficiency is low and increase with the CO content in the syngas.

1. Introduction

Syngas is a mixture of carbon monoxide and hydrogen in various compositions and can be derived from a number of sources such as coal, biomass gasification or from natural gas during steam reforming. It is used in IGCC because it is a potential clean fuel containing carbon dioxide (CO_2), water (H_2O), ammonia (NH3) and hydrogen sulphide (H_2S) constituents. However, these variations in compositions of syngas potentially affect the IGCC performance and efficiency [1].

Syngas in integrated gasification combined cycle (IGCC) applications should be free of particulates and trace metals that could damage the gas turbine. Catalytic gasification is one of the gasification technologies that produced gas containing high hydrocarbon content which is similar to natural gas. Hence, it is ideal for utilization in the combustion turbines. This is because the design of the gas turbine in IGCC was developed based on the natural gas combustion turbines. However, there are some differences between syngas and natural gas that will affect the design of combustion turbines [2]. Syngas gasification process varies in calorific value, composition, flammability characteristics and contaminants compared to natural gas. Syngas typically consists of carbon and hydrogen, whereas natural gas mainly consists of methane. As the syngas has more hydrogen content, these results in a higher flame speed and wider flammability limits. Thus, this factor make syngas produced a stable flame at leaner conditions. Besides that, it has higher combustion speed than natural gas [3].

[1] Universiti Tenaga Nasional, Jalan Ikram-Uniten, 43000, Kajang, Selangor, Malaysia

2. Research Methodology

The micro gas turbine (MGT) used is Capstone 30kW while a standard bench top gas analyser was used to run the experiment using eight different syngas with various H_2/CO ratio. Each syngas was tested from 0-30 kW load and the data of heat losses, output power, NO_x and CO emissions were recorded by the gas analyzer. The combustion efficiencies for the syngas were calculated using Equation (1) and (2). In order to calculate the combustion efficiency, the LHV values for each syngas were calculated first.

LHVgas = [30CO + 25.7H$_2$ *+ 85.4CH*$_4$ *+ 151.3C*$_2$*Hx] 0.0042 MJ/Nm*$_3$ (1)

Combustion efficiency = 100% - ([Heat Losses/ Fuel Heating Value] x 100) (2)

2.1 Experiment setup

The experiments were carried out at TNB Research IGCC Pilot Plant using Capstone 30 kW MGT. The experiment begins by connecting the syngas to the MGT and the pressure of the gas container is ensured to be between 8 to 10 psi. Thereafter, the main power of the MGT at the control panel is switched on together with the gas analyzer before running the Capstone Software to start up the MGT operation. Throughout the experiment, the flow of the gas is monitored by checking the flow meter to ensure no drop in pressure and no freezing occurs. The freezing will block the syngas from entering the MGT causing shut down which will interrupt the experiment processes. The syngas were tested for 10 minutes for each load starting from 0kW to 3kW.These steps were repeated for each syngas compositions with different H_2/CO ratio. The schematic diagram of the experiment system is shown in figure 1.

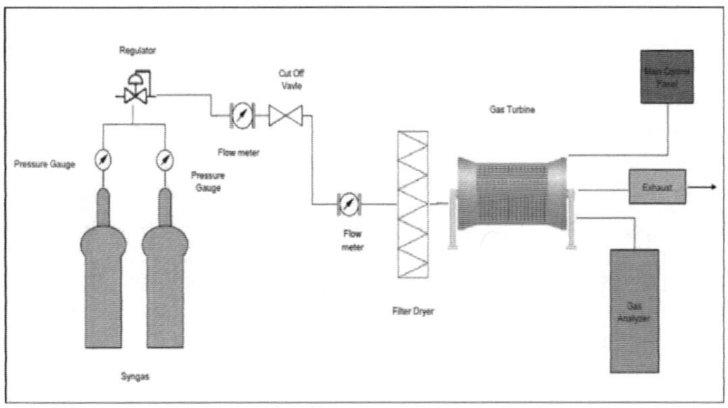

Figure 1. Schematic diagram of the experiment setup.

2.2 Syngas Composition

In this experiment, the ratio of hydrogen (H_2) and carbon monoxide (CO) in the syngas compositions was varied. The combustion efficiency and output power for each syngas were then calculated from the data collected throughout the experiments. NO_x and CO emissions were also compared between each syngas. The ratio of H_2 and CO composition variation is shown in table 1.

Table 1. Syngas compositions tested during the experiment.

Syngas	H_2 (%vol)	CO (%vol)	CH_4 (%vol)	N_2 (%vol)
1	0.6	1	60	38.4
2	0.6	2	60	37.4
3	0.6	3	60	36.4
4	0.7	1	60	38.3
5	0.7	2	60	37.3
6	0.7	3	60	36.3
7	0.8	1	60	38.2
8	0.8	2	60	37.2

3. Results and Discussion

During the experiment, the parameters recorded include the temperature profiles, the heat losses and the output power. The experiments were carried out from 0kW to 3kW for each syngas H_2/CO with 10 minutes duration for each load respectively. The collected data were used to calculate the combustion efficiency. The MGT is connected to the gas analyzer and the gas emissions such as CO (%) and the No_x (ppm) were recorded every 5 seconds.

Figure 2 shows the output power against H_2/CO ratio at load of 3 kW. The 0.6 H_2/CO has the highest output power when connected to 3kW load whereas, 0.7 H_2/CO has the lowest output power. One of the factors contributing to the higher output power is the smaller fuel heating value as it causes an increase in turbine flow [4]. Gas turbine output power is also affected by the ambient temperature. As the ambient temperature increases, less output power will be produced. The experiments 0.7 H_2/CO syngas was carried out at noon, which has a high ambient temperature as compared to other syngas ratio, which were mostly conducted in the morning. This explains the lower output power was produced by 0.7H_2/CO.

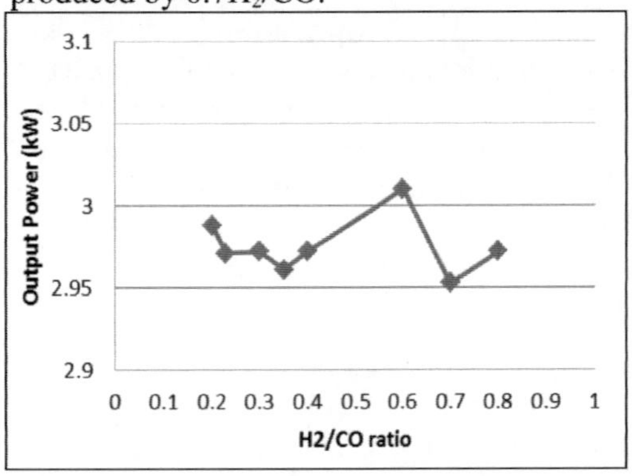

Figure 2. Output power versus H_2/CO ratio at 3 kW load for various syngas compositions.

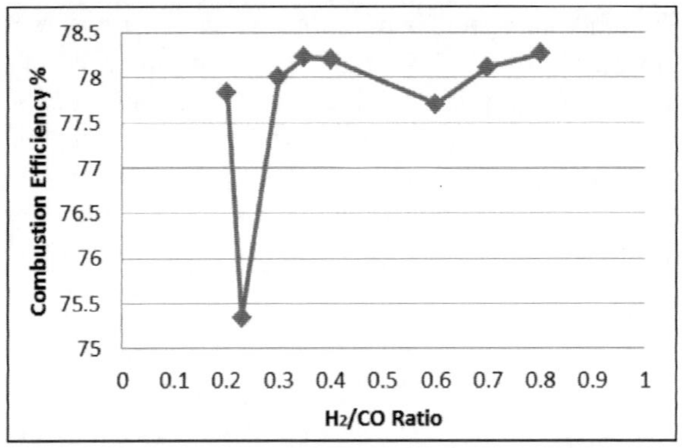

Figure 3. Combustion efficiency versus H_2/CO ratio at 3 kW load for various syngas compositions.

Combustion efficiency indicates the effectiveness of the useful energy conversion from the fuel, which is to rotate the turbine blades. From figure 3, the combustion efficiency increases as the H_2/CO ratio increases. However, there were some points particulary at 0.23 H_2/CO ratio and at 0.6 H_2/CO ratio, it decreases. In this experiment, the 0.80 H_2/CO ratio has the highest combustion efficiency of 78.30%. This H_2/CO ratio has the highest H_2 percentage volume of 0.8 (% volume) with 1 (% volume) of CO. The combustion efficiency was aided by the higher hydrogen content in this syngas H_2/CO ratio since hydrogen has very clean burning characteristics, a high flame propagation speed and wide flammability limits. As it has higher laminar combustion, it reduced the combustion duration, thus increase the combustion efficiency [5]. The syngas with higher hydrogen content elongated the flame because of the presence of more H radical. This H radical promotes the chain branching and chain propagation in which increase the reaction zone length. From the results, it shows that combustion efficiency increases as the hydrogen content increases. It is known that, hydrogen has higher moisture content that might affect the turbine component if the moisture content has reached the moisture limitations [6]. There is a drop in the graph indicating the lowest combustion efficiency of 75.34% at 0.23 H_2/CO ratio. Most of the combustion efficiency range from 78.30% to 77.71%. There was around 2.37% difference in combustion efficiency, compared to the other H_2/CO ratio. This lower combustion efficiency resulted from the higher CO content in this H_2/CO ratio, which has 0.7 H_2 (% volume) and 3 CO (% volume). However, external factors [7] such as the remaining syngas inside the turbine from the preceding experiments and ambient temperature may affect the accuracy of this result as reflected in the 2.37% difference.

Figure 4 shows the emission of NO_x (ppm) emissions for each H_2/CO ratio. The graph shows that at 0.8 H_2/CO emitted the highest NO_x and 0.2 H_2/CO has the lowest NO_x emissions. A best-fit line shows that emissions of NO_x generally increase with H_2/CO ratio. From the data recorded, it shows that NO_x emissions have a direct relationship with the hydrogen content. NO_x emissions are from the nitrogen and oxygen thermal fixation in the inlet combustion air and form the chemically-bound nitrogen conversion in the fuel.Typically, the fuel that has higher combustion efficiency will emit the highest NO_x emissions as it has the highest flame temperature. As in this experiment, 0.8 H_2/CO which records the highest efficiency also emitted the highest NO_x emissions. This is because of higher hydrogen content which relates to higher equivalent ratio. Thus, increase in the flame temperature results in higher NO_x emissions. From literature, NO_x emissions are directly proportional to the H_2 content. As the H_2 content increases,the NO_x emissions become higher [8].

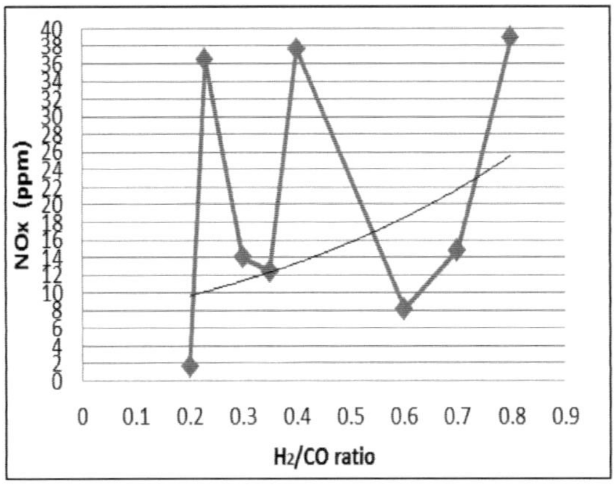

Figure 4. NO_x emissions versus H_2/CO ratio at 3 kW load for various syngas compositions.

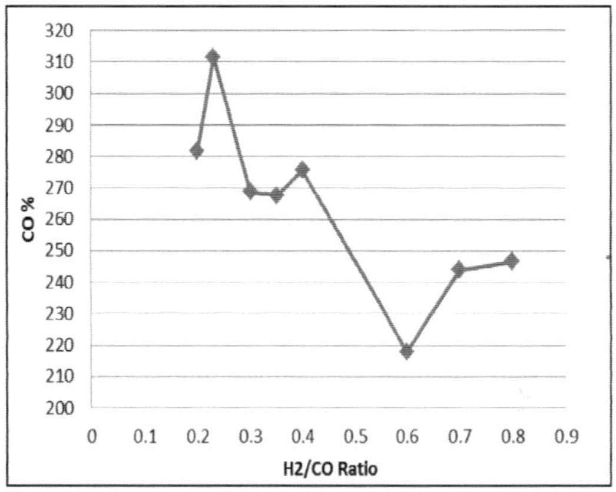

Figure 5. CO emissions versus H_2/CO ratio at 3 kW load for various syngas compositions.

Meanwhile, figure 5 shows the emissions of CO against the H_2/CO ratio. The graph shows at 0.23 H_2/CO ratio, emitted the highest CO, and at 0.6 H_2/CO emitted the lowest CO. Generally,the trend of the graph is inversely to the H_2/CO ratio. There is large dropped in CO emissions at 0.4 H_2/CO to 0.6 H_2/CO ratio. Then, the CO emissions increased at 0.7 H_2/CO onwards. CO emissions result from the incomplete hydrocarbon combustion or no combustion of CO in the fuel gas. Previous studies have proven that more CO is emitted with lower combustion efficiency and fuel with high CO content [9]. The results of this experiment also show that as the H_2/CO ratio decreases, the CO emissions increase. The smaller H_2/CO ratio indicates higher CO content. Previous studies by other researchers [10,11] also found that CO emissions result from a lower efficiency. This explains the result of this experiment at 0.23 H_2/CO, where it emitted the highest CO emissions of 311.58 ppm with the lowest combustion efficiency of 75.34 %.

4. Conclusion

In conclusion, among the eight syngas with different H_2/CO ratio, the syngas with higher hydrogen content gives out the highest combustion efficiency because it aided the combustion processes that resulted in higher laminar combustion. The syngas with higher hydrogen content also emitted more NO_x when released with high flame temperature. It is found that the NO_x emissions are maintained less than 50 ppm by using the MGT in IGCC pilot plant. These emissions are relatively low as compared to the conventional power plant gas turbine. Meanwhile, the syngas that has the lowest combustion efficiency emitted high CO emissions. CO emissions also have a direct relationship with the CO content in the syngas.

References

[1] Singh, Deepti, et al. "An experimental and kinetic study of syngas/air combustion at elevated temperatures and the effect of water addition." Fuel 94 (2012): 448-456.

[2] Hasegawa, Takeharu. "Combustion Performance in a Semiclosed Cycle Gas Turbine for IGCC Fired With CO-Rich Syngas and Oxy-Recirculated Exhaust Streams." Journal of Engineering for Gas Turbines and Power 134.9 (2012): 091401.

[3] Wang, Zhe, et al. "Syngas composition study." Frontiers of Energy and Power Engineering in China 3.3 (2009): 369-372.

[4] Aiguo, L. I. U., and W. E. N. G. Yiwu. "Effects of Lower Heat Value Fuel on the Operations of Micro-Gas Turbine." Energy and Power Engineering 1.01 (2009): 28.

[5] Azimov, Ulugbek, et al. "Effect of syngas composition on combustion and exhaust emission characteristics in a pilot-ignited dual-fuel engine operated in PREMIER combustion mode." international journal of hydrogen energy 36.18 (2011): 11985-11996.

[6] Kim, Chang Min, Tong Seop Kim, and Jeong L. Sohn. "A Study on Maximizing Power Output of IGCC Plants Considering Operating Limitations of Gas Turbine." ASME Turbo Expo 2014: Turbine Technical Conference and Exposition.2014.

[7] Kim, Young Sik, et al. "Effects of syngas type on the operation and performance of a gas turbine in integrated gasification combined cycle." Energy Conversion and Management 52.5 (2011): 2262-2271.

[8] Lee, Min Chul, et al. "Gas turbine combustion performance test of hydrogen and carbon monoxide synthetic gas." Fuel 89.7 (2010): 1485-1491.

[9] Gnanapragasam, Nirmal, Bale Reddy, and Marc Rosen. "Reducing CO_2 emissions for an IGCC power generation system: effect of variations in gasifier and system operating conditions." Energy Conversion and Management 50.8 (2009): 1915-1923.

[10] Berstad, David, Petter Nekså, and Gunhild A. Gjøvåg. "Low-temperature syngas separation and capture for enhanced efficiency of IGCC power plants." Energy Procedia 4 (2011): 1260-1267.

[11] Chiesa, Paolo, Giovanni Lozza, and Luigi Mazzocchi. "Using hydrogen as gas turbine fuel." ASME Turbo Expo 2003, International Joint Power Generation Conference.2003.

Acknowledgments

The authors would like to acknowledge technical assistance from Tenaga Nasional Berhad Research and financial support from Uniten and Titech are greatly appreciated.

Flood Forecasting via Time Lag Forward Network; Kelantan, Malaysia

Milad Jajarmizadeh[1], Lariyah Mohd Sidek[1], Hidayah Bte Basri[1], Aminah Shakira Jaffar[1]

[1]Center for Sustainable Technology and Environment, College of Engineering, Universiti Tenaga Nasional, Jalan IKRAM-UNITEN, 43000 Kajang, Selangor, Malaysia

E-mail: milad_jajarmi@yahoo.com

Abstract. Forecasting water level is one of the critical issues in Malaysia for Kelantan region. Based on the flood events in 2014, this study investigates the hourly-forecasting of water level in one station namely Kg Jenob in Kelantan. For this issue, Time Lag Forward Network (TLFN) is evaluated for forecasting the water level as dynamic model. Heuristic method in stepwise forward methodology is performed. Rainfall and water level are the input and output of the modelling respectively. For selected flood period 15/12/2014 to 30/12/2014, 8 scenarios are developed to obtain a minimum error in water level forecasting. By monitoring the error, it will show that the optimum configuration of network has 2 processors in hidden layer and 7 lags have enough contribution on the result of hourly forecasting. Transfer functions in hidden and output layers are is Tangent hyperbolic and bias. Observed and simulated data are compared with usual error criteria called Mean Square Error (MSE) and Root Mean Square Error (RMSE) which obtained 0.005 and 0.07 respectively. In conclusion, this study will be as a baseline for Kelantan to show that TLFN has promising result to forecast the flood events.

1. Introduction

Time Lag Forward Networks (TLFN) are an alternative model in hydrological prediction [1]. TLFN is classified as data driven models with short memory structures. TLFN includes local recurrent connections and they include memory layer, limit to the given data as input which they are pre-processor. TLFN is dynamic network which means that the information is based on time instead a static pattern. Therefore, TLFN can be used for prediction and forecast of time series data. For temporal prediction, TLFN is a suitable model and is under evaluations for optimization and application in different fields [1]. One of the applications of this data driven technique can be for temporal problems such as time series forecasting, system identification and temporal pattern recognition[2]. The main advantage of TLFN is the smaller network size required to learn temporal problems. Moreover, the another advantage of TLFN is their low sensitivity to noise. In this study, TLFN is investigated for hourly forecasting the water level in Malaysia, Kelantan region which is potential for flood events as Kelantan has involved with a huge flood in 2014 [3].Therefore, forecasting the fluctuation of water level is the objective of this study upon the development of different scenarios to reach optimum results.

2. Methodology

For Kelantan watershed, data are collected for Kg Jenob station in both rainfall (mm) and water level (m). The data that has been obtained in hourly time scale from 15/12/2014 to 30/12/2014 has covered the two weeks rainfall and water level in Kg Jenob station as a flood event in 2014. Figure 1 shows trend visualization rainfall and water level in Kg Jenob station. The highest rainfall has occurred in 20/12/2014 at 11:00 which is 35.6 mm. Also, the maximum recorded water level has been observed on 17/12/2014 at 8:00 which is 2.49 m. Figure 1 shows presentation of rainfall versus water level in Kg Jenob station. Rainfall has a firm fluctuation approximately. It might be a reason that water level has smooth trend. However, highest rainfall events have own impact slightly to increase the water level in upcoming water levels events.

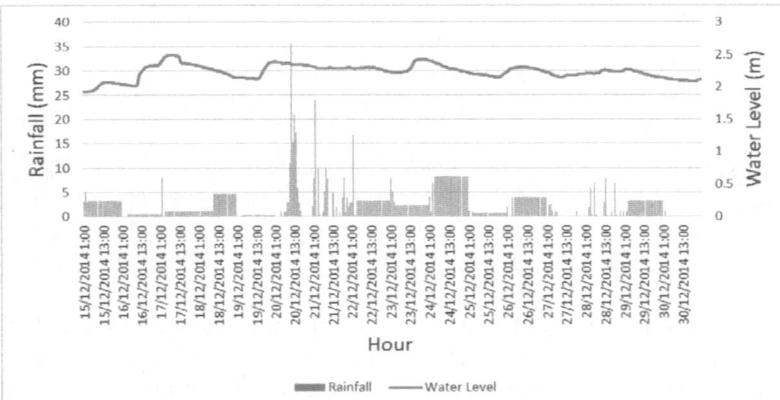

Figure 1. Presentation of rainfall versus water level at Kg Jenob station, Malaysia

The development of TLFN includes a systematic steps which are choosing of data set for learning, cross validation and test, determination of the input and output variables including scaling; choosing of the network topology and specification the required number of cells for hidden layer; training and monitoring the cross validation and testing of TLFN to find optimum results. Configuration of TLFN model for Kg Jenob station includes three layers, which are: input, hidden and output layer. The number of cells for input layer is related to the input data vector, and the cell for output layer is related to the output vector. Normally, application of one hidden layer is enough for rainfall-runoff modelling by ANNs. One of the common and broadly-used methods is the heuristic approach to find optimum configuration of ANNs. Stepwise heuristic approach keep the non-linear relationship of hydrologic phenomena and monitoring the increasing inputs help to recognition of number of lags. In this study rainfall as main element to produce the flow has been considered as input of model. It should be investigated optimum maximum lags, which is also called memory length since this will increase the memory length besides considering the complexity of the neural network by the application of a number of inputs variables (e.g. attributed lags and different variables). For Kg Jenob, 8 scenarios have been developed in step wise method. In this methodology, next hour water level is forecasted as an objective. For instance, scenario 1 is to forecast the water level at 15/12/2014 2:00, which water level derived from rainfall 15/12/2014 1:00. Consequently, scenario 8 is related with forecasting the water level at 15/12/2014 9:00, which derived from rainfalls in previous hours between 15/12/2014 1:00 to 15/12/2014 8:00 in Kg Jenob station. Table 1 illustrates developed scenarios for Kg Jenob station based on water level (Q) and rainfall (PCP).

Table 1. Developed scenarios in Kg Jenob by TLFN

Scenario	Variables
1	$Q_2 = f\{PCP_1\}$
2	$Q_3 = f\{PCP_1, PCP_2\}$
3	$Q_4 = f\{PCP_1, PCP_2, PCP_3\}$

4	Q_5 =f{ PCP_1,......, PCP_4}
5	Q_6 =f{ PCP_1,......, PCP_5}
6	Q_7 =f{ PCP_1,......, PCP_6}
7	Q_8 =f{ PCP_1, ...,PCP_7}
8	Q_9 =f{ PCP_1,......, PCP_8}

3. Result

Configuration of TLFN for Kg Jenob station includes three layers as input, hidden and output layer. The optimum TLFN network has one hidden layer and 2 neurons or process elements in hidden layer. Tangent hyperbolic and bias have been used as transfer functions in hidden and output layers respectively. Training algorithm is used back propagation with momentum term for whole network. Figure 2 shows the observed Mean Square Error (MSE) and Root Mean Square error (RMSE) for monitoring the accuracy of developed models. TLFN has been improved from scenario 1 to scenario 7 based on error optimization generally. The main reason can be related with sufficient input data to improve the pattern recognition of input data which related with output. In this study, further scenarios have not any significant improvement on result and scenario 8 has gained higher error in comparison with scenario 7 and it has been chosen the stop level. Therefore, scenario 7 has been introduced as optimum scenario. Indeed, further lags on rainfall and development of TLFN leads in a complex network which is usually considered as a disadvantage.

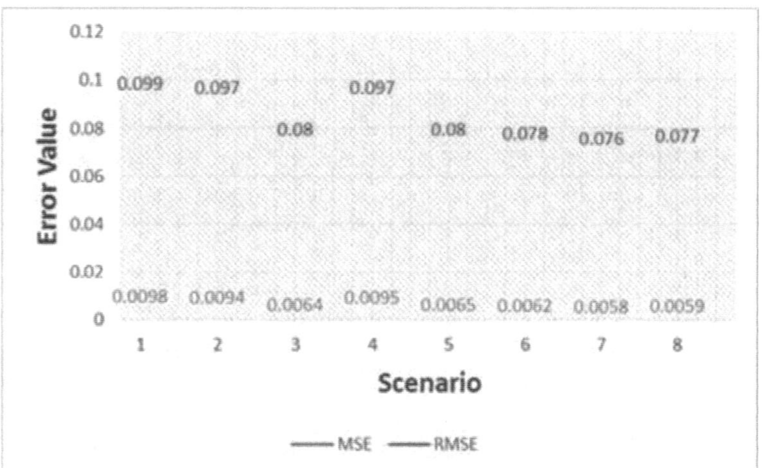

Figure 2. Error visualization of TLFN on water level forecasting in Kg Jenob

Figure 3 illustrates optimum simulation of water level for Kg Jenob station. However, the trend has under and over estimations for water level events. The minimum and maximum recorded water levels are 1.99 and 2.48 which are forecasted 2.1 and 2.37 respectively. Table 2 includes basic statistical analysis for observed and forecasted water level. Generally, discrepancies of the indices are based on over and under estimation of observed and forecasted values. For Kg Jenob station, mean, median, minimum and maximum values and standard deviation have a promising value, which are close to observed data. Although, mean, median and standard deviation have slightly underestimation.

Table 2. Basic Statistical analysis for observed and forecasted water level in Kg Jenob station

Index	Mean	Median	Min	Max	Standard deviation

Observed Water Level	2.24	2.24	2	2.4	0.1
Forecasted Water Level	2.22	2.23	2	2.3	0.07

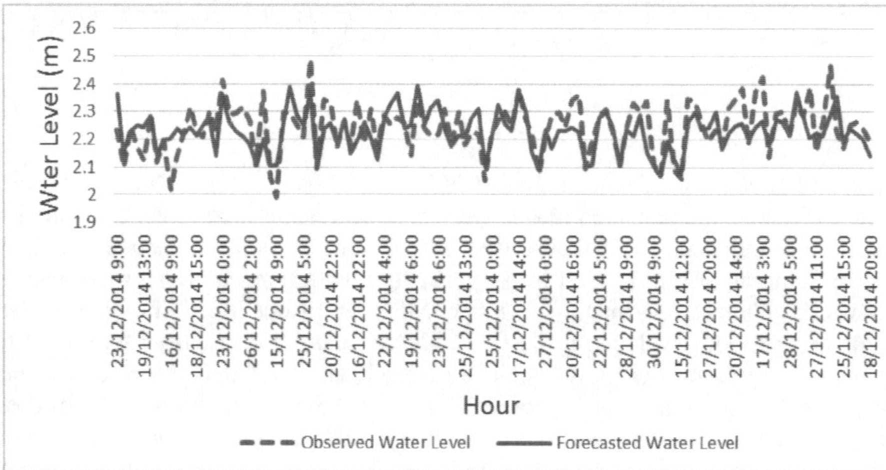

Figure 3. Trend presentation of scenario 7 as optimum forecasting of water level in Kg Jenob station

4. Conclusion

For this study, Time-Lagged Feed Forward network (TLFN) has been used for water level forecasting in kg Jenob station in Kelantan region for a flood period in 2014. For development of model is used available data as rainfall (mm) and water level (m). Development of TLFN is involved with definition of input data and related lags which is lead to 8 scenarios. Optimum TLFN is involved with one hidden layer including 2 neurons for processing. Tangent hyperbolic and bias have been used as transfer function in hidden and output layers respectively. Training algorithm is used back propagation with momentum term. Result shows that the model has progressive respond to use more lags for rainfall as inputs and using the rainfall lags bring a better memory for pattern recognition of water level in Kg Jenob station. However, using more rainfall after scenario 7 shows that there is not any significant improvement for forecasted values. It can be concluded that optimum hourly water level modeling for Kg Jenob is derived from 7 hours rainfall lags, in other word, next hour water level has high relationship with 7 hours previous rainfall. Both statistical and graphical analysis admit that TLFN can be an alternative technique in Kelantan for further evaluations in regard with forecasting the flood and water level which are beneficial as subsidiary tools for a hydrologist.

References

[1] Donate JP., Sanchez G G., Cortez P., Miguel A S. 2011. Evolving Time-Lagged Feedforward Neural Networks for Time Series Forecasting. Proceedings of the 13th Annual Conference on Genetic and Evolutionary Computation. July 12 - 16, Dublin, Ireland.

[2]Galavi, H., Mirzaei, M., Lee, T. S., & Valizadeh, N. 2013. Klang River Level Forecasting Using ARIMA and ANFIS Models. American Water Works Association (AWWA). 105 (9): 496-506.

[3] Aminah Shakirah Jaafar., Lariyah Mohd Sidek, Hanapi Mohd Noor, Hidayah Basri, Sazali Osman, Nazirul Mubin Zahari, Abdul Hafiz Mohammad, Wan Hazdy Azad Wan Abdul Majid, Jajarmizadeh M. 2015. Flood Catastrophe of Kelantan Watershed in 2014. International Symposium on Flood Research and Management. October, 5th - 6th , Shah Alam, Malaysia.

IOP Conf. Series: Earth and Environmental Science **32** (2016) 012044 doi:10.1088/1755-1315/32/1/012044

Finite Element Analysis of the Maximum Stress at the Joints of the Transmission Tower

Zarina Itam[1*], Salmia Beddu[1], Nur Liyana Mohd Kamal[1], Khaled H. Bamashmos[1]
[1]Department of Civil Engineering, Universiti Tenaga Nasional, Malaysia

E-mail: iZarina@uniten.edu.my

Abstract. Transmission towers are tall structures, usually a steel lattice tower, used to support an overhead power line. Usually, transmission towers are analyzed as frame-truss systems and the members are assumed to be pin-connected without explicitly considering the effects of joints on the tower behavior. In this research, an engineering example of joint will be analyzed with the consideration of the joint detailing to investigate how it will affect the tower analysis. A static analysis using STAAD Pro was conducted to indicate the joint with the maximum stress. This joint will then be explicitly analyzed in ANSYS using the Finite Element Method. Three approaches were used in the software which are the simple plate model, bonded contact with no bolts, and beam element bolts. Results from the joint analysis show that stress values increased with joint details consideration. This proves that joints and connections play an important role in the distributiin of stress within the transmission tower.

1. Introduction

Joint in transmission towers have been studied for several decades now due to the effects they bring to the stress distribution within the structure. Knight and Santhakumar [4] conducted tests on a full-scale quadrant of the lowest panel of a transmission tower. They pointed out that the secondary stresses caused by bolted joint could be significant enough to cause failure of leg members even under normal working-load conditions. Chan et al. [3] compared the experimental failure loads of single angle struts with those predicted by design code equations and numerical analysis results, and concluded that more reasonable ultimate load predictions can be obtained by considering both the effects of joint fixity, which increase member capacity compared to the ideal pinned-joint conditions, and eccentricity which reduces capacity compared to ideal centric loading. Engineers normally use four common methods for modeling bolted connections. The simplest kind of models for a bolted joint does not include the bolts themselves. Instead, bonded contact is defined between the joint surfaces. This model however, would not allow individual bolt loads to be extracted. Beam Element Bolt is another method for modeling bolts. In this method, beam elements will be created attached to an edge or face as a beam connection, and is generally done with a fixed joint for line bodies. The third method is Solid Element Bolt, where it allows the model to account for contact between the washer/bolt and the rest of the joint. The last method is by modeling the threads. The thread geometry with frictional contact defined between the threads can be explicitly modeled. However, these models can be extremely computationally expensive. A very fine mesh is required in the threads. In this paper, a study on the effects of stress caused by the loads applied on the tower's members will be conducted. This paper is mainly divided into two stages: The structural analysis of the tower, and the joint analysis. The structural analysis of the tower will determine the joint with the maximum stress. The joint analysis will focus on one joint to study its effects on the tower behavior.

2. Analysis of the transmission tower

2.1. Transmission tower geometry

The purpose of this stage is to analyze the whole tower to determine the stresses at the joints and the forces in each member. The effects of joint will be neglected in this stage. All the joints are assumed as pinned. STAAD Pro software will be used due to its ease of usage, enhanced features, and its efficiency in displaying results. The tower consists of 13 bays with a total height of 37.41 meter and two faces: transversal, and longitudinal face. The European Angle (L-section) was used for this model. The material used for all members is steel.

2.2. Results and discussion

In this stage, the stresses distribution in the tower under different loading condition was determined. Five loading conditions were considered; Normal Condition, Ground Wire Broken, Top Conductor + Middle Conductor Broken, Middle Conductor + Bottom Conductor Broken, and tower self-weight. The results from the analysis show that the maximum stress occurs at Node 18 (see Figure 1). Table 1 shows the stresses and internal forces of the members attached to Node 18. The force value of Members 291 and 270 are very small and can be neglected. The other forces will be considered in the joint analysis.

Table 2. Stresses and forces of members attached to Node 18.

Member	Section (mm)	Load Condition	Stress (N/mm2)	Internal Force (kN)
66	L 120 × 120 ×10	3	196	454.871
14	L 130 × 130 × 10	3	166.14	418.673
62	L 80 × 80 × 6	3	146.847	137.302
77	L 90 × 90 × 6	3	78.209	82.901
57	L 75 × 75 × 6	3	-32.685	28.600
299	L 130 × 130 ×10	3	11.091	27.951
30	L 130 130 10	3	-7.232	18.225
74	L 90 × 90 × 6	3	-1.314	1.393
270	L 75 × 75 × 6	3	0791	0.692
291	L 75 × 75 × 6	3	-0.391	0.316

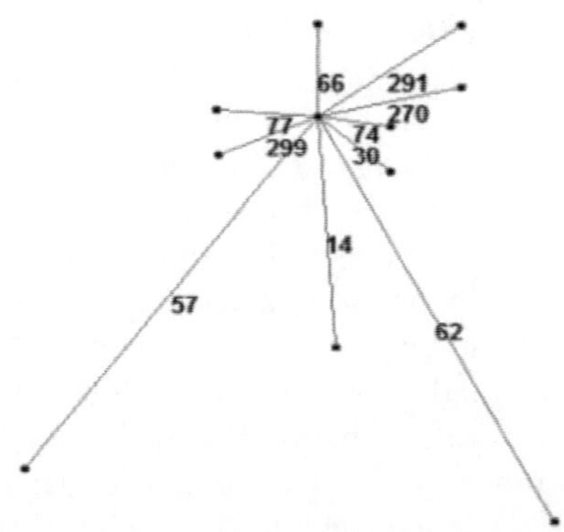

Figure 1. Node 18 and the associate members.

3. Analysis of joint

The purpose of this stage is to perform a detailed analysis for one joint of the transmission tower. Different parameters for the joint will be assumed. As given in the architectural drawing, all the connections are steel galvanised bolts and nuts as per BS4190. Finite element analysis software ANSYS was used. In this analysis, a steel plate with the thickness of 10mm and dimension of 500x500mm was used. Each member will be connected to this plate with two bolts of 16mm diameter. In this paper, three models will be considered (see Figure 2). Model 1 will consider only the plate without including the members in the design. This is the simplest design where the bolt will be represented by 16mm diameter holes. Members of the joint will be represented by its respective internal forces, as stated in Table 2. Model 2 includes a plate with the members attached to it. Similar to Model 1, bolts will be represented by 16mm diameter holes. Despite of this similarity, presence of the members attached to the plate makes it necessary to take into account the contact area between the plate and the members. Bonded contact is defined between the plate and the members. These two models are the simplest since they do not include a lot of details and there is no much setup needed for the design. At the same time, these models are considered as the least realistic models of the three models. Model 3 uses the Beam Element Bolts method. Five members will be designed and bolted to the plate. Bolts will be modeled by creating a line body and defining its cross-section in the ANSYS Design Modeler. The beam will be attached to the inner diameter of the bolt hole.

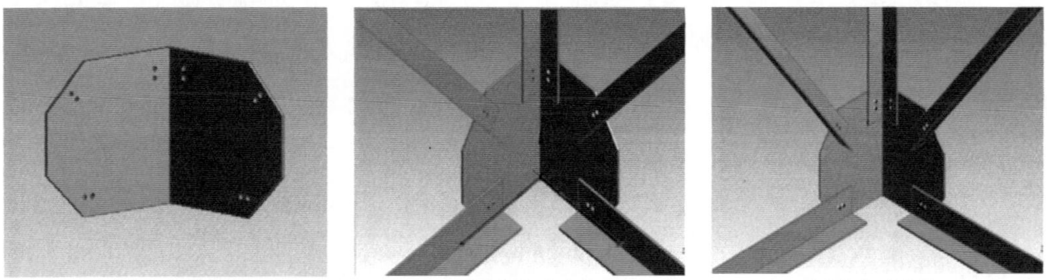

(a) Model 1 (b) Model 2 (c) Model 3

Figure 2. Tower joint model approaches.

3.1. Results and discussion

Figure 3 shows the stress distribution for the three models. It can be seen that the value of the stresses is the least at Model 1, and increases in Model 2 and Model 3. For example, in Bolt 1 the value of stresses are 141N/mm^2 for Model 1, 178.5N/mm^2 for Model 2, and 349.6N/mm^2 for Model 3. In Model 1, where the members of the joints are replaced by its forces and only the plate is considered, the maximum stress value produced at the bolts is 165N/mm^2. This result is much lesser than the value obtained by STAAD Pro where the maximum stress is 196N/mm^2. By only considering the plate and ignoring the members attached to the joint, this model becomes the least realistic between the three models. Model 2 takes into account the members attached to the plate and does not include bolts design. Thus, this model can be considered as the most similar model to that in the tower analysis. They are similar in terms of geometry and ignoring the bolts effects. For this reason, the results of this model is similar to that of STAAD Pro. The maximum bolt result in this model is 217.3N/mm^2 which is nearly 20N/mm^2 more than STAAD Pro results. This increment in the stress value is due to the presence of the holes in the members and the plate. Model 3 is the most realistic model. It considers the bolts effects in the design. The maximum stress for this model is 332.2N/mm^2.

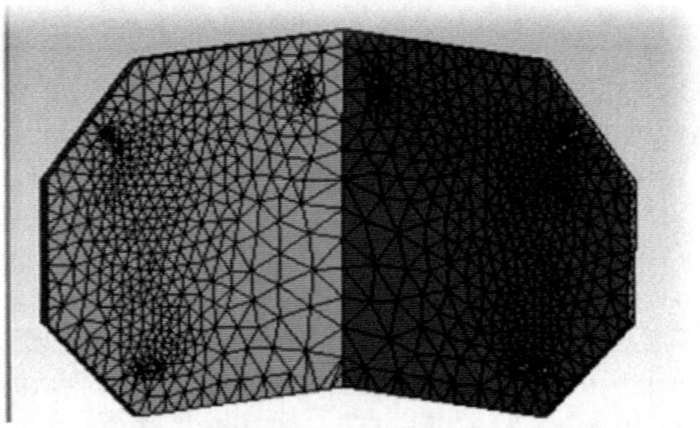

(a) Meshing Details.

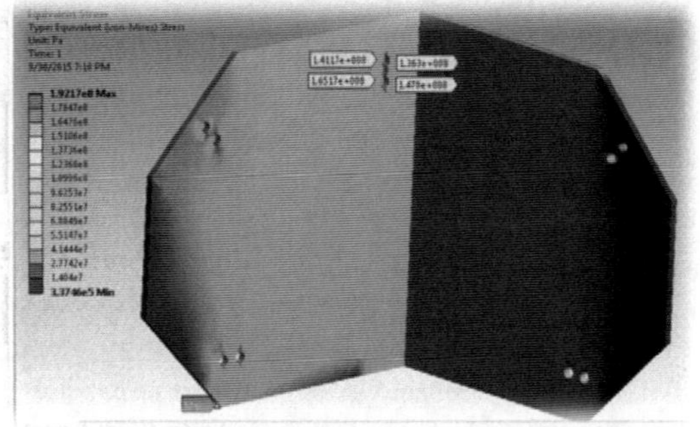

(b) Stress distribution of Model 1.

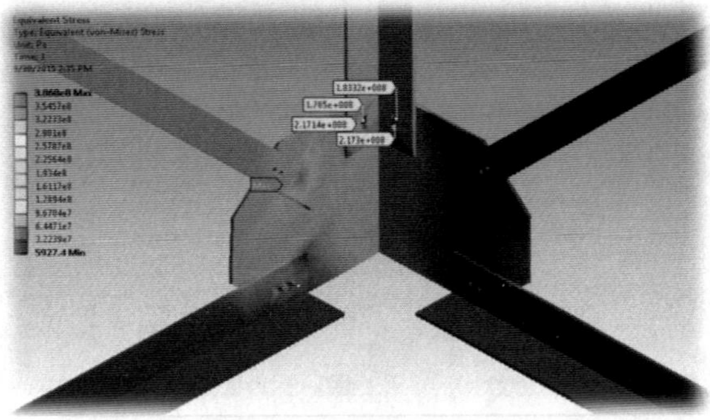

(c) Stress distribution of Model 2.

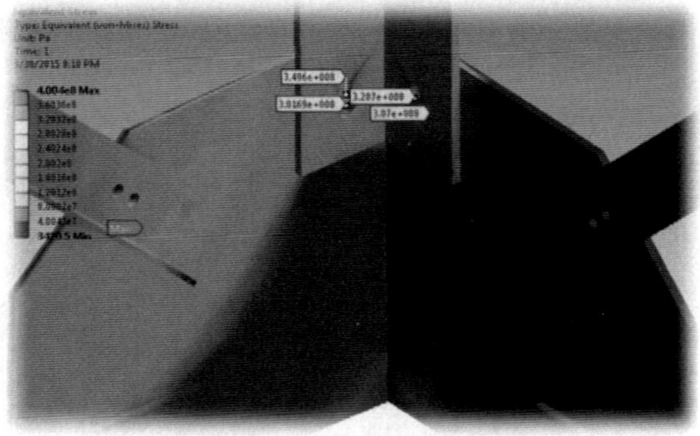

(d) Stress distribution of Model 3.

Figure 3. Stress distribution for joint models.

4. Conclusion

The summary of the findings are presented below:

- The tower has maximum stress with value of 196N/mm^2 at Node 18 of the tower.
- Model 3 shows that consideration of joint detailing in finite element simulation affects the analysis of the tower. The result of Model 3 shows a 78% stress increment compared to the tower analysis.
- As conclusion, the consideration of joint detailing in the analysis of transmission tower is effective, and can dramatically increase the stress produced at the members.

References

[1] Al-Bermani, F. G. A., Kitipomchai, S. 1992. *Nonlinear Analysis of Transmission Towers.* Engineering Structures, 14(3), 13 9- 15.

[2] Bergstrom, R N., Arena, J. R., Kramer, J. M. 1960. *Design of Self Supported Steel Transmission Towers.* Journal of the Power Division, ASCE, 86(3), 55-88.

[3] Chan SL, Cho SH. *Second-order analysis and design of angle trusses: elastic analysis and design.* Eng Struct 2008;30(3):616–25.

[4] Knight GMS, Santhakumar AR. *Joint effects on behavior of transmission towers.* J Div, ASCE 1993;119(3):689–712.

[5] Lo, D. L. C., Morcos, A., Goel, S. K. 1975. *Use of Computers in Transmission Tower Design.* Journal of the Structural Division, ASCE, 10 1(7), 1443- 1453.

The Potential of Heat Collection from Solar Radiation in Asphalt Solar Collectors in Malaysia

Salmia Beddu[1*], Siti Hidayah Abdul Talib[1], Zarina Itam[1]
[1]Department of Civil Engineering, Universiti Tenaga Nasional, Malaysia

E-mail: Salmia@uniten.edu.my

Abstract. The implementation of asphalt solar collectors as a means of an energy source is being widely studied in recent years. Asphalt pavements are exposed to daily solar radiation, and are capable of reaching up to 70°C in temperature. The potential of harvesting energy from solar pavements as an alternative energy source in replace of non-renewable energy sources prone to depletion such as fuel is promising. In Malaysia, the sun intensity is quite high and for this reason, absorbing the heat from sun radiation, and then utilizing it in many other applications such as generating electricity could definitely be impressive. Previous researches on the different methods of studying the effect of heat absorption caused by solar radiation prove to be quite old and inaffective. More recent findings, on the otherhand, prove to be more informative. This paper focuses on determining the potential of heat collection from solar radiation in asphalt solar collectors using steel piping. The asphalt solar collector model constructed for this research was prepared in the civil engineering laboratory. The hot mixed asphalt (HMA) contains 10% bitumen mixed with 90% aggregates of the total size of asphalt. Three stainless steel pipes were embedded into the interior region of the model according to the design criteria, and then put to test. Results show that harvesting energy from asphalt solar collectors proves highly potential in Malaysia due its the hot climate.

1. Introduction

Most countries currently rely on non-renewable energy sources such as coal, oil, and natural gases as an energy source. These non-renewable sources will eventually deplete and become too expensive or too environmentally damaging to retrieve. In contrast, renewable energy resources such as wind and solar energy has the tendency to become a regenerative energy with no consequence of environmental damage. Most renewable energy comes either directly or indirectly from the sun. The sun provides a cheap and abundant source of clean and renewable energy.

In the 1980's, Malaysia started implementing asphalt pavements which consists of bituminous surfacing, granular road base, drainage sub-base, and the formation subgrade replacing bituminous macadam surfacing, and design in accordance to the Arahan Teknik (Jalan) 5/85 which was adapted from American Association Of State Highway Officials (AASHTO) Road Test, subjected to specific modifications based on local experiences [1, 2]. Asphalt pavements has the capability of absorbing higher heat flux, with the capacity to produce an average of 400 to 600MJ/m^2 solar radiation per month due Malaysia's geographical profile which is at the equatorial region, thus increasing the pavement temperature [4, 5].

Hasebel et al. performed a study on the use of energy from heated pavements to produce electricity, and at the same time lower pavement temperatures during summer [9]. Measurements and modeling were used to evaluate the effect of the flow rate and temperature of the heat exchanger. It was found that the temperature of the heat exchanger fluid and the resistance of the thermoelectric modules showed a significant effect on the peak power output. Rajib et al. investigated the temperature distributions in small samples by means of finite element modeling. Tests on small and large-scale asphalt pavement samples were also performed. Many researchers have studied the thermal response in asphalt pavement especially in modern countries where sustainable development is deemed crucial. Similar to other developing countries, Malaysia is keen to promote solar energy as the alternative source to be used in the future [4]. However, since studies related to solar energy is yet to be a common research in Malaysia; therefore a wide range of data collection is needed for further development, integration, and innovation.

2. Problem statement

Asphalt pavement has gained more attention in the recent years as a potential source of renewal energy. Its surface temperature can reach up to 70°C in summer, inducing a rise in temperature of adjacent air at the surface, which is generally known as the heat island effect. This leads to an increase of power consumption due to excess use of air conditioning, which may lead to a decrease of air quality in cities. In particular, thermal oxidation rate doubles with each 10K increment in temperature. Moreover, pavements under such high temperatures are prone to suffer from rutting. Therefore, immediate actions should be taken as Malaysia is encountering a remarkable increase in energy demand and economic growth. Renewable energy is one of the best solutions to supply energy since it can be generated constantly and replenished naturally.

Asphalt solar collectors consist of pipes embedded in the pavement with an internally circulating fluid. Since solar radiation increases the thermal gradient between the pavement and the fluid circulating through pipes from its initial condition, heat transfer occurs from the pavement to the fluid. This process will continue until thermal equilibrium is reached, where both the pavement and fluid will have the same temperatures, leading to a decrease in pavement temperature and an increase in fluid temperature. This drop in asphalt temperature contributes to mitigate the heat island effect and reduce the risk of permanent deformations to the pavement. Asphalt solar collectors are usually coupled with low temperature geothermal heat pumps, resulting in reasonable efficiency and operating costs. What makes asphalt solar collectors intriguing is the potential of harnessing energy from the temperature rise from the circulating fluid in the pipes. This study aims to develop a prototype for a solar harvesting system which will prove the potential of harvesting energy from solar energy in Malaysia. Therefore, the efficiency of asphalt pavement as an energy harvester will be discovered.

Malaysia is one of the Asian countries with a warm, tropical climate. Its temperature ranges from 21°C to 35°C. The annual average daily solar irradiations for Malaysia range from 4.21 to 5.56kWh/m^2. The highest solar radiation was estimated at 6.8kWh/m^2 in August and November while the lowest was 0.61kWh/m^2 in December. Hence, the potential of Malaysia to use solar as a renewable energy resource is high. Al-Abdul et.al. [2] showed that a wide range of temperature variation results in drastic variation of asphalt concrete modulus. As temperature increases from 30°C to 60°C, the resilient modulus reduces from 5500MPa to 800MPa. Thus, to study the in-situ characteristics of various pavement designs, it is imperative to know the temperature distribution within the pavement cross-section. Recently, researchers carried out experimental slabs to investigate heat transfer characteristics of the Slab Solar Collection (SSC), heat collecting efficiency, and the influence of pipe spacing and flow rate on the SSC process [7]. It has been proven that the average heat collecting capacity is about 150 to 250W/m^2 in summer conditions. Bijsterveld et. al. conducted finite element simulations to investigate the effects of heat exchanger system embedded in asphalt pavement on the temperature distributions and stress/strains inside the pavement [1]. It was found that placing the heat exchanger tubes at shallow depths can extract more energy but results in higher stresses in the pavement, thus reduced durability of the pavement. For this reason, the effect of different materials on the thermal and structural properties of the pavement must be studied.

3. Experimental program

The model was fabricated from wood, and embedded with asphalt and stainless steel (see Figure 1). Details of the model are as the following:

- 30×30×15cm wooden box was designed for the solar asphalt pavement model.
- Bituminous mixture consists of a well-graded mixture of coarse aggregates, fine aggregates and mineral filler, bound together with petroleum bitumen using asphalt cement.
- Standard steel pipe was placed in the middle layer of the box with the length of 35cm and 2.2cm diameter. Six inlet and outlet valves sockets were added to the end of pipes in order to insert and discharge the liquid easier.

Figure 1. Asphalt solar collector model.

The model was then placed in a selected open space area, in this case, a parking lot, chosen due to its optimal environmental condition for solar radiation. Two cases has been specified:

1. *Case 1* – Testing was conducted with the frequency of 3 times a day for 5 consecutive days. The scheduled time for testing are:
   - 9 am to 10 am in the morning.
   - 1 pm to 2 pm in the afternoon.
   - 5 pm to 6 pm in the evening.

   Such times were chosen to know the actual temperature at the time of solar radiation exposure itself.

2. *Case 2* – Testing was conducted between 9 am to 3 pm, and 3 pm to 9 pm for 10 consecutive days. Temperature readings will be done after every 6 hours. This is in order to determine the temperature for transient thermal conditions.

Heat efficiency of the model is based on the fluid and air temperatures, and is calculated from the following formulation:

$$\text{Heat efficiency}=\frac{\text{Fluid temperature}(°C)\text{ - Air temperature}(°C)}{\text{Fluid temperature}(°C)}\text{x}100 \quad (1)$$

4. Results and discussion

4.1. Case One

- Results show that the percentage of heat efficiency is higher the afternoon than other times. Due to the tropical weather conditions in Malaysia, constant air temperature has been observed. Therefore, we could realize that the heat efficiency for these 5 days are similar to each other.
- The lowest heat efficiency among these five days was on the second day and in morning at 3.70% due to low and constant temperature condition, causing steady state condition within the model.
- The highest heat efficiency was on the fifth day with 25.58%. In the afternoon, the sun radiation was high enough to heat the pavement and steel, causing the heat to be captured well.

4.2. Case Two

- Results show that the percentage of heat efficiency in the first period of time which is between 9 am to 3 pm is higher than 3 pm to 9 pm. This is mostly because of the air temperature which was higher during this period.
- The lowest heat efficiency was obtained on the third day with 12.12% during the 3 to 9 pm period.
- The highest heat efficiency among the 5 days was on the fourth day with 57.69% during the 9 am to 3 pm time frame. The percentage of heat efficiency is really high and shows that the system is really effective in terms of harvesting heat and energy. The temperature that recorded is 33 °C and the fluid temperature reached to 78°C after six hours.

5. Conclusion

Based on the results obtained, the following conclusions can be made:

- Heat transfer efficiency was affected mostly by weather temperature and fluid temperature.
- The maximum heat efficiency obtained for Case 1 was on the fifth day with 25.58% efficiency in the afternoon, and for Case 2 on the fourth day between 9am to 3pm with 57.69% efficiency. These results show high efficiency of the model used for this project.

Acknowledgment

The authors would like to acknowledge the personnels of the Faculty of Civil Engineering & Earth Resources, University Malaysia Pahang for their assistance in the experimental testing. This research is financially supported by the Internal Grant RDU 130383.

References

[1] Berdahl P, Akbari H, Levinson R, Miller WA. *Weathering on Roofing Materials – An Overview.* Constr. Build. Mater. 2006;22(4), pp. 423–33.
[2] Al-Abdul Wahhab, H. and Balghunaim, F. (1994). *Asphalt Pavement Temperature Related to Arab Saudi Environment.* J. Mater. Civ. Eng., 6(1), pp. 1–14.
[3] B. K. Diefenderfer, I. L. Al-Qadi and S. D. Diefenderfer. *Model to Predict Pavement Temperature Profile: Development and Validation.* Journal of Transportation Engineering, Vol. 132, No. 2, 2006, pp. 162-167.
[4] Hermansson, A. (2000). *Simulation Model for Calculating Pavement Temperatures, including Maximum Temperature.* Transportation Research Record 1699, Transportation Research Board, Washington, D.C., pp. 134–141.
[5] Shaopeng W., Mingyu Chen., Jizhe Z. *Laboratory Investigation into Thermal Response of Asphalt Pavements as Solar Collector by Application of Small-scale Slabs.* Applied Thermal Engineering vol. 31 issue 10 July, 2011. pp. 1582-1587.
[6] X.B. Liu, S.J. Rees, J.D. Spitler. *Modeling Snow Melting on Heated Pavement Surfaces: Model Development.* Applied Thermal Engineering 27 (2007) , pp. 1115-1124.
[7] Q. Gao, M. Li, M. Yu, J.D. Spitler and Y.Y. Yan. *Review of Development from GSHP to UTES in China and other Countries.* Renewable and Sustainable Energy Reviews 13.

Light irradiance and spectral distribution effects on cyanobacterial hydrogen production

Siti Fatihah Salleh[a,1], Azlina Kamaruddin[b], Mohamad Hekarl Uzir[b], Abdul Rahman Mohamed[b], Abdul Halim Shamsuddin[a]
[a]Centre for Renewable Energy, Universiti Tenaga Nasional, 43000 Kajang, Selangor, Malaysia

[b]School of Chemical Engineering, Engineering Campus, Universiti Sains Malaysia, Seri Ampangan, 14300 Nibong Tebal, Penang, Malaysia

Abstract. Light is an essential energy source for photosynthetic cyanobacteria. Changes in both light irradiance and spectral distribution will affect their photosynthetic productivity. Compared to the light irradiance, little investigations have been carried out on the effect of light spectra towards cyanobacterial hydrogen production. Hence, this work aims to investigate the effects of both light quantity and quality on biohydrogen productivity of heterocystous cyanobacterium, *A.variabilis*. Under white light condition, the highest hydrogen production rate of 31 μmol H_2 mg chl *a* $^{-1}$ h^{-1} was achieved at 70 $\mu E\ m^{-2}\ s^{-1}$. When the experiment was repeated at the same light irradiance but different light spectra of blue, red and green, the accumulations of hydrogen were significantly lower than the white light except for blue light. As the light irradiance was increased to 350 $\mu E\ m^{-2}\ s^{-1}$, the accumulated hydrogen under the blue light doubled that of the white light. Besides that, an unusual prolongation of the hydrogen production up to 120 h was observed. The results obtained suggest that blue light could be the most desirable light spectrum for cyanobacterial hydrogen production.

1. Introduction

Cyanobacteria, also widely known as blue-green algae, are oxygenic photosynthetic prokaryotes with a long evolutionary history. The ability of cyanobacteria to fix nitrogen and to adapt to low light allow them to enter into a very wide range of symbiotic associations [1]. Cyanobacteria are also useful in the carbon sequestration outlook, since they could act as carbon sinks by capturing and transforming CO_2 into organic chemicals and many other valuable materials. In addition, some heterocystous cyanobacteria are well adapted for diazotrophic growth, and are able to synthesize hydrogen under some prerequisite conditions [2]. Both nitrogen fixation and hydrogen production are catalyzed by the O_2-sensitive nitrogenase enzymes which are confined in the heterocyst.

Like other photosynthetic organisms, cyanobacteria are able to acclimate to changes in ambient light irradiance and spectral distributions. Their responsiveness towards light treatment correlates with the changes of the contents of both light harvesting pigments and photosystems in the cells [3]. It has also been proven that for a given cell density and light irradiance, light spectra conversion gives significant

[1] Author of correspondence: Siti.Fatihah@uniten.edu.my

effect towards photosynthetic pigments composition, cell growth rates and nitrogenase activity [4-7]. However, information on its effect towards cyanobacterial hydrogen production which is also light-dependent is still limited. Considering the essentials of both light quantity and quality toward the complex cellular regulatory systems, this study aims to investigate hydrogen production performance of *A.variabilis* incubated under diversified light irradiances and wavelength distribution.

2. Methodology

2.1 Propagation of Anabaena variabilis.

Stock cultures of *Anabaena variabilis* ATCC 29413 were propagated aerobically in 500 mL Erlenmeyer flask and maintained in $BG11_o$ medium, as described elsewhere [2]. Cell concentration was monitored according to the optical density (A_{683}) using a UV spectrophotometer.

2.2 Hydrogen production experiment

Cells were collected during mid-to late-exponential growth stage and concentrated using a centrifuge. The supernatant was removed after centrifugation, and the cells pallet was washed twice with fresh $BG11_o$ medium. The concentrated cells were then diluted and suspended in 40 mL of $BG11_o$ medium inside 164 mL serum bottles, sealed with butyl-rubber stoppers and crimped aluminum rings. An anaerobic condition with 5% of CO_2 was generated prior to cell suspension. The initial innoculum concentration was standardized at ±2.32 µg chl *a* mL^{-1}. The serum bottles were incubated horizontally on an orbital shaker operated at a speed of 120 rpm. Twelve fluorescent lamps (Philips TL-D 18W/865 1SL) were placed slightly above the serum bottles, providing uniform light irradiance of 35-350 µE m^{-2} s^{-1}, measured using a quantum sensor (Licor LI-190, USA). Monochromatic lights of blue, red and green were provided using 3W RGB Power LED lights (Vollong Electronics Co., Limited, China). The calculated surface area of one side of the culture bottle (assuming it was transacted by a horizontal plane) was 45 cm^2. 200 µml of headspace gas was taken out periodically using an airtight syringe and then manually injected to a gas chromatography (Perkin Elmer AutoSystem XL) equipped with a thermal conductivity detector (TCD). Hydrogen production experiment was done in triplicates. The efficiency of photosynthetic conversion of light to hydrogen was calculated as:

$$\text{Light conversion efficiency (\%)} = \frac{\text{Energy content of } H_2}{\text{Absorption of light energy}} \times 100 \qquad (1)$$

In full expression, Equation 1 is in the form:

$$\eta_{H2}\,(\%) = \frac{\Delta G^{o}_{H2}\, R_{H2}}{E_s\, A} \times 100 \qquad (2)$$

where ΔG^{o}_{H2} is the free energy available from the heat of combustion of hydrogen which is 0.24 kJ/mmol at 25 °C [8], R_{H2} is the amount of accumulated H_2, E_s is light energy radiation, and *A* is the illumination area [9].

3. Results and Discussion

3.1 Effect of Light Irradiance

Figure 1 illustrates the effect of light irradiance towards hydrogen production rate. The maximum rate of 31 µmol H_2 mg chl a^{-1} h^{-1} was achieved at a moderate light irradiance of 70 µE m^{-2} s^{-1}. Further increase of the light irradiance beyond this value did not improve hydrogen production rate. Furthermore, the light conversion efficiency was also significantly reduced by 7 times from 70 µE m^{-2} s^{-1} to 350 µE m^{-2} s^{-1} which were 0.12% to 0.017% respectively. This implies that the light harvesting pigments had already been saturated at 70 µE m^{-2} s^{-1}.

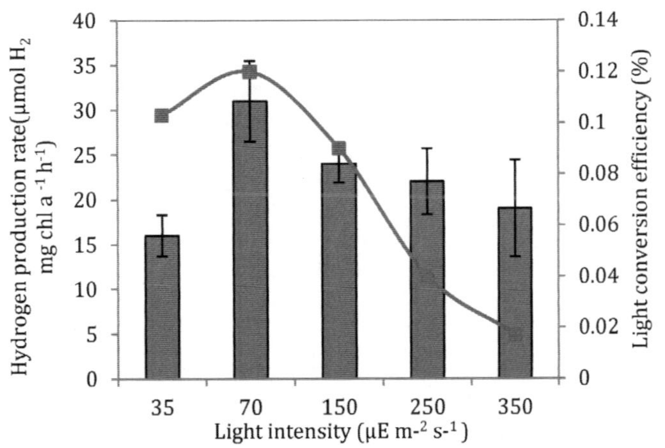

Figure 1. The effect of white light irradiance on hydrogen production rate (bar) and light conversion efficiency (■). The light conversion efficiencies were calculated using Equation 2.

3.2 Effect of Light Spectra

The hydrogen production versus time profile in Figure 2(A) depicts the effect of light spectral distribution on hydrogen production. For all light colours except for blue, hydrogen production were significantly lower as compared to control (white) light. Cells incubated under the red light condition demonstrate the lowest hydrogen production. Meanwhile, the production of hydrogen under the green light starts to drop the earliest which is at the 47th hour. The increased production of hydrogen in blue light condition might be the result of enhanced light energy harvested by the chlorophyll *a* pigments which absorbs light most strongly in the blue spectrum (430-445 nm) as well as strong stimulation of nitrogenase activity within this wavelength range [7].

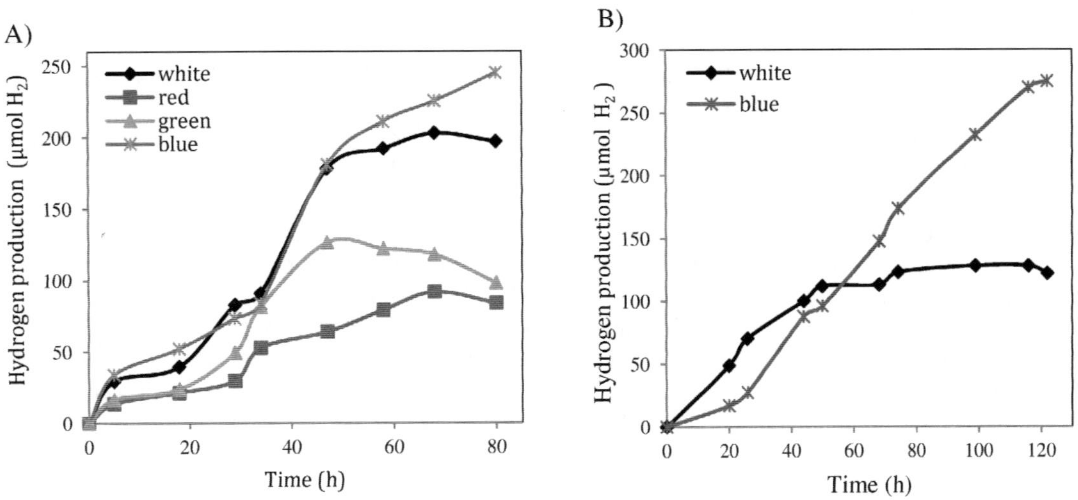

Figure 2. Accumulation of hydrogen over time under (A) moderate light irradiance of 70 µE m^{-2} s^{-1} for white (control), blue, green and red lights and (B) high light irradiance of 350 µE m^{-2} s^{-1} for white and blue lights.

The experiment was then repeated at high light irradiance of 350 $\mu E\ m^{-2}\ s^{-1}$ for both white and blue lights. As demonstrated in Figure 2(B), although the hydrogen production rate for cells incubated under blue light is initially lower than the white light during the first 44 h (21.4 μmol H_2 mg chl $a^{-1}\ h^{-1}$ compared to 24.5 μmol H_2 mg chl $a^{-1}\ h^{-1}$), the rate is however maintained throughout the experiment. Furthermore, significant prolongation of hydrogen production can be observed. The production did not cease and kept increasing even after 120 h or 5 days of incubation. The maximum hydrogen accumulation obtained under blue light condition was 275 μmol H_2, a two-fold increase compared to the white light condition.

4. Conclusion
In conclusion, the specificity of *A. variabilis* response toward variations in light conditions proved that cyanobacteria are light-sensitive and therefore, proper tunings of the light irradiance and wavelength distribution are essential in order to maximize the hydrogen productivity. The improvements in the accumulation of hydrogen and production duration suggest that blue light could be the most desirable light spectrum for hydrogen synthesis by this species.

5. Acknowledgement
The authors would like to acknowledge Universiti Sains Malaysia for funding this current research work through the Long Term Research Grant Scheme (LRGS) 203/PKT/6723003 and MyBrain15 from the Ministry of Education, Malaysia for its financial assistance to S. F. Salleh.

6. References
[1] Whitton B A, Carr N G 1982 *Cyanobateria: Current perspectives* (The biology of cyanobacteria) ed Carr N G, Whittons B A (USA: University of California Press) pp 1-8
[2] Salleh S F, Kamaruddin A, Uzir M H, Mohamed A R 2014 Effects of cell density, carbon dioxide and molybdenum concentration on biohydrogen production by *Anabaena variabilis* ATCC 29413 *Energ. Convers. Manage.*, **87** pp 599–605
[3] Kehoe D M, Gutu A 2006 Responding to color: the regulation of complementary chromatic adaptation *Annu. Rev. Plant. Biol.*, **57** pp 27-50
[4] Bazyli C 1986 The effect of light on the content of photosynthetically active pigments in plants *Phyton (Austria)*, **26** pp 1-36
[5] Mohsenpour S F, Richards B, Willoughby N 2012 Spectral conversion of light for enhances microalgae growth rates and photosynthetic pigment production *Bioresource Technol.*, **125** pp 75-81
[6] Jacob-Lopes E, Gimenes Scoparo C H, Queiroz M I, Franco T T 2010 Biotransformations of carbon dioxide in photobioreactors *Energ. Convers. Manage.*, **51** pp 894-900
[7] Staal M, Stal L J, Hekkert S T L, Harren F J M 2003 Light action spectra of N_2 fixation by heterocystous cyanobacteria form the Baltic Sea *J. Phycol.*, **39** pp 668–677
[8] Yu J, Takahashi P 2007 *Biophotolysis-based hydrogen production by cyanobacteria and green microalgae* (Communicating current research and educational topics and trends in applied microbiology) ed Mendez-Vilas A (Spain: Formatex) pp 79-89
[9] Yoon J H, Shin J H, Kim M S, Sim S J, Park T H 2006 Evaluation of conversion efficiency of light to hydrogen energy by *Anabaena variabilis Int. J. Hydrogen Energ*, **31** pp 721-727

Thermal Analysis for Improving Fuel Characteristics of Empty Fruit Bunch (EFB) Fibers

M.S Hafiz[1], K.Horst[2], M.A Md.Ariffin[3]

[1]Technical and Project Executive, Detik Aturan Sdn Bhd, Selangor, Malaysia

[2]Senior Lecturer of Energy Technology and Wood Industries, University of Applied Science, Rosenheim, Germany

[3]Managing Director, Detik Aturan Sdn Bhd, Selangor, Malaysia

E-mail: [1] sufy.ibn.hafiz@gmail.com
[2] horst.kreimes@fh-rosenheim.de
[3] mdarfizal@yahoo.com

Abstract- Agriculture waste is well known with low melting ashes. This is due to the large quantities of alkali metals are required as nutrients. Potassium is the one of the alkali metal that contributes to the low melting ash of empty fruit bunch biomass fuel. Thermal analysis was conducted in order to determine the increasing of ash fusion temperature of EFB fiber. Ash fusion temperature has four type of shape which is initial deformation, softening temperature, hemispherical temperature and fluidization temperature. EFB fiber has melting temperature 1061°C in average. To solve low melting ashes, $Ca(OH)_2$ was added with EFB fiber in order to increase the ash melting point. 2%, 4%, 6% and 8% by dry weight was added to EFB fiber. From the experiment by adding 2%, 4%, 6% and 8% of $Ca(OH)_2$, the melting temperature was increase to 1114°C, 1239°C, 1277 °C and 1315 °C respectively. As conclusion this lab scale thermal analysis test was conducted in order to allow the determination of additive percentage for larger scale in order to adapt with biomass gasifiers.

1.0 Introduction

In living plants, large quantities of alkali metals are required as nutrients and counter ions. Potassium is especially abundant in annual and short-life crops like palm oil with sodium substituting in minor amounts. In the fast growing parts of trees (small branches, twigs and leaves), the alkali content is higher compared to the trunk and larger branches. The usefulness of biomass fuels in energy production is often limited by maintenance problems such as sintering, fouling and hot corrosion. These processes, which are largely due to the presence of potassium and chlorine, reduce the capacity, efficiency and availability of thermal conversion plants using biomass. Around 90% of the alkali in biomass is present in water-soluble or ion exchangeable form and is susceptible to vaporization during heating [1].

1.3 Effect of Additive with Increasing of EFB Softening Point

Slag formation slightly form of hard foam mineral layer on gasifier will occur if the temperature exceeds the softening temperature of ash [2] with a low heat transfer coefficient. Slag formation is mainly influenced by the gasification feedstock. Feedstock with high ash content and low softening point such as palm oil EFB, will more likely to form a slag. Elements such as Ca and Mg increase the ash melting point while K and Na decrease the melting point. Chlorides and low melting alkali and alumino silicates may decrease the softening point as well [3].

According to Sorensen et al, addition of calcium compound to the biomass before or during gasification process is recommended in order to change the composition of ash leading to in compound having higher fusion point temperature. According to the inventive patent [4], it is considered reasonable to obtain $CaKPO_4$, having a melting temperature of 1560°C. Hence a goal can be to generate this compound as the major product. This means that one mol of K requires one mol of Ca.

Slaging and sintering can be avoiding by reducing the operating temperature in gasifier but lower temperature may affect the quality of flue gases. Apart from adjusting the operational temperature, another solution identified is by adding additive such as CaO to increase the melting point of the feedstock by compensating the lowering abilities of the K [5].

2.0 Methodology

The amount of additive supplemented in EFB is 2, 4, 6, and 8% mass by dry weight of EFB. The fibers and additive were mixed in a small plastic container and manually shake by hand until fibers and additive have been in a homogenously mix. Usually EFB fibers have average 12% of moisture content in natural environment. Additional water spray is not required in order to attach the additive surface to the EFB fiber surface. Sample EFB fiber with additive will be ashes in the furnace at 550 °C for 2 hours. Then the ash sample will be mould into cylindrical shape for ash fusion temperature analysis. The cylindrical shape will be observed into 4 kind of shape which is initial deformation, softening point, hemisphere temperature and fluidize temperature. The picture is taken for each 10 °C temperature increase. [6]

3.0 Result of Ash fusion temperature of EFB with addition of $Ca(OH)_2$ as additive

In this chapter, all the ash fusion result is shown by each additive percentage with amount of test. Four trials are made in order to confirm with previous test and get the average of fusion temperature for each percentage of additives.

Table 3.1 Mean value of EFB ash fusion temperature with addition of $Ca(OH)_2$

Characteristic	EFB + 0% of $Ca(OH)_2$	EFB + 2% of $Ca(OH)_2$	EFB + 4% of $Ca(OH)_2$	EFB + 6% of $Ca(OH)_2$	EFB + 8% of $Ca(OH)_2$
Initial Deformation (°C)	934	1046	1166	1237	1274
Softening Temperature (°C)	1061	1114	1239	1277	1315
Hemspherical Temperature (°C)	1170	1146	1280	1320	1320
Fluid Temperature (°C)	1185	1220	1295	1320	1320

From table 3.1, mean value of EFB ash fusion temperature is increase by addition of additive. From the initial hypothesis, by increasing ash fusion temperature of EFB, the slag and sintering occur in gasify chamber can be reduce.

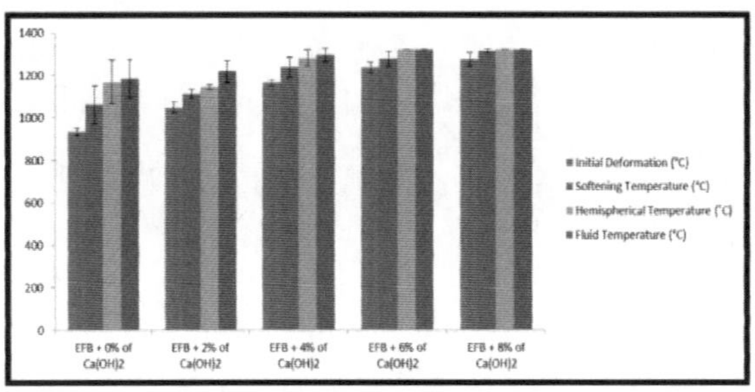

Figure 3.1 Bar column for each addition of additive (%) and ash fusion temperature (°C)

4.0 The correlation of EFB ash fusion temperature with addition of additive

In this chapter, the ash fusion temperature of EFB will be discussed. In natural form, EFB have a low ash fusion temperature. This is causes by high alkali element in EFB such as Potassium (K) and Sodium (Na). Both elements cause low softening temperature in all biomasses, especially in agricultural plants. These mineral is supply three times per year as palm oil plant growth factor or as known as inorganic fertilizer [7]. Plant will absorb all the nutrient and mineral and accumulated in straw, fruit, leave and trunk. Figure 3.1 shows the addition of additive affecting ash fusion temperature of the EFB fiber.

EFB fibers with 0% of additive have low ash fusion temperature. The mean of initial deformation is 934 ˚C with standard deviation ±16.15. Raw EFB have inhomogeneous composition that resulting high standard deviation when it reach softening temperature phase. The mean of softening temperature from four trials is 1061 ˚C with standard deviation ±91.07. This is why from previous works; slagging problem occur inside the combustion chamber even though the operating temperature is below 1000 ˚C [8]. The inhomogeneous element of EFB fiber is cause by rainy season, material handling and contamination with soil. On the other hand, ash content is increasing in biomass during growth and is affected by biomass species, origin, soil and other factor [8]. At average temperature 1170˚C ±102.8, the cylindrical shape is turn to hemisphere shape or it is called hemisphere temperature point. By continuously raising the temperature, the flow temperature is form which is half of the hemisphere shape. 1185˚C±91.15 is the flow temperature of the raw EFB which is the ashes is in fluid state.

Once again from figure 3.1, it can be seen directly from the bar column that the ash fusion temperature is increasing with addition of additive. The additive use in this study is $Ca(OH)_2$ where calcium (Ca) is the main element in increasing as fusion temperature. By addition 2% of $Ca(OH)_2$ with EFB, the initial deformation is increase with mean 1046±25°C and softening temperature is 1114˚C±20°C. While by addition 4% of $Ca(OH)_2$ the initial deformation is increase with mean 1166±15°C and softening temperature is 1239±47°C. Then by addition 6% and 8% of additive also increase the initial deformation and softening temperature compare to 2% and 4% with initial deformation and softening temperature at 1237±25°C and 1277±34°C for 6% of additive while 1274±30°C and 1315±10°C for 8% of additive.

5.0 Conclusion

As shown above, thermal analyses experiment lab scale allow the determination of additive percentage necessary for successfully running European solid biomass gasifier with Malaysian biomasses such as EFB fibers by adapting them. The lab scale test is much faster and less cost intensive than test with the original big plant.

6.0 Acknowledgement

Thank you to University of Kuala Lumpur and University of Applied Sciences Rosenheim, Germany, for giving opportunity to conduct this research at Rosenheim, in context with their common MGEEB (Master of Green and Energy Efficient Buildings) program. Then very big thank you to Stadtwerke Rosenheim, Germany for me to conduct research at their facilities. Last, but not least, we would like to say thank you to Detik Aturan Sdn.Bhd for sponsoring the raw material and transportation from Malaysia to Germany.

7.0 References

[1] A.Nurhayati and S.Fauziah (2013). The properties of wash empty fruit bunch of palm oil. Journal Science Physic, vol 24(2), 117-137

[2] Williams et al. H2 Production via Biomass Gasification, AEP Project, Task 4.1 Technology Assessments of Vehicle Fuels and Technologies, PIER Program, California Energy Commission, prepared by ITS-Davis 2007

[3] Friedrich Biedermann, Ingwald Obernberger. Ash-related Probems during Biomass Combustion and Possibilities for a Sustainable Ash Utilisation.

[4] L.H. Sorensen,J. Fjellerup and Ulrik Henriksen(2003). Method for reducing agglomeration, sintering and deposit formation in gasification and combustion of biomass. US 6615751B1.

[5] Kaltschmitt, Martin; Hartmann, Hans; Hofbauer, Hermann. Energie Aus Biomasse Grundlagen, Techniken und Verfahren 2. neu bearbeitete und erweiterte Auflage. Berlin / Heidelberg: Springer

[6] ISO 540:1995 Solid Mineral Fuel. Determination of Fusibility Ash.95.99.(2008-05-28)

[7] R. Ian and T.H Fairhust. Management of Phosphorus, Potassium and Magnesium in Mature Palm Oil. Better Crops International. Vol 13. No 1, May 1999.

[8] M.Nizam (2013). Contribution of palm oil empty fruit bunch for gasification as an energy resource. Master Thesis from Univeristy Kuala Lumpur.

International Conference on Advances in Renewable Energy and Technologies (ICARET 2016) IOP Publishing
IOP Conf. Series: Earth and Environmental Science **32** (2016) 012048 doi:10.1088/1755-1315/32/1/012048

Techno-economic analysis of bioethanol production from rice straw by liquid-state fermentation

M H M Hidayat[a], S F Salleh[a,1], T M I Riayatsyah[b], H B Aditiya[c], T M I Mahlia[a] A H Shamsuddin[a]
[a]Centre for Renewable Energy,Universiti Tenaga Nasional, 43000 Kajang, Selangor, Malaysia

[b]Department of Mechanical Engineering, Faculty of Engineering, University of Malaya, 50603 Kuala Lumpur, Malaysia

[c]Department of Mechanical Engineering, The University of Melbourne, VIC 3010, Australia

Abstract. Renewable energy is the latest approach of the Malaysian government in an effort to find sustainable alternative energy sources and to fulfill the ever increasing energy demand. Being a country that thrives in the service and agricultural sector, bioethanol production from lignocellulosic biomass presents itself as a promising option. However, the lack of technical practicality and complexity in the operation system hinder it from being economically viable. Hence, this research acquired multiple case studies in order to provide an insight on the process involved and its implication on production as well as to obtain a cost analysis of bioethanol production. The energy input and cost of three main components of the bioethanol production which are the collection, logistics, and pretreatment of rice straw were evaluated extensively. The theoretical bioethanol yield and conversion efficiency obtained were 250 L/t and 60% respectively. The findings concluded that bioethanol production from rice straw is currently not economically feasible in Malaysia's market due to lack of efficiency in the pretreatment phase and overbearing logistics and pretreatment costs. This work could serve as a reference to future studies of biofuel commercialization in Malaysia.

1. Introduction

Biofuels among other renewable and sustainable energy resources offer a worthy solution to counter depleting conventional fuel sources as well as to put a halt to climate change. As a developing nation, Malaysia's energy demand keeps increasing especially with the advancements in the industrial and transportation sector Malaysia [1]. One of Malaysia's thriving economic sectors is the agricultural sector with palm oil and paddy being the most dominant [2, 3]. What both of these have in common is the abundance of waste agricultural residue generated from harvest. Research has shown that agricultural waste such as rice straw which is a crop residue that comes from paddy harvesting has a great potential to generate as biomass recourses for the generation of biofuels [4]. Its lignocellulosic biomass could be broken down into simple sugar via hydrolysis which is then fermented into the final product, bioethanol [5, 6].

[1] Author of correspondence: Siti.Fatihah@uniten.edu.my

Rice straw can be found in abundance and currently has minimal practical uses which are as livestock feed or material for building [7]. However, despite the impressions that bioethanol production from rice straw is sustainable and environmentally benign, it has its technical limitations such as the difficulty and complexity in the operation system, low production yield and low biomass-to-product efficiency. Besides that, the produced fuels's energy properties and production cost are yet to be determined before its viability and competitiveness in the current Malaysia's economic market could be assessed. Hence, this work was dedicated to address this issue by reviewing Malaysia's biomass potential and its current renewable energy alternatives in general. Detailed analysis on the process involved in the production of bioethanol from rice straw biomass through liquid state fermentation will allow the assessment of its technical and economical feasibility.

2. Methodology

2.1 Rice straw acquisition

The rice straw is collected from paddy fields in several states in Malaysia which were; Perlis, Penang, Kedah and Perak. The energy of the process is calculated according to Equation 1;

$$E_{rsc} = E_{fc} \times F_{cc} \times A_{rsc} \quad (1)$$

where E_{rsc} is the energy of rice straw collection (J), E_{fc} is the energy of fuel for collection (J/L), F_{cc} is the fuel consumed for collection (L/ha) and A_{rsc} is the area of rice straw collection (ha). As for the cost of rice straw collection, it is as expressed in Equation 2;

$$C_{rsc} = (C_{fc} \times F_{cc} \times A_{rsc}) + (C_{dfc} \times R_c) \quad (2)$$

where C_{rsc} is the cost of rice straw collection (RM), C_{fc} is the cost of fuel for collection (RM/L), C_{dfc} is the cost of driver fee for collection (RM/rate) and R_c is the rate for collection. The amount of bale that would be produced was calculated according to Equation 3;

$$\mathrm{RSB} = \frac{RST \times A_{rst}}{R_b} \quad (3)$$

where RST is the total amount of rice straw produced (kg/ha), A_{rst} is the total plantation area (ha), RSB is the amount of rice straw bale (kg) and R_b is the bale size ratio (kg rice straw/kg bale).

2.2 Rice straw logistics

The logistics component of the production of bioethanol consists of transportation of the rice straw bale from the plantation area to either a collection center or directly to a processing plant. The energy involved was calculated according to Equation 4;

$$E_{rst} = D_t \times E_{ft} \times F_{ft} \times N_t \quad (3)$$

where E_{rst} is the energy of rice straw transportation (J), E_{ft} is the energy of fuel for transportation (J/L), F_{ft} is the fuel consumption of transportation (L/km), D_t is the distance of transportation (km) and N_t is the number of trips for transportation. The cost component is similar to the technical equation where energy is replaced with the cost of fuel shown in Equation 5;

$$C_{rst} = (D_t \times C_{ft} \times F_{ft} \times N_t) + (C_{dft} \times R_t) \quad (5)$$

where C_{rst} is the cost of rice straw transportation (RM), C_{ft} is the cost of fuel for transportation (RM/L), C_{dft} is the cost driver fee for transportation (RM/L) and R_t is the rate for transportation.

2.3 Rice straw conversion

The rice straw conversion includes the hydrolysis and fermentation otherwise known collectively as the pretreatment process of the production. The summation of these adds up to give the amount of bioethanol that can be produced. For the energy involved, it is calculated based on Equation 6.

$$E_{rsp} = M_{rs} \times \sum(E_{rs,pre} \times \eta_{conv}) \quad (6)$$

where M_{rs} is the mass of rice straw that will undergo the converting process (kg) and is a product of $RST \,.\, A_{rst}$, E_{rsp} is the energy of rice straw pretreatment (J), $E_{rs,pre}$ is the energy of rice straw pretreatment processes (J/kg) (hyrdrolysis and fermentation), and η_{conv} is the conversion efficiency.

The economic version of the pretreatment process then becomes the cost of each of the required components for the pretreatment to take place as shown in Equation 7.

$$C_{rsp} = M_{rs} \times \sum(C_{rs,pre}) \quad (7)$$

where C_{rsp} is the cost of rice straw pretreatment (RM) and $\sum C_{rs,pre}$ is the cost of all the pretreatment processes (RM/kg). Other pretreatment component such as the mechanical pretreatment will be included for processes that involves multiple types of pretreatment. The equation then becomes;

$$C_{rsp} = M_{rs} \times [(\sum C_{rs,pre}) + C_{rs,pre\ M}] \quad (8)$$

where $C_{rs,pre\ M}$ is the cost of mechanical pretreatment process (RM/kg).

2.4 Bioethanol production

The bioethanol yield is calculated according to Equation 9;

$$Y_{ethanol} = RST \times A_{rst} \times Y_{theoretical} \times \eta_h \times \eta_f \quad (9)$$

where $Y_{ethanol}$ is the bioethanol yield (L), $Y_{theoretcical}$ is the theoretical bioethanol yield from rice straw (L/kg), η_h is the hydrolysis efficiency and η_f is the fermentation efficiency.

3. Results and discussion

3.1 Bioethanol yield from rice straw

From an amount of roughly 1.3milion rice straw produced annually, the potential bioethanol yield is calculated [8]. Thus, with pretreatment efficiency of 51%, hydrolysis of 90% and fermentation of 50%, the annual theoretical yield of bioethanol is 340 million liter.

3.2 Energy consumption in bioethanol production

The total energy input was calculated as 26.67 MJ/L bioethanol produced. Comparing this value to the amount of energy released from the combustion of ethanol which is 21.2 MJ/L, it gives overall deficit of 5.44 MJ/L of energy. This shows that ethanol with its lower energy content and high energy input would not be a viable a fuel source.

3.3 Life Cycle Cost Analysis (LCCA)

The LCCA of this study was conducted to evaluate the three components of bioethanol production; Rice straw collection, logistics, and pretreatment. As demonstrated in Table 1, the major cost of the process comes from logistics.

Table 1. Cost distribution of bioethanol production

Process	RM/bale	Weightage (%)
Collection	20.44	23
Logistics	48.79	23
Production	20.47	54

3.4 Rice straw collection and pretreatment

For rice straw collection, the labor contributes to majority of the cost (63%), followed by machinery (20%), fuel (9%), and twine (8%). Meanwhile, for rice straw pretreatment, the major cost comes from sulfuric acid (RM 0.48/L) and the total costs were RM 1.51/L (with acid recovery) and RM 7.85/L (without acid recovery).

4. Conclusion

It can be concluded that rice straw presents a great potential as a biomass for bioethanol production. The theoretical bioethanol yield and conversion efficiency obtained were 250 L/t and 60% respectively. Despite having low conversion efficiency, the yield could be improved by optimizing the pretreatment process. From the economical perspective, it was found that the bioethanol production form rice straw is for the moment not viable due to expensive logistics and pretreatment components.

5. Acknowledgement

Authors would like to acknowledge this research as supported by Exploratory Research Grant Scheme (ERGS) through the Ministry of Higher Education of Malaysia (MOHE) with reference number ERGS/1/2013/TK07/UNITEN/01/01.

6. References

[1] Tye Y Y, Lee K T, Abdullah W N W, Leh C P 2011 Second-generation bioethanol as a sustainable energy source in Malaysia transportation sector: Status, potential and future prospects *Renewable and Sustainable Energy Reviews*, **15** pp 4521-4536

[2] Mekhilef S, Saidur R, Safari A, Mustaffa W E S B 2011 Biomass energy in Malaysia: Current state and prospects *Renewable and Sustainable Energy Reviews*, **15** pp 3360-3370

[3] Goh C S, Tan K T, Lee K T, Bhatia S 2010 Bio-ethanol from lignocellulose: Status, perspectives and challenges in Malaysia *Bioresource Technology*, **101** pp 4834-4841

[4] Binod P, Sindhu R, Singhania R R, Vikram S, Devi L, Nagalakshmi S, Kurien N, Sukumaran R K, Pandey A 2010 Bioethanol production from rice straw: An overview *Bioresource Technology*, **101** pp 4767-4774

[5] Mood S H, Golfeshan A H, Tabatabaei M, Jouzani G S, Najafi G H, Gholami M, Ardjmand M 2013 Lignocellulosic biomass to bioethanol, a comprehensive review with a focus on pretreatment *Renewable and Sustainable Energy Reviews*, **27** pp 77-93

[6] Park Y C, Kim J S 2012 Comparison of various alkaline pretreatment methods of lignocellulosic biomass *Energy*, **47** pp 31-35

[7] John A 2013 Alternatives to Open-Field Burning on Paddy Farms *OPTIONS*, **18** pp

[8] Nori H, Halim R A, Ramlan M F 2008 Effects of Nitrogen Fertilization Management Practice on the Yield and Straw Nutritional Quality of Commercial Rice Varieties *Malaysian Journal of Mathematical Sciences*, **2** pp 61-71

Sustainability and business: what is green corporate image?

Vathana Bathmanathan[1] **, Chikako Hironaka**[2]

E-mail: vathana@uniten.edu.my

Abstract. Green corporate image is reckon to be the driving factor in the current business setups. Stakeholder's green perception of the firm encourages growth of businesses. Organisation is moving from conventional businesses to running businesses with sustainable agenda that creates values to their brand. This paper analyses several green corporate image initiatives and concepts by various researches and shares how this can be essential for business.

1..Introduction

In recent years, more and more organisations are looking into progress and success not just through their financial balance. Success is measured through business growth, its stakeholders and how these stakeholders perceive its organisation. This paper intends to share various concepts related to sustainability in business and clarify the definition of Green Corporate Image (GCI) and its importance. Taking businesses one step further in the past decades, businesses are seen to incorporate sustainability agenda in their businesses. Sustainability is defined as the key factor to meet the current need without affecting the future generation. It is now one of the key factor of a successful business.

Kumar and Christodoulopoulou (2013) explains 'sustainability's scientific knowledge appears to be multidisciplinary in nature with many components or explained in different terms such as sustainability development, triple bottom line, green business, environmental management, corporate social responsibility and corporate citizenship'. In managing these sustainability activities, Gonzalez and Gonzalez, (2005) categorized the transformation towards environmental sustainability into three areas which are the planning and organizational practices in the management area, operational practices in the operations area and communicational practices in the marketing area. Hence, the above clearly supports the idea and importance given to sustainability in current business.

1. Sustainability and Businesses

Cramer and Karabell in their book Sustainable Excellence claims that in order for businesses to survive and thrive in this twenty first century, they not only have to measure its success in their financial aspects but instead to work towards discovering a formula to incorporate sustainability measures as part of their performance. Businesses will have to ensure they have sustainable components to achieve competitive advantage or face suffocation and die a natural death (Namkung and Jang, 2012. Chang, 2010 shares views of Bansal, 2005 and Barnet, 2007 in his paper on how companies faces a growing pressure to go green and reduce environmental effect. He further argues on consumers' green buying intention which is in line with their concern for the environment. Businesses need to add value to the society as a whole, hence include sustainability elements and shift from shareholder capitalism to responsible capitalism (V.Kumar, 2014).

There are many researchers conducted on sustainability in business. We would like to examine them and divide into 3 categories; green growth and challenge, green branding and green and sustainable reporting.

2.1. Green growth and green challenge

Too and Bajracharya, (2013) states in their research that the number of consumers preferring to purchase from companies that care about sustainability is growing. This statement is further strengthen by Namkung and Jang, (2013) in their paper by stating customers are more likely to choose a green restaurant which supports more green experiences and involvement. In the current business era, although most leading brands have moved towards developing and introducing eco-friendly products, it still faces major challenge to overcome consumer scepticism about their green operations and green attributes (Kumar and Christodoulopoulou, 2014). This is seen as greenwashing where many organisations claim to be green when they are not actually practicing it.

Another reason why it is a challenge is because the sustainability research and social sustainability attributes is difficult to measure. This is mainly because each industry faces their own and unique sustainability challenge which makes sustainability communication across various industries difficult to compare (Anne H Reilly, Katherine A.Hynan, 2014). Hooi, Hassan & Che Mat, (2012) further supports the statement by saying there is no generic approach to suit all organisation when it comes to sustainability measurement and communication.

2.2 Green branding

As explained in the earlier chapter, incorporating sustainability into branding can enhance a firm's ability to appeal to its customers that are concern about the environment. Integrating green attributes, initiatives and practices into a corporate brand improves the perception of customer that is linked to environmental commitments and concern, hence gain a competitive advantage (Too and Bajracharya, 2013). Namkung and Jang, 2013 says the same in their research. They believe that the corporate image is expressed as a function of a salient attributes of a business that drives its customers to compare with another business. Many researches support the statement that greening has become a key factor in a business survival and that businesses are much more sensitive to sustainable agenda (Ng, Butt, Khong, Ong, 2013). Therefore, there is a consistence evidence in the growing interest on marketing researches and practitioners in the firm environmental issues and their impacts towards the marketing activities and firm performance.

2.3 Green and sustainable reporting

While being seen sustainable appears to be a necessity in the current business environment, it is also crucial to communicate these green activities via active reporting to stakeholders in order to gain a competitive advantage (Hartmann, P., et. al., 2005, Aaker, 1991, Keller 2003, Namkung and Jang, 2012). Other researches Hillestad et al., (2010) explains that the reason firms give importance to environmental reporting is because they want to show the public that the top level management of the firm takes green issues seriously and this helps promote positive green image. (Nair and Menon, 2008). Green reporting is seen crucial as it has shown to improve firm performance. Even social media activities for green firms are far more then non green firms (Too and Bajracharya, 2013 and Reilly, A.Hynan, 2014).

3. Current research on Green Corporate Image (GCI)

The definition of corporate image is seen as what the stakeholders perceive the organisation as a business (Van Riel and Fombrun, 2007, Javier et.al, 2014). Corporate image is perceived as the feature of an organisation in the eye of its stakeholders. It is the desired general impression of the organisation in the minds of its stakeholders. Organisations spends its vital resources i.e. money, time, people and etc to build a strong corporate image (James Poon, 2000). Studies has shown that through this image, companies tend to secure a position in the industry and create a competitive edge for themselves (Too and Bajracharya, 2013, Ko, Hwang and Kim, 2012 and Namkung and Jang, 2012).
Several research has focused on green corporate image (GCI) across various industries. Studies have indicated diverse effect of GCI on several internal and external organizational factors such as

employee work life, top management support and commitment, firm performance, buying intentions and more (Table 1). However, these studies don't examine the relationship with other similar concepts, like sustainability, green reporting, and green branding.

Literature	**Reviews**
Nair and Menon, 2008	It is the effort firms take to create the positive green image
Hillestad et. al., 2010	Green corporate image imposes believe that decision makers takes green issues seriously
Chen (2008)	Companies which invest in efforts on environmental issue, are able to improve their corporate images and to develop new market as well as to increase their competitive advantage.
Namkung and Jang (2013),	Green practices significantly influence customer perceptions of a brand's green image and green behavioural intentions.

Table 1 : Literature on Green Corporate Image

Moving on from being sustainable, how does these effort and initiatives drives firms towards actually being sustainable and seen sustainable by stakeholders? More and more firms are keen to portray themselves as being a sustainable brand. A jargon commonly used in the marketing field as 'Corporate Image' has found its own identity to lead the industry as 'Green Corporate Image' (GCI). Green corporate image is a combination of firm's concern on the image of its business as being environmentally concerned. Green corporate image is an effect brought by Green Branding. Green branding or also known as sustainable branding are brands that consumer associates with environment and sustainable business practices. Green brands refers directly to the green product and practices of the firm where else green corporate image derives from the green perception of its stakeholders.

There are a few researches that shows a positive relationship between Corporate Image and customer satisfaction, customer loyalty (Nai and Cher, 2010, Joanna et.al., 2009) and firm performance (Javier et. al., 2014). Other researches also supports the positive relationship between green practices and improvement in firm image (Miles and Russell, 1997 and Chen, 2008). However, there is a clear gap in the study area of green practices in different industries (Poon et.al., 2000) and there isn't a clear evidence on the relationship between environmental management and firm performance (Javier et. al., 2014). The relevance of green corporate image management will depend on the industry in which the firm operates (Frooman, 1999).

4. Conclusion

In this study, we examined prior research regarding sustainability in business. There are a lot of similar concepts regarding sustainability in business. First of all, we categorized these research into 3 types. Next, we focused on 'Green Corporate Image' and clarified the relationship between GCI and other concepts. To the end, we have pointed out that the research regarding the corporate performance and GCI is limited. For further research, we would like to focus on the GCI in Malaysian companies, especially in energy industry. Energy industry is one of the leading industries in Malaysia and GCI is the significant factor for these firms to grow. This being said, there is not much research focusing on the energy industry especially in Malaysia. Therefore, there is a clear avenue for research on the green practices and firm performance in the energy industry in Malaysia. Further research should also look into how the green corporate image in the energy industry improve firm performance.

Reference

[1] Aaker, A David 2004 Leveraging the corporate brand
[2] Cramer A, Karabell Z (2010). Sustainable Excellence; The future of Business in a Fast-Changing World. New York: Rodale.
[3] Anne H. Reilly , Katherine A. Hynan 2014 *Corporate communication, sustainability , and social media : It's not easy (really) being green* Business horizons. 57, 747-758
[4] Bansal P (2005). *Evolving sustainably: a longitudinal study of corporate sustainable development* Strategic Management Journal 26(3): 197-218
[5] Barnet ML (2007) *Stakeholder influence capacity and the variability of financial returns to corporate social responsibility* Academic Management Review.33(3): 794-816
[6] Cees B.M. van Riel and Charles J. Fombrun 2007 *Essentials of Corporate Communication: Implementing Practices for Effective Reputation Management* (London; New York : Routledge)
[7] Chen, Y, 2008, *The driver of Green innovation and Green image – green core competence* Journal of Business Ethics Volume 81, Issue 3, pp 531-543
[8] Eunha J., Jang S (Shawn), Day J , Ha S 2014 *The impact of eco-friendly practices on green image and customer attitudes: An investigation in a café* Int Jour of Hospitality Mgmt p.10-20
[9] Eunju Ko , Yoo Kyung Hwang , Eun Young Kim 2013 *Green marketing' functions in building corporate image in the retail setting* Journal of Business Research p.17091715
[10] Frooman, J., 1999 Stakeholder influence strategies
[11] Gonzalez-Benito, J., Gonzalez-Benito, O., 2005 *A study of the motivations for the environmental transformation of Companies*
[12] Hart, S., Dowell, G., 2010 *A natural resource-based view of the firm: fifteen years after* Marketing Intelligence & Planning, Vol. 23 Iss: 1, pp.9 - 29
[13] Hillestad, T., Xie, C Haugland, S.A., 2010 *Innovative corporate social responsibility : the Founder's role in creating a trustworthy corporate brand through 'green innovation'.*
[14] James Poon Teng Fatt, Meng Wei, Sze Yuen and Wee Suan 2000 *Enhancing corporate image in organisations* Management Research News. 23 Iss 5/6, 28-54.
[15] Javier Amores-Salvado, G. M.-d.-L. (2014). *Green corporate image: moderating the connection between environmental product innovation and firm performance.*
[16] Joanna Minkiewwicz, J. E. (2011). Corporate Image in the leisure services sector. Journal of Services Marketing. 25/3, 190-201.
[17] Keller, Kevin Lane 2003 *Strategic Brand Management: Building, Measuring, and Managing Brand Equity*
[18] Keoy Kay Hooi, Fadzil Hassan and Masnizan Che Mat 2012 *An Exploratory Study of Readiness and Development of Green University Framework in Malaysia*
[19] Kumar, Leela, 2014 *The Impact of Corporate Social Responsibility on Sustainable Development*
[20] Linda Too Bhishna Bajracharya 2015 *Sustainable Campus: engaging the community in sustainability* Int. Journal on Sustainability in Higher Education Vol 16 Iss 1 p 57-71
[21] Morgan P. Miles and Gregory R. Russell ,1997 *ISO 14000 total quality environmental management: The integration of environmental marketing, total quality management, and corporate environmental policy* Journal of Quality Management Vol 2, Iss 1, 1997, P 151–168
[22] Chang N.J and Fong C.M 2010 *Green product quality, green corporate image, green customer satisfaction, and green customer loyalty* African Journal of Bus. Mgmt Vol 4(13)p2336-44
[23] Nair, S.R., Menon, C.G., 2008 *An environment marketing system-a proposed model based on Indian experience.*
[24] Pui Fong Ng, Muhammad Mohsin Butt, Kok Wei Khong And Fon Sim Ong, 2014. *Antecedents of Green Brand Equity.*
[25] V.Kumar and Angeliki Christodoulopoulou 2014 *Sustainability and branding: An integrated perspective*
[26] Young N, Jang S(Shawn) 2013 *Effects of restaurant green practices on brand equity formation : Do green practices really matter?*

IOP Conf. Series: Earth and Environmental Science **32** (2016) 012050 doi:10.1088/1755-1315/32/1/012050

Energy Reporting Practices among Top Energy Intensive Industries in Malaysia

N E Tasrip[1], N Mat Husin[1] and B Alrazi[1]
[1]Department of Accounting, Universiti Tenaga Nasional, Malaysia

E-mail: elyana_dec@yahoo.com

Abstract: This study content analyses the energy content in the corporate report of top 30 Malaysian energy-intensive companies. Motivated by the gap among prior corporate social responsibility and environmental reporting studies in respect of energy, this study provides evidence of Malaysian companies' initiative to reduce energy consumption. While the evidence suggests that not all 30 companies have reported energy-related information, the findings provide an overview on the response of energy intensive companies in relation to Malaysian government initiatives on energy.

1. Introduction

Global warming and climate change are two critical issues that are expected to give significant impact to humankind. Scientific and political communities have long debated on tackling these issues which include mitigating the emissions of greenhouse gas (GHG) emissions, particularly with respect to energy use [1]. It is reported that Malaysia is the third-largest energy consumer in the ASEAN region and its per-capita energy consumption is currently relatively high for the region, at 61% of the OECD average [2].

With the increase of energy consumption, Malaysia is undertaking a comprehensive planning to develop and formulate an Energy Policy for Malaysia (2013 – 2050) that encompasses all relevant sectors as well as incorporating current issues and future challenges related to energy. Prior to this, the Tenth Malaysia Plan (10th MP) 2011-2015 emphasizes on green technology that vested in National Green Technology Policy and new renewable energy act and feed-in tariff mechanism. As national guidelines and policies deliver a common message to mitigate the emissions, there is a call for business leaders to also confront the issue of climate change [3]. This is because industries are dependent on fossil fuel for their production and manufacturing activities which contribute to the GHG emission [4].The industrial sector uses more energy than any other end-use sectors, currently consuming about 37% of the world's total delivered energy [5].

With the visibility of efforts to shift Malaysia towards becoming a more energy efficiency intensive country and the contribution of industrial sector in energy usage, it is an obligation of energy intensive industries to demonstrate their commitment in supporting government initiatives particularly in relation to energy. Therefore, it is the aim of this study to explore the extent of energy disclosure made by the energy intensive sectors among public listed companies in Malaysia. A self-developed index is used to measure the quantity of energy related items being reported in the annual reports, stand-alone reports and also the corporate websites.

2. Regulatory Environment and Energy Reporting

Malaysian government has embarked on various strategies to enhance its' energy-based initiatives. The most prominent strategy is the introduction of various energy based policies. Among those are the introduction of National Energy Policy (1979), National Depletion Policy (1980), National Biofuel Policy (2006), National Green Technology Policy (2009), and National Renewable Energy Policy (2010). Currently, a final draft of National Energy Efficiency Action Plan is introduced in January 2014. The plan outlines strategic actions that are needed for an effective and efficient implementation of energy efficiency programs. The introduction of National Green Technology Policy (2011-2015) provide details on the strategy to restructure the energy subsidies on natural gas and electricity used by industry sector as well as discounted energy pricing mechanism for selected industrial users.

With several energy policies being implemented by the government that acts on behalf of the public, this means a company is now being bequeathed with legitimacy based on its ability to operate within the bounds imposed by the society in order to enjoy continued access to products and resource markets [6]. Corporate reporting, therefore, has become one of the strategies used by companies to educate the public or to communicate the changes that have been made, in response to the pressure made by the government/public to contribute to energy-based initiative. Nonetheless, with financial reporting being the center of attention since the existence accounting, the reporting of non-financial based information in Malaysia has only been made compulsory in year 2007 with the mandatory requirement of Corporate Social Responsibility (CSR) reporting to all Malaysian public listed companies (PLCs) [7].

As reporting becomes a mean for companies to legitimize their action, it lead to an apparent link between accounting research and legitimacy theory that revolves around the annual report and related disclosures [8]. Prior studies on climate change reporting (see for example [9]; [10]), CSR reporting (see for example [11]; [12]) and environmental reporting (see for example [13]; [7]) are among studies that have evaluated companies disclosure, in response to pressure resulting from social contract, in different forms of corporate reports. Most of these studies have also incorporated energy based information as part of the information being evaluated. A CSR reporting study conducted by [14], for example, reported only small minority (1 percent) of companies in industrial sector disclose information related to on energy. Nonetheless, with energy-based information being incorporated with other types of sustainability information, the analysis can be regarded as being very limited. To the knowledge of this study, there is only one study that put a specific focus on energy reporting and it is conducted by [15]. This study which is based in South African developed a guideline for the development and implementation of effective energy reporting. Unfortunately, [15] only focuses on preparing a special energy report for a company' energy management system that may not be able to represent overall energy initiatives implemented by the company. Taking into consideration the importance of energy conservation issue in Malaysian and the lack of focus on energy reporting per se, it is the objective of this study to provide a close examination of energy information reported by top energy intensive companies in Malaysia.

3. Sample Selection and Research Methods

Energy-based disclosure is examined for the 30 largest Malaysian public listed companies, by market capitalization, focusing only on energy intensive sectors. Only the largest Malaysian companies were chosen to conform to the view suggested by [16] that large companies are more likely to be more progressive and innovative because they have the financial resources that enable this type of reporting. Energy intensive sectors include all companies under iron and steel, cement, pulp and paper, chemical, textile, aluminum, oil and gas, electricity and heat, ceramic, glass, wood, coal mining and transport.

The study adopts content analysis method to assess energy-based information in the three main types of corporate report i.e. annual reports, stand-alone sustainability reports and corporate websites. Both annual and sustainability reports were for the 2014 financial year. Meanwhile, the analysis for the corporate websites disclosure was performed in the month of November, 2014. The website analysis was limited up to three levels of the homepage/sitemap, unless further links indicate the disclosure of energy information beyond the third level. This is reasonable as it is expected that stakeholders will spend less time going through various sections of the website to evaluate company environmental policy and performance [17]. The extent of disclosure was then measured using self-developed index that is modified from the Global Reporting Initiative (GRI) index. Each of the energy-based information found was then classified into several categories i.e. monetary, non-monetary, declarative, to provide more insight on how the information was reported.

4. Findings and analysis

Table 1 describes the number of energy-related items found in the 30 Malaysian energy intensive companies' annual report year 2014, sustainability report year 2014 and websites. A total of 143 energy-related items were found in the companies' annual report making it the most popular venue of reporting. This is contributed by the fact that it is mandatory for Malaysian PLCs to produce this report annually. Nonetheless, taking into consideration the analysis was conducted on energy-intensive industries, the non-existence of energy-related information in 5 (only 25 companies currently report the energy-based information) of the companies could provide an indication the non-existence of activities related to energy-based initiative or companies unwillingness to report such information. The analysis also shows website as the next popular venue followed by sustainability report. It is important to highlight that only 2 companies chose to produce a separate report on CSR.

Table 1: Quantity of energy reporting by forms of reporting

No.	Items	Annual Report				Sustainability Report				Websites			
		TQ	M	NM	D	TQ	M	NM	D	TQ	M	NM	D
1	Energy consumption of organization	37	9	19	9	2	0	2	0	0	0	0	0
2	Initiative to reduce energy consumption	47	0	0	47	13	0	2	11	14	0	0	14
3	Energy governance & ethics	18	0	0	18	6	0	0	6	2	0	0	2
4	Energy based achievement & awards	9	0	0	9	6	0	0	6	2	0	0	2
5	Reduction of energy consumption	12	1	7	4	9	0	7	2	0	0	0	0
6	Renewable energy	20	0	0	20	11	0	0	11	3	0	1	2
	Total scores	**143**	**10**	**26**	**107**	**47**	**0**	**11**	**36**	**21**	**0**	**1**	**20**
	Total companies	**25**				**2**				**15**			

Note: TQ: Total Quantity; M: Monetary; NM: Non-Monetary; D: Declarative

For each type of corporate report, item 'initiative to reduce energy consumption' has been recorded as the most reported item. This indicates Malaysian companies' initiative to support their government agenda in becoming energy-efficiency nation. YTL Corporation, for example, reported the following information in their annual report:

'Overall, YTL has a two-fold strategy in pursuing green practices Group-wide. Firstly, we have a monitoring system put into place to measure and track energy usage. Secondly, we are pursuing and steadily increasing renewable energy usage, wherever possible'

In addition to item related to energy efficiency initiative, information related to RE has also being among the most reported item indicating companies' initiative to support or possibly invest in RE project. Malaysian Airport, for example, reported the following information in their sustainability report:

"One mode of renewable energy that we have embarked on is solar power system at KL International Airport (KLIA)."

Analysis on different categories of reporting shows most of the information are declarative in nature. The subjective nature of declarative information may have contributed to its popularity as it will make it harder for the information to be refuted or audited. Nonetheless, it is important to highlight the existence of companies who willingly produce quantitative information which include monetary information. Malaysia Marine and Heavy Engineering Holdings Berhad, for example, reported the following information in the annual report:

"Almost 100% of spent lubricating oil, spent hydraulic oil, electric/ electronic and acid battery waste in the yards were sent for recovery as value waste. This resulted in a saving of approximately RM0.5 million in overall 3R activity in MHB"

5. Conclusion

This study began by discussing Malaysian government initiative to strengthen its energy-related policies and the need for the industrial sector as one of the largest users of energy to support these initiatives. In response to this issue, this study is motivated to initiate a first study, within the context of Malaysia, to explore the extent of energy-based information in the corporate reports of top 30 Malaysian energy-intensive companies.

In summary the observed statistics shows that 25 out of 30 the companies have voluntarily provided energy-based information with most of the information are declarative in nature. While the figure represents 83% of total sample, the fact that the study was conducted on energy intensive industries could be alarming as it signal the possibility that some of energy-intensive companies are not that willing to invest or support government energy-based initiatives. While findings the exact factors leading to these findings is beyond the scope of this study, the fact that this issue is currently being discussed at international level and the companies will continue using energy to sustain its operation, the need for energy-based disclosure is still there if the companies want to legitimize their position among the society. It is also the role of the regulators to facilitate provision of such information without comprising the need of various parties including the companies.

References

[1] Reddy B S and Assenza G B 2009 The great climate debate. *Energy policy 37*(8) 2997-3008
[2] International Energy Outlook 2014 *World energy and economic outlook*
[3] Lee K H, Min B and Yook K H 2015 The impacts of carbon (CO 2) emissions and environmental research and development (R&D) investment on firm performance *Int. J. of Prod. Econ.* 167 1-11
[4] Soni M and Bhanawat S S 2015 Strategies Adopted by Indian Corporate Sectors to Mitigate the Carbon Risk: An Analysis. *The IUP J. of Acc. Res. & Audit Prac.* 14(2) 43-52
[5] Abdelaziz E A, Saidur R and Mekhilef S 2011 A review on energy saving strategies in industrial sector *Ren. & Sus. En. Rev.* 15150-168
[6] Campbell, D., Craven, B., and Shrives, P. 2003 Voluntary social reporting in three FTSE sectors: A comment on perception and legitimacy. *Acc. Aud. & Account. J.* 16(4) 558–81
[7] Sulaiman M, Abdullah N and Fatima A H 2014 Determinants of Environmental Reporting Quality in Malaysia *Int. J. of Econ., Mgmt. & Acc. 22*(1) 63-90
[8] Tilling, M. V. and Tilt, C. A. 2010 The edge of legitimacy – Voluntary social and environmental reporting in Rothmans' 1956–1999 annual reports *Acc. Aud. & Account. J.* 23(1), 55–81.
[9] De Aguiar T R S & Bebbington J 2014 Disclosure on climate change: Analysing the UK ETS effects *Acc. F.* **38** (4) 227-40
[10] Ahmad N N N and Hossain D M 2015 Climate Change and Global Warming Discourses and Disclosures in the Corporate Annual Reports: A Study on the Malaysian Companies *Proc. Soc. & Bhvr, Sci.* 172, 246-53
[11] Othman R and Ameer R 2009 Corporate social and environmental reporting: Where are we heading? A survey of the literature. *Int. J. of Discl. & Gov.* 6(4), 298-320
[12] Saleh M Zulkifli N and Muhamad R 2010 Corporate social responsibility disclosure and its relation on institutional ownership: Evidence from public listed companies in Malaysia *Mgrl. Aud. J. 25*(6) 591-613
[13] Ienciu I A, Popa I E and Ienciu N M 2012 Environmental Reporting and Good Practice of Corporate Governance: Petroleum Industry Case Study *Proc.Econ. & Fin. 3* 961-67
[14] Ahmad N N N Sulaiman and Siswantoro 2003 Corporate social responsibility disclosure in Malaysia: An analysis of annual reports of KLSE listed companies *Int.J. of Eco., Mgmt. & Acc.*11(1) 51-86
[15] Maneschijn, R., Vosloo, J.C., and Pelzer, R. 2014 *Effective Multi-Level Energy Reporting In South African Industry* Int. Conf. on the Ind. & Com. Use of En. 19-20 Aug. 2014
[16] Guthrie, J., Petty, R., and Ricceri, F. 2006 The voluntary reporting of intellectual capital: Comparing evidence from Hong Kong and Australia. *J. of Int. Cap.* 7(2) 254–71
[17] De Villiers C and Van Staden C J 2011 Where firms choose to disclose voluntary environmental information. *J. of Acc. and Pub. P.* 30(6), 504-25

Rapid Economic Growth and Natural Gas Consumption Nexus: Looking forward from Perspective of 11th Malaysian Plan

H A Bekhet[1], T Yasmin[2]
[1]Graduate Business School, College of Graduate Studies
Universiti Tenaga Nasional (UNITEN), Malaysia
[2]PhD Student, College of Graduate Studies
Universiti Tenaga Nasional (UNITEN), Malaysia
Email: profhussain@uniten.edu.my, yasmin_t35@yahoo.com

Abstract. The present study investigates the relationship between economic growth and energy consumption by incorporating CO_2 emissions, natural gas consumption and population in Malaysia. Annual data and F-bound test and granger causality have applied to test the existence of long run relationship between the series. The results show that variables are cointegrated for long run relationship. The results also indicate that natural gas consumption is an important contributing factor to energy demand and hence economic growth in case of Malaysia. The causality analysis highlights that the feedback hypothesis exists between economic growth and energy consumption. While, conservative hypothesis is validated between natural gas consumption and economic growth which implies that economic growth will push natural gas consumption policies in future. This study opens up new direction for policy makers to formulate a comprehensive natural gas policy to sustain environment for long span of time in case to achieve 11th MP targets.

Keywords: *Economic Growth; Energy Consumption; CO_2 Emissions; Natural Gas Consumption; Population; Malaysia.*

1. Introduction

There is close interaction between economic growth and energy consumption with pace of development. Due to increase energy consumption the resources are depleting and resulting in environmental challenges. In recent years, natural gas pipeline development is driven by market supply and demand. It has estimated that roughly 70% of natural gas flows across the globe are transported to market destinations within the country of production. While an additional 20% flows cross international borders through pipelines, and nearly 10% is moved to market destinations as liquefied natural gas (LNG). Total natural gas consumption is expected to grow at 18% annually between 2007 and 2035. The key end-use sector that is driving continued growth in natural gas demand is the electric power sector, in which natural gas has largely replaced oil and is successfully competing with coal [1]. Also, the growth of the renewable energy is enhancing due to Cross-State Air Pollution Rule, which leads to tradeoffs between the fuels used for electricity generation. For the past years during 2000 to 2012, electricity generation from natural gas-fired plants rose more than doubled due to fuel switching [2], [3].The increasing demand of natural gas is relatively linked to electricity generation process. The fact is that natural gas generates relatively less CO_2 emissions than other fossil fuels e.g. coal. With this objective, many countries are exploring the options for better use of natural gas as an alternative energy source [4, 5].

In the Southeast Asian market as a net key exporter of LNG to global markets is also facing various challenges. For example, in case of Indonesia and Malaysia, due to geographical mismatch between the location of gas resources and demand has created the situation in which they are simultaneously importers and exporters of LNG [6]. Specifically in Malaysia, the fuel mix for power generation relies heavily on natural gas and coal. The share of coal and natural gas consumption for electricity generation recorded an annual growth rate 16.3% and 7.5% for the (1992-2013) period respectively [7]. Looking various renewable energy policies it has highlighted by [8] that the 10th Malaysia Plan (10MP) focused on attaining a sustainable level of economic growth that will enable the country to attain a developed nation status by 2020. They suggested that natural gas consumption has an indirect effect to the Malaysian economic growth but there is need of efficient exploitation of natural gas reserves of the country to meet future demand. Taking into consideration these policies, the objective of current paper is to look ahead the natural gas planning in coming years for fuel mix and emission reduction. According to the above discussion, this study raises the further question of what could be the contributions of natural gas consumption towards boosting the Malaysian economic growth prospects to aid in the attainment of the 11th Malaysian Plan. The growth hypothesis asserts that energy consumption is vital in economic growth both direct, as an input in the production process and indirectly, as a complement to labor and capital inputs. Moreover, this hypothesis suggests that an increase in economic growth causes an increase in energy consumption [9], [10], [11], [12], [13]. There are some studies used time series and panel data analysis in natural gas consumption perspective such as: [14], [15] for Taiwan; [16] for US; [17] for Iranian economy; [18] for US, [4] panel of 67 countries, [19] for the G7 countries; [20] for Bangladesh and [21] for Pakistan. Furthermore, [22] reinvestigated the relationship between natural gas consumption and economic growth Malaysia for the 1971–2012 period. They discovered natural gas consumption, foreign direct investment, capital formation and trade openness have positive influence on economic growth in Malaysia.

The above discussion clearly indicates that there is still room of research on the relationship between natural gas consumption and economic growth not only in the existing literature but also in case of Malaysia. So, this study is a good effort to fill up the gap in the existing literature by taking into consideration various policy perspectives. The rest of the paper is organized as follows: section 2 states data and methodological framework; section 4 discusses the empirical results; and the last section concludes the paper with policy implications.

2. Data Source and Methodology

The present study aims to investigate the relationship between economic growth and energy consumption while incorporating CO_2 emissions, natural gas consumption and population. We follow the methodological framework of aggregate growth model like [23], [24], [25], and [26] to examine the relationship between economic growth, energy consumption, CO_2 emissions, natural gas consumption and population. Annual time series data from 1980 to 2013 for above variables have been obtained from the [27], [28].The cointegration approach, the ARDL bounds testing approach developed by [29] is used to explore the existence of long-run equilibrium relationship between the variables. This approach has several advantages [30], [31]. By using the appropriate order, it is possible to simultaneously correct the serial correlation in residuals and the problem of endogenous regressors [32]. We have transformed all the series into logarithmic form as shown in Eq. (1).

$$\Delta LY_t = \alpha_o + \sum_{i=1}^{q} \beta_1 \Delta LY_{t-i} + \sum_{i=0}^{q} \beta_2 \Delta LEC_{t-i} + \sum_{i=0}^{q} \beta_3 \Delta LCO_{2t-i} + \sum_{i=0}^{q} \beta_4 \Delta LNG_{t-i} + \sum_{i=0}^{q} \beta_5 \Delta POP_{t-i} + \lambda_1 LY_{t-1} \quad (1)$$
$$+\lambda_2 LEC_{t-1} + \lambda_3 LCO_{2t-1} + \lambda_4 LNG_{t-1} + \lambda_5 POP_{t-1} + \mu_t$$

Equation (1) presents two segments of results. The first part indicates the short run relationship such as βs; and second λs to explore the long run associations among the variables in the current study.

Lastly, we have applied multivariate Granger Causality test to check the direction among research variables (Eq. 2).

$$\begin{bmatrix} \Delta LY \\ \Delta LEC \\ \Delta LCO_2 \\ \Delta LNG \\ \Delta LPOP \end{bmatrix}_t = \begin{bmatrix} \alpha_1 \\ \alpha_2 \\ \alpha_3 \\ \alpha_4 \\ \alpha_5 \end{bmatrix} + \sum_{s=1}^{d} \begin{bmatrix} \beta_{11} & \beta_{12} & \beta_{13} & \beta_{14} & \beta_{15} \\ \beta_{21} & \beta_{22} & \beta_{23} & \beta_{24} & \beta_{25} \\ \beta_{31} & \beta_{32} & \beta_{33} & \beta_{34} & \beta_{35} \\ \beta_{41} & \beta_{42} & \beta_{43} & \beta_{44} & \beta_{45} \\ \beta_{51} & \beta_{52} & \beta_{53} & \beta_{54} & \beta_{55} \end{bmatrix} \begin{bmatrix} \Delta LY \\ \Delta LEC \\ \Delta LCO_2 \\ \Delta LNG \\ \Delta LPOP \end{bmatrix}_{t-s} + \begin{bmatrix} \mu_1 \\ \mu_2 \\ \mu_3 \\ \mu_4 \\ \mu_5 \end{bmatrix} ECT_{t-1} + \begin{bmatrix} \varphi 1 \\ \varphi 2 \\ \varphi 3 \\ \varphi 4 \\ \varphi 5 \end{bmatrix}_t \quad (2)$$

3. Empirical Results

Augmented-Dickey-Fuller (ADF) is used to test the unit root properties of the variables. It finds that all series contain unit root problem at their levels, *I(0)* and stationary at 1st difference*I(1)*. The calculated F-statistics is 8.33which is greater thanupper critical bund at 1% level, provided by [33]. The study highlights that the long-run results are robust, these findings are presented in Table 1 and they show that natural gas has negative and significant impact on economic growth. The empirical result display that a 1% increase in energy consumption will lead to increase the economic growth by 0.41% in long run and 0.50% in short run and both of them are significant at 1% level of significance. In addition, a 1% increase in emissions is positively linked with economic growth at 0.42% and0.179% in long and short run respectively. Whereas, the natural gas consumption shows a decline with economic growth in long run by 0.1019%, but positive in the short run. Also, the study discuss that how population was found to have a positive and significant link with economic growth as 1.19% in long run and 0.067% in short run. These findings are consistent with [34] and [35], who highlighted this relationship in Malaysian economy.

Table 1: Long run and Short run coefficients.

Long run Coefficents				**Short run Coefficents**			
Dependent Variable = LY				**Dependent Variable = ΔLY**			
Variable	Coefficient	T-Statistic	Probability	Variable	Coefficient	T-Statistic	Probability
Constant	-11.350	-5.467	0.000	Constant	0.007*	6.052	0.000
LEC	0.413	2.324	0.038	ΔEC	0.505	2.122	0.041
LCO_2	0.422	8.871	0.000	ΔCO_2	0.179	3.147	0.004
LNG	-0.102	-6.089	0.000	ΔNG	0.007	0.212	0.834
LPOP	1.199	7.663	0.000	ΔPOP	0.066	2.0714	0.046
				ECM_{t-1}	-0.096**	-2.177	0.036

Furthermore, the lagged error term ECM_{t-1}, confirms the significant long-run relationship between the variables (at 5% level). Moreover, Figure 1, shows that there is bidirectional causality between economic growth and energy consumption; energy consumption and natural gas and economic growth and population. Whereas, unidirectional causality from economic growth to natural gas consumption; energy consumption to CO_2 emissions and population to CO_2 emissions.

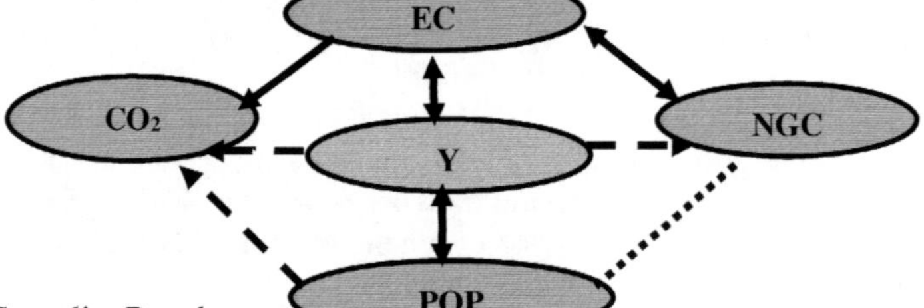

Figure 1: Granger Causality Result.
Note: ←→ shows bidirectional, - → unidirectional and ···· no causality.

4.Conclusion and Policy implications

The paper investigated the relationship between economic growth and energy consumption by using CO_2 emissions, natural gas consumption and population in a multivariate framework in case of Malaysia over the 1970–2013 period. It has applied the ARDL bounds testing approach for testing the stationary properties by unit root tests and explores short and long run relationship. The direction of causality between the variables is investigated by applying the Granger causality. The findings depicts that economic growth pushed energy demand and CO_2 emissions but declines natural gas consumption in the case of Malaysia. These findings propose various policy implications in this context. It is very important to have careful planning for renewable energy (RE) as an alternative energy source. It can further enhance to support the continuous increase of energy demand complemented with nominal efforts on energy efficiency measures. Furthermore, various efforts will be undertaken to ensure security of natural gas supply which is significant to the reliability of electricity supply. Key efforts include increasing natural gas capacity through the laying of additional pipelines and installing another receiving terminal in Kerteh, Terengganu to receive additional gas volume from Malaysia-Thailand Joint Development. One of the factors contributing to the unbalanced energy mix is the highly subsidized natural gas, which is the preferred fuel for the electricity subsector as it incurs the least cost. It has proposed in 11MP that due to increased natural gas demand in near future it is important to develop future market price for industries, secondly, there is fuel switching plan and lastly look new resources to fill the gas demand and supply gap [36].

References

[1] Energy Information Agency (EIA) 2014 *Leidos, Inc.*, Under Contract to EIA.
[2] Environmental Protection Agency (EPA) 2014: http://www.epa.gov/airtransport/CSAPR.
[3] Energy Information Agency (EIA) 2013: Retrieved from: https://www.iea.org/publications/.
[4] Apergis N and Payne J E 2010 *J. Applied Energy* **87** 2759–2763
[5] Salehnia N, Falahi M A., Seifi A and Adeli M H M. 2013 *J. of Nat. Gas Sci. and Eng.* **14** 238-249.
[6] International Energy Agency (IEA) 2013 Retrieved from: https://www.iea.org/publications.
[7] Energy Commission Malaysia 2014: Retrieved from: https://www.ec.malaysia/publications.
[8] Rafindadi A A and Ozturk I 2015 *J.Renewable and Sustainable Energy Rev.* **49** 1221–1232
[9]Shahbaz M Khraief N Uddin G S and Ozturk I 2014 *J. Renewable and Sust. Energy Rev.* **34** 325-336
[10] Bölük G and Mert M 2015 *J. Renewable and Sustainable Energy Rev.* **52** 587-595.
[11] Baek J and Kim H S *J. Energy Econ.* **36** 744–749
[12] Al-mulali U, Tang C F and Ozturk I 2015 *J. Renewable and Sustainable Energy Rev.* **50** 918-924.
[13] Jebli M B, Youssef S B and Ozturk I 2016 *J. Ecological Ind.* **60** 824-831
[14] Yang H Y 2000 *J. Energy Econ.* **22** 309–317
[15] Lee C C Chang C P 2005 J. Energy Econ. **27** 857–872
[16] Ewing B T Sari R and Soytas U 2007 *J. Energy Policy* **35** 1274–1281.
[17] Zamani M 2007 J. Energy Econ. **29** 1135–1140
[18] Sari R Ewing B T and Soytas U 2008 *J. Energy Econ.* **30** 2302–2313
[19] Kum H Ocal O and Aslan A *J. Renewable and Sustainable Energy Rev.* **16** 2361–2365.
[20] Das A McFarlene AA and Chowdhury M *J. Renewable and Sustainable Energy Rev.* **22** 269–274
[21] Shahbaz M Lean H H and Farooq A 2013 *J. Renewable and Sustainable Energy Rev.* **18** 87–94
[22] Solarin S and Shahbaz M 2015 *J. Renewable and Sustainable Energy Rev.* **42** 835–845
[23] Hamdi H Sbia R and Shahbaz M 2014 *J. Econ. Modelling* **38** 227-237.
[24] Farhani S Shahbaz M Arouri M and Teulon F 2014 *J. Energy Policy* **66** 677–684.
[25] Ang J B 2008 *J. of Policy Modeling* **30** 271-278.
[26] Soytas U and Sari R 2009 *J. Ecological Econ.* **68** 1667–1675
[27] World Bank Malaysia 2015 Retrieved from: http://data.worldbank.org/country/malaysia
[28]Energy Information Administration (EIA) 2015: Retrieved from: http://data.eia.org/
[29]Pesaran M H Shin Y and Smith R J 2001 *J. of Applied Economics* **16** 289–326
[30] Inder B 1993 *J. of Economics* **57** 53–68.
[31] Engle R F and Granger C W J 1987 *J. Economics* **55** 251–76
[32] Pesaran H M Shin Y and Smith R P 1999 *J. of the American Statistical Ass.* **94** 621–34
[33] Narayan P K 2005 *J. Applied Econ.* **37** 1979–1990
[34] Bekhet H A and Yasmin T 2013a. *J. Earth and Environmental Sci.* **16** 01 2015
[35] Bekhet H A and Yasmin T 2013b *J. of Social Human Science and Eng.* **7** 50-55
[36] Eleventh Malaysian Plan 2015

A proposed model of factors influencing hydrogen fuel cell vehicle acceptance

N H Noor Imanina[1], Tan Kwe Lu[2] and A R Fadhilah[2]
[1]College of Graduate Studies, Universiti Tenaga Nasional, Putrajaya Campus, Jalan IKRAM-UNITEN, 43000 Kajang, Selangor, Malaysia.
[2]College of Engineering, Universiti Tenaga Nasional, Putrajaya Campus, Jalan IKRAM-UNITEN, 43000 Kajang, Selangor, Malaysia.

E-mail: noorimanina@hotmail.com, tan@uniten.edu.my, fadhilah@uniten.edu.my

Abstract. Issues such as environmental problem and energy insecurity keep worsening as a result of energy use from household to huge industries including automotive industry. Recently, a new type of zero emission vehicle, hydrogen fuel cell vehicle (HFCV) has received attention. Although there are argues on the feasibility of hydrogen as the future fuel, there is another important issue, which is the acceptance of HFCV. The study of technology acceptance in the early stage is a vital key for a successful introduction and penetration of a technology. This paper proposes a model of factors influencing green vehicle acceptance, specifically HFCV. This model is built base on two technology acceptance theories and other empirical studies of vehicle acceptance. It aims to provide a base for finding the key factors influencing new sustainable energy fuelled vehicle, HFCV acceptance which is achieved by explaining intention to accept HFCV. Intention is influenced by attitude, subjective norm and perceived behavioural control from Theory of Planned Behaviour and personal norm from Norm Activation Theory. In the framework, attitude is influenced by perceptions of benefits and risks, and social trust. Perceived behavioural control is influenced by government interventions. Personal norm is influenced by outcome efficacy and problem awareness.

1. Introduction

Hydrogen fuel cell vehicle (HFCV) is the new era in automotive industry. HFCV uses hydrogen as the fuel which combines with oxygen in the air to produce electricity for the propulsion and emits only water and heat. Although there are issues on the carbon emission during the production of hydrogen fuel, there is no doubt that HFCV has a zero tailpipe carbon emission [1].

Recently, Honda has launched its third generation of fuel cell sedan, the Honda Clarity in Japan which is expected to be on sale in the first quarter of 2016. In Malaysia, Datuk Seri Idris Jusoh, the Education Minister launched the first Malaysian-made HFCV, a golf buggy at the National University of Malaysia (UKM) in December 2014. UKM was given RM7 million by the Education Ministry to lead the project to develop a zero emission vehicle using indigenous fuel cells.

Since HFCV is still new and probably unknown to Malaysian, it is important to understand to what extent public will accept this new sustainable technology and what are the predictors of such intention. This is to give an insight on how the technology should be introduced and implemented and to give the idea to design communication about the technology to the public later [2]. To the authors' knowledge, there is still no published study on the acceptance model of HFCV in Malaysia.

2. Literature review

The proposed model is developed base on two psychological theories of acceptance/behaviour and factors from other empirical studies.

2.1. Acceptance

There are three types of acceptance distinguished, that are, socio-political acceptance (public reaction to national or international policy-making), citizen acceptance (public responses towards the adoption of a technology in one's living areas as a results of others) and consumer acceptance (public's reactions to the availability of innovations on the market) [3]. The authors also proposed the term 'acceptance' for "actual behaviour in reaction to the technology". In this study, the acceptance is limited to consumer acceptance and we defined acceptance as behaviour towards HFCV. However, HFCV is still undergoing technical developments and is still not available in automotive market in Malaysia. Therefore, we could not measure the actual behaviour such as actual acceptance. Instead, we proposed to measure the intention to accept (intention to support, purchase or using) which is a direct predictor of the acceptance [2] of HFCV and study the determinants of such intention.

2.2. Theory of planned behaviour (TPB). TPB stated that intention influences behaviour via three determinants, namely, attitude, subjective norm and perceived behavioural control [4]. Attitude towards behaviour is defined as the extent to which an individual has a good or bad evaluation of the behaviour [4][5]. Subjective norm refers to perceived social pressure to perform or not perform the behaviour while perceived behavioural control refers to the perceived ease or difficulty of performing the behaviour [4].

2.3. Norm activation theory (NAT). Pro-social and pro-environmental behaviours, both involve giving up own self-interest for the sake of others [6]. According to [7], altruistic behaviour, that is behaviour that benefits others than oneself, is induced by personal norms. Personal norms reflects "feelings of moral obligation to perform or refrain from specific actions" [8]. These personal norms are influenced by two main factors; the problem awareness and outcome efficacy [9]. The problem awareness refers to the awareness that not performing a specific behaviour has certain negative consequences [9][10]. Outcome efficacy can be defined as the extent to which a person feels in control over the problem [10] or whether one thinks that the new technology/relevant behaviour will actually reduce energy problems [5]. Many studies include this norm activation model in predicting pro-environmental behaviours [2][11][12][13].

2.4. Perceptions of benefits and risk. Attitude is influenced by perceived benefits and risks of the technology/behaviour [5]. Studies showed that perceptions towards the technology significantly influenced usage intention [14][15] and acceptance [16][17]. Therefore, we proposed that these perceptions of benefits and risks influenced acceptance through attitude in the model. These perceptions is then influenced by social trust [18][19].

2.5. Social trust. As HFCV is still not in the Malaysia's automotive market, we assume that the public have little or no knowledge about HFCV. The technology and innovation of alternative energy is a complex topic that only fully known by experts and actors responsible in the area [20]. People will base their attitudes toward new technology through social trust when there is insufficient knowledge or resources [19]. Social trust refers to "willingness to rely on those who have the responsibility for making decisions and taking actions related to the management of technology, the environment, medicine, or other realms of public health and safety" [19]. Through trust, public would not need to fully understand the details of the benefits and risks and rely on the information provided by the trusted parties to make decision [21]. Several studies showed that trust in those responsible for the technology influences perceived benefit positively and perceived risk negatively [2][18][19].

2.6. Government interventions. The availability of resources and opportunity which results in less perceived impediment will cause a higher PBC in an individual [4]. This resources and opportunity refers to 'facilitating condition' as used by [22]. Government intervention has become a facilitating condition to the adoption of new technology and has a significant influence in several studies related to vehicle acceptance [14][23]. Therefore, we proposed government interventions such as tax exemption and subsidies as the facilitating resources to be the antecedent of perceived behavioural control.

3. The proposed model of acceptance

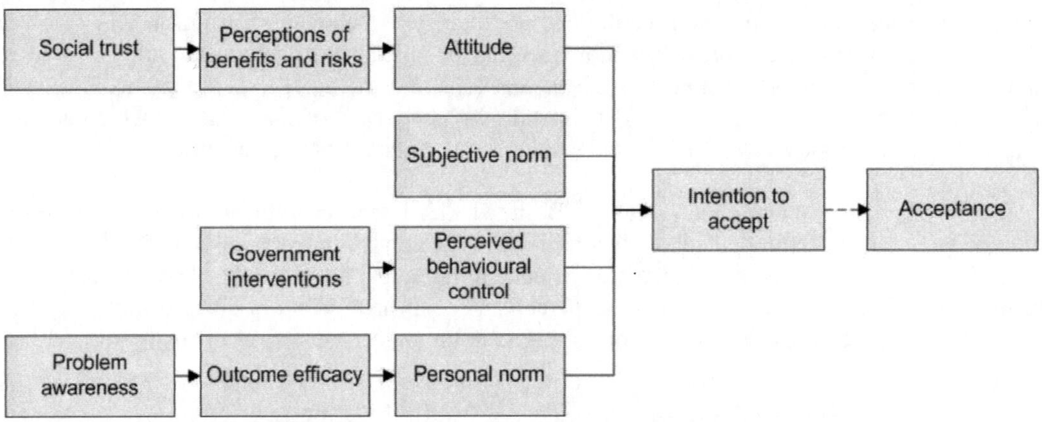

Figure 1. The proposed model of factors influencing HFCV acceptance.

Table 1. Factors and their definitions.

Factor	Definition
Intention to accept	Intention to support, purchase or use HFCV.
Attitude	The degree to which a person has a favorable or unfavorable evaluation of accepting HFCV.
Perceptions of benefits and risks	The extent to which a person thinks that HFCV has certain advantages and disadvantages.
Social trust	Willingness to rely on those who have the responsibility for making decisions and taking actions related to HFCV.
Subjective norm	The extent to which a person thinks that important people to him/her believes he/she should support, purchase or use HFCV.
Perceived behavioral control	The perceived ease or difficulty to support, purchase or using HFCV.
Government interventions	The extent to which a person thinks the government will intervent in the penetration of HFCV especially in financial help.
Personal norm	A feeling of moral obligation to accept HFCV.
Outcome efficacy	The perceived effectiveness of supporting, purchasing or using HFCV in reducing energy problems.
Problem awareness	Awareness (having realization) that not accepting HFCV has a certain negative consequences.

Figure 1 depicts the proposed model. There are ten latent variables to be measured with exception on acceptance. The reason is explained in 2.1. The lines with arrows show the directions of the relationships between the factors and the dashed line shows the relationship that will not be tested. The detailed definition of each factor is shown in Table 1.

4. Conclusion

This study will be a significant effort in the implementation of HFCV in Malaysia. Malaysia's government will be able to have an insight about the public acceptance of this new type of transport, HFCV, which will help in the decision-making and policy-making of the implementation. From the significant factors, the government could organize and plan better for the introduction and implementation of HFCV. In return, this will enhance the mitigation of the local environmental problems and energy insecurity caused by transport industry. The automotive industry will also benefit from this study. The results of this research will be helpful for automotive manufacturers in Malaysia in their decision-making to produce and sell HFCV and would also convey the public concerns regarding HFCV such as the costs, risks and benefits.

5. Reference

[1] Pollet B G, Staffell I and Shang J L 2012 *Electrochimica Acta* **84** 235-49
[2] Huijts N M A, Molin E J E and Van Wee B 2014 *J. Environmental Psych.* **38** 153-66
[3] Huijts N M A, Molin E J E, Chorus C G and Van Wee B 2011 *Transition Towards Sustainable Mobility: The Role of Instruments, Individuals and Institutions* ed H Geerlings et al (Hampshire: Ashgate)
[4] Ajzen I 1991 *Organizational Behavior and Human Decision Processes* **50(2)** 179-211
[5] Huijts N M A, Molin E J E and Steg L 2012 *Renewable and Sustainable Energy Reviews* **16(1)** 525-31
[6] Abrahamse W, Steg L, Gifford R and Vlek C 2009 *Transportation Research: Traffic Psych.and Behaviour* F **12(4)** 317-24
[7] Schwartz S H 1977 *Advances in Experimental Social Psych.* **10** 221-79
[8] Schwartz S H and Howard J A 1981 *Altruism and Helping Behavior* 189-211.
[9] Huijts N M A, De Groot J I M, Molin E J E and Van Wee B 2013 *Transportation Research: Policy and Practice* A **48** 63-74
[10] Steg L and Groot J 2010 *British J. Social Psych.* **49(4)** 725-43
[11] Klöckner C A 2013 *Global Environmental Change* **23(5)** 1028-38
[12] Onwezen M C, Antonides G and Bartels J 2013 *J. Economic Psych.* **39** 141-53
[13] Han H 2014 *J. Environmental Psych.* **40** 462-71
[14] Sang Y N and Bekhet H A 2015 *J. Cleaner Production* ***92*** 75-83
[15] Mansor N, Yahaya S N, Nizam N Z and Hoshino Y 2014 *Int. J. Business Administration* **5(1)** 27
[16] Matsumoto Yasuo, Takanashi Hirokazu, Uemura Yoshimitsu and Kai Takami 2008 *Environmental science* **21(6)** 435-49
[17] Kang M J and Park H 2011 *Energy Policy* **39(6)** 3465-75
[18] Siegrist M 2000 *Risk Analysis* **20(2)** 195-204
[19] Siegrist M, Cvetkovich G and Roth C 2000 *Risk Analysis* **20(3)** 353-62
[20] Perlaviciute G and Steg L 2014 *Renewable and Sustainable Energy Reviews* **35** 361-81
[21] Huijts N M, Midden C J and Meijnders A L 2007 *Energy Policy* ***35*(5)** 2780-89
[22] Chu P Y, Hsiao N, Lee F W and Chen C W 2004 *Government Information Quarterly* ***21*(2)** 219-34
[23] Hong Y H, Khan N and Abdullah M M 2013 *Australian J. Basic and Applied Sciences* **7(8)** 347-54

Acknowledgement
This work was supported by the Long-term Research Grant Scheme (LRGS) funded by the Education Ministry of Malaysia (01201401 LRGS).

Liquid cooled plate heat exchanger for battery cooling of an electric vehicle (EV)

M M Rahman[1], H Y Rahman, T M I Mahlia, J L Y Sheng
Department of Mechanical Engineering, College of Engineering,
Universiti Tenaga Nasional, Jalan IKRAM-UNITEN, 43000 Kajang, Selangor, Malaysia

E-mail: mujibur@uniten.edu.my

Abstract. A liquid cooled plate heat exchanger was designed to improve the battery life of an electric vehicle which suffers from premature aging or degradation due to the heat generation during discharging and charging period. Computational fluid dynamics (CFD) was used as a tool to analyse the temperature distribution when a constant surface heat flux was set at the bottom surface of the battery. Several initial and boundary conditions were set based on the past studies on the plate heat exchanger in the simulation software. The design of the plate heat exchanger was based on the Nissan Leaf battery pack to analyse the temperature patterns. Water at different mass flow rates was used as heat transfer fluid. The analysis revealed the designed plate heat exchanger could maintain the surface temperature within the range of 20 to 40°C which is within the safe operating temperature of the battery.

1. Introduction

Recently, there has been a growing demand in newer, cleaner and greener technology where it aims to replace the conventional fossil fuel as primary energy source. The automotive industries are the major innovators in developing new technologies for greener and cleaner vehicles as engineers race to provide their solutions in discovering new energy source in powering the automotive industries. Electric vehicle (EV) has become one of the fast growing innovations in the automotive industries. Auto makers have gone into developing the electric vehicle technology. Vehicles which do not pollute the environment show a very promising sign for the future development in the automotive industries. However, the electric vehicle is still at a juvenile stage where several models that were manufactured mainly for experimental purposes and market research. Auto makers continue to improve their respective models to be made into the mass market for consumers.

The electric vehicle (EV) that is sold in Malaysian market namely the Nissan Leaf uses passive air-cooling as its battery thermal management system (BTMS) which means when the battery is charging, it would not be cooled as the cooling relies on the movement of the car. Hence, the electric vehicle suffers from the loss of driving range and having to replace the battery pack after a short period of time as the heat is not removed from the battery pack compartment. Moreover, in hot climate like Malaysia, the requirement for an active cooling system for the batteries is even greater as the heat could not be transferred out as easily as compared to colder countries. Hence, a battery thermal management system (BTMS) has to be designed for the Nissan Leaf for better adaptation for the hot countries like Malaysia.

[1] Corresponding author: Prof. Dr. Md. Mujibur Rahman.

There are a few studies that were done on the battery thermal management system [1-3] which shown that having a liquid cooling system in the battery pack possess a few advantageous over forced air cooling. Figure 1 shows the setup of the liquid cooled plate heat exchanger assembled together with the battery pack. As the battery pack is very congested, the total height of the liquid cooled plate heat exchanger is at 5 mm.

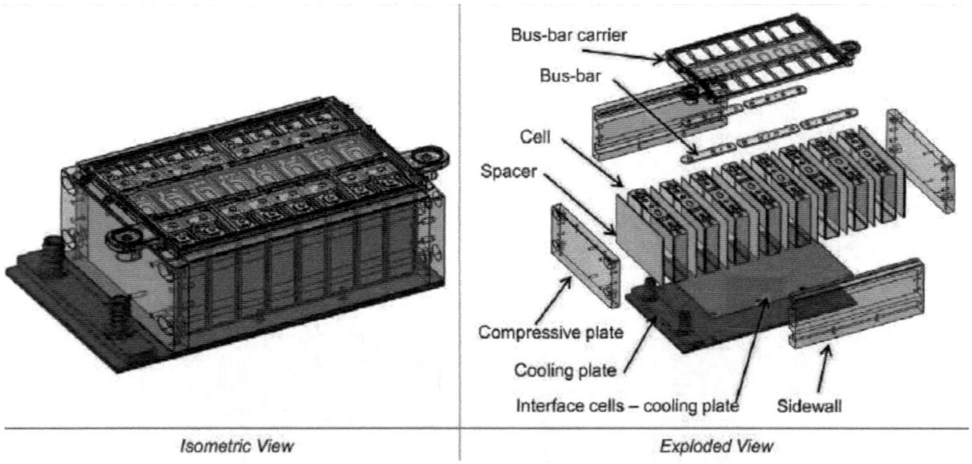

Figure 1 Placement of the liquid cooled plate heat exchanger [3]

Another study on thermal behaviour of the flow channel of the liquid cooled plate is done by Jarrett and Kim [4] in where they studied the optimum design of the electric vehicle cooling plates. They have considered three factors which determine the optimum operating conditions, i.e., pressure drop, average temperature, and temperature deviation where it is subject to a constant heat flux of 500 W/m^2. Another research [5] showed that the mass flow rate of the coolant plays an important role in temperature uniformity and the average temperature. By increasing the mass flow rate, the surface of the plate will eventually reach temperature uniformity where temperature deviation across the surface is close to zero.

2. Methodology

The electric vehicle used as the reference for this design is the Nissan Leaf since it employs a passive air cooling system which relies on the movement of the vehicle to cool the battery. The battery does not have a forced air or liquid cooling when it is charging in its stationary state. The battery used specifically for this case study is same as used in the Nissan Leaf which is is a prismatic cell made from its own manufacturer in Japan.

From the data that is obtained from Nissan, the prismatic battery cell that is used provides a power output of 24 kilowatt per hour and the voltage output of 403.2 volts gives the battery pack to run the 80 kilowatt AC synchronous electric motor a driving range of 84 miles which is roughly 140 kilometers in one single charge. The battery pack contains a total of 48 modules and each module contains 4 prismatic cells equaling a total of 192 cells in the whole battery compartment. The dimensions of the battery pack are listed on the table 1.

Table 1 The dimensions of the prismatic cell, modules and pack of the Nissan Leaf

	Quantity	Length (mm)	Width (mm)	Thickness (mm)
Battery pack	1	1570.5	1188.0	264.9
Battery module	48	303.0	223.0	35.0
Battery cell	192	290.0	216.0	7.1

The battery pack of the Nissan Leaf is placed below the passengers and driver. As according to the manufacturer, the placement of the battery pack at the center of the vehicle will lower the center of gravity for the car as to ensure stability of the car. The battery modules are arranged in such manner where 24 modules are placed vertically and the remaining 24 batteries lie among each other on two sides of the battery pack. Due to the volumetric constraints of a conventional battery pack, ensuring that the terminals of the battery module are not disrupted when the plate is installed is vital to avoid incomplete contact at the terminals which may results in power loss or current leakage. Hence, the height of the liquid cooled plate heat exchanger has to be small [3]. The height of the plate in total will be 5 mm with the radius of the inlet to be set at 2 mm. and the height of the plates to be at 0.5 mm. As the width is much bigger than the height of the plate, it can be assumed that the conduction takes place only in one direction which is the normal to the surface plate.

2.1 Computational Fluid Dynamics (CFD) Model Generation

Figure 2 shows the top view of the flow channel of the liquid cooled plate exchanger.

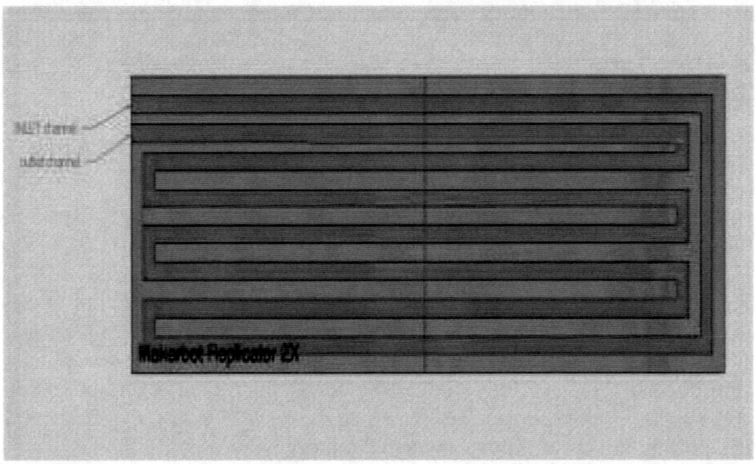

Figure 2 Top view of the single channel design of the liquid plate heat exchanger

The pre-processor, i.e., GAMBIT was used to model the desired geometry before it was exported into the CFD software. Moreover, it is used to mesh the entire surface of the geometry so that the geometry can be plotted out in the CFD software. After the geometry of the plate is modeled out, the meshing is then done on the plate model. A 2-dimensional approach to the liquid plate heat exchanger was done as the thickness of the plate was only 5 mm. Furthermore, since the length of the plate is longer than the thickness of the plate which is 840 mm and 5 mm, respectively, the ends and side thermal effects are negligible. Hence, the thermal analysis from the computational fluid dynamics simulation can be done accurately with two dimension. The use of 2 dimensional simulation can also reduce the computer resources allowing a quicker convergence of the residuals when simulation is done. Once the meshes are generated, the cooling plate is then exported into FLUENT. A uniform heat flux of 500 W/m^2 on the surface of the cooling plate was applied which follow the studies reported in [4]. The coolant flow rates were varied from 0.01 kg/s to 0.05 kg/s. The pressure at the outlet channel was set as zero.

3. Results and discussion

Contour curve of typical temperature distribution at a mass flow rate of 0.03 kg/s is presented in Figure 3 and the outlet temperatures of the water at different mass flow rates are presented in Figure 4.

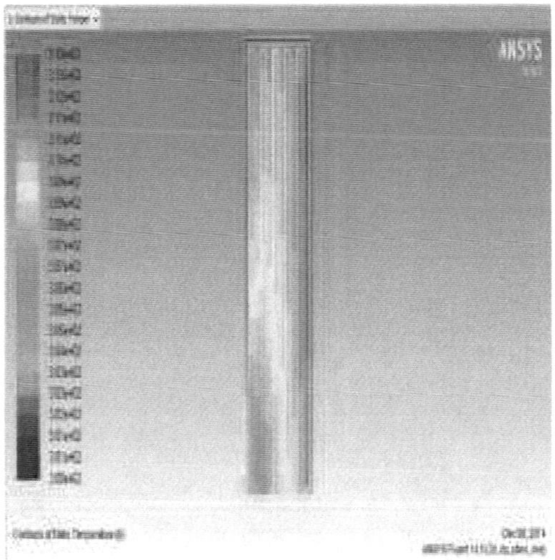

Figure 3 Temperature contour at a mass flow rate 0.03 kg/s

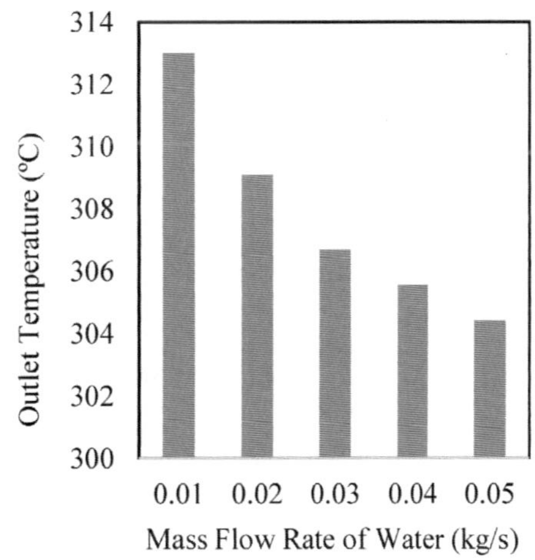

Figure 4 Outlet temperature of water at different mass flow rates

It is clear from Figures 3 and 4 that as the mass flow rate of the cooling fluid increases, the outlet temperature also decreases. As the water flows into the channel duct, heat is exchanged from the top surface of the plate into the duct walls. As the water flows along the channel, heat is transferred due to the temperature difference. Higher mass flow rate decreases the water outlet temperature since water was in contact with the duct surface for a short time hence less hat is swept away by the water.

4. Conclusions

A liquid cool plate heat exchanger was designed to remove excessive heat generated by the battery pack during the operation of an electric vehicle in Malaysian environment. Water was considered as working fluid. The heat transfer analysis was conducted by simulating the flow of water through the heat exchanger by CFD. It was observed that a mass flow rate of water about 0.02 kg/s could remove the excessive heat generated which could maintain the temperature in an allowable level.

5. Acknowledgement

This research is financially supported by Ministry of Education (MOE) Malaysia through the research projects 20140117 FRGS and 10012013 ERGS.

6. References

[1] Pesaran A A, Kim G H and Keyser M 2009 EVS-24 *24th International Batterie, Hybrid and Fuel Cell Electric Vehicle Symposium,* **NERL/CP-540-45779** 1-7.

[2] Pesaran A A, Keyser M 2001 Proc, *The Sixteenth Annual Battery Conference on Applications and Advances*, 219-225.

[3] Smith J, Hinterberger M, Hable P and Koehler J 2014 *Journal of Power Sources*, **267** 784-792.

[4] Jarrett A, Kim I Y 2013 *Journal of Power sources*, **245** 644-655.

[5]. Rahman M M, Hua T. J and Rahman H Y 2014 *Applied Mechanics and Materials*, **663** 213-217.

Is MLQ Instrument Applicable to Verify the Blue Ocean Leadership Traits?

Wan Noordiana Wan Hanafi[1], **Salina Daud**[2], **Nur Lyana Baharin**[3]

E-mail: dianahanafii@gmail.com[1],salina@uniten.edu.my[2],nurlyana@uniten.edu.my[3]

Abstract. Blue Ocean Leadership aims to achieve an organizational increase in leadership strength quickly and at low cost, which translates into high performance for the business. This is why in building a sustainable business and community in the organization, leadership plays a very important role. The purpose of this paper is to review past researches and to identify whether the "Multifactor Leadership Questionnaire" (MLQ From 5x) is applicable to be used to measure blue ocean leadership traits. It is expected that the instruments proposed in this paper will contribute to the new findings in the leadership study.

1. Introduction

One of the most crucial components of management today would be leadership (Weihrich et al., 2008), and even nowadays, according to Salleh and Grunewald (2013), leadership has a high demand in different organizations and a great skill of leadership is needed by the Chief Executive Officer in order to successfully run the organization effectively. Lok and Crawford (2004) stated that the success and failure of an organization are contributed by the styles and practices shown by the leaders. Furthermore, in an organization these days, organizational leaders are found to be practicing different styles of leadership and behaviour at work (Oshagbemi & Ocholi, 2006).

Blue Ocean Strategy is known for providing companies with a new way of thinking about where and how to compete to drive profitable growth. Meanwhile, Blue Ocean Leadership is a leadership style created by Chan Kim and Renée Mauborgane (2014) that uses many of the same principles and critical frameworks as those of blue ocean strategy. The basic idea of Blue Ocean Leadership is to achieve an organizational increase in leadership strength quickly and at low cost, which translates into high performance for the business.

Studies on blue ocean leadership are important for today's organization, and thus, it is important to find the right leadership styles to measures the blue ocean leadership traits. The objective of this paper is to enhance the "Multifactor leadership Questionnaire" (MLQ From 5x) instrument that was developed by Bass and Avolio (2004) to be used to measure the blue ocean leadership traits.

2. Leadership styles

Transformational leadership has captured the attention of many scholars' interests (Antonakis, Cianciolo, & Sternberg, 2004; Basham, 2010; Northouse, 2010; Sarros & Santora, 2001), and has begun to be the central model in leadership. According to Bass and Riggio (2006) and Sivanathan and

[1] To whom any correspondence should be addressed.

Fekken (2002), transformational leader delivers a vision and a sense of organizational mission; inspire pride, respect and trust among their followers. Pearce and Sims (2002) stated that the dimensions which are common to transformational leaders are charismatic leadership (or idealised influence), inspirational motivation, intellectual stimulation, idealized behaviors and idealized traits.

Transactional Leadership on the other hand, was referred by Bass (1989), Avolio, Waldman and Yammarino (1991) and Burke, Stagl, Salas, Pierce and Kendall (2006) as concerned managers who get things done by making and satisfying promises to employees. Northouse (2010) stated in his study that a transactional leadership is different compared with transformational leadership because a transactional leader does not individualize the needs of subordinates or focus on their self-improvement. Transactional leaders employ three factors which are: (1) Contingent reward, (2) Management by exception: active, and (3) management by exception: passive (Bass & Avolio, 1994).

Blue Ocean Strategy provides organization with a new way of thinking about where and how to compete to drive profitable growth. Blue Ocean Leadership uses many of the same principles and frameworks to help organization engage the previously disengaged employee by changing the organization leadership profile. Thus, the basic idea of Blue Ocean Leadership is to achieve an organizational increase in leadership strength quickly and at low cost, which translates into high performance for the business. This is done by transforming unengaged employees into engaged ones. The following table summarize the above leadership styles traits (Nur Lyana, Salina, Aerni, & Mohd Yunus, 2015)

Table 1: Leaders Traits for Transformational, Transactional and Blue Ocean Leadership

Traits	Transformational	Transactional	Blue Ocean Leadership
Visionary	Leader sets vision	Vision is not clear	The vision is to always challenge the ordinary
Contingent Reward	Intrinsic satisfaction	Extrinsic satisfaction	Team gratification
Courage	Empowerment	Initiation by leader	Let's go (togetherness)
Idealized Influence	Use of charisma	Use of power	Promotes new ideas & thinking
Inspiration	High inspiration	Leads the way	Walks with the team
Intellectual Stimulation	High intellectual stimulation	Low intellectual stimulation	Engagement in creativity & innovation
Passion	Beyond organizational goals & objectives	Organizational goals & objectives	Always connected to reality
Strategic Thinking / Planners	Long term planners	Short term planners	Game changer
Focus	Vision and goals of organization	Values, qualities and behavioral style	Acts and activities leaders need to undertake
Collaborate	Executive & senior level	With subordinates	All levels in organization
Innovate	Encouragement	Slow or no innovation	Innovates to survive
Willingness to change	Moderate to fast	Only when problem exists	Do or die
Communication	Two-way	Top-down	360 degrees

3. Multifactor Leadership Questionnaire (MLQ) instrument

The Multifactor Leadership Questionnaire (MLQ) was first published in 1985 (Bass, 1985) and over the past twenty years, MLQ has been revised several times by scholars (Antonakis et al., 2003; Avolio & Bass, 2004; Bass et al., 2003; Judge & Piccolo, 2004). According to Bass and Avolio (2004), the MLQ scale showed high internal consistency and factor loadings and the reliabilities of MLQ were suggested to be a useful tool for future researches in leadership arena. Ozaralli (2003) in his study has also stated that MLQ is "the best validated measure of transformational and transactional leadership" (p.338).

Table 2: Multifactor Leadership Questionnaire (MLQ) dimensions

Transformational	Transactional	Laissez-fair
• Idealized influence (IA) • Idealized behaviour (IB) • Inspirational motivation (IM) • Intellectual Stimulation (IS) • Individual Consideration (IC)	• Contingent Reward (CR) • Management by exception: active (MBEA) • Management by exception: passive (MBEP)	• Laissez-Faire (LF)

The above MLQ dimensions were adapted from Bass and Avolio (1999) which measure leadership and non-leadership dimensions. This non-leadership is known as *laissez-faire* (LF) leadership and reflects the absence of leadership and avoidance of intervention. There is no attempt to make agreements with followers, to motivate them, to set standards or to give feedback. In sum, as can be seen in Table 2 above, the MLQ claims to measure nine distinct leadership dimensions, which are grouped into three broader categories: transactional, transformational, and non-leadership.

6. Future Study and Conclusion

After acknowledging the MLQ criticisms by refining several versions of the instruments, the version of the MLQ, the Form 5X (Bass & Avolio, 1999) is successful in effectively capturing the full leadership factor of the transformational leadership theory. Table 3 below was developed to enhance the MLQ by including the Blue Ocean Leadership traits.

Table 3: Improvising the Multifactor Leadership Questionnaire (MLQ) Dimensions

Transformational	Transactional	Laissez-fair	Blue Ocean Leadership Attributes
• Idealized influence (IA) • Idealized behaviour (IB) • Inspirational motivation (IM) • Intellectual Stimulation (IS) • Individual Consideration (IC)	• Contingent Reward (CR) • Management by exception: active (MBEA) • Management by exception: passive (MBEP)	• Laissez-Faire (LF)	• Eliminate • Raise • Reduce • Create

The findings from this study are expected to contribute to the new dimension of leadership study which could help an organization to increase its leadership strength quickly and at low cost, which translates into high performance for the business.

7. References

[1] Antonakis, J., Avolio, B. J., & Sivasubramaniam, N. (2003). Context and leadership: an examination of the nine-factor fullrange leadership theory using the Multifactor Leadership Questionnaire. The Leadership Quarterly, 14(3), 261-295

[2] Antonakis, J., Cianciolo, A. T., & Sternberg, R. J. (2004). The nature of leadership. Thousand Oaks, CA: Sage Publications

[3] Avolio, B. J., Jung, D. I., Murry, W., Sivasubramaniam, N. and Garger, J. (2003). Assessing shared leadership: Development of a Team Multifactor Leadership Questionnaire. In Pearce, C.L., and Conger, J. A. (Eds.), Shared Leadership: Reframing the Hows and Whys of Leadership Thousand Oaks: Sage. pp. 143-172.

[4] Avolio, B.J., Waldman, D.A., & Yammarino, F.J. (1991). Leading in the 1990's; the four I's of transformational leadership. Journal of European Industrial Training, 15(4),8-16.

[5] Basham, L. M. (2010). Transformational and Transactional Leaders in Higher Education. S.A.M. Advanced Management Journal, 77(2), 15-23,37,12.

[6] Bass BM, Avolio BJ (1994). Improving organizational effectiveness through transformational leadership. California: Sage

[7] Bass, B. M. & Avolio, B. (2004). Multifactor leadership questionnaire technical report. Center for Leadership Studies. Redwood City, CA: Mind Garden.

[8] Bass, B. M. (1985). Leadership and performance beyond expectations. New York: Free Press.

[9] Bass, B. M. (1989). From transactional to transformational leadership: Learning to share the vision. Organizational Dynamics. 19-31

[10] Bass, B. M. and Avolio, B. J. 1997. Full range leadership development: Manual for the multifactor leadership questionnaire, CA, Mind Garden

[11] Bass, B. M., & Avoilio, B. J. (2003). The multifactor leadership questionnaire report. Palo Alto, CA: Mind Garden

[12] Bass, B., & Riggio, R. E. (2006).Transformational Leadership (2nd ed.). Mahwah, N.J.: Lawrence Erlbaum.

[13] Boehnke, K., Bontis, N. Distefano, J., & Distefano, A. (2003). Transformational Leadership: An Examination of Cross-national Differences and Similarities. Leadership and Organization Development Journal, 24(1/2), 5-17

[14] Burke, C.S., Stagl, K.C., Salas, E., Pierce, L., & Kendall, D. (2006). Understanding team adaptation: A conceptual analysis and model. Journal of Applied Psychology, 91, 1189-1207.

[15] Gallup Inc., (2013). State of the American Workplace. Washington. Hollensen, S, 2013. The Blue Ocean that disappeared – the case of Nintendo Wii. Journal of Business Strategy, Volume 34 No. 5, 25-35.

[16] Judge, T. A., & Piccolo, R. F. (2004). Transformational and Transactional Leadership: A Meta-Analytic test of their relative validity. Journal of Applied Psychology, 89(5), 755-768

[17] Kim, W. C. & Mauborgne, R. (2014). How to Develop and Select Your New Leadership Profiles, Blue Ocean Leadership Series IV.

[18] Lok, P. and Crawford, J. (2004). The effect of organizational culture and leadership style on job satisfaction and organizational commitment – A cross-national comparison. Journal of Management Development, 23 (4), 321-338.

[19] Northouse, G. P. (2007). Leadership: Theory and Practice. Thousand Oaks, Cal.: Sage Publications, Inc

[20] Northouse, P. G. (2010). Leadership: theory and practice (5th ed.). Thousand Oaks, Calif.: Sage Publications.

[21] Nur Lyana, B., Salina, D., Aerni, I. and Mohd Yunus, M. (2015). Conventional leadership versus blue ocean leadership. Proceedings of National Symposium and Exhibition on Business and Accounting 2015, Universiti Tenaga Nasional.

[22] Oshagbemi, T. & Ocholi, S.A. (2006). Leadership styles and behavior profiles of managers. Journal of Management Development, 25(8), 748-762. doi: 10.1108/02621710610684231

[23] Ozaralli, N. 2003. Effects of transformational leadership on empowerment and team Effectiveness , Leadership & Organization Development Journal, 24(6) ,335-344

[24] Pearce, C.L. & Sims, H.P. Jnr. (2002). Vertical versus shared leadership as predictors of the effectiveness of change management teams: An examination of Aversive, Directive, Transactional, Transformational, and Empowering leader behaviours. Group Dynamics: Theory, Research and Practice, 6 (2), 172-197.

[25] Salleh, M., & Grunewald, D. (2013). Organizational Leadership—The Strategic Role of the Chief Exec. Journal of Leadership, Accountability, and Ethics, 10(5), 9–20.

[26] Sarros, J. C., & Santora, J. C. (August 01, 2001). The transformational-transactional leadership model in practice. Leadership and Organization Development Journal, 22, 8, 383-394.

[27] Sivanathan, N., & Fekken, G. C. (2002).Emotional Intelligence, Moral Reasoning and Transformational Leadership. Leadership and Organizational Development Journal, 23(3/4), 198-204.

[28] Weihrich, H., Cannice, M.V. and Koontz, H. (2008) Management (12th ed.). New Delhi: Mc Graw Hill

What Is Green Growth Strategy for Government Link Company?

Maryam Jamilah Asha'ari[1], Salina Daud[2], Hasmaizan Hassan[3]

Abstract. Disasters around the world are very extreme because of the global warming and climate change. Malaysia firms have to play their role in handling the challenging of environmental problems in order to sustain. The feature of the new strategy which is green growth strategy has been identified. The study focuses on the features of the green growth strategy which discuss on the keys to sustaining the strategy, marketing emphasis, production emphasis, product line, basis of competitive advantage and strategic target. Business had contributed to the industrialisation era positively or negatively and therefore there is a must for business people to use the best strategy in reducing the environmental risks. By 2020, Malaysia will achieve the target in becoming an advanced economy by applying the right strategy. The objective of this paper is to propose a feature for new strategy known as green growth strategy. Future study is to propose to conduct an empirical analysis to confirm the green growth strategy features.

1.0 Introduction

In facing of environmental and economic challenges nowadays, national have put a lot of efforts in promoting a new source of growth which is green growth. OECD (2012) stated that the policy of sustainable and economic is a green growth matter. In order to sustain, business play a significant role. Anuar (2012) believe that the impact that we experience today is because of the contributing of the business in terms of society, economy and environment.

According to Fayers (1999), the environment is harm due to the growing physical evidence of unsustainable future at the current rate of natural resources consumption. Disasters around the world are very extreme because of the global warming. In decision making, there is a must for the business people to analyse carefully the impact of the environment and society (Amran, 2010). Disaster risks and inclusivity can be enhanced as well as the economic opportunities can be expanded by successful green growth (Eleventh Malaysia Plan).

The objective of this paper is to propose the new business strategy besides five generic competitive strategies which is green growth strategy. This new strategy supports the Eleventh Malaysia plan to become an advanced economy by 2020 as this strategy applies green growth practiced in achieving a resilient, low-carbon, efficient in term of resource and socially-inclusive manner. Besides that, in preparing the nation for challenges in the future, green growth is a development trajectory which considers the dimensions of sustainable development which is social, environment and economy (Eleventh Malaysia Plan).

---

1 To whom any correspondence should be addressed.
2 To whom any correspondence should be addressed.
3 To whom any correspondence should be addressed.

2.0 Five Generic Competitive Strategies

According to Ritika (2013), for an industry to create a defendable position, competitive strategies involve taking defensive or offensive actions. Broad differentiation, focused low-cost provider, low-cost provider, focused differentiation strategy and best-cost provider are the five generic competitive strategies (Thompson, Strickland and Gamble, 2011). Ritika (2013) believe that generic strategies help organization in competing rivals and help in coping with the competitive forces.

Rather than being one of competitors with comparatively low costs, firms can achieve low-cost leadership. It can be achieves when the company becomes the industry's lowest-cost provider (Thompson, Strickland and Gamble, 2011). The services or products creation that is unique throughout its industry at the broad market is aimed as differentiation (Ritika, 2013). According to Thompson, Strickland and Gamble (2011), whenever customer's preferences and needs are too diverse to be fully satisfied by sellers or by a standardized product with identical capabilities it is the time when differentiation strategies are attractive.

In incorporating upscale attributes, the competitive advantage of a best-cost provider is lower costs than competitors. For whose products that have similar upscale attributes, it is putting the company in a position to underprice competitors. Focused low-cost provider is a focused strategy that is based on low cost. By serving customers a lower price than the competitor and in the market niche that is targeted, this strategy aims at securing competitive advantage (Thompson, Strickland and Gamble, 2011).Thompson, Strickland and gamble (2011) stated that a focused differentiation strategy is a strategy keyed to differentiation that aims at securing a competitive advantage. Differentiation of competitive advantage with a product offering carefully design to appeal to the needs of a narrow, well-defined group of customer needs and preferences that is unique.

3.0 Green Growth Strategy

Former President of Republic Korea (2008) defines green growth as "a new national development paradigm for job creation and new growth by using green and clean technology" (World Water Council, 2015). Green growth means ensuring the natural assets continue to provide the environmental services and resources on which our wellbeing relies and fostering the development and economic growth (OECD, 2011). UNEP (2011) believed that the concept of green growth is significantly reducing ecological scarcities and environmental risks. Green growth is also significantly improving social equity and improved human well-being.

Example of green growth strategy that can be practiced by company is by producing green products (OECD, 2015). OECD (2009) states that green products or environmentally products are product that having packaging that is eco-friendly which has the long term goal of natural preservation with low detrimental environmental impact at all stages of its life cycle. These products are bio-degradable, non-toxic chemical, recyclable and reusable.

4.0 Features of Company Strategy

From above discussion, below is the summary of the business strategy for all business strategy discuss in this paper.

Table 1. Features of company strategy

Features	Low-Cost Provider	Broad Differentiation	Best-Cost Provider	Focused Low-Cost Provider	Focused Differentiation	Green Growth
1. Strategic target	A broad cross-section of the market.	A broad cross-section of the market.	Value-conscious buyer.	A narrow market niche where buyer needs and preferences are distinctly different.	A narrow market niche where buyer needs and preferences are distinctively different	Environment conscious buyers
2. Basis of competitive advantage	Lower overall costs than competitors.	Ability to offer buyers something attractively different from competitors.	Ability to give customers more value for the money.	Lower overall cost than rivals in serving niche members.	Attributes that appeal specifically to niche members.	Ability to give customers more value for the money
3. Product line	A good basic product with few frills (Acceptable from quality and limited selection)	Many product variations, wide selection; emphasis on differentiating features.	Items with appealing attributes; assorted upscale features.	Features and attributes tailored to the tastes and requirement of niche members.	Features and attributes tailored the tastes and requirements of niche members.	Product variations that is friendly environment
4. Production emphasis	A continuous search for cost reduction without sacrificing acceptable quality and essential features.	Build in whatever differentiating features buyers are willing to pay for; strive for product superiority.	Build in upscale features and appealing attributes at lower cost than rivals	A continuous search for cost reduction while incorporating features and attributes to niche member preferences.	Custom made products that match the tastes and requirements of niche members	Build in environmentall y safe product features
5. Marketing emphasis	Try to make a virtue out of product feature that lead to low cost	Tout differentiating features Charge a premium price to cover the extra costs of differentiating features.	Tout delivery of best value	Communicate attractive features of a budget-priced product offering that fits niche buyers' expectations.	Communicate how product offering does the best job of meeting niche buyers' expectations.	Communicate attractive features of a green product that fits the buyers' expectations.
6. Keys to sustaining the strategy	Economical prices/good value. Strive to manage costs down in every area of the business.	Stress constant innovation to stay ahead of imitative competitors.	Unique expertise in managing costs down while incorporatin g upscale features and attributes.	Stay committed to serving the niche at lowest overall costs.	Stay committed to serving the niche better than rivals;	Stay committed to produce green product in order to sustain

5.0 Conclusion and future study

This paper would like to propose among the green strategy features which are strategic target, basis of competitive advantage, product line, production emphasis, marketing emphasis and keys to sustaining the strategy. The green growth strategy that is proposed in the generic competitive strategies is focusing on the strategy in producing green goods and services. The product line of green growth strategy is to produce product variations that are friendly environment which is green products. According to OECD (2013), the key policies to be considered are fostering green technology research and development (R&D), innovation and identifying the potential for green specialization. The green product specialization thus can reduce the environmental risks towards the economy, environment and social as potential sources for growth.

The strategic target of green growth strategy is value-conscious buyers. Tsay (2010) stated that customer that perceiving green product in having superior quality product seems to pay higher premium for the higher income group customers. Green purchase decision could be triggered by emotions exhibited towards the enhancement of the responsibility of the individual and the environmental protection (Rex and Baumann, 2007). Buyer behavior towards buying green products is impacted by the desire for knowledge regarding the product attributes and novelty (Tanners and Kast, 2003). Therefore, firms have to apply the green growth strategy in order to sustain and compete with the rivals.

For future study, the future study is to propose to conduct an empirical analysis to confirm green growth strategy features as above.

References

[1] Amran, A., &Haniffa, R., (2010). Evidence in Development of Sustainability Reporting: a Case of a Developing Country.
[2] Anuar (2012). Sustainability Practices and Corporate Performance: A Case of Malaysia Companies.
[3] Eleventh Malaysia Plan (2015). Pursuing Green Growth for Sustainability and Resilience.
[3] Fayers, C. (1999). Environment and Investment: The Role of Personal Investment Choice in Creating Sustainability.*Sustainable Development*, **7**, 64-76
[4] Hammer, S. et al. (2011). Cities and Green Growth: a Conceptual Framework. *OECD Regional Development Working Papers*, OECD Publishing.
[5] Gurel, O. (2015). Determinants of Green Supplier Selection. *Social and Behavioral Science*, **181**, 131-139
[6] Jamet, S. and J. Coffee-Morlot (2009). Assessing the Impacts of Climate Change: Literature Review.*OECD Economics Department Working Papers*, No. 691, OECD, Paris..
[7] OECD (Organisation for Economic Co-operation and Development) (2012). Green Growth and Developing Countries: A summary for Policy Makers.
[8] OECD (2013).Innovation-driven Growth in Regions: The Role of Smart Specialisation
[9] OECD (2015). Green Growth.
[10] Rex, E., Baumann, H., 2007. Beyond Ecolabels: What Green Marketing Can Learn From Conventional Marketing. J. Clean. Prod. **15** (6), 567e576.
[11] Ritika, T. (2013). Porter's Generic Competitive Strategies. *Journal of Business and Management,* **15**, 11-17.
[12] Tanner, C., Kast, S.W., 2003. Promoting Sustainable Consumption: Determinants of Green Purchases by Swiss Consumers. Psychol. Mark. 20 (10), 883e902.
[13] Thompson, Strickland and Gamble, (2011). Crafting and Executing Strategy (17th Edition). McGraw-Hill Higher Education..
[14] Tsay, Y.Y., (2010). The impact of Economic Crisis on Green Consumption in Taiwan
[15] World Water Council (2015). Water and Green Growth: Beyond the Theory for Sustainable Future

IOP Conf. Series: Earth and Environmental Science **32** (2016) 012056 doi:10.1088/1755-1315/32/1/012056

Institutionalising Human Governance Determinant: Steering Organizations towards Sustainability

Zulkefli Muhamad Hanapiyah, Salina Daud, Wan Mohammad Taufik Wan Abdullah
College of Business and Accounting, University Tenaga Nasional, Muadzam Shah, 26700, Pahang, Malaysia

E-mail: jacky.mucklow@iop.org

Abstract. This paper discusses past researches on human governance elements. Eight elements of human governance are proposed in this paper: leadership, integrity, religiosity, spirituality, culture, recruitment, training and internal control policy. Empirical study shall be conducted in the future study to confirm the eight elements of human governance proposed in this paper. It is expected that these elements may enhance the human governance practice in the organizations.

1. Introduction

Human plays an important role in an organisation at every level of management. This is because most of organisations' activities such as documentation, organizing and managing, are conducted by the employees. These activities will be influenced by conflicts whether conflict in their workplace or conflict in family (Greenhaus & Beutell, 1985, p. 77). Human governance is about internal, inside-out and value-based conviction to guide human to behave. It focuses on the axiology covering the traits of value, religion, belief system, culture and ethics in order to encourage the trust culture, which produces high ethical values and moral conduct (Salleh & Ahmad, 2010). According to the World Bank (2007), human governance is defined as the manner in which public officials and institutions acquire and exercise the authority to shape public policy and provide public goods and services.

2. Literature review

Leadership

Leadership plays an important role in promoting and managing organisation development by influencing the subordinates both directly or indirectly. According to Pierce and Newstrom (2011), they are individuals who are capable of taking ambiguous situations, interpreting these situations, and framing for the followers an understanding of the situation and what needs to be done to move forward.

Integrity

According to Transparency International (TI) (2000), it is acknowledged that political corruption can be remedied when citizens demand integrity and accountability from their leaders. It showed that, the level of integrity of employees is lower at work place.

Religiosity

The term religion refers to faith in a divinely created order of the world. It is an agreement with which is the means of salvation for a community and thus, each individual has a role in that community. This notion is principally applied by Judaism, Christianity and Islam (Samanta, 2011).

Spirituality

Meanwhile, Falk (2001) and Mead (2006) stated some notable pieces in the literature on the connection between religion and governance and religion and corruption (Paldam, 2001; Woodberry & Shah, 2004), which addressed the overlap portions of spirituality and religion. However, there is a large and significant portion of spirituality that lies outside the bounds of religion.

Culture

Additionally, economists narrowed down the definition of culture which is, to share values and beliefs governing interaction among individuals (Huntington, 2000; Greif, 2004; Fernandez, 2008; Barr & Serra, 2010). In all definitions, culture is "shared" among group members, and indeed, the group may be defined by these shared values. This is another angle to represent human governance framework.

Recruitment

According to Taylor (2008) and Rees and French (2010), recruitment is the process whereby an organisation collects applications for a position and generates a pool of potential suitable employees, while selection involves using techniques or different methods to assess the applicants and decide who is best suited to the available position, given management goals and legal requirements.

Training and Development

Apart from that, Bates and Davis (2010) emphasized on the usefulness of training programme which is possible only when the trainee is able to practise the theoretical aspects learned in the training programme in actual work environment. They highlighted the use of role playing, cases, simulation, mediated exercises, and computer based learning to provide exposure to a current and relevant body of knowledge and real world situations.

Internal Control Policy

Lastly, effective internal control system provides information to management about the entity's progress, or lack of progress toward the achievement of their objectives (Changchit et al., 2001; Jokipii, 2010; Nilniyom & Chanthinok, 2011; Vijayakumar & Nagaraja, 2012; Verdina, 2011). Thus, it is also vital to be included in the human governance proposed framework.

3. Human Governance

Human governance is a religious-based governance that guides the view of nature, thought, decision making and the behaviour of the leaders. Human governance looks at the axiology, encompassing the traits of values, religion, belief system, culture, and ethics in order to foster the culture based on trust where human within the organization was viewed as the soul of the organization (Salleh & Ahmad, 2010).

4. Gap in the literature

There are several terms and definitions that apply consistently where further undermine public administration debate (Oman & Arndt, 2013). Since 2008, the development of human governance by researcher and practitioners had focused on broadening and achieving the objective. Apart from having corporate governance, human governance is really needed which emphasizes more about the spirit of the law in order to guide the human.

5. Exploratory Study Framework

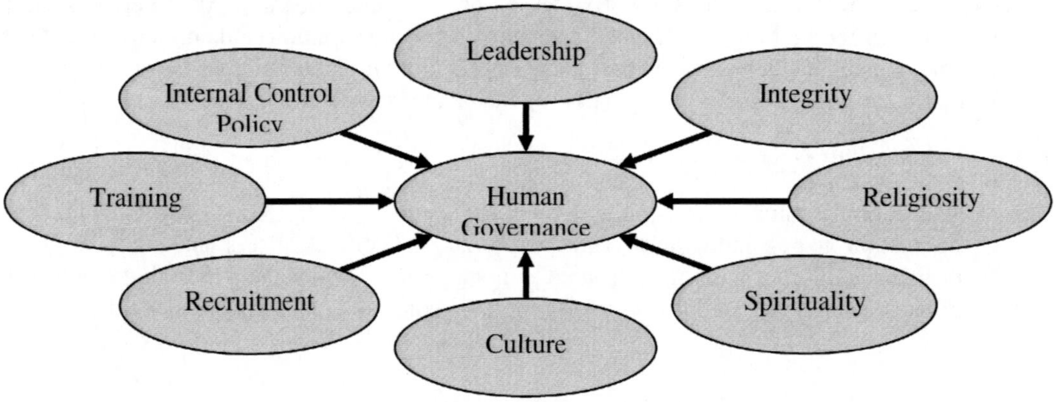

The above theoretical framework is mainly to determine the influencing determinants that represent human governance. This is supported by the rotten apple theory that justifies that corruption depends on leadership, individual characteristic, ethical and moral failure, and also the culture of organization (Chandrasekaran & Raza, 2015).

6. Conclusion

The proposed framework intends to measure each of the elements on how they influence human governance. Elements that have a relationship with human governance will be used in another study in order to combat corruption.

7. Acknowledgement

This research wishes to acknowledge the Ministry of Education (MOE) for the fund granted through the Fundamental Research Grant Scheme (FRGS). Further, the researcher also requests to acknowledge The Malaysian Anti-Corruption Commission (MACC) which agreed to cooperate by participating in the focus group discussion and survey which will be conducted as part of the data collection activity.

References

[1] Barr, A., Serra. D., 2010. "*Corruption and culture: An experimental analysis*." Journal of Public Economics **94**, 862-869
[2] Bates, Donald L. and Davis, Tammy J. (2010), "*The Application Bridge: A Model for*

Improving Trainee Engagement in the Training Process", International Journal of Management, Vol. **27**, No. 3, Part 2, December, pp. 770-776.

[3] Chandrasekaran, V. and Raza, M.A., 2015. *Micro Analysis of Police Corruption In India*: Global Journal for Research Analysis, **4(2)**: 96-97.

[4] Changchit, C., Holsapple, C. W. & Madden, D. L. (2001). *Supporting managers' internal control evaluations: An expert system and experimental results*. Decision Support Systems, **30**, 437–449.

[5] Corruption Perceptions Index 2010. *Transparency International.*

[6] Falk, Richard, "On Human Governance. Toward a New Global Politics", University Park PA, 1995, p. 14

[7] Fernandez, R. 2008. "Culture and economics" in Steven N. Durlauf, Lawrence E. Blume (Eds.), The New Palgrave Dictionary of Economics (Second Edition), Palgrave Macmillan

[8] Glance, A. 2006. *Effectiveness of internal control. Local Government*: Results of the 2006- 07 Audits, 19 – 26.

[9] Greenhaus, J.H., & Beutell, N.J. 1985. *Sources of conflict between work and family roles. Academy of Management Review, **10**,* 76-88.

[10] Greif, A. 1994. "*Cultural beliefs and the organization of society: a historical and theoretical reflection on collectivist and individualist societies*." The Journal of Political Economy, **102**, pp. 912–950.

[11] Huntington, S.P. 2000. "*Cultures count*." In L.E. Harrison, S.P. Huntington (Eds.), *Culture Matters: How Values Shape Human Societies*, Basic Books, NY.

[12] Jokipii, A. 2010. *Determinants and consequences of internal control in firms: a contingency theory based analysis.* J Manag Gov, **14**, 115–144.

[13] Mead, W. R. 2006. "God's Country?", Foreign Affairs, vol. 85, issue 5, pp. 24-43.

[14] Nilniyom, P. & Chanthinok, K. 2011. *Accounting system innovation and stakeholder acceptance of Thai listed firms: Mediating internal control effectiveness*. Review of Business Research, **11**, 26 – 37

[15] Oman, C. P., & Arndt, C. 2013. *Measuring Governance*, **(39)**.

[16] Paldam, M. 2001. "*Corruption and Religion Adding to the Economic Model*", Kyklos, vol. **54**, pp. 383-414.

[17] Pierce, Jon & Newstrom John (2011); *Leaders and Leadership Process; Reading, Self-Assessments and Applications*, (6th edition): McGraw-Hill Education, New York

[18] Pierce, Jon & Newstrom John (2011); *Leaders and Leadership Process; Reading, Self-Assessments and Applications*, (6th edition): McGraw-Hill Education, New York

[19] Rees, G. & French, R., 2010. *Leading, Managing and Developing People*. 3rd ed. s.l.:CIPD.

[20] Salleh, A. and A. Ahmad, 2010. *Human Governance: Bringing the meaning of integrity in the life of professional accountants.*

[21] Samanta, Subarna K. 2011. "*Corruption, Religion and Economic Performance* in OPEC

[22] Taylor, S., 2008. *People Resourcing* 4th Edition. London: The Chartered Institute of Personnel and Development.

[23] Verdina, G. 2011. *Risk management as a tool for securing internal control in the process of study program implementation at higher education institutions*. Economics and management, **16**, 987- 991.

[24] Vijayakumar, A. N. & Nagaraja, N. 2012. *Internal control systems: Effectiveness of internal audit in risk management at public sector enterprises*. BVIMR Management Edge, **5(1),** 1-8.

[25] Woodberry, R. D. & T. S. Shah 2004, "*The Pioneering Protestants*", Journal of Democracy, Vol. **15**, issue 2, pp. 47-62.

[25] World Bank 2007, p. 38.

GSM Web-Based Centralized Remote Wireless Automatic Controlling and Monitoring of Aquafeeder

C L Wong [1], **A Idris**[1] **and Z Hasan**[1]

[1] College of Engineering, Universiti Tenaga Nasional, Selangor, Malaysia.

Abstract. This project is about producing a prototype to feed fishes at fish ponds of remote location with the use of GSM mobile phone. An automatic fish feeder is an electric device that has been designed to give out the right amount of pellets at the designed time. In this project, the automatic feeder designed consists of photovoltaic solar cells that are used to generate electricity and storing it into batteries. Solar charge controllers can be used to determine the rate of which current is drawn and added from the batteries. GSM cellular communication is used to allow user to control from a distance. Commands or instructions are sent to the operating system which in return runs the servomotor and blower by blowing certain amount of fish pallets into the pond to feed the fishes. The duration of the feeding processes is fixed by the user, hence the amount of fish food pallets released are precisely the same for each time. This technology is especially useful for fish farmers where they can remotely feed their fishes.

1. Introduction

1.1 Problem Statement

Fish farmers run daily fish feeding operation manually with paid help. This is particularly difficult for larger farm with various types of fish and ages. An independent system that runs automatically to replace the conventional feeding method by sending instruction via Web server and GSM mobile phone. The server will send an SMS to the remote automatic feeder equipped with GSM receiver, and this will trigger the servo motor to operate and release the food.

1.2 Project Objectives

The objectives are as such:

- To design a GSM web-based centralized remote wireless automatic controlling and monitoring of aquafeeder.
- To build a prototype of the aquafeeder.

2. Literature Review

2.1 Photovoltaic System

A photovoltaic system, also known as a PV system is a display of components designed to use the Sun as the power source to supply usable electric power for a wide range of usage [11]. A solar array or photovoltaic array are made out of solar panel which consist of several photovoltaic modules, to generate usable direct current (DC) electricity from converting solar radiation (sunlight). An inverter is usually installed into the

PV system to produce alternating current (AC) by converting the DC power generated [12]. The alternating current then can be used to power loads such as motors and lights. In PV array the modules are connected in series to obtain the desired voltage then the individual strings are joined parallel so that more current can be produced by the system.

2.2 Current aquafeeder

The device comes in three main sections, namely: mobile phone, controller and action unit.

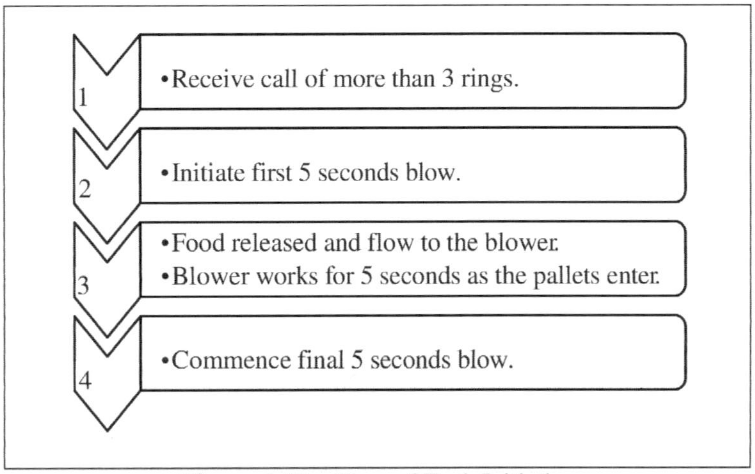

Figure 1. The Flow of Each Initiation

By using Global System Mobile (GSM), the feeding operation can be done with just a call away. Since mobile telephone is readily available, the user does not need to carry heavy and unnecessary device. The use of this technology enables remote activation [12]. When feeding time, the user can control and switch the device ON or OFF.

Figure 2. Wavecom GSM Modem

Figure 3. Maxim Max 232 for RS-232 to TTL converter

3. Methodology

3.1 Study the Present Prototype

The circuit consists of:

- Arduino microcontroller.

- Power Management Unit (PMU).
- Auger and Blower controller module.
- TTL to RS-232 converter and GSM Modem.

3.2 Familiarization with Arduino

Arduino is a single board microcontroller. This controller enables the application of interactive devices; especially devices that interact with environment by sensors and actuators. The hardware consists of an open-source hardware board designed around an 8-bit ATMEL AVR microcontroller, or a 32-bit Atmel ARM. Current models feature a USB interface, 6 analog input pins, as well as 14 digital I/O pins which allow the user to attach various extension boards. There are plenty of models in the market to choose from based on their suitability to our objective.

4. Result and Analysis

4.1 Call and SMS

These methods of initiation are as easy as calling your contacts. You only need to dial or send an SMS to the number used on the Arduino. A feedback will be sent to the user's number upon each successful feeding.

Figure 6. Methods of Initiation (Call and SMS)

Figure 7. Website page initiation.

4.2 Website

The steps are as below:

1. Visit website: **coe3.uniten.edu.my/coe3members/** from any device with internet access.
2. Login by entering username and provided.
3. Search for Fish feeder at the list on the left section of the page.
4. Choose which pond and click on the INITIATE FEEDING button to start the feeding process.
5. Receive an SMS indicating that the feeding is complete.

5. Conclusion

The device comes in three main sections, namely: mobile phone, controller and action unit. The system is set to STANDBY mode and can be triggered by a missed call and received SMS. Any number can be used to make this call. Once the system receive more than three ringing, the blower will start first and work for about 5 seconds. This is to clear what ever there might be in the channel. After the initial blow, fish pallets will be released and blow out for about 10 seconds. Another 5 seconds blow is allowed to clear the pipe hole. The completed prototype consists of the frame, Arduino, mechanical components and electronic components. All these are interconnected but electrical wiring. The Arduino act as the 'brain' of the system and will trigger the system of the aquafeeder. Currently, the prototype is located at Electrical Engineering Lab, COE (BH, UNITEN).

6. References

[1] DOF (Electronic Materials)
[2] Davidson J, Good C, Barrows F T, Welsh C, Kenney P B and Summerfelt S T 2013 *Aquaculture Engineering* vol **52** pp 45-57
[3] Smeltzer D C 2014
[4] Halford A C 2014
[5] Zimmerman D 2014
[6] Marsden H *Electronic Systems News* vol **1985** no 1 pp 7-8.
[7] Zhang N, Wang M and Wang N 2006 *Computers and Electronics in Agriculture* vol **50** pp 1-14
[8] Zhao C J, Zhang H H, Fan H P and Wang Z W 2011 *Real-time Remote Monitoring and Warning System* pp 160-163
[9] Sverrisson J A F 2014 *Renewables 2014* pp 49
[10] Northern Arizona Wind & Sun (Electronic Materials)
[11] *Photovoltaic System*
[12] Mohd S S 2011 *Solar powered remote auto feeding with GSM mobile phone* Putrajaya
[13] Ellis G 2004 *Control System Design Guide* pp 303–339
[14] Hall S 2012 2012 *Branan's Rules of Thumb for Chemical Engineers* pp 118–133
[15] Elliott T V and Pierce F J 2008 *Regional and On-Farm Wireless Sensor Networks for Agricultural Systems in Eastern Washington* vol **61** pp 32-43
[16] N N and H M 2012 *Procedia Engineering* vol **30** pp 737-741
[17] Halford A *C Automatic fish feeder* (Patent, US, 6082299 A 15 April 1999)
[18] Kang F W C 2007 *Automatic Fish Feeder for Cultivation Pond* Universiti Teknikal Malaysia Melaka
[19] Pierce T F J 2008 *Regional and On-Farm Wireless Sensor Networks for Agricultural Systems in Eastern Washington* vol **61** pp 32-43
[20] Oancea D 2011 *GSM Infrastructure Used for Data Transmission* p 4
[21] Yang X M J C 2010 *Application of Extension Method to Selecting of Solar Power Systems* pp 160-163

The effect of dilute acid pre-treatment process in bioethanol production from durian (*durio zibethinus*) seeds waste

K A Ghazali[a], S F Salleh[a, 1], T M I Riayatsyah[b], H B Aditiya[c], T M I Mahlia[a]
[a]Centre of Renewable Energy,Universiti Tenaga Nasional, 43000 Kajang, Selangor, Malaysia

[b]Department of Mechanical Engineering, Faculty of Engineering, University of Malaya, 50603 Kuala Lumpur, Malaysia

[c]Department of Mechanical Engineering, The University of Melbourne, VIC 3010, Australia

Abstract. Lignocellulosic biomass is one of the promising feedstocks for bioethanol production. The process starts from pre-treatment, hydrolysis, fermentation, distillation and finally obtaining the final product, ethanol. The efficiency of enzymatic hydrolysis of cellulosic biomass depends heavily on the effectiveness of the pre-treatment step which main function is to break the lignin structure of the biomass. This work aims to investigate the effects of dilute acid pre-treatment on the enzymatic hydrolysis of durian seeds waste to glucose and the subsequent bioethanol fermentation process. The yield of glucose from dilute acid pre-treated sample using 0.6% H_2SO_4 and 5% substrate concentration shows significant value of 23.4951 g/L. Combination of dilute acid pre-treatment and enzymatic hydrolysis using 150U of enzyme able to yield 50.0944 g/L of glucose content higher compared to normal pre-treated sample of 8.1093 g/L. Dilute acid pre-treatment sample also shows stable and efficient yeast activity during fermentation process with lowest glucose content at 2.9636 g/L compared to 14.7583g/L for normal pre-treated sample. Based on the result, it can be concluded that dilute acid pre-treatment increase the yield of ethanol from bioethanol production process.

1. Introduction

Bioethanol is a renewable energy source which offers a sustainable solution to the rise in the global energy demand and depleting oil reserves. The current trend of producing bioethanol from food crops such as corn, barley, wheat and sugarcane has led to concern over food security. Therefore, lignocellulosic biomass which is defined as lignocellulosic material with cheap and abundant properties derived from plant is deemed as a better feedstock in the second generation of biofuel [1]. Durian is a seasonal fruit which belongs to *Bambaceae* family, found typically in the South East Asia. Malaysia as one of the biggest producer and exporter of durian generates large amount of agricultural wastes with an average of 90,141 metric tonnes of waste produced annually from 2008 to 2010 [2]. Currently, durian seeds are not being utilised for commercial purposes and usually thrown away as a waste product. Nevertheless, the raw seeds contain high percentages of carbohydrate (43.6%) [3] and therefore, can potentially be used as biomass feedstock for bioethanol production.

---

[1] Author of correspondence: sitifatihah.salleh@gmail.com

The bioethanol production process of lignocellulosic biomass starts from pre-treatment, hydrolysis, fermentation, distillation and finally obtaining the final product, ethanol [4]. Different pre-treatment methods are available to prepare the raw materials for enzymatic hydrolysis. One of them is the chemical pre-treatment method which mainly uses acids and bases of different strengths and under different conditions to effectively break down the hemicellulose [5]. This is so that the cellulose becomes easily accessible to enzymes during the subsequent hydrolysis process where it is converted to glucose. In this work, dilute acid was used during the pre-treatment process of durian seeds wastes since it is less corrosive than concentrated acid and less inhibitive to the yeast during the fermentation process. The effects of this pre-treatment process towards the production of sugar and bioethanol were investigated.

2. Methodology

2.1. Preparation of raw material

Durian seeds waste was obtained from local fruits stalls around Kajang, Selangor. The seeds were later washed with plain water and dried in the oven at 70°C for 3 hours to remove excess moisture from the outer layer of the seeds. Then, the outer skins of the seeds were peeled to remove the lignocellulosic material as only the inner part contain high amount of carbohydrates useful for ethanol production. The peeled seeds were then shredded by vegetable grater, followed by drying at 80°C for 4 hours and grinding until it become granulated-powdery durian seeds.

2.2. Dilute acid pre-treatment of durian seeds

Four sets of samples, each with 100 ml of H_2SO_4 of different concentrations 0.019 M, 0.056 M 0.113M and 0.167 M were prepared inside conical flasks. 5 g of physically pre-treated durian seeds, were weighed and transferred into respective conical flasks. The prepared flasks were then sealed and heated in an autoclave at 130°C for a retention time of 30 minutes and left to cool down to room temperature. 100 ml of prepared 0.038M, 0.113M, 0.225M and 0.338M sodium hydroxide solutions were added to each conical flask respectively for neutralization purpose.

2.3. Enzymatic hydrolysis of dilute acid pre-treated durian seeds

0.025g of sodium chloride was added to the acid pre-treated samples for stabilization. Then, α-amylase enzymes (Sigma-Aldrich) was added into the samples and incubated in a shaking incubator for 90 minutes.

2.4. Fermentation of reducing sugars

Nutrients, *Saccharomyces cerevisiae* yeast extract, ammonium chloride (NH_4Cl) and potassium dihydrogen phosphate (KH_2PO_4) with a ratio of 1:0.2:0.4 gram/100mL solution were added to the acid pre-treated and enzyme hydrolysed sample. The samples were then autoclaved at 121 °C for 15 minutes. After autoclaving, *Saccharomyces cerevisiae* was introduced into all the samples, except the one which acted as the control for the experiment. Then, the samples were incubated using an incubator shaker for up to 48 hours at a setting of 160 rpm and 37 °C.

2.5. Analysis of sugar content using DNS method

1 ml solution was collected from each samples into test tubes containing 1 ml of 0.1% DNS reagent [6]. A control solution was prepared by mixing the DNS reagent with 1 ml of RO water. The test tubes were heated in a water bath at 90 °C for 5 minutes. The corresponding absorbance values (A_{540}) were measured using a spectrophotometer.

IOP Conf. Series: Earth and Environmental Science **32** (2016) 012058 doi:10.1088/1755-1315/32/1/012058

2.6. Analysis of ethanol content by High Performance Liquid Chromatography (HPLC)

20 μL samples were filtered by using 0.50 μm fibre meshes and later injected into the HPLC. The running time of HPLC was set at 20 minutes. The values resulting from the HPLC were shown in the peak area unit and they were converted into the percentage of ethanol concentration through a standard curve made from several known ethanol concentrations.

3. Results and discussion

3.1. Glucose yield from acid pre-treatment on durian seeds

The total reduced sugar yield after 90 min from different concentration of acid was plotted in Figure 1. The highest sugar yield was at 0.113 M which was 23.5 g/L. Further increase in acid concentration gave no meaningful increase in sugar yield.

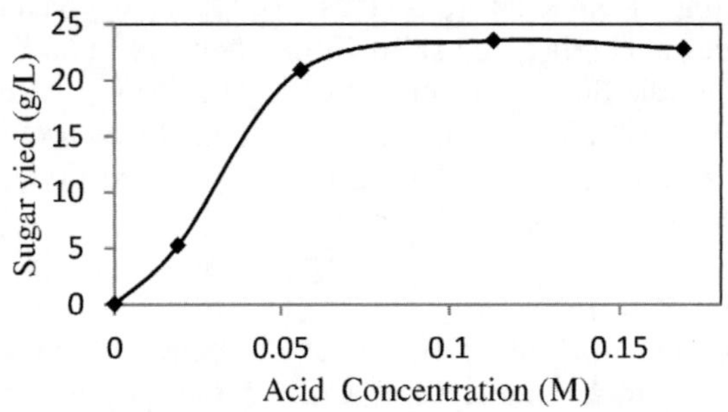

Figure 1. Effect of different acid concentrations on sugar yield

3.2. Glucose yield from enzymatic hydrolysis on durian seeds

The sample containing durian seeds that were pre-treated with 0.113 M H_2SO_4 was carried forward for the enzymatic hydrolysis process. The hydrolysis was also repeated with durian seeds without acid pre-treatment (only mechanical pre-treatment) as a control. Figure 2 shows the results obtained by each sample. Durian seeds without acid pre-treatment showed almost negligible sugar yield of 5.56 g/L at 90 min. On the other hand, additional acid pre-treatment enhances the yield from enzymatic hydrolysis significantly to 50.1 g/L at 90 min.

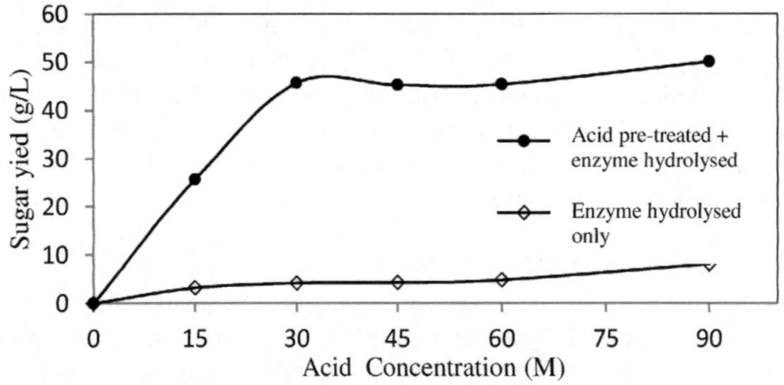

Figure 2. Effect of additional acid pre-treatment on sugar yield from enzymatic hydrolysis

3.3. Analysis of ethanol produced

Figure 3 compares the fermentation profiles of the enzyme hydrolysed with additional acid pre-treatment sample and enzyme hydrolysed sample. The reduction in the glucose content signifies the amount of glucose being fermented by yeast to produce ethanol. The fermentation activity of the yeast is significantly lower than the acid pre-treated sample, from initial measure of 23.1 g/L to 13.9 g/L glucose after 30 hours of incubation. On the other hand, the yeast in acid pre-treated with an enzymatic hydrolysis sample takes only 24.2 hours to consume 34.5 g/L of glucose. From the HPLC method, the enzyme hydrolysed with additional acid pre-treatment sample produced 0.47% of ethanol.

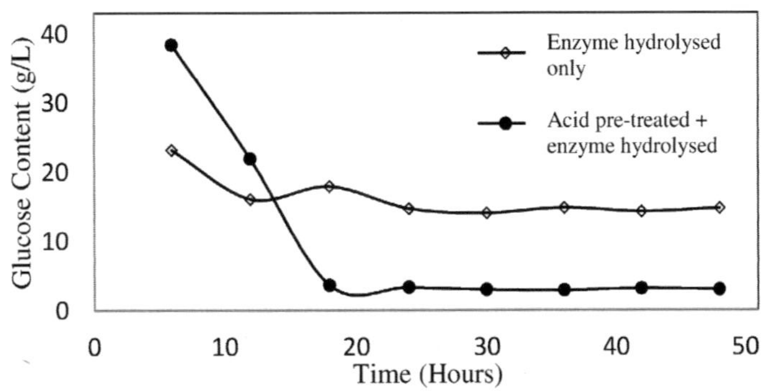

Figure 3. Fermentation profiles of the enzyme hydrolysed with additional acid pre-treatment sample and enzyme hydrolysed sample.

4. Conclusion

The pre-treatment of durian seeds with dilute H_2SO_4 (0.113 M) has significantly enhanced the yield of sugar obtained in enzymatic hydrolysis (50.1 g/L) and ethanol in the fermentation process (0.47%).

5. Acknowledgement

Authors would like to acknowledge this research as supported by Exploratory Research Grant Scheme (ERGS) through the Ministry of Higher Education of Malaysia (MOHE) with reference number ERGS/1/2013/TK07/UNITEN/01/01.

6. References

[1] Gomez L D 2008 Sustainable liquid biofuels from biomass: the writing's on the walls *New Phytol*, pp 473-485
[2] Omar A A M F Z Zakaria Z Long H D Abdullah A G 2011 Durian skin machine *6th MARDI Science & Technology Exhibition*, p 75
[3] Brown M J 1997 *Durio* - a bibliographic review (Canada: International Plant Genetic Resources Institute)
[4] Mussatto S I Dragone G Guimarães P M R Silva J P A Carneiro L M Roberto I C Vicente A Domingues L Teixeira J A 2010 Technological trends, global market, and challenges of bio-ethanol production. *Biotechnol Adv* , pp 817–830
[5] Zheng Y Pan Z Zhang R 2009 Overview of biomass pretreatment for cellulosic ethanol production *Int J Agric & Biol Eng*, **2** (3) p 51
[6] Miller G L 1959 Use of dinitrosalicylic acid reagent for determination of reducing sugar *Anal Chem*, pp 426-428

Comparison of an Electromagnetic Energy Harvester Performance using Wound Coil Wire and PCB Coil

MSM Resali, H Salleh

Centre for Renewable Energy, Energy Harvester Research Group, Universiti Tenaga Nasional, UNITEN, Putrajaya Campus, Jalan IKRAM-UNITEN, 43000, Kajang, Selangor, Malaysia.

Mohdsofwan@gmail.com, hanim@uniten.edu.my

Abstract. This paper presents the performance of two types of electromagnetic energy harvester, one using manually wound coil wire (EH-EC) and the other one using printed circuit board (PCB) coil (EH-EP). The objective of the study is to measure the corresponding output voltage and power by varying the number of coils and the position of the magnet. The experiment was conducted at a fix 50 Hz of frequency and at 0.25g of acceleration. The EH-EP was found to be more effective than the 350 turns of the wound coil wire, with maximum power of 26 μW. Overall, the performance of the EH-EC showed better result with maximum power of 125 μW for 1050 turns when compared to the EH-EP.

1. Introduction

Recently, due to small electronics devices, there is a high demand of low power applications using energy harvesters. [1]. Energy harvesting can be useful in many areas of application such as biomedical, building monitoring, health and remote sensing. The harvested energy can be generated from many sources such as from vibration, light, heat [2]. There has been many research focusing on vibration based energy harvester since it is a promising technology to produce high energy output when the source of vibration is available [3-5].

There are three types of harvester that researcher work to harvest energy from the vibration to convert vibration energy to the electrical energy which are electrostatic, electromagnetic and piezoelectric energy harvester. Electromagnetic energy harvester is one of the efficient harvester to convert the mechanical energy to electrical energy [6, 7]. For this study, two types of electromagnetic energy harvester to harvest energy from the vibration and will be described in the next following section below.

2. Mathematical model

The basic equation of the electromagnetic energy harvester based on the Faraday's law of electromagnetic induction as per equation (1):

$$V_{emf} = -\frac{d\phi}{dt} \qquad (1)$$

Where V_{emf} is the voltage induced in the conductor proportional to the time rate of change of the magnetic flux ϕ of the circuit. For the practical of the electromagnetic energy harvester with the fix and permanent magnet, the V_{emf} generated of the coil with N turns and the average flux ϕ, the equation is expressed:

$$V_{emf} = -N\frac{d\phi}{dt} \tag{2}$$

For vibration based electromagnetic harvester, the relation of the motion between coil and permanent magnet in the direction of x-direction , the voltage V_{emf} can be expressed as a product of a flux linkage gradient and the velocity of movement. The equation can be expressed in equation:

$$V_{emf} = -N\frac{d\phi}{dx}\frac{dx}{dt} \tag{3}$$

To measure the power generated by the electromagnetic energy harvester is by connecting load resistance R_L to the harvester. In the frequency domain, the power generated by the harvester can express as in given by [8]:

$$Q(\omega) = \frac{K_{em}m\omega^2 Y_0}{\left((R_i + R_L) + jL\omega\right)\left(-m\omega^2 + j\omega C + K\right)} \tag{4}$$

The open circuit voltage V_{OC} of the power output P_{opt} of the electromagnetic generator as can be expressed as follow:

$$V_{OC} = \frac{K_{em}m\omega^3 Y_0}{\sqrt{\left(\omega_0^2 - \omega^2\right) + \left(2\zeta_m\omega_0\omega\right)^2}} \tag{5}$$

$$P_{opt} = \frac{R_L\left(K_{em}m\omega^3 Y_0\right)^2}{2\left((R_L + R_i)^2 + (L\omega)^2\right)\left(\left(\omega_0^2 - \omega^2\right) + \left(2(\zeta_e + \zeta_m)\omega_0\omega\right)^2\right)} \tag{6}$$

3. Design and experiment

There are 2 types of electromagnetic energy harvester in this paper. The first one is labelled as EH-EC as shown in figure 1, where the electromagnetic energy harvester is generated by using coil wire at the free end of cantilever beam and NdFeB magnet is located near to the cantilever beam. For EH-EC, the magnet is fixed while the cantilever beam is vibrating. The second type , labelled as EH-EP as shown in figure 2, where electromagnetic energy harvesters is generated by using coils from printed circuit board (PCB). For EH-EP, the NdFeB magnet is attached at the free end of the cantilever beam and the coil from PCB is fixed.

The effect of number of coil turns on the performance of the EH-EC energy harvester was evaluated. The performance indicator was the output voltage and power. Three different number of coil turns were proposed, 350 turns (A), 700 turns (B) and 1050 turns (C). Meanwhile, for the EH-EP, the performance of the energy harvester was compared to the location of the NdFeB magnet at the free end of the cantilever.

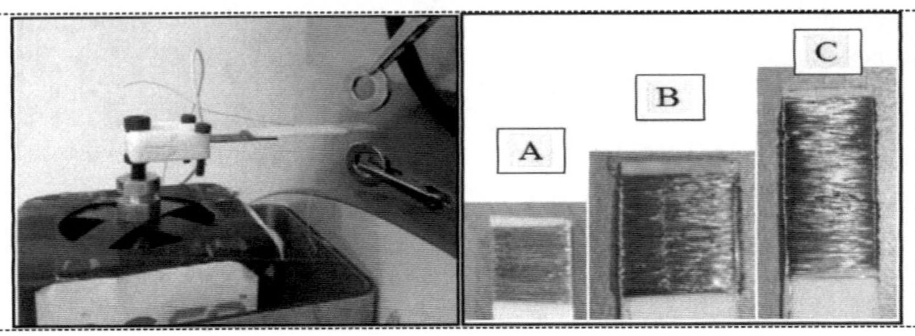

Figure 1. EH-EC electromagnetic energy harvester.

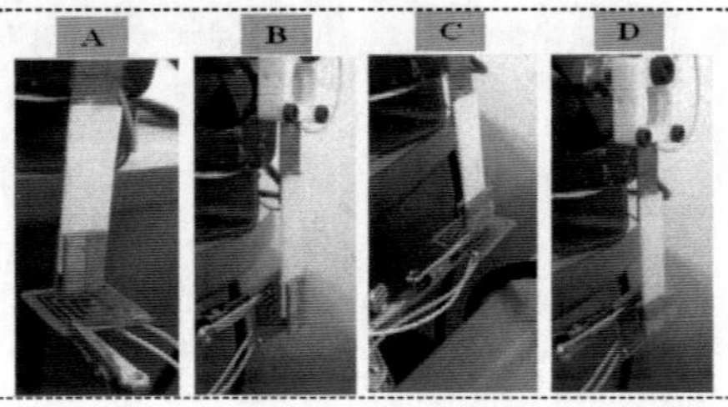

Figure 2. EH-EP electromagnetic energy harvester.

The experimental setup for the electromagnetic energy harvester for EH-EC and EH-EP is shown in figure 3. The objective of the experiment is to measure the output voltage and power generated of the EH-EC and EH-EP at the frequency 50 Hz and 0.25g of acceleration. The function generator was set to produce the frequency at 50 Hz and connected to the amplifier before start to vibrate the shaker. The computer was installed with the Labview software and equipped with the vibration analysis to monitor the current value of the frequency and acceleration of the vibration of the shaker. The data of vibration is sent by DAQ card and connected to the laser vibrometer and pointed on the electromagnetic energy harvester EH-EC and EH-EP. The value of the open circuit voltage was measured by connecting the output terminal to the digital multimeter and for the power output; the magnitude voltage was measured by using digital multimeter at the load resistance.

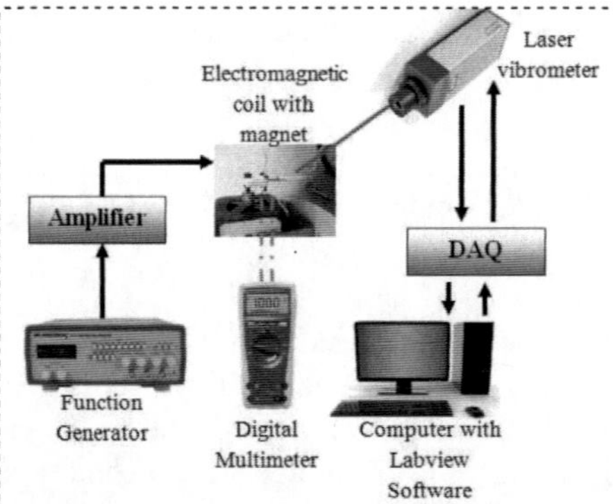

Figure 3. Schematic diagram for the experimental setup

4. Result and discussion

4.1. EH-EC energy harvester

The summary of the result for EH-EC is shown in Table 1. In this experiment, the load resistance is fixed at 50 KΩ, 70 KΩ and 100 KΩ. The open circuit voltage and power output is increased when the number of turns for coil is increased. The EH-EC for the 1030 turns showed the highest value of the open circuit voltage and power output. When the number of turns is increased, the weight at the free end of the cantilever beam increased and as a result, it increased the value of velocity of movement.

Table 1. The measure output of the EH-EC electromagnetic energy harvester

	350 turns (A)	700 turns (B)	1050 turns (C)
Frequency (Hz)	50	50	50
R_L (KΩ)	50	70	100
V_{OC} (V)	1.5	3.1	5
P_{opt} (μW)	22.5	68.6	125

4.2. EH-EP energy harvester

Table 2 shows the summary of the measured value for EH-EP electromagnetic energy harvester. The open circuit voltage and output power was measured based on the 4 different position of the magnet at the free end of cantilever beam as shown in figure 2. The EH-EP is fixed but the magnet is moving with the same position of the motion of the cantilever beam. The different of the position will produce for the different value of the flux magnet generated and give different result of the value for open circuit voltage and power generated by EH-EP electromagnetic energy harvester.

Table 2. The measure output of the EH-EP electromagnetic energy harvester at different position

	A	B	C	D
Frequency (Hz)	50	50	50	50
R_L (KΩ)	50	50	50	50
V_{OC} (V)	1.5	5.1	4.95	0.8
P_{opt} (μW)	22	26	24.5	6.4

5. Conclusion

The effect of using two types of coil for an electromagnetic energy harvester were presented in this paper. The first type was manually wound coil wire (EH-EC) and the second type was PCB board coil (EH-EP). The EH-EC showed increasing number of turns for coil affected the output power and produce high velocity of motion. The highest value of power was 125 μW for 1050 turns. In the EH-EP arrangement, the position of the magnet at the free end of the cantilever beam affected the output power. More power generated when more area of magnet exposed to the surface of EH-EP. The EH-EP was found to be more effective than the 350 turns of the wound coil wire, with 26 μW at position B. However, overall the EH-EC showed better result compared to the EH-EP.

6. References

[1] Paradiso, J.A., Starner, T. 2005 Energy scavenging for mobile and wireless electronics *IEEE Pervasive Comput.* pp. 18-27.
[2] Carmo, J.P., Concalves, L.M., Correcia, J.H,. 2010 Thermoelectric microconverter for energy harvesting systems *IEEE Trans. Ind. Electron.*, pp 861-867.
[3] Mohammed Dhia Salim, Hanim Salleh, Dhia Shaker Salim, 2013. A low frequency tunable hybrid generator, *Microsystem Technologies, 19 (11), 1839-1844.*
[4] *Kherbeet, A.S. Salleh, H. Salman, B.H. Salim, M.* .2015. Vibration-based piezoelectric micropower generator for power plant wireless monitoring application. *Sustainable Energy Technologies and Assessments. Volume 11, Pages 42-52.*
[5] Md Ralib, A.A. Nordin, A.N. Salleh, H. Othman, R. 2012. Fabrication of aluminium doped zinc oxide piezoelectric thin film on a silicon substrate for piezoelectric MEMS energy harvesters. *Microsystem Technologies. Volume 18, Issue 11, pp 1761-1769.*
[6] Seong, T.O.K., Salleh, H., Nurashikin, A. 2014. Optimization of resonator design for vibration-based electromagnetic energy harvester. *Applied Mechanics and Materials* .471. Noise Vibration and Comfort. p355. 2014.ISBN-13: 978-3-03785-930-8.
[7] Wang, P.-H., Dai, X.-H., Zhao, X.-L., Niu, L.-J. 2009 Electromagnetic self powered low-level vibration energy scavenger with microelectroplated nickel resonator *IET Electron. Lett.*, pp. 832–833.
[8] E. Bouendeu, A. Greiner, P. J. Smith, and J. G. Korvink, Design synthesis of electromagnetic vibration-driven energy generators using a variational formulation, *J. Microelectromech. Syst.*,.

IOP Conf. Series: Earth and Environmental Science **32** (2016) 012060 doi:10.1088/1755-1315/32/1/012060

Effect of Synthesis Methods Using Renewable PODFA on Structural Characteristics of Metal-Organic Framework (MOF-5)

S Z Othman[1], H Misran, W A Affendy and N F Mahadi
Nanoarchitechtonic Laboratory, College of Engineering, Universiti Tenaga Nasional, Jalan IKRAM-UNITEN, 43000 Kajang, Selangor, Malaysia

E-mail: zubaidah@uniten.edu.my

Abstract. In this work, metal-organic framework (MOF-5) syntheses were successfully synthesized with three different methods namely i) modified direct-mixing method ii) microwave assisted method and iii) autoclave assisted method without using surfactants. Palm oil derived fatty alcohol (PODFA), dodecyl alcohol (C8) was used as renewable non-surfactant template. Four prominent peaks were observed in all three samples attributable to the reflection planes of (200), (220), (400) and (420) of a cubic structure at ca. 2θ =6.9°, 9.6°, 13.7° and 15.2°. The morphology of MOF-5 produced using modified direct-mixing method was faceted structure with average size of ca. 1 μm. In addition, the morphology of MOF-5 produced using microwave assisted synthesis was cubic structure with average size of ca. 8.34 μm. However, the morphology of MOF-5 using autoclave assisted method was interpenetrated cubic with average size of ca. 22.65 μm.

1. Introduction

Metal-organic frameworks (MOF) are a class of microporous organic-inorganic hybrid materials made of metal clusters interconnected with organic linkers with a geometrically well-defined structure [1]. The organic linkers were typically a single, di-benzene or tri-benzene rings. MOF materials has unique structure with permanent but flexible porosity. It has extremely high surface area and large pore volume compared to other mesoporous materials such as mesoporous silica [1, 2]. MOF materials are gaining extensive interest in recent years due to these advanced characteristics that were making them suitable materials in gas-storage, gas separation, gas sensing and as catalysts [2]. In particular, the research of MOF-5 materials are gaining increasing interest because it was relatively easy to be produced among all other MOFs and has simple cubic structure. This is because of its organic linker that consists of only a single benzene ring [1]. The compound serves as a prototype for an extensive family of metal-organic frameworks in which oxocentered Zn_4 tetrahedra (similar to the type encountered in basic zinc acetate) that were connected through linear organo-dicarboxylates to give a cubic network structure. As the synthesis were carried out in a solution assembly reaction, the pores within the structure are filled with guest solvent molecules. The studies on various and efficient synthesis methods of MOF materials suggested to be the major role in advancing MOF materials by improving their structural properties, surface area, and single crystals size as well as overall time duration in the synthesis process. The synthesis method of MOFs includes direct-mixing method [3],

diffusion method, liquid-crystal templating (using surfactant) method [4], ultrasonic-assisted method and microwave-assisted method [5].
Therefore, this paper attempts on various synthesis approach using renewable nonsurfactant template of fatty alcohol derived from biomass, namely palm oil with C12 carbon chain (dodecyl alcohol). The methods include modified direct-mixing method, microwave-assisted and autoclave-assisted method. Then, the effect various synthesis methods on the structural characteristics of MOF-5 were investigated and studied.

2. Materials and Method

2.1 Materials
Zinc nitrate (98% $Zn(NO_3)_2.6H_2O$, Sigma-Aldrich), 1,4-benzenedicarboxylic acid (98% H_2BDC, Merck), N,N dimethylformamide (99.8% DMF, Merck), triethylamine (99% TEA, Merck) were used. Palm oil derived fatty alcohol (PODFA) dodecyl alcohol ($C_{12}H_{26}O$) was purchased from Emery Oleochemicals. All materials were used in the as-received condition.

2.2. Synthesis Approach
i) Direct-Mixing Method
$Zn(NO_3)_2.6H_2O$ and H_2BDC were dissolved in DMF under mild stirring condition for 10 minutes. Next, the dodecyl alcohol and TEA were added into the solution. Then, the solution was stirred for four hours. After that, the solution was centrifuged and washed twice with DMF. The sample was aged for two days, prior to drying. The resultant powder was dried overnight and denoted as DMOF-5.[6]
ii) Microwave-Assisted Synthesis
$Zn(NO_3)_2.6H_2O$ and H_2BDC were dissolved in DMF under mild stirring condition for 10 minutes. Next, the dodecyl alcohol was added into the solution. The solution was stirrer for another 10 minutes. Then, the solution was poured into microwave digestive bomb and heated in microwave (Sharp, Model R290HS) for 15 minutes. After that, the digestive bomb was cooled in room temperature. Then, the solution were centrifuged and washed twice with DMF. The sample was aged for two days, prior to drying. The resultant powder was dried overnight and denoted as MMOF-5.
iii) Autoclave-Assisted Synthesis
$Zn(NO_3)_2.6H_2O$ and H_2BDC were dissolved in DMF under mild stirring condition for 10 minutes. Next, the dodecyl alcohol and TEA were added into the solution. The solution was poured into Teflon autoclave. The autoclave was heated at 100°C for 24 hours. Then, the solution was heated in oven at 100°C in oven for 24 hours. After that, the autoclave was cooled in room temperature. Later, the solution was centrifuged and washed twice with DMF. The sample was aged for two days, prior to drying. The resultant powder was dried overnight and denoted as AMOF-5.

2.3. Characterizations
The X-ray diffraction analyses were conducted using Shimadzu XRD-6000 with CuK_α radiation at $2\theta = 5°- 40°$ with 30 kV and 20 mA. The scan speed was 2°/min with scan step of 0.02°. SEM images were obtained with LEO 1455 VP SEM with 20 kV. Nitrogen adsorption were done on a Quantachrome Autosorb iQ at 77 K. Samples were degassed at 373 K overnight prior to adsorption.

3. Results and Discussion
Figure 1 shows the X-ray diffraction patterns of MOF-5 synthesized with three different methods. All samples exhibited four prominent peaks attributable to the reflection planes of (200), (220), (400) and (420) at 6.9°, 9.6°, 13.7° and 15.2° of a cubic crystal structure. These peaks were also observed in MOF-5 prepared by previous groups [1, 7]. These results suggested the successful formation of MOF-5 using PODFA with 12 carbon chains. Moreover, higher relative intensity of (220) reflection peaks compared to (200) reflection peaks suggested the pore filling effect given rise from the effect of DMF

molecules inside the pore framework during washing by solvent-lattice interactions [7,8]. This interactions was weaker than those observed in zeolitic materials [7, 8].

The results suggested that oven-drying in 24 hours at 373 K was not able to remove the solvent. The peak splitting at 9.7° was due to distortion of the cubic symmetry and resulting with tetragonal crystal formation [8]. Small peaks were observed at ca. 2θ = 32°, 34° and 36° attributable to the ZnO crystals originating from the starting precursor [9]. Additional peak at ca. 8.9° (starred) was observed for samples prepared using direct-mixing method and autoclave-assisted method suggesting that the partial hydrolysis of metal clusters and organic linkers with water molecules during the synthesis or due to exposure to water molecules in the environment [10]. Therefore, samples prepared with microwave-assisted method exhibited better stability compared to other methods. Figure 2 shows the SEM images of dried MOF-5 synthesized using different methods. The SEM image of DMOF-5 suggested faceted crystal structure. The microwave assisted resulted with cubic crystal structure. The autoclave assisted method resulted with interpenetrated cubic structure. The non-interpenetrated structure offer larger pores, larger total pore volume and lower density [11]. The average crystal sizes were ca. 1 μm for DMOF-5, 8.34 μm for MMOF-5 and 22.6 μm for AMOF-5 respectively. The modified direct-mixing synthesis suggested smallest crystals size compared to other methods. The

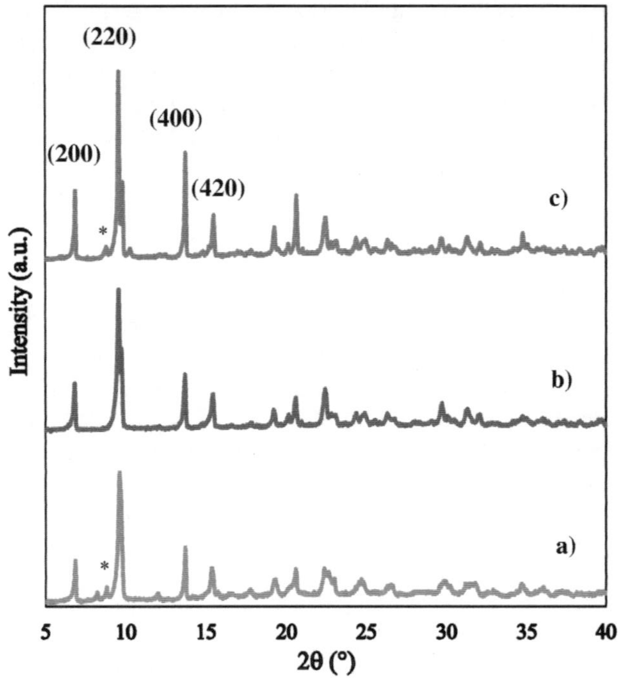

Figure 1. X-ray diffraction analyses of MOF-5 synthesized at different condition a) DMOF-5, b) MMOF-5 and c) AMOF-5.

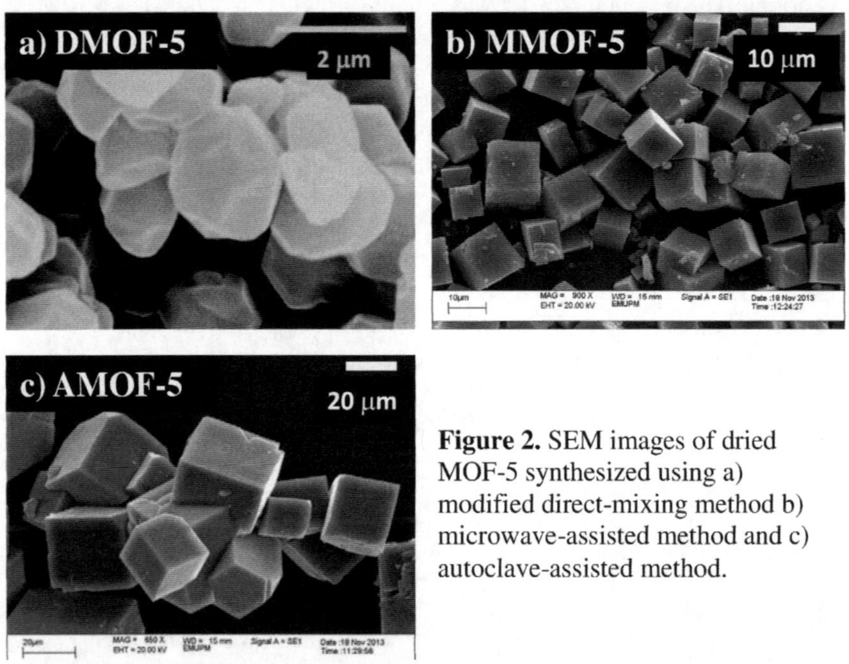

Figure 2. SEM images of dried MOF-5 synthesized using a) modified direct-mixing method b) microwave-assisted method and c) autoclave-assisted method.

autoclave assisted technique resulted with largest MOF-5 crystal with interpenetrated structure because longer heating duration at 100° for 24 hours. The long heating duration allowed crystal growth to larger size for AMOF-5. The microwave heating allowed uniform and faster nucleation which lead to homogeneous nucleation and reducing the crystallization time compared to conventional oven heating [5]. Figure 3 shows the nitrogen adsorption isotherm and its corresponding Horwath-Kawazoe (HK) pore size distribution of DMOF-5. The isotherm was Type 1 typical of microporous materials with Langmuir surface area of ca. 348 m2/g and HK pore size distribution of ca. 0.52 nm suggesting successful formation of MOF-5 with microporous structure.

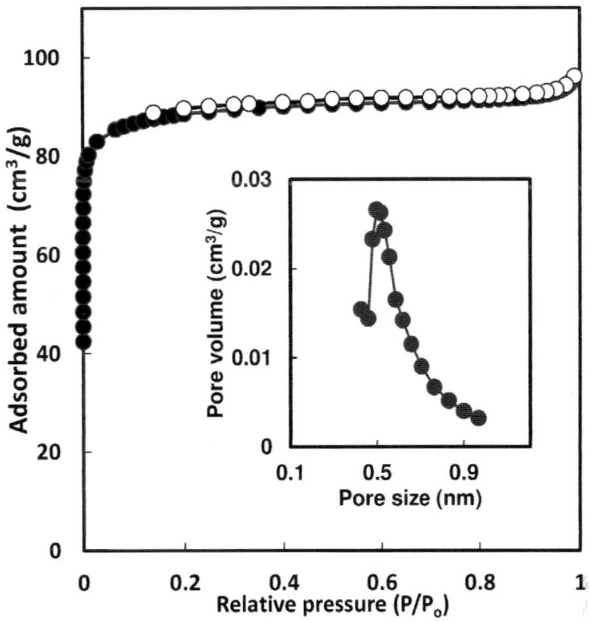

Figure 3. Nitrogen adsorption isotherm of MOF-5 prepared using microwave-assisted method and the corresponding H-K pore size distribution (inset).

2. Conclusions

All three syntheses methods were applicable to synthesis MOF-5 with dodecyl alcohol as non-surfactant template. The syntheses methods influenced crystallite formation. The microwave assisted technique resulted with cubic crystals with average crystal size of 8.34 μm, while interpenetrated cubic was obtained using autoclave assisted technique. The smallest average crystal size was obtained by modified direct-mixing technique with faceted morphology.

3. References

[1] Li H, Eddaoudi M, Keeffe M O' and Yaghi O M 1999 *Nature* **402** 276
[2] Mueller U, Schubert M, Teich F, Puetter H, Schierle-Arndt K and Pastre J, 2006 *J. Mater. Chem.* **16** 626–636
[3] Tranchemontagne D J, Hunt J R and Yaghi O M. 2008 *Tetrahedron* **64** 8553–8557
[4] Huang L, Wang H, Chen J, Wang Z, Sun J, Zhao D, and Yan Y, 2003 *Micropor. Mesopor. Mater.* **58**, 105-114
[5] Choi J S, Son W J, Kim J, Ahn W S, 2008 *Micropor. Mesopor. Mater.* **116** 727–731
[6] Misran H, Othman S Z, Manap A, Pauzi N I M, and Ramesh S, 2014 *Sci. Adv. Mater.* **6** 1–7
[7] Münch A S, Lohse M S, Hausdorf S, Schreiber G, Zacher D, Fischer R A, Mertens F O R L, 2012 *Micropor. Mesopor. Mater.* **159** 132–138
[8] Hafizovic J, Bjørgen M, Olsbye U, Dietzel P D C, Bordiga S, Prestipino C, Lamberti C and Lilerud K P, 2007 *J. Am. Chem. Soc.***129** 3612-3620
[9] Zhang L and Hu Y H, 2011 *Phys. Lett. A* **375** 1514–1517
[10] Kaye S S,Dailly A,Yaghi O M, and Long J R, 2007 *J. Am. Chem. Soc.* **129**, 14176-14177
[11] Jiang H L, Makal T A and Zhou H C, 2013 *Coord. Chem, Rev.* **257** 2232-2249

Acknowledgments
The authors thank Ministry of Education for FRGS (20130106FRGS and 20130104FRGS) and ERGS Scheme (20130901ERGS) and Universiti Tenaga Nasional for facilities.

A virtual laboratory for the simulation of sustainable energy systems in a low energy building: A case study

M Breen[1], A O'Donovan[1], M D Murphy[1*], F Delaney[1], M Hill[2], P D O Sullivan[1]

[1]Dept. of Process Energy and Transport Eng, Cork Institute of Technology, Ireland
[2]Dept. of Electrical and Electronic Eng, Cork Institute of Technology, Ireland

*E-mail: MichaelD.Murphy@cit.ie

Abstract. The aim of this paper was to develop a virtual laboratory simulation platform of the National Building Retrofit Test-bed at the Cork Institute of Technology, Ireland. The building in question is a low-energy retrofit which is provided with electricity by renewable systems including photovoltaics and wind. It can be thought of as a living laboratory, as a number of internal and external building factors are recorded at regular intervals during human occupation. The analysis carried out in this paper demonstrated that, for the period from April to September 2015, the electricity provided by the renewable systems did not consistently match the building's electricity requirements due to differing load profiles. It was concluded that the use of load shifting techniques may help to increase the percentage of renewable energy utilisation.

Introduction

The reduction of energy consumption and embodied carbon emissions is a major global challenge. With buildings accounting for a third of global energy consumption [1], upgrading existing buildings and improving building design is crucial in reducing carbon emissions. A holistic approach needs to be taken that considers both supply level and demand level reduction needs in order improve this sector [2]. Reducing electricity demand in buildings can be achieved through the use of energy efficient equipment and lighting, while the reduction of heat loss through a buildings envelope can reduce the amount of thermal energy demand [3–5]. From a supply perspective the use of clean renewable energies such as solar photovoltaics and wind power can reduce the amount of carbon intensive grid supplied electrical energy, leading to reduced carbon emissions. Regarding the investment in and use of these renewable systems; many studies have been carried out to assess financial and environmental implications. Examples of such studies include work carried out in [6–8]. In the European Union, the importance of renewable technologies has increased in recent years as member countries are obliged to meet the goals of the 20-20 by 2020 initiative. The aims of this initiative are to reduce greenhouse gas emissions by 20% when compared with 2005 levels, to increase energy efficiency by 20% and to increase the share of renewable technologies in energy use by 20% by the year 2020 [9]. This paper aims to provide an overview of the National Build Energy Retrofit Test-bed's virtual laboratory at Cork Institute of Technology, Ireland, while also providing an example of the type of analysis which can be carried out using this platform. This sample simulation was conducted using data from April to September 2015, and analysed electricity use in the building during this period.

Materials and Methods

1.1. Building description

The National Build Energy Retrofit Test-bed (NBERT) [10] (figure 1) is located in the zero2020 building on the main campus of Cork Institute of Technology (CIT). The zero2020 building is an example of a deep building retrofit where part of CIT's existing main campus building, which was designed in 1974, was upgraded to have a higher level of energy performance [11]. The retrofit project saw both the installation of a variety of sustainable energy systems (figure 2) and upgrades to the existing buildings envelope which led significant energy consumption savings [12]. As an example of a living laboratory the NBERT allows researchers the opportunity to develop and calibrate virtual laboratories to study; microgrid applications, thermal comfort, and demand side management.

Figure 1. Zero2020 building retrofit test-bed

Figure 2. Sustainable energy systems used by the zero2020 building

Figure 3. Weather station and instrumentation used in the building

1.2. Models used for sample analysis

To simulate the output of the 12kWp photovoltaic system on the rooftop of the zero2020 building during a five month period from April 2015 to September 2015, a widely used photovoltaic model [13] was employed. This model consists of a current source with one diode and no series or shunt resistances. It uses the following equations to determine the voltage, current and power of a photovoltaic module at its maximum power point;

$$V_m = \frac{nN_s k_B T}{q} \ln\left(\frac{nN_s k_B T}{qI_o}\frac{I_{sc}}{V_{oc}}\right) \tag{1}$$

$$I_m = I_{ph} + I_o - \left(\frac{nN_s k_B T}{q}\frac{I_{sc}}{V_{oc}}\right) \tag{2}$$

$$P_m = V_m I_m \tag{3}$$

Where P_m = Power at maximum power point (W); V_m = Voltage at maximum power point (V); I_m = Current at maximum power point (A); n = Diode quality coefficient; N_s = Number of cells in series per module; k_B = Boltzmann constant (J/K); T = Cell Temperature (K); q = Charge on the electron (C); I_o = Saturation Current (A); I_{sc} = Short Circuit Current (A); V_{oc} = Open circuit voltage (V); I_{ph} = Photocurrent (A).

The inputs to this model included irradiance and ambient temperature data recorded by the for the period specified, as well as specifications of the photovoltaic array. A more detailed description of the materials and methods used, as well as model validation can be found in [14].

To simulate the output of the on-site Proven 2.5 wind turbine (2.5 kWp), a power curve was created by fitting a high order polynomial to data provided by the manufacturer. This curve is shown in Figure 4 and is described by the following equation;

$$P = au^6 + bu^5 + cu^4 + du^3 + eu^2 + fu + g \tag{4}$$

Where: P = Turbine power output (W), u = Wind speed (m/s), a = -0.0025, b = -0.1136, c = 1.717, d = -11.19, e = 55.16, f = -85.2, g = 11.25.

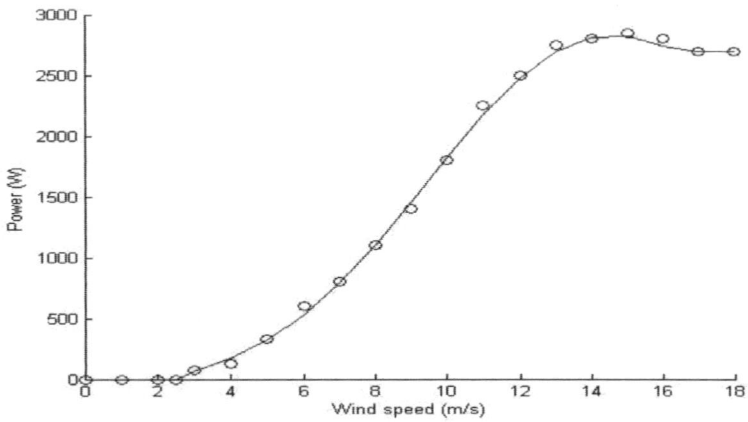

Figure 4. Power curve for modelled wind turbine

The National Build Energy Retrofit Test-bed's electricity use is logged at hourly intervals and can be divided into three categories; heating, lighting and general services. The electricity consumed by the building during the five month period from April 2015 to September 2015 was used in the sample analysis presented. In order to examine multiple electricity pricing scenarios for the five month period, three electricity tariffs were used. The first tariff used was a flat tariff, which consisted of a constant 24 hour rate of €0.19 per kWh consumed. The second tariff used was a Day/Night (DN) tariff, which consisted of a price of €0.20 from 8:00 to 0:00 and a price of €0.10 from 0:00 to 8:00. The third tariff used was a time of use (TOU) tariff, which consisted of a price of €0.15 from 8:00 to 0:00, a price of €0.14 from 0:00 to 8:00, and a price of €0.22 from 17:00 to 19:00.To explore the possibility of being paid for exporting electricity, a feed-in tariff (FIT) of €0.10 per kWh was also included in the analysis. As demand side management can yield monetary benefits for varying pricing tariffs [15] two control algorithms were investigated with regard to purchasing and selling electricity; "Grid priority" and "Load priority". Grid priority implied that all electricity produced by the PV system and/or wind turbine was sold. Load priority implied that only the excess electricity produced by the PV system and/or wind turbine (i.e. remaining electricity after the building load has been met) was sold.

Results and Discussion

Table 1. Averaged external conditions for the period monitored

	Temperature (°C)	Wind Speed (m/s)	Irradiance (W/m^2)
April	9.3	2.7	207
May	11.1	3.0	189
June	14.1	2.5	231
July	14.6	2.8	181
August	14.5	2.2	161
September	13.3	1.9	121

Table 2. Building electricity cost (€) over the five month period when using PV, Wind and a combination of both. Negative figures indicate a profit was made (from selling electricity produced by PV/Wind).

	PV			Wind			PV and Wind		
Priority	Flat	DN	TOU	Flat	DN	TOU	Flat	DN	TOU
Grid	121.53	9.03	-62.95	894.31	781.81	709.83	71.54	-40.96	-112.94
Load	414.66	302.16	230.18	932.76	820.26	748.28	403.12	290.62	218.64

As can be seen in Table 2, the grid priority algorithm provided substantially lower monetary costs than those incurred using the load priority algorithm. It is interesting to note that the costs incurred when using a wind turbine were similar for both the grid priority and load priority algorithms, whereas the costs incurred when using a PV system varied significantly depending on the algorithm used. The size of the PV system (12kW) compared to that of the wind turbine (2.5kW) was the reason for this difference in cost. In Figure 5, the amount of the building's electricity demand which was met by the PV system, wind turbine, and a combination of both is shown. In total, 59% of the demand was met by the PV system, 8% by the wind turbine, and 61% by the combination of PV and wind. The building demand over the five month period totalled 4.97 MWh. Interestingly, the PV system produced 8.23 MWh during this period, which would have been more than enough to meet the building's demand. However, much of this was exported to the grid as excess electricity, due to the power demand of the building not corresponding to the power output from the PV system. The building demand peaked in the morning between 8:00am and 9:00am during the period analysed, as this was the time of maximum occupancy. At this time the PV system was producing little electricity as its peak output occurred during the middle of the day. If load shifting were to be introduced, the PV system output could meet 100% of the building's demand, while also exporting over 3 MWh of electricity to the grid.

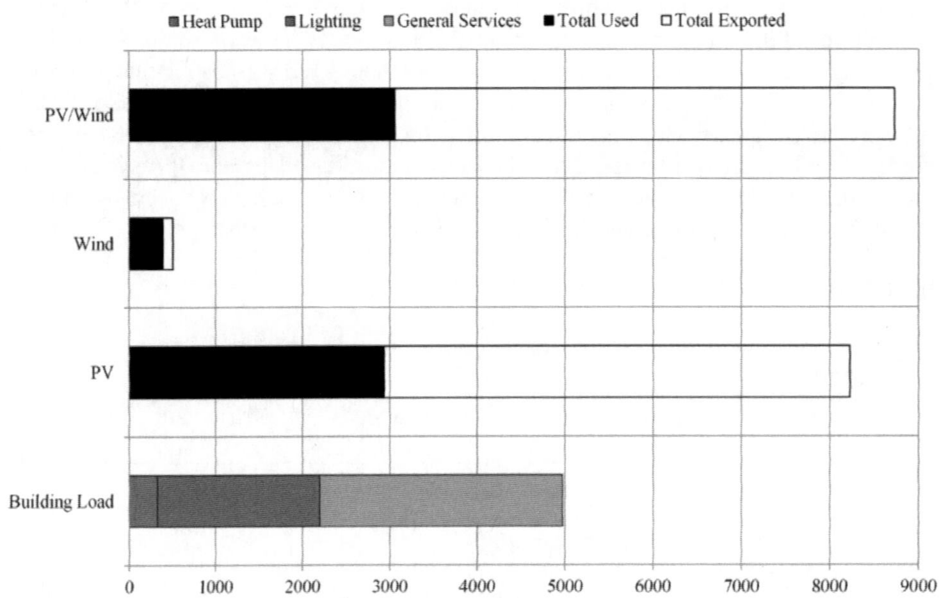

Figure 5: Illustration as to how much of the building's electricity demand is met by PV, wind and a combination of both during the five month period analysed.

Conclusion

In this study a virtual laboratory for the purpose of simulating renewable energy system in the National Build Energy Retrofit Test-bed at the Cork Institute of Technology was presented. This test-bed was a suitable application due to the amount and variety of data, both internal and external, which is available for analysis. The simulations conducted in the virtual laboratory demonstrated that while the photovoltaic system provided sufficient electricity to meet the building's demand, it was not possible to do so due to contrasting load profiles. It was deduced that load shifting techniques may allow the building's electricity requirements to be met, while also providing 3MWh excess for export to the national grid.

References

[1] International Energy Agency (IEA), *World Energy Outlook 2014*. 2014.

[2] A. Allouhi, Y. El Fouih, T. Kousksou, A. Jamil, Y. Zeraouli, and Y. Mourad, "Energy consumption and efficiency in buildings: current status and future trends," *J. Clean. Prod.*, pp. 1–13, 2015.

[3] S. H. Hong, T. Oreszczyn, and I. Ridley, "The impact of energy efficient refurbishment on the space heating fuel consumption in English dwellings," *Energy Build.*, vol. 38, no. 10, pp. 1171–1181, 2006.

[4] Z. Zhou, S. Zhang, C. Wang, J. Zuo, Q. He, and R. Rameezdeen, "Achieving energy efficient buildings via retrofitting of existing buildings: a case study," *J. Clean. Prod.*, 2015.

[5] L. Liu, P. Rohdin, and B. Moshfegh, "Evaluating indoor environment of a retrofitted multi-family building with improved energy performance in Sweden," *Energy Build.*, vol. 102, pp. 32–44, 2015.

[6] T. K. Boomsma, N. Meade, and S.-E. Fleten, "Renewable energy investments under different support schemes: A real options approach," *Eur. J. Oper. Res.*, vol. 220, no. 1, pp. 225–237, 2012.

[7] A. Menegaki, "Valuation for renewable energy: A comparative review," *Renew. Sustain. Energy Rev.*, vol. 12, no. 9, pp. 2422–2437, 2008.

[8] D. R. Thiam, "Renewable decentralized in developing countries: Appraisal from microgrids project in Senegal," *Renew. Energy*, vol. 35, no. 8, pp. 1615–1623, 2010.

[9] COM(2008) 30, "Communication from the Commission to the European Parliament, the Council, the European economic and social Committee and the Committee of the Regions - 20 20 by 2020 - Europe's climate change opportunity," *Eur. Comm.*, pp. Brussels, 1–12, 2008.

[10] "NBERT," 2015. [Online]. Available: http://messo.cit.ie/nbert.

[11] P. D. O. Sullivan, F. Delaney, M. O' Riain, T. Clancy, J. O' Connell, and D. Fallon, "Design and Performance of and External Building Envelope Retrofit Solution for a Grid Optimised Contcrete Structure: A Case Study," in *30th International Manufacturing Conference*, 2010, pp. 1–12.

[12] CIT, "Zero2020energy.com," 2014. [Online]. Available: http://www.zero2020energy.com/.

[13] Saloux, E., Teyssedou, a., & Sorin, M. (2011). "Explicit model of photovoltaic panels to determine voltages and currents at the maximum power point". *Solar Energy,* vol. 85, no. 5, pp. 713-722, 2011.

[14] M. Breen, M. D. Murphy, J. Upton, "Development and validation of photovoltaic and wind turbine models to assess the impacts of renewable generation on dairy farm electricity consumption," in *ASABE 2015*, pp. 1–11, 2015.

[15] M. D. Murphy, M. J. O'Mahony, and J. Upton, "Comparison of control systems for the optimisation of ice storage in a dynamic real time electricity pricing environment". *Applied Energy* Vol 149, pp. 392-403, 2015.

IOP Conf. Series: Earth and Environmental Science **32** (2016) 012062 doi:10.1088/1755-1315/32/1/012062

Numerical analysis of single particle impact in the context of Cold Spray: a new adhesion model

P Profizi[1,3], A Combescure[1,2] and K Ogawa[3]
[1]Laboratoire de Mécanique des Contacts et des Structures, INSA de Lyon, UMR-CNRS 5259, 18-20 Rue des sciences, 69621 Villeurbanne, France
[2]AREVA-SAFRAN chair
[3]Fracture and Reliability Research Institute, Tohoku University, 6-6-1-11 Aramaki Aoba-ku, Sendai, 980-8573, Miyagi, Japan

E-mail: paul.profizi@insa-lyon.fr

Abstract. A new adhesion model for numerical simulation of single particle impact in the context of Cold Spray is introduced. As in other studies, cohesive forces are put between the particle and substrate to account for adhesion. In this study however, the forces are put only when a local physical criterion is met. The physical phenomenon most often attributed to Cold Spray adhesion is a shear stress instability. The Johnson-Cook material law is used with a shear damage softening law to enable strong localization at the interface without the need for an extremely fine mesh. This localization is then detected as a drop in local yield stress value by the algorithm, which then implements a local cohesive force. The evolution of this cohesive force is defined by an energy dissipative cohesive model, using a surface adhesion energy as a material parameter. Each cohesive link is broken once all its associated surface energy is dissipated. A criterion on the damage value is also used to break a cohesive bond prematurely, to account for the effect of erosion at higher speeds. This model is found to reproduce the Cold Spray-like adhesion behavior with observed critical and maximum speeds.

1. Introduction

The Cold Spray process consists in coating an object with a thin layer of a metallic or amorphous material by projecting very small particles at high speed in a pressurized carrier gas. It is used industrially to create a coating on a piece of material to give it new surface characteristics, to create a new piece altogether in an additive manner like in 3D printing, but also to repair damaged structures and surfaces easily[1]. The first experiments showed that the projected particles adhere only for certain pairs of particle-substrate materials and within a specific velocity range [1,2]. Many hypotheses, such as the presence of a shear stress instability at the interface upon impact, have been formulated to explain these experimental observations [3]. They were based on the subsequent observation of the microstructures and their association with numerical results giving the histories of temperatures and other parameters at the interface [4].

These numerical results relied mainly on explicit dynamic finite element calculations [5]. However, most finite element calculations are unable to simulate the very significant crushing of soft particles against a rigid substrate or the deep penetration of hard particles into a softer substrate, so they often resort to a remeshing method [3]. Some calculations are based on coupled Euler/Lagrange formulations: these lead to better results when one of the two objects is subjected to very high strains[6].

Solid SPH methods also appear to lead to good predictions in the case of highly distorted meshes. They also enable local material tearing to be easily simulated [7,8]. Nevertheless, the common point of these simulations is that while striving for a precise evaluation of the material behaviour, they do not include any adhesive forces at the interface during the computation, and instead conclude on adhesion based on a post-computation criterion looking at the evolution of different variables [9].

To the author's knowledge, only a few finite element or SPH fast dynamics numerical simulations use tools to account for adhesive forces during impact [7,10]. In [10], a cohesive zone defined by a cohesive stress value is put on the whole contact surface, and the particle is said to rebound when the contact surface drops to zero.

The new adhesion model presented in this paper differs from this approach as it involves: an adhesion activation criterion based on shear stress instability, an adhesion deactivation criterion based on surface energy dissipation, and an adhesion deactivation criterion based on erosion of the elements. It follows a study on adhesion in dynamics impacts, using the same model but without any criteria[11].

2. Numerical model

2.1. Geometry, limit conditions and mesh

A ball of radius R_b=12.5µm impacts a substrate of sides L_s≈73.7µm and height H_s≈33.1µm (approximate values due to the SPH packing method) at a chosen initial velocity. To simulate a semi-infinite substrate, elastic impedances are put on every side but the impacted one.

The SPH method is used to mesh both parts in a hexagonal compact lattice. The SPH particle radius is R_{SPH}=1.25µm. A complete description of the model can be found in [11] and more explanations on the SPH method for solids can be found in [11–13]. The coarse mesh used, for quick computations, contains 15776 elements in the substrate and 737 elements in the impacting particle.

2.2. Material law

The material law used for both parts is the common Johnson-Cook law, which takes into account equivalent plastic strain hardening, equivalent plastic strain-rate hardening and thermal softening. Here adiabatic heating is considered with 90% of plastic work converted in heat. A damage softening coefficient D is added to give the final equation (1) and get better localization at the interface, and to account for element erosion and improve general behaviour [10]. For stability purposes, the maximum allowed damage softening coefficient is set as 90% with no element deletion.

$$\sigma_y = \left[A + B\left(\varepsilon_{eq}^{pl}\right)^n\right]\left[1 + C\,ln\left(\frac{\dot{\varepsilon}_{eq}^{pl}}{\dot{\varepsilon}_{ref}^{pl}}\right)\right]\left[1 - \left(\frac{T - T_{ref}}{T_{melt} - T_{ref}}\right)^m\right](1 - D) \quad (1)$$

The material is aluminum for both parts with the parameters shown in table 1.

Table 1. Material parameters used for Aluminum

E (MPa)	Ro (kg/m³)	nu	A (MPa)	B (MPa)	C	n	M	$\dot{\varepsilon}_{ref}^{pl}$ (s⁻¹)	T_{ref} (°K)	T_{melt} (°K)	K (J/°K)
70.9e9	2710	0.3	148.4e6	345.5e6	0.001	0.183	0.895	1	300	916	904

The shear criterion defined in [14] is used for damage initiation with parameters too long to discuss here but given in input files for the "*Progressive failure analysis of thin-wall aluminum extrusion under quasi-static and dynamic loads*" example available online. A linear increase of damage until a maximum plastic equivalent deformation of 2 gives damage evolution. Discussion on the choice of the damage model is not of interest for this specific short paper but its extreme influence on the results should not be forgotten and is currently under investigation.

2.3. Adhesion model

A cohesive stress is applied between two elements if they have been in contact, have both met the activation criterion, and are separating. The activation criterion is a 30% drop in the local yield stress value. The Lagrange multiplier given by the contact algorithm is used for the cohesive stress value. The work associated with this stress is updated at every step and compared to the maximum work allowed, defined by the product of an adhesive surface energy G_c=30J.m^{-2}and the surface associated with an SPH particle $\pi * R_{SPH}^2$. When this limit is reached, the bond is broken and the cohesive stress is no longer applied. The bond is also broken if one of the elements of the adhesive link has reached a maximum allowed damage value for adhesion MAXD.

3. Behaviour of the model

Four sets of computations were carried out, all with a final simulated time of around 0.177μs found to be high enough for energy variations to settle down. For each simulation are observed: the number of elements having been in contact, the number of adhesive couples having been activated, the number of adhesive couples having dissipated all of their assigned energy, the number of adhesive couples broken by damage, and the number of adhesive couples still active.

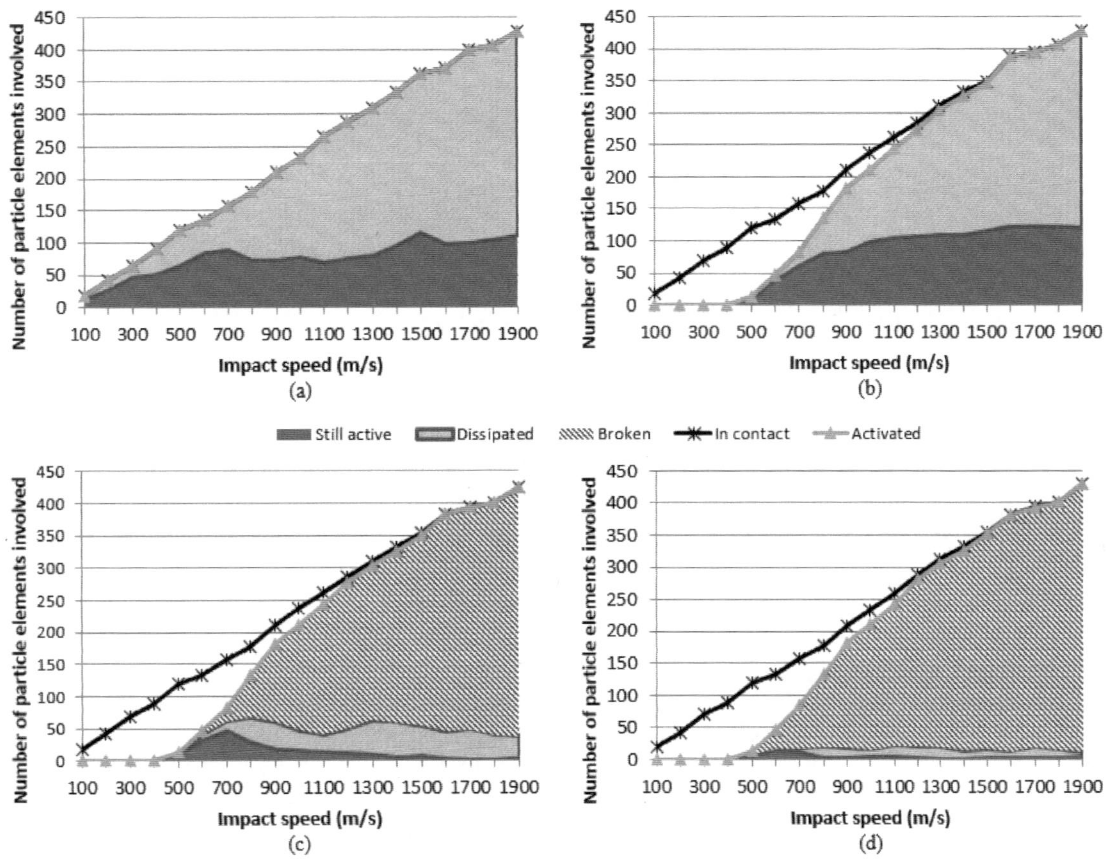

Figure 1. Model behaviour for different sets of parameters

Figure 1 shows the variation of these values with impacting speed for four scenarios. In scenario (a) no activation trigger nor erosion trigger were used, so all contacting elements undergo adhesion and some little adhesion is still present even for low speeds of 100m.s^{-1}, making it difficult to conclude on a critical speed. Scenario (b) shows the effect of the activation criterion based on a drop in local yield stress value. The activated surface is no longer the same as the contacting surface, resulting in adhesion only above 500m.s^{-1}, which can be seen as the critical velocity of the Cold Spray process. The surface under adhesion then converges to the surface in contact at high speeds, meaning that the activation criterion is met on an increasing part of the contacting surface. The number of adhesive links still active at final time however still stabilizes with increasing speed.

In scenario (c) the adhesion erosion criterion is used with a value of MAXD=0.5. The behaviour at low speeds is the same as in scenario (b), but once above 600m.s^{-1}, an increasing part of the adhesive couples is broken before they can dissipate all their assigned energy. The presence of the deletion criterion results in a drop in the number of active adhesive links at final time with increasing impact speed, to finally get a maximum impact speed for adhesion at around 1700m.s^{-1}. The relation between contacting surface and adhesive surface is however unaffected. Scenario (d) uses a more strict criterion of MAXD=0.4, resulting in nearly no final adhesion over the whole impact speeds range.

The notion of final adhesion is here chosen as at least one adhesive couple being still active, but while convenient this choice is of course not realistic and the notion of "minimal final adhesive surface for sufficient particle adhesion strength" will later be implemented.

4. Summary and further work

A new cohesive model for particle adhesion during the Cold Spray process was introduced. This model where the occurrence of adhesion is based on physical mechanisms is shown to exhibit a sticking velocity window as in experiments, with a different behaviour than models lacking these criteria.

Current and further work includes: the study of the cases of copper/copper, copper/aluminum and aluminum/copper to ascertain the capability of the SPH model to correctly represent the high deformation behaviours for different couples of material; a mesh-dependency study; a complete parametric study to find the range of influence of the different parameters involved. Different criteria based on different physics could also be implemented for the onset and breaking of local adhesive links.

References

[1] Papyrin A N, Kosarev V F, Klinkov S, Alkhimov A P and Fomin V 2007 *Cold Spray Technology*

[2] Gilmore D L, Dykhuizen R C, Neiser R a., Roemer T J and Smith M F 1999 Particle Velocity and Deposition Efficiency in the Cold Spray Process *J. Therm. Spray Technol.* **8** 576–82

[3] Assadi H, Gärtner F, Stoltenhoff T and Kreye H 2003 Bonding mechanism in cold gas spraying *Acta Mater.* **51** 4379–94

[4] Luo X-T, Li C-X, Shang F-L, Yang G-J, Wang Y-Y and Li C-J 2014 High velocity impact induced microstructure evolution during deposition of cold spray coatings: A review *Surf. Coatings Technol.* **254** 11–20

[5] Li W Y, Zhang C, Li C J and Liao H 2009 Modeling aspects of high velocity impact of particles in cold spraying by explicit finite element analysis *J. Therm. Spray Technol.* **18** 921–33

[6] Xie J, Nélias D, Walter-Le Berre H, Ogawa K and Ichikawa Y 2015 Simulation of the Cold Spray Particle Deposition Process *J. Tribol.* **137** 041101

[7] Vergani L, Guagliano M, Manap A, Okabe T and Ogawa K 2011 Computer simulation of cold sprayed deposition using smoothed particle hydrodynamics *Procedia Eng.* **10** 1145–50

[8] Lemiale V, King P C, Rudman M, Prakash M, Cleary P W, Jahedi M Z and Gulizia S 2014 Temperature and strain rate effects in cold spray investigated by smoothed particle hydrodynamics *Surf. Coatings Technol.* **254** 121–30

[9] Grujicic M, Zhao C ., DeRosset W . and Helfritch D 2004 Adiabatic shear instability based mechanism for particles/substrate bonding in the cold-gas dynamic-spray process *Mater. Des.* **25** 681–8

[10] Yildirim B, Fukanuma H, Ando T, Gouldstone A and Müftü S 2014 A Numerical Investigation Into Cold Spray Bonding Processes *J. Tribol.* **137** 011102

[11] Profizi P, Combescure A and Ogawa K 2016 SPH modeling of adhesion in fast dynamics : application to the Cold Spray process *Comptes Rendus Mec.*

[12] Liu M B and Liu G R 2010 *Smoothed Particle Hydrodynamics (SPH): an Overview and Recent Developments* vol 17

[13] Maurel B, Combescure A and Potapov S 2012 A robust SPH formulation for solids *Eur. J. Comput. Mech. Eur. Mécanique Numérique*

[14] Simulia D 2014 ABAQUS Analysis User's Manual, Volume III: Materials *Provid. Stand*

Institutional Governance Framework for Determining Carbon-related Accounting Practices: An Exploratory Study of Electricity Generating Companies in Malaysia

B Alrazi[1] and N Mat Husin
Department of Accounting, College of Business Management and Accounting, Universiti Tenaga Nasional, MALAYSIA

E-mail: Bakhtiar@uniten.edu.my

Abstract. Electricity industry is the major contributor of the global carbon emissions which has been scientifically identified as the main cause of climate change. With the various initiatives being implemented at the international, national, and industry levels, companies in the electricity industry are currently facing immense pressure from various stakeholders to demonstrate their policies, initiatives, targets, and performance on climate change. Against this background, accounting system is argued to be able to be play important roles in combating climate change. Using institutional governance as the underlying framework, we have identified several governance mechanisms as the determining factors for companies to have a systematic accounting system related to carbon emissions. The factors include environmental management system certification, environmental organization, publication of stand-alone sustainability reports, the use of GRI guidelines, environmental strategic planning, governance quality, and participation in CDP surveys and emissions trading scheme. We explored this issue in the context of major electricity generating companies in Malaysia and found that except for certified environmental management system, the other governance mechanisms are still lacking. The findings suggest that companies in Malaysia, in particular, from the electricity industry are not well prepared in facing risks related to climate change.

1. Introduction

Electricity industry is the largest contributor of carbon dioxide (CO_2) emissions [1]. In 2011, the industry was responsible for 42 percent of energy-related CO_2 emissions. In fact, the emissions from the industry had increased by 70% as compared to the level in 1990. The issue of CO_2 emissions is significant for several reasons. Firstly, the report by Intergovernmental Panel on Climate Change [2] has identified greenhouse gas (GHG) emissions build-up (of which 77% is CO_2 emissions) as the main cause of climate change. Secondly, various initiatives have been established at the international, regional, and national levels to mitigate climate change issues including the ratification of Kyoto Protocol, the implementation of European Union Emissions Trading Scheme (EU ETS), and the establishment of investor groups like Carbon Disclosure Project (CDP). These developments have made companies from the electricity industry at the centre of debates and discussion. Thus, it is of paramount importance for electricity companies to demonstrate carbon accountability to their

[1] To whom any correspondence should be addressed.

stakeholders (including shareholders, creditors, the government, and society). In essence, they need to show the public that they are taking necessary measures to mitigate CO_2 emissions and climate change issues.

Against this background, accounting system is argued to be able to be a powerful tool to manage carbon emissions as well as other environmental issues [3]. In the context of the environment, environmental accounting could mean "all areas of accounting that may be affected by the business response to environmental issues, including new areas of accounting" [4, p. 7]. These include accounting for contingent liabilities/risks, cost analysis, the development of new accounting information systems and auditing. Some of the potential applications of environmental (management) accounting include: to estimate ecological strengths and weaknesses and improvements needed; to control negative impacts on the environment; to assess eco-efficiency; to provide a foundation for both internal decision making and external communication; and to help promote ecologically sustainable development [5]. Empirical findings also suggested that environmental costs (i.e., both capital and operating) [6] and reward systems [7] are associated with better environmental performance. This relationship could be extended to climate change in which companies with accounting system that captures CO_2 emissions could have a better carbon performance.

In order to establish an accounting system that will be able to capture and manage climate change issues, there is a need for supporting mechanisms - both internally and externally. Institutional governance framework essentially argues that corporate competitiveness relies on both: (1) firm resources, capabilities, and strategies, and (2) governments and institutions interventions [8]. In the context of carbon-related accounting practices, companies need to have the necessary internal mechanisms to allow them to develop such a system. These might include financial resources and non-financial resources (e.g., management systems, leadership, and governance). Additionally, they must be pressure imposed by parties external to the organizations including from the governments, industry associations, non-governmental organisations, media and public at large.

Hence, the objectives of the study are twofold. Firstly, it aims to identify several governance mechanisms that are essential to help companies in their climate change initiatives. Specifically, we reviewed institutional governance framework as proposed by [8] and other related literature. Secondly, it explores the implementation level of institutional mechanisms identified among the major electricity generating companies in Malaysia. By so doing, this study adds to the body of knowledge by identifying institutional governance mechanisms imperative in determining the carbon-related accounting practices, which in turn, can be implemented by organizations.

The remaining sections of the paper are structured as follows. The next section provides the review of literature on institutional governance framework. It is followed by the discussion on sample and findings. The final section concludes, highlights the limitation, and provides some recommendations.

2. Institutional governance framework

Institutional governance system can be defined as "the configuration of state and private organizations and institutional arrangements that impact and create mechanisms by which economic and social outcomes in nations are produced" [9, p. 416]. The framework has been developed in response to the limitations inherent in the arguments put forth in the strategic management literature and political economy literature in regards to the national industry and corporate competitiveness [8]. The proponents of strategic management literature posit that the generation of competitive advantage is firm-centric in which it is driven by companies' internal resources, capabilities, and strategies and industry characteristics and structures. By contrast, the proponents of the political economy literature emphasize on the important role of national governments and institutions in fostering competitive capabilities (hence, state-centric). Therefore, institutional governance system framework proposes the integration of four institutional arrangements for national industrial competitiveness, namely market governance, corporate governance, state governance, and joint governance [see further [8]].

Institutional governance implies that for environmental initiatives to be effective, there must be internal corporate governance in place and the pressure from the market and state [8]. In this regard, a study by [10] found that the voluntary greenhouse gas emissions disclosure (and its credibility) among the largest Australian companies are positively associated with environmental management system, compliance to GRI recommendations, voluntary participation in CDP survey, corporate governance quality, company size, and environmental sensitivity of the industry. This indicates the importance of appropriate internal organizational structure [see [11]] to manage, monitor, measure, and report on environmental issues.

Meanwhile, in another study by [12] on the web-based environmental reporting of 278 public listed companies in Malaysia found that the environmental disclosure is positively associated with the certification of environmental management system, the establishment of a specific organizational structure, the inclusion of environmental issues in the strategic planning, the publication of a stand-alone sustainability report or a report in conformity with GRI guidelines, and the quality of corporate governance.

Therefore, based on the above discussion, it can be posited that the proclivity of companies to have a carbon-related accounting system is dependent upon several factors. Carbon-related accounting system will in turn improves corporate carbon performance. Figure 1 depicts the conceptual model and Table A1 in the appendix describes each of the variables.

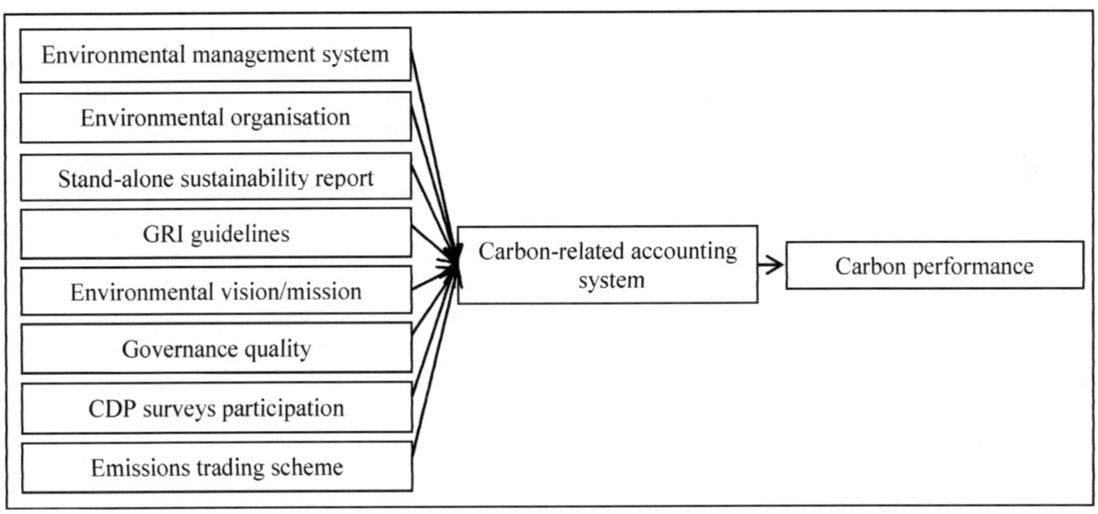

Figure 1. The conceptual model

3. Sample and Findings

Table 1 provides the sample included in this exploratory study. The list of electricity companies in Malaysia was obtained from the Energy Commission website [13], after matching the subsidiaries with the parent companies for data availability reason. Units generated (GWj) were estimated based on [13], while percentage of operations was determined using the data reported in the 2014 annual reports. The data for institutional governance mechanisms were obtained through content analysis of annual reports, stand-alone sustainability and corporate websites for the 2007-2014 period.

Table 1. Summary of the findings for each company

Companies	*EMS*	*ORG*	*SAR*	*GRI*	*MISS*	*GOV*	*CDP*	*ETS*
Eden Inc. Bhd	√	X	X	X	X	X	X	X
Malakoff Corporation Bhd	√	X	X	X	X	X	X	X
Mega First Corporation Bhd	X	X	X	X	X	X	X	X
MMC Corporation Bhd	√	X	X	X	X	√	X	X
Tenaga Nasional Bhd	√	X	X	X	√	√	X	X
YTL Corporation Bhd	√	√	√	X	√	X	X	X
YTL Power International Bhd	√	X	X	X	X	√	X	X

Based on Table 1 above, we found that all but one companies have their environmental management system (*EMS*) being certified, particularly ISO14001. Additionally, three companies are included as among the Top MCGI companies (*GOV*), while two companies with vision and/or mission statement (*MISS*) addressed environmental issues. None of the sample companies prepared their report in accordance to GRI (*GRI*), participated in CDP surveys (*CDP*), and involved in any emission trading scheme (*ETS*). Finally, only one company published stand-alone sustainability reports (*SAR*) and has a specific committee on environmental issues. If these factors were imperative for the development of carbon-related accounting system, it can be inferred that Malaysian electricity companies are generally are not well prepared in facing risks associated with climate change.

4. Conclusion and Direction of Future Research

This paper reviews institutional governance framework and discusses the application of the framework in the context of carbon-related accounting practices. In essence, it argues that in order for companies to establish a systematic accounting system, they need to first have established several institutional mechanisms including certification of environmental management system, establishing a specific environmental committee, publication of stand-alone sustainability reports, the use of GRI guidelines, inclusion of environmental issues in the vision and/or mission statement, good corporate governance and participation in CDP surveys and emissions trading scheme. Since carbon-related accounting practice is a new concept, the institutional mechanisms identified earlier could be indicators for the readiness of companies in Malaysia to deal with climate change risks. Using a sample of electricity generating companies as a case, we found that except for environmental management system certification, other mechanisms are still lacking among the sample companies. Considering that the industry is the main contributor to CO_2 emissions, this situation is rather alarming and regulatory intervention seems to be needed to ensure that the aim to reduce national emissions by 40% by the year 2020 will materialize (as compared to the 2005-emission level).

Several research can be proposed based on this paper. Firstly, researcher could identify the extent of existing accounting system used by the companies has incorporated carbon and climate change issues. This could be in the form of carbon disclosure (reporting in the annual reports and stand-alone sustainability reports, and on the website) or actual carbon accounting system using questionnaire survey or interview methods. Secondly, using primary data collection methods would also reveal as to whether the factors identified in the paper are able to explain the carbon-related accounting practices. The sample could be extended to other industries to increase the generalizability of the findings.

References

[1] IEA 2013 *CO_2 Emissions from Fuel Combustion: Highlights*
[2] IPCC 2007 *Climate change 2007: Synthesis report*
[3] Alrazi B, De Villiers C and Van Staden C 2015 *J. Clean. Prod.* **102** 44

[4] Gray R. and Bebbington J 2001 *Accounting for the environment*
[5] Burritt R, Hahn T and Schaltegger S 2002 *Aust. Acc. Rev.* **12** 39
[6] Rao P, Singh A K, O'Castillo O L, Intal P S and Sajid A 2009 *Bus. Strategy Environ.* **18** 14
[7] Epstein M J and Wisner P S 2005 *Adv. Manag. Acc.* **14** 115
[8] Griffiths A and Zammuto R F 2005 *Acad. Manag. Rev.* **30** 823
[9] Griffiths A, Haigh N and Rassias J 2007 *Euro. Manag. J.* **25** 415
[10] Rankin M, Windsor C and Wahyuni D 2011 *Acc. Aud. Accountability J.* **24** 1037
[11] Adams C A, 2002 *Acc. Aud. Accountability J.* **15** 223
[12] Alrazi B, 2014 *Proceedings of Nat. Sympo. Exh. on Bus. Acc.*
[13] Suruhanjaya Tenaga 2015 *IPPs Directories*
[14] Melnyk S A, Sroufe R P and Calantone R 2003 *J. Oper. Manag.* **21** 329
[15] Sumiani Y, Haslinda Y and Lehman G 2007 *J. Clean. Prod* **15** 895
[16] Vafeas N and Nikolaou V 2001 *Adv. in Pub. Int. Acc.* **8** 195
[17] Bansal P and Roth K 2000 *Acad. Manag. J.***43** 717
[18] McKendall M, Sanchez C and Sicilian P 1999 *The Inter. J. Org. Anal.***7** 201
[19] Elijido-Ten E 2007 *Asian Rev. Acc.,* **15** 164
[20] Kent P and Chan 2009 *Corp. Own. Cont.* **7** 399
[21] Kolk A, Levy D and Pinkse J 2008 *Euro. Acc. Rev.* **17** 719
[22] CDP 2006 *Climate change report: Carbon Disclosure Project*
[23] Bebbington J and Larrinaga-González C 2008 *Euro. Acc. Rev.* **17** 697

Appendix

Table A1. Summary of each institutional governance variables.

No	Variables	Description
1	*Environmental management system certification (EMS)*	• Assist in developing appropriate environmental policies and goals, their execution, and monitoring their effectiveness; identifying legislative requirements and prioritizing environmental impacts; fostering and managing employees' commitment and awareness; enhancing employees' skills; establishing a management process to review and audit; and maintaining appropriate communication with relevant internal and external parties [14] • Necessitates the development of environmental targets, improvement in processes on a continuous basis, and participation in regular audits by the certifying organization [15]. • Requires communication to both internal and external stakeholders [10].
2	*Environmental organization (ORG)*	• Including having a specific manager/committee/department dealing with environmental issues • Enables to monitor corporate environmental policy [16], oversee its ecological impacts [17] [18], and continuously monitor and manage relationship with key stakeholders [19] [20]. • More likely to see the importance of environmental reporting to their public, government, and other stakeholders [10] and thus demand accountability from management so as to protect reputation [18].
3	*Publication of stand-alone sustainability reports (SAR)*	• Voluntary in nature; various labels including Corporate Citizenship Report, Corporate (Social) Responsibility Report, Sustainable, Development Report, Sustainable Value Report, and Sustainability Report • Contents are usually based on some stakeholder engagement activities or reporting guidelines/standards
4	*GRI guidelines (GRI)*	• Contains a total of 30 indicators on the environment and at least 5 indicators related to climate change

No	Variables	Description
		• Subject to external assurance of the report (optional)
5	*Environmental strategic planning (MISS)*	• Having a specific vision/mission statement on the environment or vision/mission statement incorporates environmental issues. • Provide objectives, guiding principles and values, which are important in directing its strategic decision making and actions [20].
6	*Governance Quality (GOV)*	• Quality of governance system, measured by whether the company is included as among the Top MCGI index • More likely to better align stakeholder expectations with corporate goals and objectives. Failure in managing the environmental issues properly could result in financial disaster which in turn could hurt the shareholders' returns.
7	*Participation in CDP surveys (CDP)*	• A voluntary effort to encourage standardized reporting procedures for companies to provide investors relevant information relevant about the business risks, opportunities from climate change [21] • The CDP represents several large investors with assets of over $US55 trillion and is supported by institutional investors. These investors are concerned about the risks associated with climate change, thus are calling for more information about how companies are addressing the challenges of climate change.
8	Participation in emissions trading scheme (*ETS*)	• Requires the development of corporate accounting and technical expertise to manage and to report on CO_2 emissions [21] • Participants of an ETS are required to submit verified emissions data annually to a central registrar [22] • Causes investors, policy makers, and the general public to be more interested in corporate CO_2 emissions management, cash flow implications, the potential impacts the ETS has on the long term company viability [23], and the extent to which the ETS has shifted company focus on managing other significant environmental impacts.

Time Series Analysis of Heavy Metal Concentrations along the Watershed Gradient in Cameron Highlands: Geospatial Approaches.

S H Haron[1], B S Ismail[1], M R Mispan[2], N F Abd Rahman[3], K Khalid[4], M Z Abdul Rasid[5], L M Sidek[6]
[1]Faculty of Science and Technology, Universit Kebangsaan Malaysia, Malaysia
[2]Malaysian Agricultural Research and Development Institute (MARDI)
[3]Faculty of Civil Engineering, Universiti Teknologi MARA, Shah Alam, Malaysia
[4]Faculty of Civil Engineering, Universiti Teknologi MARA Jengka, Pahang, Malaysia
[5]Faculty of Engineering, Universiti Putra Malaysia, Malaysia
[6]Centre for Sustainable Technology and Environment (CSTEN), Department of Civil Engineering, Universiti Tenaga Nasional, Malaysia

E-mail: sthumaira@yahoo.com

Abstract. Heavy metal, particularly cadmium, lead, and arsenic, constitute a significant potential threat to human health. Some metals are extremely toxic to humans and the toxic heavy metals of greatest concern include cadmium, lead, and arsenic. The objective of the study conducted was to determine the accumulation and distribution status of heavy metal cadmium (Cd) in the sediment of Bertam River from September 2014 to February 2015 in the agricultural areas of Cameron Highlands, Malaysia. The sediment samples were collected randomly in three replicates from ten sampling points in the agricultural areas of Cameron Highlands. The heavy metals in the sediment were extracted using the wet acid method and the sample concentrations are then tested for metal concentrations by the spectrography method using Inductively Coupled Plasma (ICP) spectrography. Inverse distance weighting (IDW) was used to create a map of metal concentrations for a point on the polygon dataset spatial interpolation. There is an increasing trend of Cd from the upstream to downstream stations along Bertam River during the rainy season. The activity range of Cd is 0.07 to2.83 μg/g during the rainy season, whereas, during the dry season, Cd activity ranged from 0.26-0.83μg/g.

1. Introduction

The presence of heavy metals traces in the atmosphere, soil, and water can cause serious problems to all organisms and the ubiquitous bioavailability of these heavy metals can result in bioaccumulation in the food chain that can be highly dangerous to human health [2, 3, 5, and 6]. Traces of heavy metals, such as arsenic, cadmium, lead, chromium, nickel, and mercury are key environmental pollutants, particularly in areas with high anthropogenic pressure [3]. The objectives of this study were to examine the status of heavy metal concentrations cadmium (Cd) in agricultural areas and to evaluate the factors that contribute to its accumulation.

The study was conducted along Bertam River in Cameron Highlands, Pahang. The river flows from Forest Reserve in the north, passing through the valley to the south and ends at Sultan Abu Bakar Dam. Bertam River is also the source of drinking water for the residents nearby. Therefore, conservation and ongoing monitoring is necessary to ensure that the river is free from pollution.

2. Materials and Methods

2.1 Sampling

Along the Bertam River, ten sampling points were selected with the location and activities of the surrounding land use recorded, as shown in Figure 1 and Table 1. Sediment samples were collected using a stainless steel scoop and replicated 3 times from each station from September 2014 to February 2015. The sediment samples were carefully wrapped using aluminium foil and put in a labelled polythene zipper bag before storing them in a cooler box.

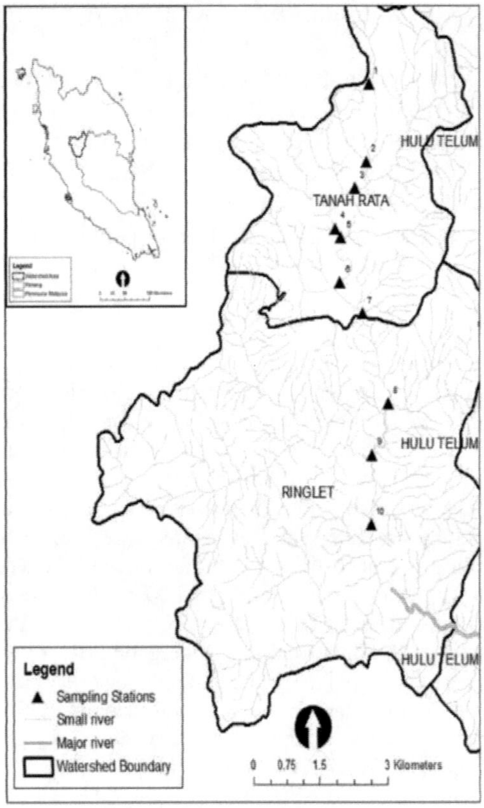

Figure 1: Sampling Location

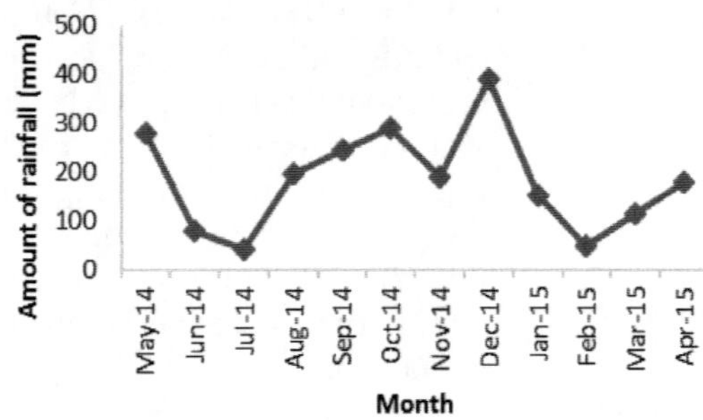

Figure 2: Rainfall trends from May 2014 to April 2015

Table 1: Land use and coordinate for sampling points

No.	Longitude	Latitude	Land use
1	N04°30'17.0"	E101°23'14.7"	Forest
2	N04°29'30.1"	E101°23'12.5"	Orchard
3	N04°29'14.4"	E101°23'4.3"	Rangeland
4	N04°28'49.9"	E101°22'50.2"	Forest
5	N04°28'44.7"	E101°22'54.5"	Urban area
6	N04°28'18.0"	E101°22'54.0"	Forest
7	N04°27'58.4"	E101°23'10.1"	Urban area
8	N04°27'5.4"	E101°23'29.1"	Orchard
9	N04°26'34.40"	E101°23'17.20"	Orchard
10	N04°25'52.70"	E101°23'16.40"	Lake

The samples were then filtered and determination of Cd was carried out using an air–acetylene flame atomic absorption spectrophotometer (AAS) Perkin-Elmer Model 4100. Output data are presented in µg/g dry weight (dw) basis. Calibration curves were analysed using multiple-level calibration standards by concentrations and calculated. 1000-mg/l stock solution was prepared for each heavy metal (MERCK Titrisol). Samples were stored in metal-free plastic tubes at room temperature until the determination of heavy metals was conducted using ICP-MS (PerkinElmer, USA) [2]. Inverse distance weighting (IDW) was use to create a map of metal concentrations along the watershed gradient for a point on the polygon dataset spatial interpolation.

2.2 *Rainfall Data*

Monthly rainfall data of Cameron Highlands from May 2014 to April 2015 were obtained from the Department of Irrigation and Drainage (DID) as shown in Figure 2. During dry season (January to April 2015) sampling period, the rainfall recorded in Cameron Highlands was between 150-170 mm. Meanwhile, during the rainy season (September to December 2014), a scale ranging from 243-388 mm of total rainfall was recorded.

3. Results and Discussion

The concentrations of heavy metal Cd in the water at the 10 sampling stations areas shown in Figure 3. This study shows that the concentration of Cd in Bertam river basin is between 0.07 to 2.83µg/g. The concentration of heavy metals in water were compared with the Guidelines for Air Quality, United States [4] which allowed or maximum concentration (CMC) and a constant concentration(CCC) to the surface water without causing significant impact to aquatic wildlife. The results showed that the Cd concentrations found in all stations are low and do not exceed the recommended maximum concentration (EPA CMC 2µg/g) except for S10 during the wet season.

The concentration of Cd showed an increasing trend at all locations (Figure 3) from July 2014 to November 2014 especially at station 10, which is located at Sultan Abu Bakar Dam (0.07µg/g, lowest value; 2.83µg/g, highest value). This could be due to the authorities wanting to ensure that the water level is lower than the intake depth to facilitate excavation of sediment in the lake. The results also showed that S10 recorded the highest range and mean of concentration activity of Cd as obviously seen in Fig.3. This is because S10 is located near the mouth of the river connected by a dam and indirectly there are additional sediment inputs from agriculture, farming and industries in the upstream areas of small rivers.

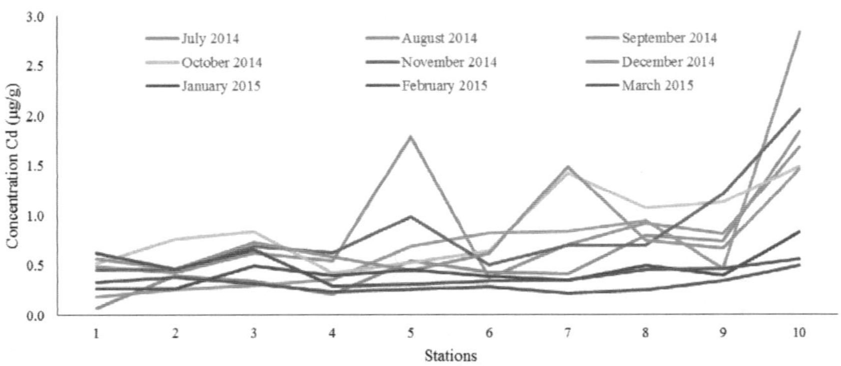

Figure 3: Concentration of Cadmium from July 2014 to March 2015

Other than S10, during dry season, S5 (1.79 µg/g) and S7 (1.49 µg/g; 1.42 µg/g) concentrations of Cd are higher than the other locations where the same pattern of Cd occurred. The concentration was slightly higher than other stations, indicating there is an accumulation of Cd due to the circulation of water at the outlet of the river. All stations show the similar concentration profile for Cd with the exception of S5, S7 and S10, which showed a higher concentration (Figure 3). Cd is insoluble in water and is usually associated with solid material [7]. Due to the additional input from a small river that connects to S10, Cd in the form of suspended solids will gather at the river junction. The confluence of two rivers will increase the Cd content in the sediment itself. The trend of Cd concentrations that were higher at S5 and S7can also be observed at S10 can be attributed to the same cause.

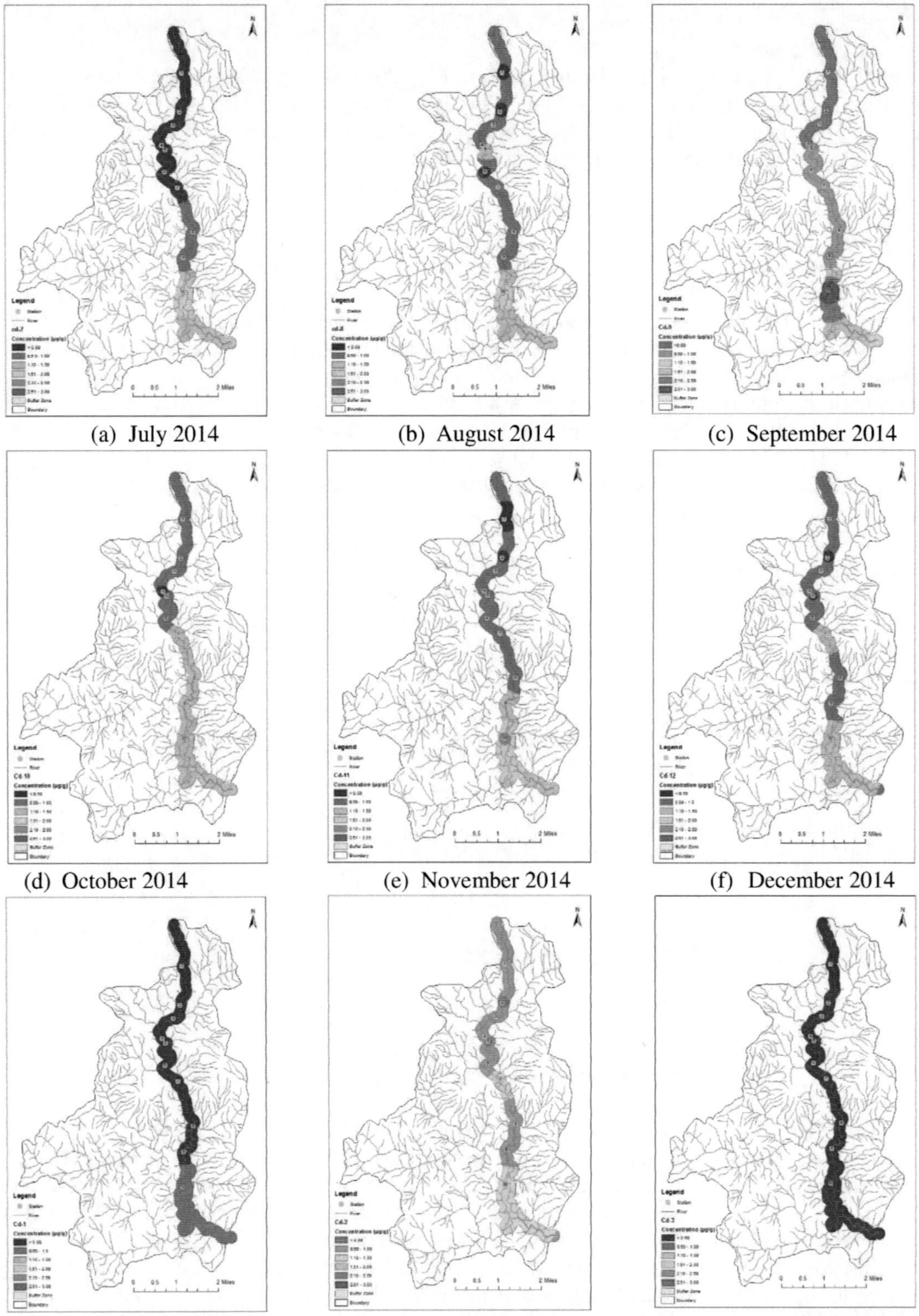

(a) July 2014 (b) August 2014 (c) September 2014

(d) October 2014 (e) November 2014 (f) December 2014

(g) January 2015 (h) February 2015 (i) March 2015

Figure 4: Concentrations of heavy metal (Cd) along the watershed gradient from July 2014 to March 2015

The concentrations of heavy metal (Cd) along the watershed gradient from July 2014 to February 2015 are as shown from figures 4 to 10 (Figure 4). As the slope decreased along Sungai Bertam, the concentration of Cd increased during the rainy season especially at S10. This might be due to heavy rainfall, increased agricultural activities where all the sediment inputs of pesticides and fertilizers enter the river. Moreover, land use features also contributed to sediment input for Cd. As in this study, S10 is located at the lowest point, which is also known as the Sultan Abu Bakar Dam, where all the runoff sediment accumulated and this will lead to increased concentrations of Cd in the event of heavy rain. As for the dry season, the trend decreased starting from January to April 2015.

4. Conclusion

This study concluded that in general, concentrations of heavy metal, Cd in the water of the research area of Bertam River Basin is low except for S10 where there is an increasing trend, especially in areas where there are anthropogenic activities such as agriculture and industry. There is an increasing trend of Cd from upstream to downstream stations of Bertam River during the rainy season. Determination of Cd was established by using alpha spectrometry. The activity range of Cd is from 0.07 to 2.83µg/g during rainy season, while for dry season the Cd activity ranged from 0.26 to 0.83µg/g.

Acknowledgements

The authors are grateful to the Department of Agriculture, Cameron Highlands (through the Ministry of Agriculture), Ministry of Education (My Brain 15), Malaysian Agricultural Research and Development Institute (MARDI- Grant No. 003) and SWAT Network of Malaysia for their financial and technical support.

References

[1] Hirata S 1985 Phosphorus and metals bound to organic matter in coastal sediments - An investigation of complexes of P, Cu, Zn, Fe, Mn, Ni, Co and Ti by inductively coupled plasma-atomic emission spectrometry with sephadex gel chromatography *Marine Chemistry* **16** 23-46

[2] Ghazali A R, Razak N E A, Othman M S, Othman H, Ishak I, Lubis S H, Mohammad N, Hamid Z A, Harun Z, Kamarulzaman F and Abdullah R 2012 Study of Heavy Metal Levels among Farmers of Muda Agricultural Development Authority, Malaysia *Journal of Environmental and Public Health* **2012**

[3] Waseem A, Arshad J, Iqbal F, Sajjad A, Mehmood Z and Murtaza G 2014 Pollution Status of Pakistan: A Retrospective Review on Heavy Metal Contamination of Water, Soil, and Vegetables *BioMed Research International* **2014** 29

[4] Environmental Protection Agency United State 2004. National Recommended Water Quality Criteria. Office of Science and Technology, Washington, DC.

[5] Ismail B S, Farihah K, Khairiah J 2005. Bioaccumulation of heavy metals in vegetables from selected agricultural areas *Bulletin of Environmental Contamination and Toxicology* **2005** 74 320-327

[6] Haron, S H and Ismail B S 2015. Comparison pesticide residue levels in the surface of Bertam River in Cameron Highlands, Pahang *AIP Conference Proceedings* **2015** 020015

[7] Martínez-Aguirre A, Garcia-León M and Ivanovich M 1995 U and Th speciation in river sediments *Science of The Total Environment* **173–174** 203-9

Gross Pollutant Traps: Wet Load Assessment at Sungai Kerayong, Malaysia

M R Mohd Shah[1], N M Zahari[1], N. F Md Said[1], L M Sidek[1], H Basri[1], M S F Md Noor[1], M M Mohammad Husni[1], Milad Jajarmizadeh[1], ZA Roseli[2], N Mohd. Dom[2]

[1] Centre for Sustainable Technology and Environment, Universiti Tenaga Nasional, Jalan IKRAM-UNITEN 43000 Kajang, Selangor, Malaysia.
[2] Humid Tropics Centre, Department of Irrigation and Drainage, Jalan Ledang, Off Jalan Duta, 50480 Kuala Lumpur, Wilayah Persekutuan Kuala Lumpur, Malaysia.

Email: rashidshah92@gmail.com

Abstract. The purpose of this project is to carry out assessment on the effectiveness and performance of Gross Pollutant Traps (GPTs) stormwater quality control in the urban areas. The study aims to provide a management and planning tool for effective management of the gross pollutants in the urban areas specifically in River of Life (ROL) project. ROL project is a Malaysian Government initiative under the Economic Transformation Program. One of the program in the greater Klang Valley is to transform Klang River into a vibrant and livable waterfront by the year 2020. The main river in ROL catchment is Sungai Klang (upper catchment), with main tributaries Sungai Gombak, Sungai Batu, Sungai Jinjang, Sungai Keroh, Sungai Bunus, Sungai Ampang and Sungai Kerayong. This paper objective is to study the gross pollutant wet load at Sungai Kerayong 1 and Sungai Kerayong 2 which is located at the downstream location of the ROL project. The result shows that Sungai Kerayong 2 produced higher gross pollutant wet load (8025.33 kg/ha/yr) than Sungai Kerayong 1 (4695.12 kg/ha/yr). This could be due to high contributions amounts of gross pollutant traps from residential area, the degree of develop area, and also the location of the river itself related to climate and rainfall.

1. Introduction

Urbanization often relates to deterioration of stormwater quality due to many factors such as uncontrolled pollution and waste disposal. It leads to improvement of socio economic life in one area, but also brings wide range of challenges to the environment Therefore, management of water quality impacts in urban areas need to be address in order to protect our environment. Gross pollutants are defined as discarded materials larger than 5 mm and include litter and debris, and coarse sediments are particles with grain sizes greater than 0.5 mm [1]. Visible street waste such as anthropogenic litter and organic matter (sediments, leaves and grass clippings) are classed as gross pollutants and it can exhibit varying degrees of physical and material properties such as firmness, shape, size and density [2]. The main purpose of gross pollutant traps (GPTs) is to eliminate gross pollutants washed into the stormwater system before the stormwater enters the receiving waters [3]. In Malaysia, the use of GPT was suggested in Manual Saliran Mesra Alam (MSMA) where it located at the end of every drain to trap gross pollutants before entering main river system [6]. At the same time, it is important to understand that tropical climate always related to situation where high rainfall intensity occurs in short duration which is cause the performance of GPT cannot be recognise properly [7].

2. Study Area

Figure 1 shows the locations of Sg Sungai Kerayong included within the locations of ROL project. It is located at the downstream of the project. Type of land use for Sungai Kerayong are residential, commercial and mix development. Sungai Kerayong catchment area with 61km2 [4] is located in Wilayah Persekutuan Kuala Lumpur and has been one of the major tributaries of Klang River [5]. The residential area forms the largest fraction at which covering 50% of the total catchment. This location has a year-round warm tropical climate and sunny with heavy rainfall, especially during the Southwest Monsoon from April to September [4].

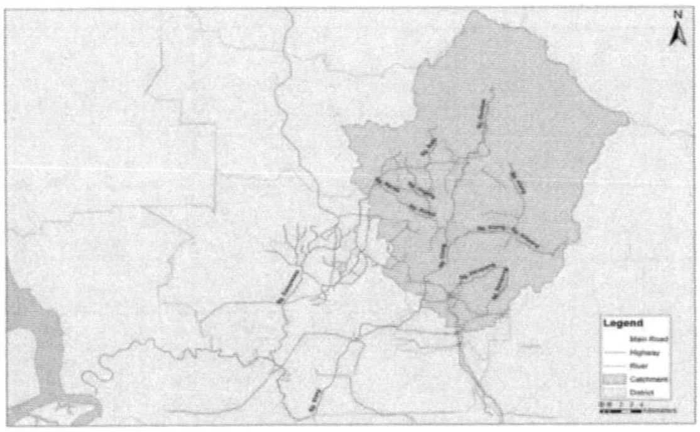

Figure 1: Location of ROL Project and Sungai Kerayong

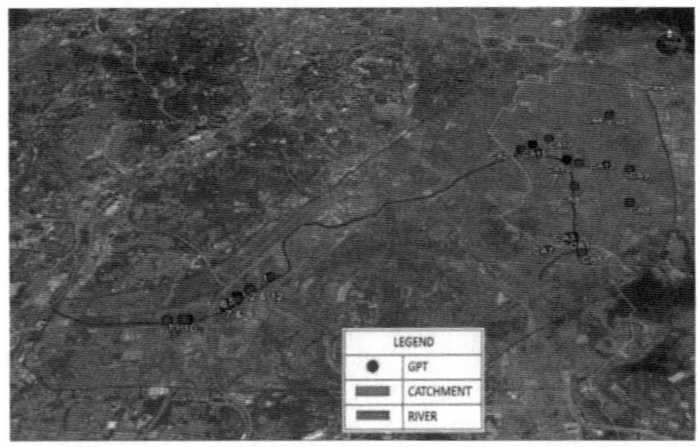

Figure 2: Location of Proprietary GPTs for Sungai Kerayong

3. Type of GPT Used

In this study area, there are 2 types of GPT used at Sungai Kerayong which are Continuous Deflective System (CDS) and Cleans All (CA). CDS installed at Sungai Kerayong 1 (12 GPT) meanwhile CA installed at Sungai Kerayong 2 (20 GPT). For CDS, the gross pollutants and sediments are separated from stormwater by centrifugal forces, which also prevent the screen from becoming blocked by debris and other items such as plastic bags. Meanwhile, for CA type, the underground gross pollutant and sediment trap that utilise a screened basket to separate the gross pollutants followed by a deeper sump for silt retention.

Figure 3: Continuous Deflective Separation (CDS)

Figure 4: Cleans All (CA)

4. Results and Discussion

In order to evaluate the performance of GPTs system in this study, the results obtained from data collection process was analysed. Gross pollutants wet load were obtained from GPTs maintenance data obtained from Pejabat Lembangan Sungai Klang (PLSK). The data was analysed based on catchment types of land use. The wet load data analysed for Sungai Kerayong 1 and Sungai Kerayong 2 were from April 2015 until December 2015 respectively.

Table 1: Summary of Gross Pollutants Wet Load for Sungai Kerayong 1 & 2 (April – December 2015)

Catchment	Type of Land Use	Total No. of GPTs	Wet Load (kg/ha/months) (Land Use)	Wet Load (kg/ha/yr) (Land Use)	Wet Load (kg/ha/yr/GPT) (Land Use)	Total Wet Load (kg/ha/months) (Catchment)	Total Wet Load (kg/ha/yr) (Catchment)
Sg Kerayong 1	R	1	94.90	126.53	126.53	3521.34	4695.12
	C	8	2784.50	3712.67	464.08		
	M	3	641.94	855.91	285.30		
Sg Kerayong 2	R	13	4346.59	5795.45	445.80	6019.00	8025.33
	M	7	1672.41	2229.88	318.55		

R = Residential, C = Commercial, M = Mix Development

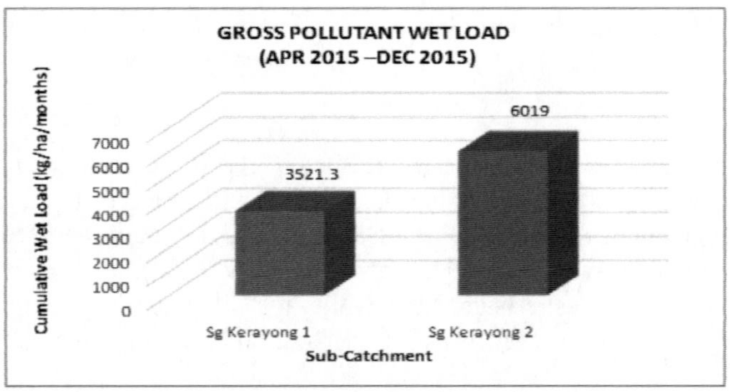

Figure 4: Gross Pollutants Wet Load for Sungai Kerayong 1 & Kerayong 2

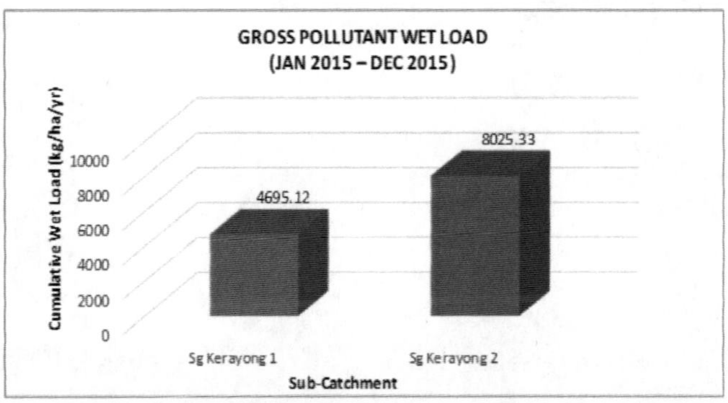

Figure 5: Gross Pollutant Wet Load trapped in a year for Sungai Kerayong 1 & Kerayong 2

The result shows cumulative gross pollutant wet load for Sungai Kerayong 1 is 3521.34 kg/ha, with a total number of 12 GPTs installed in the catchment. Sungai Kerayong 2 with 20 GPTs resulted in 6019 kg/ha within 9 months. In terms of year, total wet load for Sg Kerayong 1 is 4695.12 kg/ha/yr and Sungai Kerayong 2 8025.33 kg/ha/yr. Figure 4 and 5 illustrate the gross pollutant wet load for the captured by 32 GPTs in Sungai Kerayong. Sungai Kerayong is a highly urbanized area with 74 % of imperviousness, which explained the amount of gross pollutant wet load captured in the GPTs [4]. From the result, the highest gross pollutant wet loads is in Sungai Kerayong 2. It occurred caused by high contributions amounts of gross pollutant traps from residential area and also the location of the river itself at which the conditions of climate and rainfall.

5. Conclusion

The wet load assessment at Sungai Kerayong 1 and Sungai Kerayong 2 have been conducted successfully. Result shows that Sungai Kerayong 2 produced higher gross pollutant wet loads than Sungai Kerayong 1 due to the degree of development area, climate and rainfall conditions. The assessment was conducted based on the catchment type of land used. It can be concluded that it is important to identify the amount of gross pollutant emanated from catchment in order to ensure optimum number of GPTs installed to trap gross pollutants before entering river system. This should be subjected to further discussion and deliberation.

Acknowledgement

The authors are grateful to Humid Tropics Centre Kuala Lumpur, Department of Irrigation and Drainage, Malaysia for their financial and technical support.

References

[1] Allison RA, Chiew FH, McMahon TA. *Stormwater Gross Pollutants*. Report 97/11. Australia: Co-operative Research Centre for Catchment Hydrology, eWater CRC, University of Canberra; 1997.

[2] Jehangir T. Madhani, R. J. (2014). *The Capture and Retention Evaluation of a Stormwater Gross Pollutant*. Ecological Engineering, 1-4.

[3] Fitzgerald, B. and Bird, W. (2010). Literature Review: *Gross Pollutant Traps as a Stormwater Management Practice*. Auckland Council Technical Report 2011/006.

[4] Norlida, M.D., Ismail, A. and Rozi, A. (2012) *Dissolved Organic Carbon Production and Runoff Quality of Sungai Kerayong*, International Journal of Engineering & Technology IJET-IJENS Vol: 12 No: 04 Kuala Lumpur, Malaysia.

[5] Ismail Abustan, A. H. (2008). 11th International Conference on Urban Drainage, Edinburgh, Scotland, UK. *Determination of Rainfall-Runoff Characteristics in An Urban Area: Sungai Kerayong Catchment, Kuala Lumpur*, 10.

[6] DID. (2012), Chapter 10, *Storm Water Management Manual For Malaysia 2nd Edition*, Department of Irrigation and Drainage (DID), Malaysia.

[7] Lariyah, M., Nor, M., Khairudin, K., & K.H. Chua. (2006). *Development of Stormwater Gross Pollutant Traps (GPT's) Decision Support System for River Rehabilitation in National Conference on Water for Sustainable Development Towards a Developed Nation by 2020, Guoman Resort, Port Dickson*. 1-8.

Evaluation of Gross Pollutant Wet Load in Sungai Sering, Malaysia

M S F M Noor[1], L M Sidek[1], H Basri[1], N M Zahari[1], N F M Said[1], ZA Roseli[2], N Mohd. Dom[2]

[1]Center for Sustainable Technology and Environment, College of Engineering, Universiti Tenaga Nasional, Jalan IKRAM-UNITEN, 43000 Kajang, Selangor, Malaysia
[2]Humid Tropics Centre, Department of Irrigation and Drainage, Jalan Ledang, Off Jalan Duta, 50480 Kuala Lumpur. Wilayah Persekutuan Kuala Lumpur, Malaysia.

Email: siddiqfarouq@gmail.com

Abstract: The gross pollutant wet load of Sungai Sering catchment, a tributary of Sungai Klang was evaluated. The catchment represents an urban catchment with rapid growth rate due to urbanization process. Urbanization frequently relates to drop of stormwater quality due to many factors such as uncontrolled pollution and improper waste disposal. The study aims to improve the understanding on the wet load trapped in 10 Gross Pollutant Traps (GPTs) in the study area. The amount of wet load analysed based on data collected in the duration from August to December 2015. There is a linear relationship between the amount of gross pollutant wet load and rainfall depth.

1. Introduction

Gross pollutants are the size of debris and litter which is greater than 5 mm and often targeted first for the removal with many structural measures with varying results. These accumulated pollutants are not only aesthetically unattractive, but also demonstrate environmentally threatening and devastating effect to the natural equilibrium, and impede hydraulic performance of the urban drainage system [1].
The rising awareness about the degradation of river water quality by gross pollutants has led to the implementation of gross pollutant management strategies as a holistic approach for water quality improvement [2]. The integration of both structural and non-structural measures is important to ensure the effectiveness of gross pollutant management. Structural measures are constructed in-transit treatments which separate and contain pollutants. The introduction of gross pollutant traps before entering the receiving water such as pond, wetland and river is based on the concept of "control-at-source" with the objective to control stormwater quantity and quality [2].
As a high density population area, Sungai Klang basin is exposed to severe environmental issues due to socio economic activities and rapid land development. Therefore, it is important to understand the factors causing accumulation of gross pollutants in order to manage them effectively, especially in urban waterways [2]. Sungai Sering is one of the main tributaries of Sungai Klang and it was surrounded mostly with residential area. In residential areas, a bulk volume of pollutant could be grass clippings with only small volumes of plastic, bottles and cans [3]. Studies have indicated that a significant proportion of gross pollutants discharged to waterways are generated by residential land, as this type of development constitutes a significant proportion of the land use in most catchments [4].

2. Study Area

The study area located in the state of Selangor and lies in north east of Kuala Lumpur. Sungai Sering is located at the upper stream of Sungai Klang (Figure 1). The total number of GPTs installed included in this study are 10 which has been divided into three types namely Continuous Deflective Separation (CDS), Downstream Defender (DD), and CleansAll (CA). The rainfall depth for this area is extracted from rainfall station 3114005, located at KM 10 Ulu Kelang, Ulu Kelang Height.

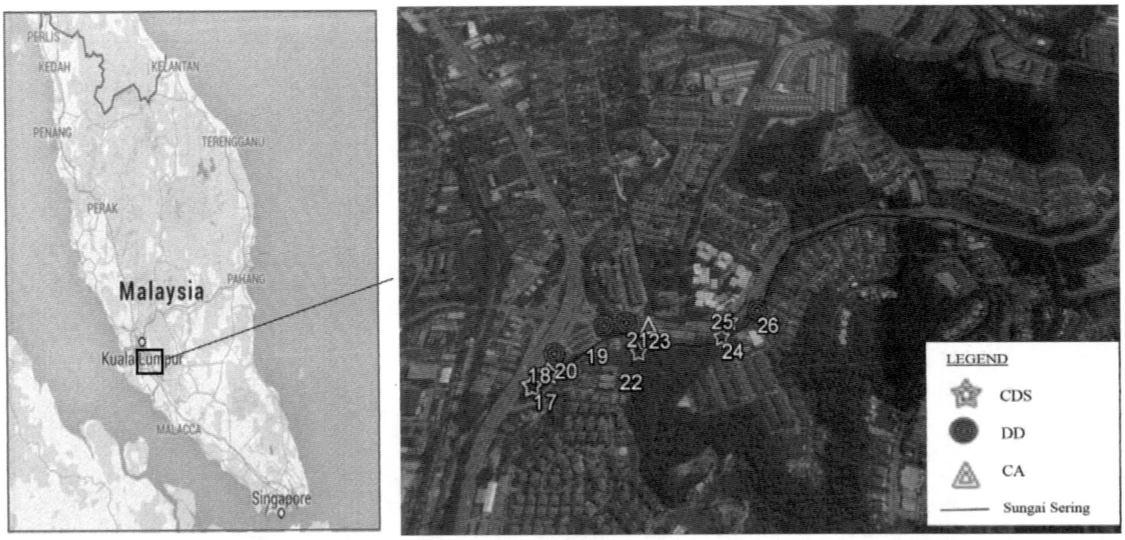

Figure 1: Sungai Sering located in the state of Selangor (left) and the location of 10 GPTs in Sungai Sering catchment (right).

3. Type of GPT and Wet Load Removal Mechanism

Generally, gross pollutant traps can be divided into two types; conventional and proprietary [5]. Conventional gross pollutant trap is a common trap that can be installed at any types of catchment area, while proprietary gross pollutant traps are usually manufactured according to specific site characteristic such as catchment area and rainfall intensity [5]. The type of GPT in Sungai Sering catchment is conventional [6]. Further classification of gross pollutant traps are based on its removal mechanism. It is necessary to identify the removal mechanism of GPTs in order to suit with the purpose of their installation at a particular site [6].

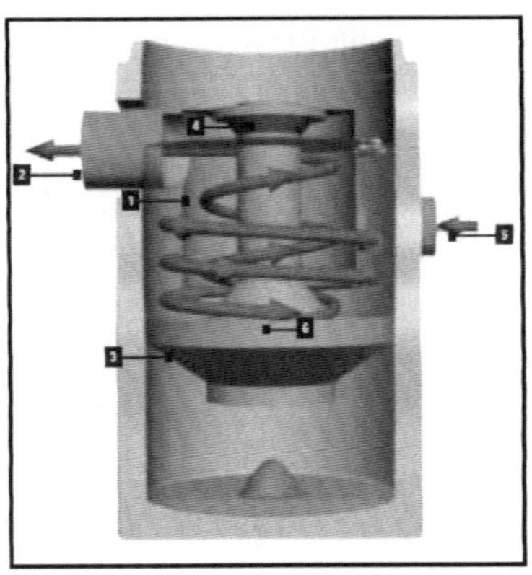

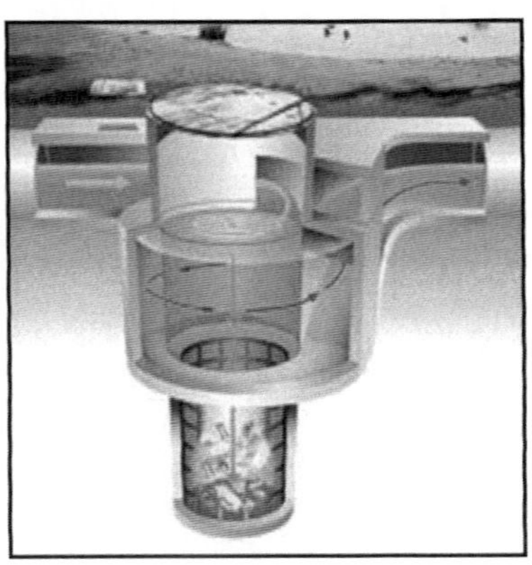

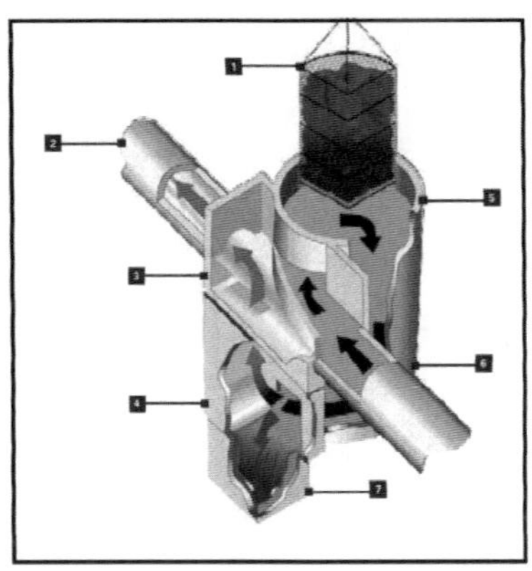

Figure 2: Cross Section of DD [5]. **Figure 3**: Cross Section of CDS [5]. **Figure 4**: Cross Section of CA [5].

4. Results and Discussion

Figure 5 indicated the wet load data for 10 units of GPTs installed in Sungai Sering. The highest wet load recorded was from GPT/SS/DD/23 in December with the weight of 65 kg meanwhile the lowest data recorded was from GPT/SS/CDS/17 in August with load of 11 kg. For cumulative wet load data in each month, December recorded the highest amount of wet load with the total of 354 kg meanwhile the lowest is in August with 204 kg. The reason of highest wet load recorded in December is because of rainy season. Therefore, the rainfall data was measured to evaluate the relationship between rainy season and incremental of wet load in that particular month.

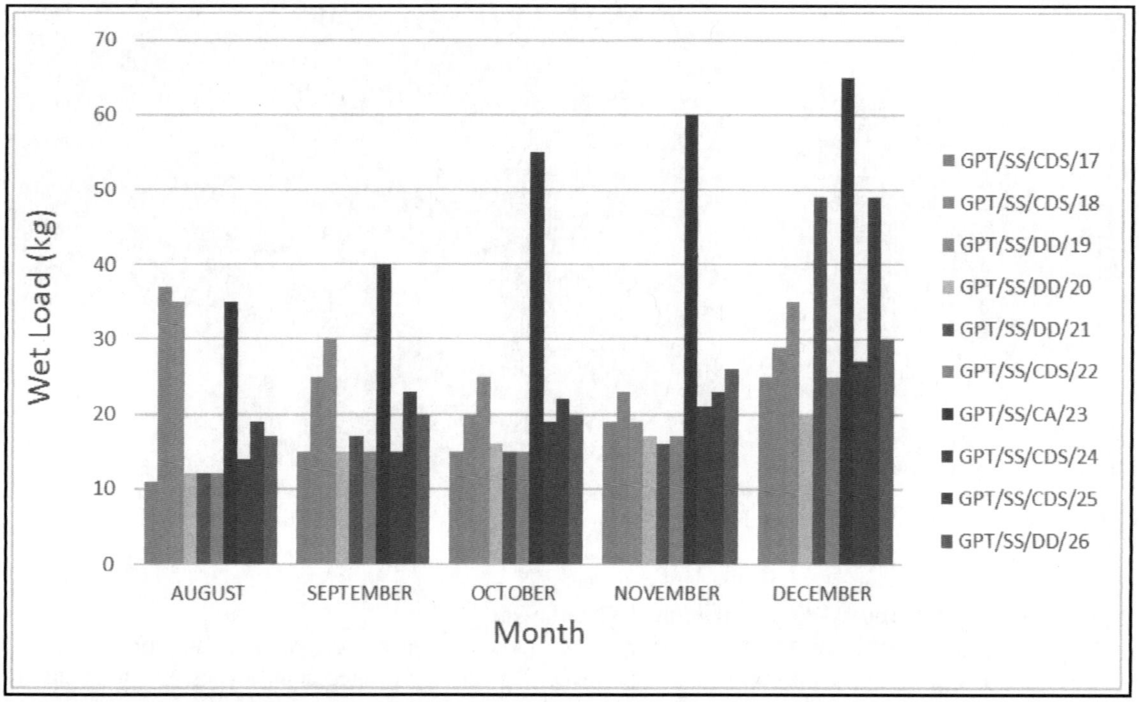

Figure 5: Wet load for GPTs in Sungai Sering catchment area from August to December 2015

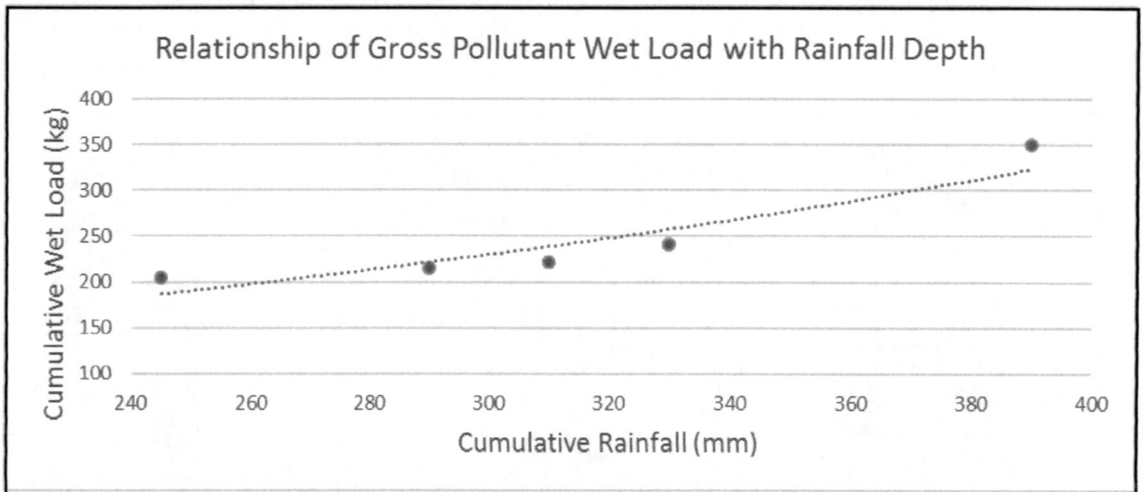

Figure 6: Relationship of Gross pollutant load with rainfall depth.

Figure 6 shows the rainy season that has started in November had contributed to higher wet load trapped in the GPT. This is because the higher flow rate of water has been channelled into the drainage line and eventually carried the wet load into the GPT.

Besides that, the reason for the apparent differences amount of wet load among the GPT could be due to other factors such as location and population. The GPTs located at downstream of the river and in high density residential area trapped higher wet load compared to the GPTs at upstream and low-density residential area. For example, GPT/SS/CDS/18 and GPT/SS/DD/19 recorded cumulative wet load of 134 kg and 144 kg respectively which are higher when compared to GPT/SS/CDS/25 and GPT/SS/DD/26 where the amount of wet load are only 133 kg and 113 kg respectively.

5. Conclusion

An analysis of wet load trapped in the 10 Gross Pollutant Traps (GPTs) that were installed in Sungai Sering has been conducted successfully. Overall, all GPTs have trapped 1233 kg of wet load in five months during the period from August to December 2015. It is vital to study on the performance of GPT to trap the wet load as it can reduce the wet load from being discharged into Sungai Klang; and thus able to improve water quality of the river. Moreover, the reasons of wet load is higher on particular month have been identified. The relationship between the rainfall depth and amount of wet load had proven that the higher rainfall depth contributed to the higher wet load trapped in the GPT. Furthermore, the reasons why higher wet load recorded were because location of GPT and high-density residential area. As a conclusion, the implementation of GPT in the main tributaries of Sungai Klang such as Sungai Sering should be broadened.

Acknowledgements

The authors wish to express our gratitude to Humid Tropics Centre Kuala Lumpur, Department of Irrigation and Drainage of Malaysia (DID) in the collaboration of this study related to the project "River of Life".

References

[1] Armitage, N. (2006). *The Removal of Urban Solid Waste from Stormwater Drains,* Department of Civil Engineering, University of Cape Town, South Africa

[2] Asian Development Bank (2007), *Malaysia: Klang River Basin Environment Improvement and Flood Mitigation Project.*

[3] Q. Zhang, J.L. Zhao, G.G. Ying, Y.S. Liu, C.G. Pan (2014), Emission estimation and multimedia fate modelling of seven steroids at the river scale basin scale in China, Environ. Sci. Technol., 48 7892-7992.

[4] Armitage, N.P., Rooseboon, A., Nel, C. and Townshend, P., (1998). *The Removal of Urban Litter from Stormwater Conduits and Streams*, Report To Water Research Commission Of South Africa, Report No. 95/98.

[5] Caltrans (2002). *Continuous Deflective Separation Units - Operations, Maintenance & Monitoring CTSW-RT-02-037*, California Department of Transportation Sacramento.

[6] Fitzgerald, B. and Bird, W. (2010). Literature Review: Gross Pollutant Traps as a Stormwater Management Practice. Auckland Council Technical Report 2011/006.

Hydrological Analysis for Inflow Forecasting into Temengor Dam

MI Najid, LM Sidek, B Hidayah, ZA Roseli

Centre for Sustainable Technology and Environment, Universiti Tenaga Nasional, Jalan IKRAM-UNITEN 43000 Kajang, Selangor, Malaysia.

E-mail: Idham2521@gmail.com

Abstract. These days, natural disaster such as flood is the main concern for hydrologists. One of solutions in understanding the reason of flood is by prediction of the event sooner than normal occurrence. One of the criteria is lead time or travel time that is important in the study of fresh waters and flood events. Therefore, estimation of lead or travel time for flood event can be beneficial primary information. The objective of this study is to estimate the lead time or travel time for outlet of Temengor dam in Malaysia. Tenaga Nasional Berhad (TNB) Sungai Perak dam operation has the main contribution on decision support for early water released and flood warning to authorities and locals resident for in the down streams area. For this study, hydrological analysis carried out will help to determine which years that give more rainfall contribution into the reservoir. Rainfall contribution of reservoir help to understanding rainfall distribution and peak discharge on that period. It also help for calibration of forecasting model system for better accuracy of flood hydrograph. There may be various methods to determine the rainfall contribution of catchment. The result has shown that, the rainfall contribution for Temengor catchment, is more on November in each year which is the monsoon season in Malaysia. TNB dam operational decision support systems can prepare and be more aware at this time for flood control and flood mitigation.

1. Introduction

Hydrological phenomena such as flood become regular as it happened almost every year in different states in Malaysia [1]. However, rainfall forecasting is one of the key that plays important role in regard with water resources. These days, one of the concerns for hydrologists is natural inflows to reservoirs. For flood mitigation, engineers estimate flood discharge and volume to assess of flood storage dam in order to determine channel capacity. Moreover, there is a need to understand the rainfall pattern and attributed temporal variation, evaporation and infiltration for estimation of irrigation storage and requirement. In addition, estimation of flow from the catchment, losses due to evaporation and seepage are necessary. Mentioned issues are related with hydrological analysis. Indeed, hydrological analysis is involved with flood forecasting, water resources modelling, real-time monitoring and management of water resources system and flood risk mapping. Inflow forecasting is a commonly used technique in hydrology to predict inflow based on forecasted weather and real-time data. It is currently used for flood prediction and hydropower reservoirs. The advantages of inflow forecasting systems are related with flood management, efficient water allocation. Inflow forecasts are one of the key inputs in the short-term scheduling of hydroelectric generation, and therefore, an inflow forecast module is an essential part of an operational Decision Support System (DSS) for hydro scheduling. In this study, inflows forecasting is one of the key inputs in the short-term scheduling of hydroelectric generation, and therefore an inflow forecast module is an essential part of an operational Decision Support System (DSS) for hydro scheduling. The objective of the study is to ensure that the longest possible forecast time is achieved for developing a flood forecasting system that will be characterized by reflecting and addressing the needs of TNB.

2. Study Area

The Perak Hydroelectric Scheme consists of four dams on Sungai Perak (Figure 1). Hydroelectric starts with the Temengor Dam at the upstream and continues by Bersia Dam, Kenering Dam and Chenderoh Dam toward the downstream.

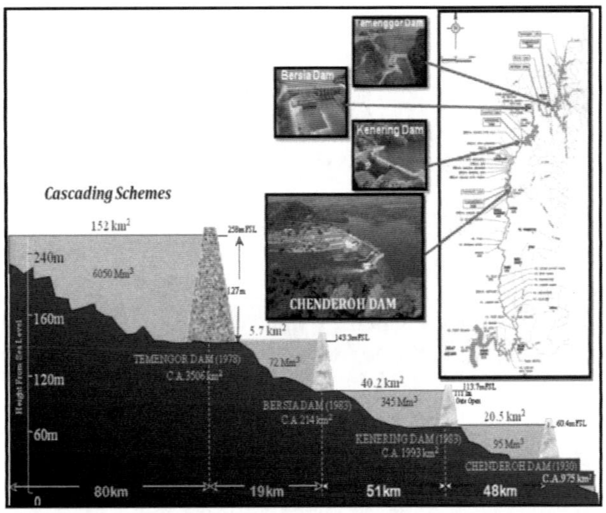

Figure 1: Sungai Perak Hydroelectric Scheme

3. Methodology

In this study, rainfall analysis has been implemented due to transformation of precipitation into channel flow. Usually, a favourable practice is to use hydrological model to represent watershed processes. Rainfall time series analysis will give some figure on rainfall contribution on reservoir, while double mass curve analysis is to verify the time series consistency. Therefore, it is considered to calculate the 7 Tenaga Nasional Berhad (TNB) rainfall stations and 1 Department of Irrigation and Drainage (DID, Malaysia) rainfall station at Temengor catchment area. One hour data of rainfall from the total of 8 stations is used to define the annual average ~~of~~ rainfall.

3.1. Double Mass Curve

The double-mass curve is used to find the harmony and consistency of hydrologic data with comparison of a reference station with selected stations in the catchment [4]. In this study, the cumulative rainfall and the average cumulative rainfall of considered stations are plotted. To give more definite results, the accumulations of rainfall data can be plotted against the accumulations of the average of many records of other nearby stations which are influenced by the same meteorological condition is less affected by an inconsistency in the record of any one station. Figure 2.a and 2.b shows the double mass curve analysis for TNB station 9061 and average accumulative rainfall at other six stations at Temengor catchment.

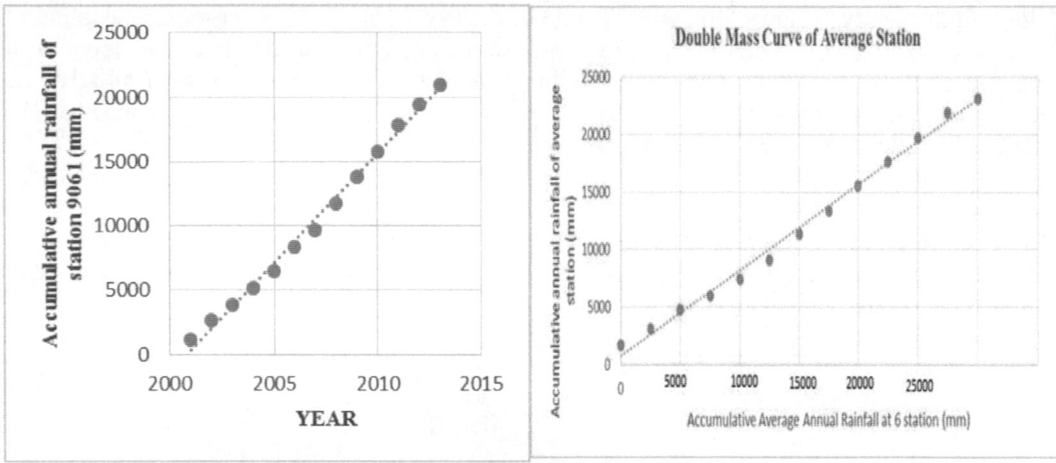

Figure 2.a: Individual station double mass curve analysis for station 9061, **2.b**: Double mass curve analysis for station 9061 against other selected 6 stations

3.2. Rainfall Trend Analysis

The hourly recorded rainfall data is collected in time series for all DID and TNB stations. Figures 3.a and 3.b present 1 hour time series for Empangan Temengor (Station No: 9061) and Kg. Sg. Tiang (Station No: 9120)

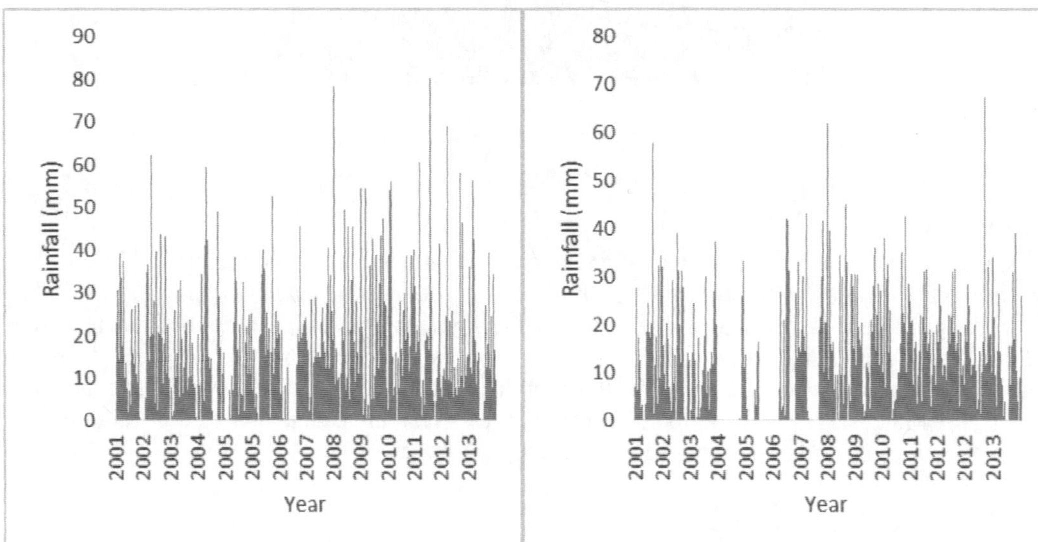

Figure 3.a: 1 Hour Time Series for Empangan Temengor **3.b**: 1 Hour Time Series for Sg. Tan. Hain

4. Results and Discussion

Due to recognition of rainfall into the reservoir, monthly average rainfall is calculated from each station to monitor high precipitation months. Table 1 shows the example of rainfall contribution data into reservoir for station 9119 Sg. Tan. Hain. From the data, November 2009 has contributed the most rainfall with average 435.5 mm which has a relative contribution roughly 22.7%. In January 2010, the lowest rainfall contribution were recorded with the average of 2.8 mm per month which has a relative contribution roughly 1.5%.

Table 1: Percentage of rainfall contribution into reservoir

Date	Rainfall (mm)	Percentage of Rainfall Contribution (%)
July 2009	170	8.85
Aug 2009	192	9.99
Sept 2009	201.5	10.49
Oct 2009	194.5	10.12
Nov 2009	435.5	22.66
Dec 2009	142	7.39
Jan 2010	28	1.46
Feb 2010	37.5	1.95
Mar 2010	28.5	1.48
Apr 2010	73	3.80
May 2010	208.5	10.85
Jun 2010	210.5	10.95

5. Conclusion

From the rainfall data analysis, the estimation of rainfall contribution for Temengor catchment is possible. Double mass curve can help to check the consistency of hydrologic data at the upstream stations. Rainfall contribution time forecast can be estimated at that rainfall stations by statistical rainfall analysis. The rainfall contribution for Temengor catchment is more on November each year which is the monsoon season in Malaysia.

6. References

[1] I Hafiz, N D M Nor, L M Sidek, H Basri, F K, M N Hanapi, Livia L (2013). *Application of Integrated Flood Analysis System (IFAS) for Dungun River Basin.* IOP Conf. Series: Earth and Environmental Science 16 (2013)

[2] World Meteorological Organization (WMO-No. 1072) (2011). Manual on Flood Forecasting and Warning

[3] Faridah Othman and Mahdi Naseri (2011) *Reservoir Inflow Forecasting Using Artificial Neural Network,* International Journal of the Physical Sciences Vol.6(3), pp 434-440, 4 February, 2011

[4] James K.Searcy and Clayton H. Hardison (1960). *Manual of Hydrology: Part 1. General Surface-Water Techniques,* Geological Survey Water-Supply Paper 1541-B.

[5] Deepesh Machiwal and Madan K. Jha, (2006). *Time Series Analysis of Hydrology Data for Water Resources and Planning and Management.* Journal Hydrol. Hydromech, 54, 2006

[6] Government of Malaysia Department of Irrigation and Drainage (DID Malaysia) (2009). Did Manual, Volume 4 – Hydrology and Water Resources

IOP Conf. Series: Earth and Environmental Science **32** (2016) 012068 doi:10.1088/1755-1315/32/1/012068

Decision Support System (DSS) for MSMA Integrated Stormwater Management Ecohydrology for Sustainable Green Infrastructure

L M Sidek[1], H A Mohiyaden[1], H Haris[1], H Basri[1], Z C Muda[1], Z A Roseli[2], M D Norlida[2]

[1]Centre for Sustainable Technology and Environment (CSTEN), Department of Civil Engineering, Universiti Tenaga Nasional, Malaysia

[2]Humid Tropics Centre, No 2 Jalan Ledang, Off Jalan Duta, 50480 Kuala Lumpur

E-mail: Hairun@uniten.edu.my

Abstract. Rapid urbanization has known to have several adverse impacts towards hydrological cycle due to increasing impervious surface and degradation of water quality in stormwater runoff. In the past, urban waterways have been confined to narrow river corridors with the channels canalised and concrete and other synthetic materials forming the bed and banks of the river. Apart from that, stormwater pollutants such as litter, debris and sediments in drainage system are common problems that can lead to flooding and the degradation of water quality. To solve this problem, implementing stormwater Best Management Practices (BMPs) proves very promising due to its near natural characteristics and multiple effects on the drainage of stormwater runoff in urban areas. This judgment of using BMPs depends on not only relevant theoretical considerations, but also a large amount of practical experience and the availability of relevant data, as well. To fulfil this task, the so-called Decision Support System (DSS) in MSMA Design Aid and Database system are able to assist engineers and developers in management and improvement of water quantity and quality entering urban rivers from urban regions. This system is also helpful when an expert level judgment procure some repetitive and large amount of cases, like in the planning of stormwater BMPs systems for an entire city catchment. One of the advantages of an expert system is that it provides automation of expert-level judgement using availability of checking tools system.

1. Introduction

Malaysia faces serious challenges in urban water management. In terms of stormwater management, even with the introduction of Urban Stormwater Management Manual (MSMA) in 2001 [1], many new technologies have yet to be studied in depth especially the interrelation between water quantity and quality. Rapid urbanization has known to have several adverse impacts towards hydrological cycle due to increasing impervious surface and degradation of water quality in stormwater runoff. Solids in stormwater runoffs adversely influence waterways located in or near urban area. [2] Stormwater pollutants such as litter, debris and sediments in drainage system are common problems, which can lead to flooding and the degradation of water quality.

In Malaysia, MSMA has introduced various storm water Best Management Practices (BMPs) as treatment control. BMPs proposed to be install at the downstream end of drains or engineered waterways [3]. The installation of storm water BMPs at downstream end of drains or engineered waterways has become common practice since the introduction of the new MSMA [3,4]. This structural measure of urban stormwater is based on the concepts of "control at source" with the objective to control stormwater quantity and quality [4].

To fulfill this task, the so-called an integrated checking aid tools called Design Aid and Database System, is proposed and introduced. One of the advantages of a Design Aid is that it provides automation of expert-level judgment. This is extremely helpful when an expert-level judgment is needed repeatedly for a large amount of cases, like in the planning of stormwater BMPs for an entire city catchment.

Thus, MSMA Design Aid and Database System is capable to optimize the performance of storm water BMPs as it is strongly dependent on specific site criteria including type of land use, hydrological data and maintenance frequency with the objective to assist engineers and local authorities to select the most appropriate strategy for storm water BMPs trapping stormwater pollutants in urban area during construction and post-construction. This software can provide the detail design of each selected stormwater BMPs based on MSMA 2nd Edition. It is also aim to expand the sources for managing stormwater pollutants in order to rehabilitate the river system and prepare a budget allocation of using storm water BMPs in terms of installation cost and maintenance cost annually including the Life Cycle Cost analysis. MSMA Design Aid and Database System module framework covers six main components; Database, Hydrological component, Design Quantity Control, Design Quality Control, Conveyance component and Checking Tools component. Overall, this project describes a self-developed design aid tool and database, as well as a case study for the selection of stormwater BMPs measures for the selected area in Kuala Lumpur.

2. Materials and Methods

2.1 Module Framework MSMA Design Aid and Database System

MSMA Design Aid and Database System module framework covers six main components; Database, Hydrological component, Design Quantity Control, Design Quality Control, Conveyance component and Checking Tools component. Figure 1 shows MSMA Design Aid and Database Modules.

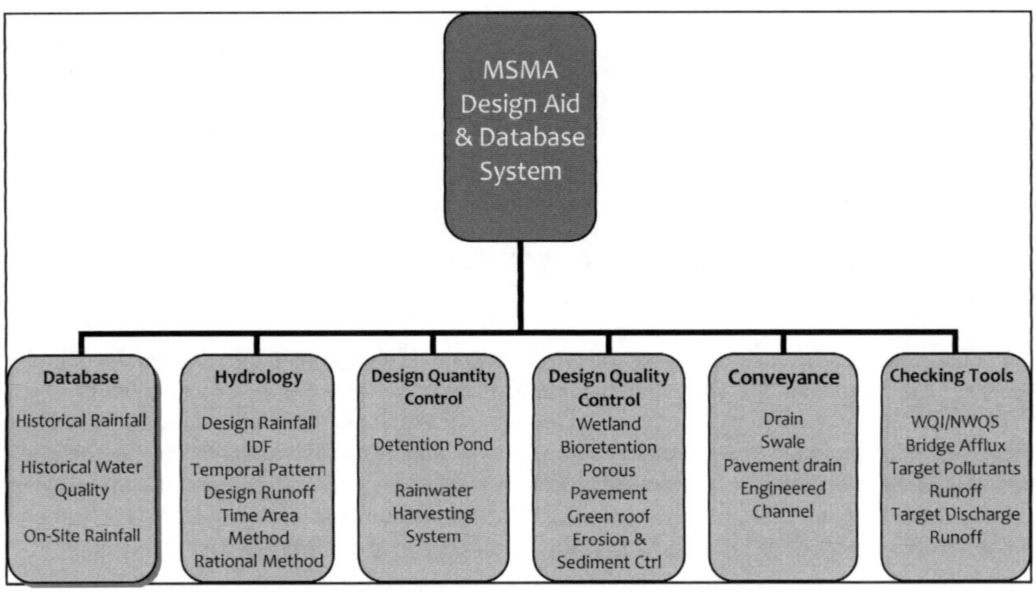

Figure 1: MSMA Design Aid and Database Modules

Figure 1 shows every component of MSMA Design Aid and Database Modules which consists of minimum two sub-components. Database component is divided into Historical Rainfall data (HTC and UNITEN Rainfall station), Historical Water Quality Data (HTC and UNITEN Rainfall station) and On-site Rainfall data Collection (HTC). Basic aims in Hydrological Module are to calculate the Design Rainfall and Design Runoff. Methods available in this module are Design Rainfall, IDF Curves, Temporal Pattern and Design Runoff using Time Area Method and Rational Method. In design Quantity control design menu, it consists of Detention Pond design calculation and Rainwater Harvesting System design calculation. In Design Quality Control module, it consists of several BMPs design preference; Wetland, Bioretention, Porous pavement, Green Roof, Infiltration facilities and Erosion and Sediment Control. For Conveyance Module, MSMA software provides design calculation for all types of drain available in MSMA Guidelines including pavement drain and swales. There are many options available in Checking Tools Module including target pollutants runoff, target discharge runoff, Water Quality Index (WQI) Calculation and Bridge Afflux.

2.2 MSMA Design Aid and Database System Database Work Flowchart

Database work flowchart consisting of three main phases as shown in Figure 2. The first phase is a data input MSMA Design Aid and Database System e.g rainfall data and water quality data. The second phase namely Model of is consisting of Data Entry Procedure, Data Record, Data search and Database management, Data analysis and search result and analysis. This model will integrate with User Input and External Model. The Third phase is aimed to depict the outputs in the form of report, tables, drawing, statistical report.

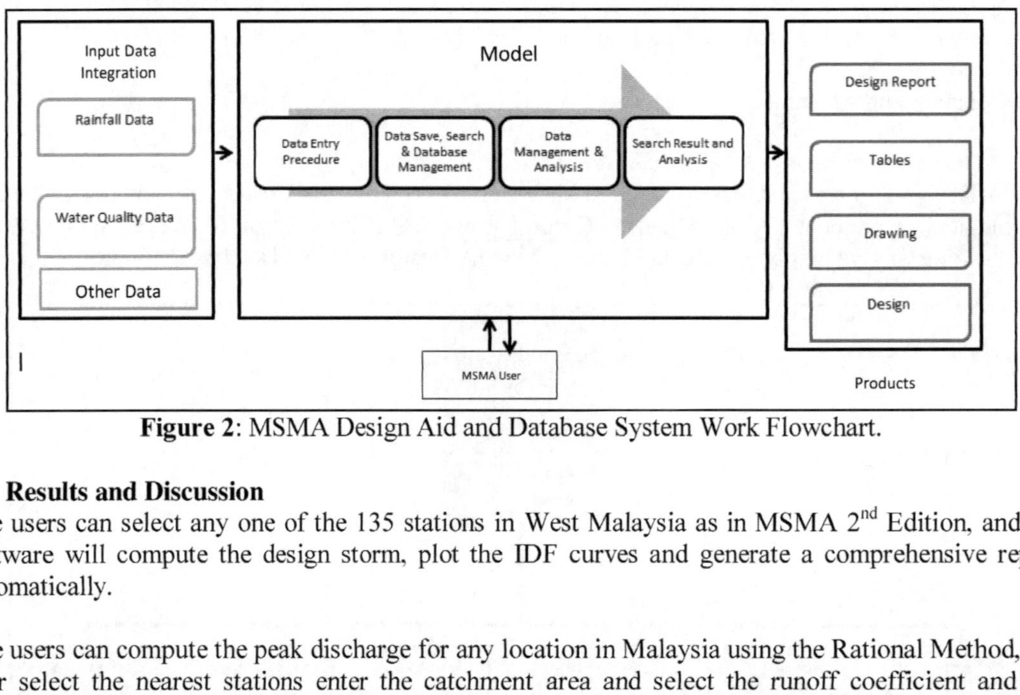

Figure 2: MSMA Design Aid and Database System Work Flowchart.

3. Results and Discussion

The users can select any one of the 135 stations in West Malaysia as in MSMA 2nd Edition, and the software will compute the design storm, plot the IDF curves and generate a comprehensive report automatically.

The users can compute the peak discharge for any location in Malaysia using the Rational Method, the user select the nearest stations enter the catchment area and select the runoff coefficient and the software will compute the design discharge and generate a report automatically. The same applies to the Time-Area Method where user can select the nearest station and the software will automatically generate the design storm data, and use them in the convolution procedure to compute the discharge. The users can calculate the pollutant loading for various landuse e.g. residential, commercial, industrial, roads and highways land use areas, which consists of up to 17 pollutant parameters.

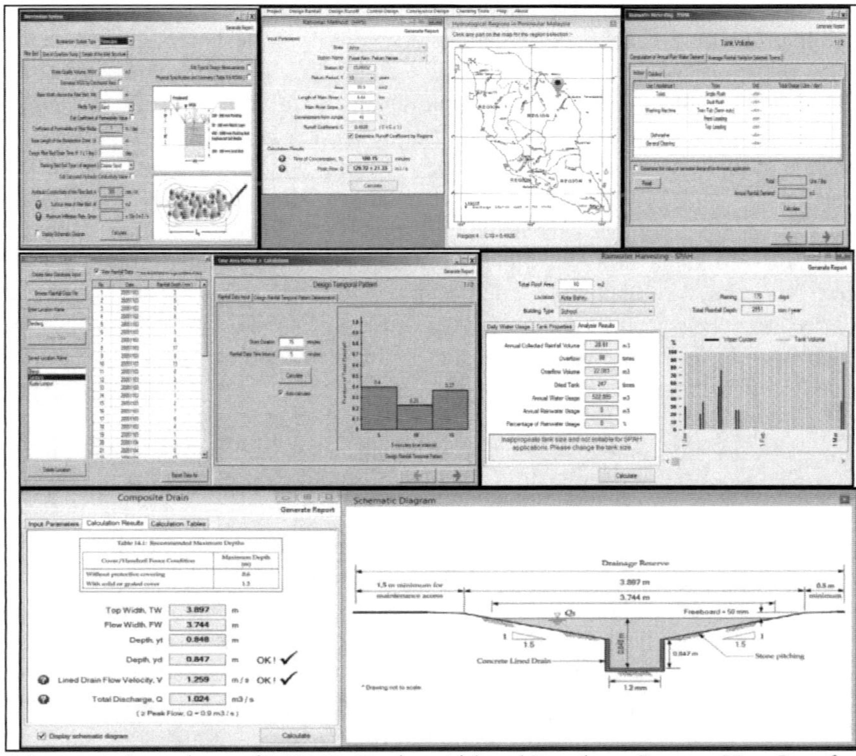

Figure 4: Snapshots of MSMA Design Aid and Database System User Interface

This software complies with editable function of the Event Mean Concentrations (EMC) values for each pollutant parameter and also supports more than 50 sub-catchments in one calculation process and generate automatically. The users can estimate the water quality volume before any assessment can be made of pollutant loads depending on landuse type. In this software, it consists of two calculation methods due to land use area either by percentage or by hectare based on 40 mm rainfall depth over the equivalent impervious contributing catchment area for water quality treatment. The detail design for each components in MSMA 2nd Edition can be done automatically in the software namely Quantity Design Control (e.g. detention pond, rainwater harvesting), Quality Design Control (e.g. wetland, water quality pond, bio retention, green roof, infiltration facilities, porous pavement) and Conveyance Design (e.g. open drain, pipe drains, composite drain, engineered channel, swale, curb opening inlet and road gutter). For detention pond design, the software allows the users to choose one out of three methods of computing the inflow hydrograph for reservoir routing i.e., Rational Method, Time-Area Method and user defined hydrograph.

The software computes the design storm for the hydrograph computation and single detention pond automatically. For Rainwater Harvesting (RWH) design, the software allows the users to choose one of two methods of designing RWH i.e. MSMA 2nd Edition method (Chapter 6) or SPAH method (JKR). The rainwater harvesting is a technique of collecting rainfall as a supplementary source of water supply for household, commercial, landscape and recommended to install for new development. The software can design the culvert based on the design discharge from one of three methods of computing peak discharge as follows: Rational Method, Time-Area Method and user defined discharge. The software provides checking tools for the users including Bridge Afflux Estimator, WQI Calculator and Statistical Analysis as an option for the users to evaluate the performance and reliability of the design.

MSMA Design Aid and Database System is interactive process software, which involves a number of different stakeholders such as water resource engineers, urban planners, economists, landowners and the public from, planning to designing stages. Therefore, this increases the pressure of making appropriate decisions when there are contrasting opinions from different stakeholders. MSMA Design Aid and Database System is suggested to upscale into Model driven Decision Support Systems (DSS) as in environmental/stormwater modelling are becoming more popular presently rather than the stand along modelling tools, which provide only a direct outcome. The major reason for that is DSS is a platform that combines both comprehensive modelling along with other important aspects such as social, technical and economic considerations that are becoming equally important in catchment modelling. However, along with the advancement of model driven DSS, some areas need be addressed in regards to increasing their accuracy and effectiveness. MSMA Design Aid and Database System can be further enhance by using the remote sensing and GIS technologies for large scale distributed catchment modelling and urban planning activities for processing, analyzing and visualizing digital spatial data. This also requires development of GIS and remote sensing mapping databases, developing links between models and the spatial software and enabling of the import and export of data sets between two platforms which itself needs to be further researched in the future.

MSMA Design Aid and Database System can be further improved by using Web based simulation as it provides a more integrated approach in modelling due to the ability to access a wider range of spatially distributed data sets. Future research studies need to conduct for selection of the best optimization algorithms for particular MSMA Design Aid and Database System selection processes. These optimization algorithms should also consider model runtimes and data handling capacities [7]. It is important for sustainability of MSMA Design Aid and Database System for the developer to consider the initial purchase price, upgrades and costs for on-going training and maintenance of the program are important criteria in model selection. Computers are routinely upgraded which directly affects programs due to different system requirements. Some programs may have a lower initial price but higher maintenance costs [8]. During urban stormwater modelling of a project, things can go wrong or data errors occur [9]. A hands-on software development is therefore an important part of the complete MSMA Design Aid and Database System software package.

4. Conclusion

The MSMA Design Aid and Database System have potential to be very useful software. The upgraded software development in the future will make MSMA Design Aid and Database System become portal based type decision support system. Further development in future will make MSMA Design Aid and Database System has the capability to integrate with GIS system and adopted by government agencies and engineering consultants in Malaysia.

Acknowledgements

The authors would like to express their gratitude to Universiti Tenaga Nasional and Fundamental Research Grant Scheme (U-GN-CR-15-02) funded by Humid Tropics Centre (HTC) Kuala Lumpur for supporting this research.

References

[1] Basri, N. (1999). An expert system for the design of composting facilities in developing countries. PhD Thesis, Department of Civil Engineering, University of Leeds

[2] Cohn, A. G., & Thomas, J. (1987). Artificial intelligence and its applications: John Wiley & Sons, Inc. USA

[3] DID. (2000). Urban Stormwater Management Manual, Volume 13, Chapter 34, Malaysia

[4] DID. (2011). Urban stormwater management manual for Malaysia: Department of Irrigation and Drainage Malaysia

[5] Gasching, J., Reboh, R., and Reiter J. (1981). Development of knowledge-based system for water resources problems: SRI international, Menlo Park, California

[6] Roesner, L.A., Amy Pruden And Elizabeth M.Kidner (2007). Improved Protocol For Classification And Analysis Of Stormwater Borne Solid, Water Environment Research Foundation, Colorado State University

[7] Saunders, M., Sullivan, T., Nash, B., Tonnessen, K., & Miller, B. (2005). A knowledge-based approach for classifying lake water chemistry. Knowledge-Based Systems, 18(1), 47-54

[8] Mohd. Sidek, L., Zakaria, N.A, Ab Ghani, A., Desa, M.N. & Othman, N. (2006). An Assessment Of Stormwater Management Practices In Malaysia, Jurutera IEM, p. 8-19

[9] Allison, R., Chiew, F and Mcmahon, T (1997). Stormwater Stormwater Pollutants Industry Report 97/11, Cooperative Research Centre For Catchment Hydrology, Monash University, Victoria

Hydrology Analysis and Modelling for Klang River Basin Flood Hazard Map

L M Sidek[1], N.E Rostam[1], B. Hidayah[1], ZA Roseli[1], W.H.A.W.A Majid[2], N.Z Zahari[1], S.H.M. Salleh[1], R.D.R. Ahmad[3], M.N. Ahmad[3]
[1]Centre for Sustainable Technology and Environment, College of Engineering, Universiti Tenaga Nasional, Kajang, Selangor, Malaysia.

[2]Department of Irrigation and Drainage, Ampang, Selangor, Malaysia.

[3]TNB Research Sdn Bhd, Kajang, Selangor, Malaysia.

E-mail: lyeena14@gmail.com

Abstract. Flooding, a common environmental hazard worldwide has in recent times, increased as a result of climate change and urbanization with the effects felt more in developing countries. As a result, the explosive of flooding to Tenaga Nasional Berhad (TNB) substation is increased rapidly due to existing substations are located in flood prone area. By understanding the impact of flood to their substation, TNB has provided the non-structure mitigation with the integration of Flood Hazard Map with their substation. Hydrology analysis is the important part in providing runoff as the input for the hydraulic part.

1. Introduction

Globally, floods are frequent natural disaster that cause severe damage to both lives and property. It is estimated that of the total economic loss caused by all kinds of disasters, 40% are due to flooding [1]. In Malaysia, floods are the most important natural hazard in terms of population affected, frequency, areal extent, and socio-economic damage [2]. As the largest utility company in Malaysia, Tenaga Nasional Berhad (TNB) is facing a major threat due to the rising severity and frequency of flood poses to their supply performance, inflicting unexpected revenue loss and high repair and maintenance cost. During the flood event on December 2014, TNB's losses are RM10million in the state of Kelantan. [3]. This has given lots of pressure to TNB for the early recovery from the flood impact. With the facts that the nature disaster such as flood cannot be stopped, the mitigation measure becomes the most critical part for TNB in protecting their assets.

Flood Hazard Map (FHM) is one of the non-structure approach practice by Department of Irrigation and Drainage (DID) under the flood mitigation scheme. Up to 2013, there are 28 FHMs for selected river basin available from the total of 189 [4] main river basin. These FHMs are develop by using combination of Hydrological and Hydraulics modelling. This map is developed by using different Average Recurrence Interval (ARI).The accuracy of the modelling is highly depending on the quality and accessibility of the data. With the cooperation of FHM in TNB system, it will help in identifying the risk of flooding to all their existing substations and planning for their upcoming substation.

This study is focusing on development of FHM in Klang River Basin. With an estimated population of over 3.6 million, (about 21% of the national population) and with a growth rate of almost 5% per year, the basin has experienced the highest economic growth in the country [5, 6]. Klang River basin is bonded with 2 states which are Selangor and Wilayah Persekutuan Kuala Lumpur. Major cities such as Kuala Lumpur grew at the confluence of the Klang and Gombak rivers and is exposed to flood risk [4]. Situated in a flood plain, Kuala Lumpur has been experiencing a number of major flood hazards. The development of FHM ~~map~~ will give a huge impact to the TNB itself in improving the resilient their substation from flooding.

This paper is focusing on the hydrological analysis aspect of the development of Klang River Basin FHM. In this analysis involved three important parts; design storm, catchment characteristic and calibration. All calculation is adopted from Urban Stormwater Management Manual for Malaysia (MSMA) 2nd edition that has been publish by Department of Irrigation and Drainage. [7]

2. Objective

The main objective of hydrologic analysis is to generate runoff from all the tributaries/sub-catchment using rainfall runoff model. This runoff value will become the main input to the river and flood plain model which carries out the flood routing. This subsection discusses on the process for the development of rainfall runoff model. The rainfall runoff model is based on the Soil Conservation Service (SCS) unit hydrograph method. This method was chosen since it is capable to produce runoff hydrograph and suitable for unsteady flow simulation which requires river and flood plain analysis.

3. Methodology

The preparation of hydrologic data is the major part to produce an accrued flood hazard map that reflects the real situation. This analysis will involve the rainfall estimation and sub-catchment of Klang River basin. As a result, the data that been analysed will become the input for the Infoworks RS to develop the runoff hydrograph. The river catchment had been develop in the model include the river alignment by using all the nodes and river line. In the Infoworks RS, the parameter that need to be include for the hydrological modelling are rainfall intensity, types of soil concentration and river alignment. In this paper the SCS hydrograph method.

The design storm involves the rainfall data from DID rain-gauge stations, while the catchment characteristic involve the time of concentration (t_c) for each of the sub catchment in Klang river basin. The calibration is used to compare the actual rainfall event with the simulation done based on the available data. In the simulation, there are few assumptions done to provide an accrued model that reflects the real situation.

Based on MSMA, there are 135 numbers of rain-gauge stations that have developed IDF curves throughout Peninsular Malaysia at various return periods. Fifteen of the rainfall stations are within the Sungai Klang catchment and Table 1 lists the rainfall stations adopted in this study while Equation 1 is use to develop the rainfall intensity for various storm duration.

$$i = \frac{\lambda T^k}{(d+\theta)^n} \qquad (1)$$

where: i = Rainfall Intensity
T = Design return period
d = Duration of storm
λ,k, θ,η = Derived Intensity Duration Frequency (IDF) parameters

Table 1: The list of rain gauge station used in hydrological modelling

No.	Station No	Station Name	Latitude	Longitude	λ	κ	θ	η
1	3015001	Puchong Drop, KL	03° 01' 10.3"	101° 35' 50.7"	69.65	0.151	0.223	0.88
2	3116003	Ibu Pejabat JPS	03° 09' 05"	101° 41' 05"	61.976	0.145	0.122	0.818
3	3116004	Ibu Pejabat JPS1	03° 08' 50"	101° 41' 15"	64.689	0.149	0.174	0.837
4	3116005	SK Taman Maluri	03° 11' 50"	101° 38' 10"	62.765	0.132	0.147	0.82
5	3116006	Ladang Edinburgh	03° 11' 00"	101° 38' 00"	63.483	0.146	0.21	0.83
6	3216001	Kg.Sungai Tua	03° 16' 20"	101° 41' 10"	64.203	0.152	0.25	0.844
7	3216004	SK Jenis Keb. Kepong	03° 14' 04"	101° 39' 38"	73.602	0.164	0.33	0.874
8	3217001	Ibu Bek.KM16,Gombak	03° 16' 05"	101° 43' 45"	66.328	0.144	0.23	0.859
9	3217002	Emp.Genting Kelang	03° 14' 10"	101° 45' 10"	70.2	0.165	0.29	0.854
10	3217003	Ibu Bek.KM11,Gombak	03° 14' 10"	101° 42' 50"	62.609	0.152	0.221	0.804
11	3217004	Kg. Kuala Seleh, H.Klg	03° 15' 30"	101° 46' 05"	61.516	0.139	0.183	0.837
12	3217005	Kg. Kerdas, Gombak	03° 14' 45"	101° 42' 55"	63.241	0.162	0.137	0.856
13	3317001	Air Terjun Sg. Batu	03° 20' 05"	101° 42' 15"	72.992	0.162	0.171	0.871
14	3317004	Genting Sempah	03° 22' 05"	101° 46' 15"	61.335	0.157	0.292	0.868
15	3117070	JPS Ampang	03° 09' 11"	101° 44' 56"	65.809	0.148	0.156	0.837

4. Results and Discussion

The main objective for this paper is to provide the overall element in developing hydrological analysis for FHM Klang River. Table 2a and 2b are the results of Equation 1. This value will be an input together with the catchment characteristic of the sub-catchment for Klang river basin. The value of t_c for this study is gathered from Kirpich Method. The Kirpich equation is normally used for natural basins with well-defined channels. By using the Infoworks RS, the rainfall runoff hydrograph is produce. Figure 1 shows the calibration done for this model and the details location with the events is shown in Table 3.

The first calibration is at Sungai Damansara at TTDI Shah Alam 3 (Figure 1 (a)). It can be concluded that the results from the simulation is in agreement with the observation, the peak water level is lower than observed. The simulated peak is at 7.67m whereas the observed peak is 7.84m resulting in 0.17m variant. The overall shape of the simulated hydrograph is also a good fit, which gives confidence that the model is representing the complete runoff process. Based on the calibration Sungai Klang at Jalan AU3 (Figure 1(b)) it can be concluded that the results from the simulation is in agreement with the observation, albeit, the peak water level is slightly lower than observed. The simulated peak is at 46.813m whereas the observed peak is 46.830m resulting in 0.017m variant. The overall shape of the simulated hydrograph is also a good fit, which gives confidence that the model is representing the complete runoff process.

While for calibration no 3 (Figure 1(c)) ~~shows~~ the overall results show a good match between the simulated and observed water level. However, a close inspection shows that there can be some significant differences between the model and observed peak water levels.

Table 3. The list of rainfall and water level station for model calibration

No	Location Of Calibration	Duration	List of Rainfall Station	Water Level Station
1.	Sg Damansara @ TTDI Jaya	13th & 15th October 2007	1) Air Itam Puchong (No. 2915117) 2) Rumah Pam Rantau Panjang (No. 3013001) 3) UITM Shah Alam (No. 3014091) 4) Setia Alam (No. 3114085) 5) Ibu Pejabat JPS Wilayah Persekutuan (No. 3116003) 6) Ladang Edinburgh Site 2 (No. 3116006)	1) Sg. Damansara TTDI Jaya (No. 3015490)
2.	Sg Klang at AU3	3rd & 6th September 2008	1) IBMBS @ Kg. Kemensah 2) Sg Klang @ Jln AU3 3) Sg Klang @ Empangan Klang Gate 4) Kuala Seleh @ Empangan Klang Gate	1) Sg. Klang @ Jln AU3 (No. 3015490)
3.	Sg Gombak at Jalan Sultan Ismail – SMART hydrological stations	1st May- 30th June 2011	1) Apartment UIA @ Jln Gombak 2) Sg Gombak @ Jln Changkat Tmn Greenwood 3) Pusat Pengajian Luar UM @ Jln Gombak 4) Sg Batu @ Jln Chendurah 5) Tmn Sri Sinar	1) Sg. Gombak @ Jalan Sultan Ismail 2) Sg. Gombak @ Jln Changkat

Table 2a. Rainfall intensity (mm/hr) for various storm durations and ARIs at Station Puchong Drop, Kuala Lumpur

Duration (hr)	Yearly Return Period, ARI (mm/hr)				
	5	10	20	50	100
0.25	171.6	190.6	211.6	243.0	269.8
0.5	118.1	131.2	145.7	167.3	185.7
1	74.4	82.6	91.7	105.3	116.9
3	44.0	48.8	54.2	62.3	69.1
6	17.8	19.7	21.9	25.2	27.9
12	9.8	10.9	12.1	13.9	15.4
24	5.4	6.0	6.6	7.6	8.4
48	2.9	3.3	3.6	4.2	4.6
72	2.1	2.3	2.5	2.9	3.2

Table 2b. Total rainfall (mm) for various storm durations and ARIs at Station Puchong Drop, Kuala Lumpur

Duration (hr)	Yearly Return Period, ARI (mm/hr)				
	5	10	20	50	100
0.25	42.9	47.6	52.9	60.7	67.5
0.5	59.1	65.6	72.8	83.6	92.9
1	74.4	82.6	91.7	105.3	116.9
3	87.9	97.6	108.4	124.5	138.2
6	106.6	118.4	131.5	151.0	167.6
12	117.7	130.7	145.2	166.7	185.1
24	129.0	143.2	159.0	182.6	202.8
48	140.7	156.3	173.5	199.3	221.3
72	148.0	164.3	182.4	209.5	232.6

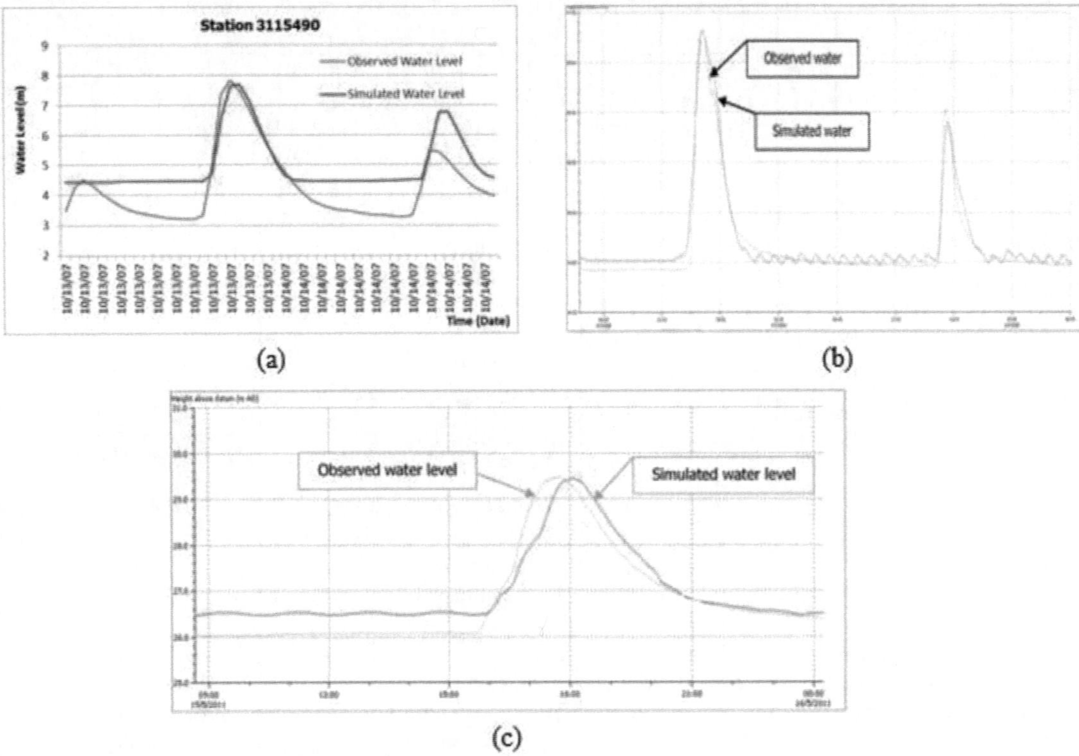

Figure 1. Result of calibration for water level output at: (a) Sg Damansara@TTDI Jaya (b) Sg Klang at AU3 (c) Sg Gombak at Jalan Sultan Ismail

5. Conclusion

As a conclusion, this hydrological analysis has shown a reliable result based on the calibration that has been done in the overall Klang River Basin. The production of rainfall runoff will be used in the hydraulic analysis to provide the complete FHM.

Acknowledgment

This paper was written as part of project Research of Floods on TNBD's PMU/PPU/SSU/PE and The Proposed Mitigation Measures. This project is collaboration between Universiti Tenaga Nasional (UNITEN) under the Centre for Sustainable Technology and Environment (CSTEN) with Tenaga Nasional Berhad-Research (TNBR) for Tenaga Nasional Berhad – Distribution (TNBD). The authors would like to thank the Department of Irrigation and Drainage for all the data given and all the people that have contributed to this project.

References

[1] Feng LH, Lu J (2010). The practical research on flood forecasting based on artificial neural networks. Expert Syst Appl 37:2974–2977
[2] Pradhan B, Youssef AM (2011) A 100-year maximum flood susceptibility mapping using integrated hydrological and hydrodynamic models: Kelantan River Corridor, Malaysia. J Flood Risk Management
[3] Malaymail online, 2015. Chairman of Kelantan's Flood Disaster Operations Committee: TNB

suffered losses RM10m in flood hit area. [online] Available at: http://m.themalaymailonline.com[Accessed Feb. 2015].

[4] Noorazuan, M.H *Flash Flood, Imperviousness and Urban Drainage in Malaysia: The Geographical Perspective.* **105**

[5] Department of Irrigation and Drainage 2013 *Flood Management Approaches.* [online] Available at: http://www.water.gov.my [Accessed 25 June 2015]

[6] Ketua Pengarah Jabatan Pengairan & Saliran Malaysia 2007 *Draft Text Speech: Flood and Drought Management in Malaysia* **6**

[7] Urban Stormwater Management Manual for Malaysia (MSMA) 2nd Edition updated 23 May 2012.

[8] Comes T., Van de Walle B., Measuring Disaster Resilience: The Impact of Hurricane on Critical Infrastructure Systems, *Proceeding of the 11th International ISRCRAM Conference University Park Pennsylvania, USA, May 2014.* **200**

A Review on Flood Events for Kelantan River Watershed in Malaysia for Last Decade (2001-2010)

Aminah Shakirah J[1] L M Sidek[1] B Hidayah[1] Nazirul M.Z[1] M. Jajarmizadeh[1] F.C. Ros[1], ZA Roseli[1]

[1]Centre for Sustainable Technology and Environment (CSTEN), Universiti Tenaga Nasional, Jalan IKRAM-UNITEN 43000, Kajang Selangor, MALAYSIA

E-mail: Aminahshakirah91@gmail.com

Abstract. Malaysia is located at tropical zone and high precipitation area that frequently hit by flood events when it is near monsoon season. This hydro hazard has been one of the main concerns for governmental and non-governmental sectors. High floods lead in financial damages and they are related with human's life. Kelantan watershed is one of the challenging watersheds which mostly suffer from flood events and heavy rainfall events. Flood in Kelantan watershed is related with monetary misfortunes and lives. Clearly, flood have significant influence on various water sectors such water supply, agriculture, human health and eco-systems therefore study of this topic and presentation of available of any data and information can be a valuable baseline for upcoming research in vulnerable case studies. In this study, Kelantan watershed is selected because it is prone to flooding and urban areas classified as vulnerable districts. This overview is discussed on the last decade (2001-2010) floods events in Kelantan.

1. Introduction

Malaysia has abandon resources of water resources by receiving sufficient amount of rain yearly. The average annual rainfall for Peninsular Malaysia is 2,400 mm, and for Sarawak and Sabah is 3,800 mm Moreover, flood is related with wet season and two groups of factors fresh waters to flood circumstances such as moderate intensity, long-duration rainfall covering a wide area and high intensity, short-duration localised rainfall. In Malaysia, flood reports shows that seasonal pattern has an important role to produce flood events in east coast and the southern part of Peninsular Malaysia, Sabah and Sarawak which flood occurs mostly during December to January by contribution of northeast monsoon. Increasingly, as we move towards the year 2020, the country is expected to face serious challenges related to flood and drought management. Per capita availability of water will greatly decrease as a result of a growing population and greater per capita use of water for a better quality of life, urbanization and industrialization. Malaysia has been involved with a long time by flood disaster since 1920. The country has experienced major flood events in the years of 1926, 1963, 1965, 1967, 1969, 1971, 1973, 1979, 1983, 1988, 1993, 1998, 2005. In 1886, Kelantan was faced with a severe flood with gale-force winds. Later in 1926 the worst flood affected most of Peninsular Malaysia. Next, during 1967 huge flood circumstances happened across Kelantan, Terengganu and Perak river basins. In 1971, a catastrophic flood swept across many parts of the country such as Pahang. Recently, in December 2006 and January 2007 flood was main concern of governmental issues for Johor. Finally, flood occurred in Kelantan in 2014 which made a catastrophic phenomenon for who lives in urban area. Nevertheless, the objective of this study is the review of flood event and analysis of flood events in last decade (2001-2010) in Kelantan.

2. Material and Method

2.1 Case Study

Kelantan covers an area of 15,099 km2 in Malaysia and it is located on the north eastern region of Peninsular Malaysia (Figure 1). Kelantan comprises of ten administrative districts, which are Kota Bharu, Tumpat, Machang, Pasir Puteh, Bachok, Jeli, Tanah Merah, Pasir Mas, Kuala Krai and Gua Musang. Flood prone river at Kelantan state are such as Sungai Nenggiri, Sungai Galas, Sungai Pergau, Sungai Kelantan, Sungai Golok, Sungai Kemasin, Sungai Pengkalan Chepa, Sungai Pengkalan Datu and Sungai Semerak. Therefore, Kelantan state deserves to have study for flood review and analysis based on availability of data owing to monsoon season and heavy frequent rainfall with high intensity. Usually, monitoring the trends of the flood in Kelantan and identification of various extreme rainfall and runoff events are recorded by Department of Irrigation and Drainage (DID) Malaysia. Moreover, the water levels are frequently exceeding the danger level. Hence, summarization of raw data and primary analysis help for decision making on later studies

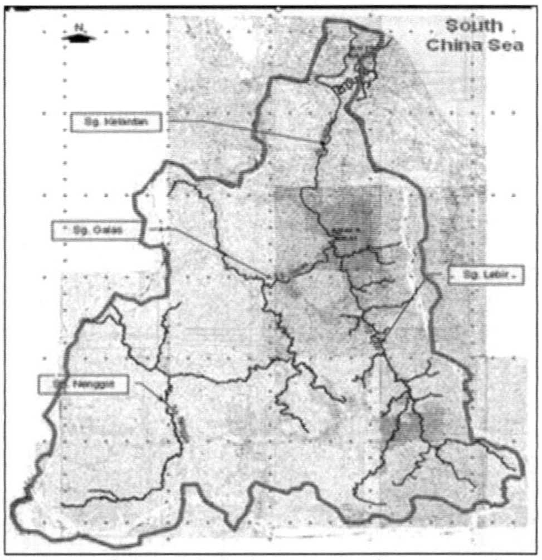

Figure1: Map of Sungai Kelantan catchment and main rivers

2.2. Primary Knowledge and Analysis

Figure 2 illustrates the prone flooded area from year 2000 – 2010 in Kelantan. It is clear that downstream ~~of~~ especially in Sungai Kelantan Catchment, north of Kelantan state is badly affected more than the upstream area. Therefore, urban areas in northern part of watershed are more vulnerable by flooding. The Kelantan River has large tributaries, namely the Lebir, Galas, Pergau and Nenggiri rivers. The two principal tributaries, the Lebir and Galas rivers merge immediately upstream of Kuala Krai after which the Kelantan river meanders over a wide and flat coastal plain before discharging into the South China Sea. Figure2 shows eastern part of Kelantan has more contribution with flood in comparison with west, it might be related with Lebir and Galas rivers which joint before downstream of Kelantan. Moreover, Northeast monsoon starts from early of November and ends in the end of March. The climate of the study area is generally hot and wet throughout the year with little variation. In addition, it is also characterized by uniform high temperature, high relative humidity and heavy rainfall. The catchment is influenced strongly by the Northeast monsoon which brings along heavy rainfall in the months of November to December. The long duration of the rain events cause prolonged flooding along some stretches of the river. Based on the DID flood report, during the Northeast Monsoon season in 2009, Kelantan has been involved with two phase flood on the 5th - 11th November 2009 (first phase) and 20th - 26th November 2009 (second phase). The flooded area include territorial at Kota Bharu, Tumpat, Pasir Mas, Pasir Puteh, Bachok, Tanah Merah, Machang, Kuala Krai and Jeli.

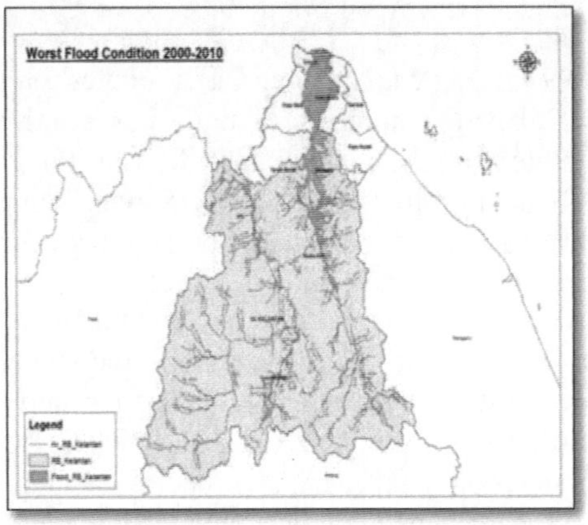

Figure 2: Flooded area at Kelantan in last decade

3. Results and Discussion

3.1. Extreme Hydrological Events at Kelantan

In Nov 2009, Kelantan has suffered from several waves of flood due to influence Northeast monsoon. The 1st wave of flood started from 5 Nov until 10 Nov 2009. Highest 24-Hr rainfall of 358.0mm was recorded at Kusial on 6 Nov 2009. Maximum annual stream flow for Sungai Kelantan in 2009 was recorded on 23 Nov with amount of 8014 m^3/s. At that time, Sungai Kelantan was loaded with large amount of rainfall water introduced by 2nd wave of flood (20-28 Nov 2009). Figure 3 and Figure 4 shown below are the rainfall hyetograph for Sungai Kelantan at Kusial from 5 – 10 Nov 2009 during the 1st wave and from 20 – 25 Nov 2009 during the 2nd wave of the flood event.

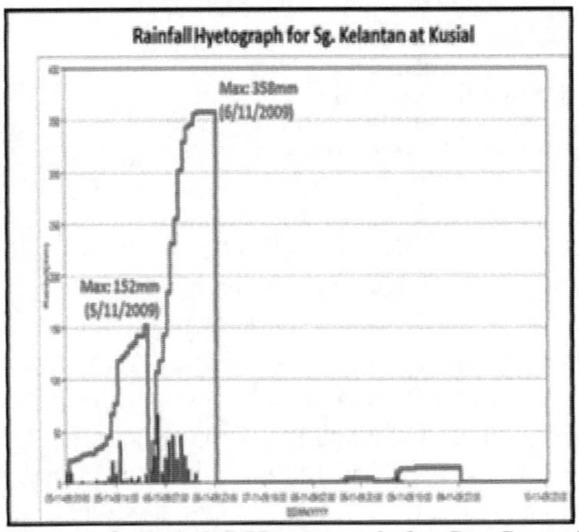

Figure 3: Rainfall Hyetograph for Sg. Sungai Kelantan at Kusial from 5 – 10 Nov 2009 (1st wave)

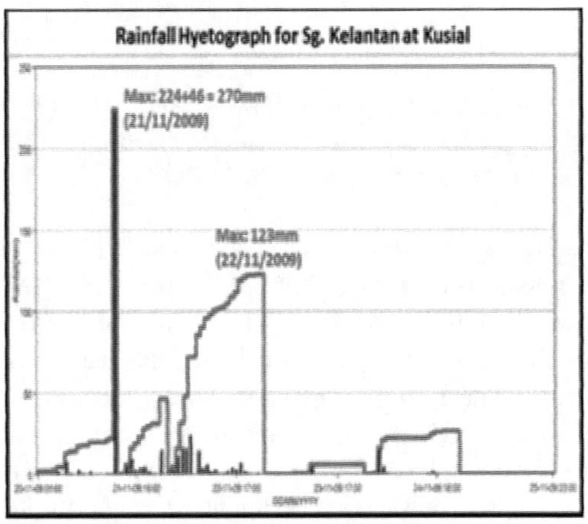

Figure 4: Rainfall Hyetograph for Sg. Sungai Kelantan at Kusial from 20 – 25 Nov 2009 (2nd wave)

Figure 5 and 6 shows the stage hydrograph for Sungai Kelantan at Kusial during 1st and 2nd wave of flood in Nov 2009. It is observed that the maximum water level recorded during 1st wave was 18.06m on 7/11/2009 (01:00) which exceeded danger level (16.0 m) of about 2.06 m. The maximum water level recorded during 2nd wave was 18.66m on 23/11/2009 (08:00) which exceeded danger level (16.0 m) of about 2.66 m.

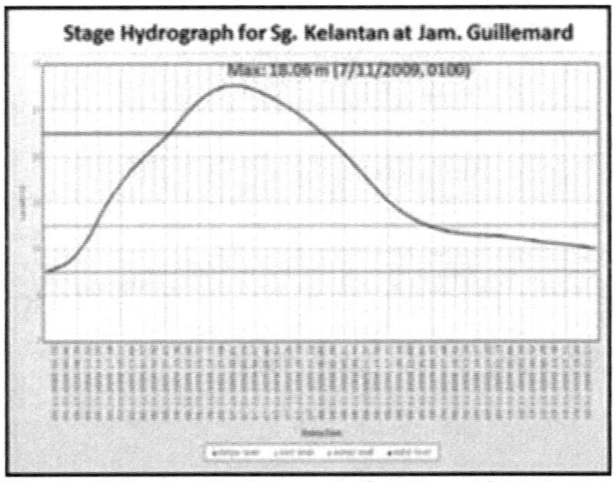

Figure 5 Stage Hydrograph for Sungai Kelantan at Kusial from 5 – 10 November 2009 (1st wave)

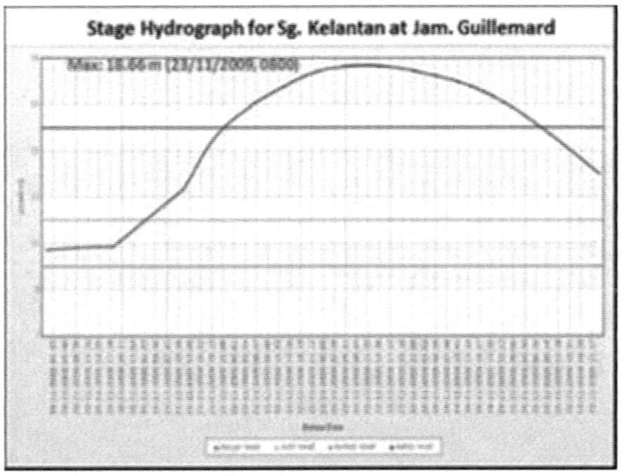

Figure 6 Stage Hydrograph for Sungai Kelantan at Kusial from 20 – 25 November 2009 (2nd wave)

4. Conclusions

Extreme storm events are attributed to several contributory factors. Among the contributory factors, intensity/duration of precipitation is widely adopted in relating to hydrological extreme events. Assigning arbitrary thresholds of precipitation or flow amplitude is deemed as the onset of an extreme hydrological event. The thresholds could be chosen either on the basis of hydrological parameters related to recurrence interval or in relationship to such levels that crossing them leads to significant economic losses. As conclusion, the extreme rainfall event is showing increasing trend as well as the pattern of flood event is showing the increasing trend. The analysis of the hydrological flood event will be used as an input into the flood forecasting model. Hence, it is crucial to compile and analyze major hydrological extreme events which could serve as a reference in the planning and management of water resources projects especially for introducing additional safety factors in the design of water resources projects.

Acknowledgement

This research is a Long-term Research Grant Scheme supported by Ministry of Higher Education Malaysia (MOHE) collaboration with Japan International Cooperation Agency (JICA) under project "Research and Development for Reduction Geo-Hazard Damage in Malaysia Caused by Landslide and Flood". This project also owes to the government sectors Department of Irrigation and Drainage (DID) Malaysia and International Centre for Water Hazard and Risk Management (ICHARM) Japan.

References

[1] Chan N W 1995 Final Report A Contextual Analysis of Flood Hazard Management in Peninsular Malaysia 7-8
[2] Chia C W 2004 Buletin Ingenieur Managing Flood Problems in Malaysia 22 38-43.ISSN 0128-4347
[3] Department of Irrigation and Drainage (DID) 2009 *LAPORAN Banjir 2008- 2009*. Unit Hidrologi Jabatan Pengairan dan Saliran Negeri Kelantan
[4] A.M. Jafri, Z. Hashim, M.L.Kawas, Z.Q Chen, N.Ohara Development of Atmospheric Based Flood Forecasting and Warning System for Selected River Basins in Malaysia
[5] A.M Jafri 2011 National Conference on the Impacts of Climate Change on Water Resources and their Consequences to Major Economic Sectors Implementation of Climate Change Adaptation Programmes for Water Resources Development Projects
[6] Ong S H 2011 National Symposium on Climate Change Adaptation Goals and Aspirations for Coastal and Water Resources in the Context of Climate Change Adaptation

A Factorial Analysis Study on Enzymatic Hydrolysis of Fiber Pressed Oil Palm Frond for Bioethanol Production

F S Hashim[1,2], H W Yussof[2], M A K M Zahari[2], R M Illias[2] and R A Rahman[3]

[1] Faculty of Chemical and Natural Resources Engineering, Universiti Malaysia Pahang, Lebuhraya Tun Razak, 26300 Kuantan, Pahang
[2] Faculty of Chemical and Natural Resources Engineering, Universiti Malaysia Pahang, Lebuhraya Tun Razak, 26300 Kuantan, Pahang
[3] Department of Bioprocess Engineering, Faculty of Chemical Engineering, Universiti Teknologi Malaysia, 81310 UTM Skudai, Johor.

E-mail: syazwana.fatin@gmail.com

Abstract. Different technologies have been developed to for the conversion of lignocellulosic biomass to suitable fermentation substrates for bioethanol production. The enzymatic conversion of cellulose seems to be the most promising technology as it is highly specific and does not produce substantial amounts of unwanted byproducts. The effects of agitation speed, enzyme loading, temperature, pH and reaction time on the conversion of glucose from fiber pressed oil palm frond (FPOPF) for bioethanol production were screened by statistical analysis using response surface methodology (RSM). A half fraction two-level factorial analysis with five factors was selected for the experimental design to determine the best enzymatic conditions that produce maximum amount of glucose. FPOPF was pre-treated with alkaline prior to enzymatic hydrolysis. The enzymatic hydrolysis was performed using a commercial enzyme Cellic CTec2. From this study, the highest yield of glucose concentration was 9.736 g/L at 72 hours reaction time at 35 °C, pH 5.6, and 1.5% (w/v) of enzyme loading. The model obtained was significant with p-value <0.0001. It is suggested that this model had a maximum point which is likely to be the optimum point and possible for the optimization process.

1. Introduction

According to Malaysian Innovation Agency in year 2011, the palm oil sector correspondingly generates the largest amount of biomass waste that is estimated around 80 million dry tons in 2010. It is expected to increase around 100 million dry tons by 2020 [1]. This huge amount of biomass waste may leads to serious disposal problem. Therefore, many studies have been conducted in turning this biomass waste into fermentable sugars by various methods and pretreatment and as a substitute for non-renewable energy sources. Oil palm frond (OPF) is currently been considered as one of the most sustainable lignocellulosic biomass (LCB) sources which are valuable and profitable to the agriculture industry. Presently OPF is used in local beef and dairy industry as source of fiber or as an ingredient in a complete fodder for ruminant animals in terms of pallets, silage or cubes [2]. Recent study showed

[1] To whom any correspondence should be addressed.

that OPF was able to produce renewable sugars by simply pressing fresh OPF to obtain the juice [3]. The remaining from the pressing process, which is fiber pressed oil palm frond (FPOPF) was normally disposed. Due to its abundant availability and low cost materials, FPOPF has the potential as an alternative and renewable LCB sources for the production of many value-added products such as bioethanol. Alkaline pretreatment is the most effective pretreatment in increasing biomass digestibility by reducing lignin content. Sodium hydroxide (NaOH) was used in alkaline pretreatment because it can causes swelling of the biomass structure by increasing the internal surface of the cellulose and decreasing degree of polymerization that allows the lignin to broken down [4]. The use of enzymes in the hydrolysis of LCB is more effective that acid hydrolysis because enzymes are highly specific, can work at mild process condition and does not produce substantial amounts of unwanted byproducts [5]. However, most of the previous research focused on using purified enzyme which is expensive rather than commercial industrial enzymes. This study aimed to find the best condition in the enzymatic hydrolysis of pretreated FPOPF to produce glucose using Cellic CTec2. A half fraction two-level factorial analysis with five factors was studied for their effect to the enzymatic hydrolysis. This study is crucial for the optimization stage later because each main effects had the tendency to affect one another as the experiment runs.

2. Materials and Method

Oil palm frond (OPF) was obtained from a local palm oil plantation at Kuantan, Pahang. The OPF was pressed to remove juice by using sugarcane press machine. Then, fiber pressed oil palm fronds (FPOPFs) were dried under the sun for three days and then mechanically shredded into pieces using grinder. Subsequently, the shredded FPOPF were sieved to obtain fractions with a particle size less than 1 mm. Dried FPOPF was stored in sealed plastic bag at room temperature. Enzymatic hydrolysis was carried out using Cellic CTec2, a commercial enzyme preparation kindly provided by Novozymes (Denmark).

2.1 Biomass characterization

Characterization of FPOPF was done according to the analytical procedure by [6][7][8]. Extractives were determined in two steps of Soxhlet extraction. Next, the chlorination method (ASTM standard) was performed to determine the cellulose and hemicellulose content by removing the lignin [7][8]. The hemicellulose content was determined by subtracting the holocellulose content with the α-cellulose. Ash content was determined by using muffle furnace at 575 ± 25°C for 24 ± 6 hours (NREL standard). The sample was then cooled inside a desiccator before the measurement of sample mass using analytical balance.

2.2 Alkaline pretreatment

FPOPF sample was soaked in NaOH solutions with concentration 4.42% (w/v). The sample was treated at 100°C for an hour [9]. Then, the treated OPF was washed thoroughly with de-ionized water until turned to neutral before dried in the oven at 105°C and stored. The dried sample was used for enzymatic hydrolysis.

2.3 Two-level factorial analysis experimental setup

To identify the factors that affect the enzymatic hydrolysis, several factors were tested earlier and obtained from literature study. Five independent variables which are agitation speed (A), percentage (w/v) of enzyme loading (B), pH (C), temperature (D) and time (E) were selected for factorial designs of 2^{5-1} in order to screen their effect on the response of glucose yield as shown in Table 1. All experiments were carried out in triplicate and the averages were taken as responses.

2.4 Enzymatic hydrolysis

Enzymatic hydrolysis was carried out in 250 mL Erlenmeyer flask using 1.5 grams of pretreated FPOPF in the presence of sodium acetate buffer at different pH (pH 4 – pH 5.6) and enzyme loading

(1.5% – 6% w/v). The enzyme loading was calculated based on a gram of Cellic CTec2 was loaded for every 20 gram of substrate. The reaction mixture (30 mL) was incubated at different temperature (35°C – 65°C), reaction time (3 – 72 hours), and agitation speed (50 rpm – 200 rpm). Then, the hydrolysates were filtered for further analysis.

2.5 Glucose analysis

The hydrolysate was analysed by 3,5-dinitrosalicyclic acid (DNS) method with glucose as a standard. 3 mL of sample was added to 3 mL of DNS reagent and placed in boiling water bath for 10 min. The mixture was cooled to room temperature and 1 mL of potassium sodium tartrate was added to maintain the colour of mixture. Then, reducing sugar was analysed using UV-Vis spectrophotometer with absorbance at 575 nm.

Table 1. Experimental design of factor screening process using 2^{5-1} factorial design with respond.

Run	Agitation speed, rpm	Enzyme loading, %	pH	Temp., °C	Time, hour	Glucose, g/L
1	50	1.5	5.6	65	72	7.372
2	50	1.5	5.6	35	3	7.414
3	200	6.0	4.0	65	3	0.850
4	200	1.5	5.6	35	72	9.736
5	50	6.0	4.0	35	3	0.276
6	200	1.5	4.0	35	3	1.608
7	50	1.5	4.0	65	3	0.251
8	50	1.5	4.0	35	72	3.949
9	200	6.0	5.6	35	3	7.482
10	200	1.5	5.6	65	3	8.167
11	50	6.0	5.6	35	72	7.647
12	200	6.0	4.0	35	72	3.337
13	50	6.0	5.6	65	3	8.153
14	50	6.0	4.0	65	72	4.624
15	200	1.5	4.0	65	72	4.179
16	200	6.0	5.6	65	72	8.629

Table 2. Characterization of raw FPOPF and pretreated FPOPF.

Characterization	Raw FPOPF, %	Pretreated FPOPF, %
Extractives	10.68	4.84
Cellulose	25.19	45.42
Hemicellulose	24.50	14.01
Lignin	35.58	25.53
Ashes	2.59	1.94
Moisture content	1.45	8.26

3. Results and Discussion

3.1 Biomass characterization

The result on the composition of FPOPF is shown in Table 2. The composition of extractives, hemicellulose, lignin and ashes in raw FPOPF were higher than the pretreated FPOPF. The differences of lignin composition from 35.58% to 25.53% showed that delignification was occurred during the alkaline pretreatment. The cellulose content in pretreated FPOPF was nearly double than raw FPOPF. It is proved that the alkaline pretreatment process has altered the structure of biomass by removing the lignin in order to make cellulose more accessible to the enzymes and later converted to fermentable sugars [10].

3.2 Factor affecting enzymatic hydrolysis

The effect of agitation speed, percentage (w/v) of enzyme loading, pH, temperature and time were evaluated using factorial analysis that consists of 16 run as shown in Table 1. The highest glucose concentration was obtained at 9.736 g/L where the hydrolysis conditions at 72 hours reaction time with temperature of 35°C, pH 5.6, and 1.5% (w/v) of enzyme loading. This finding was similar to the work by [11]. Thus, the presence of a sufficient amount of enzyme was important in order to obtain higher glucose yield during enzymatic hydrolysis. According to [12], the enzyme cost significantly contributed to the cost of biomass conversion process. Thus, by minimizing the enzyme dosage accordingly, this can reduce the cost of biomass conversion process. Eventually, in this present work, 1.5% (w/v) of enzyme loading was enough to perform a cellulose conversion during enzymatic hydrolysis. So, the cost of glucose in bioethanol production can be further reduced.

In percentage of contribution, the most important factor which gives the highest percentage contribution for enzymatic hydrolysis is pH with 82.18% followed by reaction time of 9.25%. Meanwhile, temperature shows the lowest contribution which is 0.024%. Other variables that less contributed to the effect of the enzymatic hydrolysis were agitation speed and percentage of enzyme loading (w/v) by having a value of 0.73% and 0.11% respectively.

From the ANOVA of this experimental design, the model obtained was significant as suggested by the model F value at 229.97 and low probability value (p-value <0.0001). The R^2 value obtained in this model was 0.9962, which is in good agreement with the adjusted R^2 value of 0.9919. The high R^2 value of 0.9962 indicates that the model was well adapted to the response. Equation 1 shows the response surface quadratic model for glucose production which can be presented in terms of coded factors as in the following equation:

$$\text{Glucose} = 5.23 + 0.27X_1 - 0.10X_2 + 2.85X_3 + 0.049X_4 + 0.95X_5 - 0.32X_1X_2 + 0.39X_2X_4 - 0.68X_3X_5 \quad (1)$$

Where, X_1 = agitation speed, X_2 = enzyme loading, X_3 = pH, X_4 = temperature and X_5 = time

Based on the empirical model, the coefficients of X_1 to X_5 are small compared to the interception. This gives an indicator of the existent of the design plateau as the other coefficient for the factors are lower than the interception at 5.23. This plateau showed that the design had a maximum point which is likely to be the optimum point and possible for the optimization process later.

4. Conclusion

The best condition for the *enzymatic* hydrolysis process of FPOPF *using* Cellic CTec2 was at pH 5.6, 35°C, enzyme loading of 1.5%, at 72 hours and 200 rpm. The model obtained had a maximum point which is likely to be the optimum point and possible for the optimization process later. Enzymatic hydrolysis using the commercial enzyme is looking more promising for the industrial application in the production of bioethanol. More work need to done in recycling the enzyme to further reduce the biomass conversion process.

5. Acknowledgements

The authors gratefully acknowledge financial support from the RACE grant RDU151304 and Faculty of Chemical and Natural Resources Engineering, University Malaysia Pahang (UMP) that made this work possible. The authors also wish to acknowledge the Ministry of Education Malaysia for sponsoring postgraduate study via MyBrain15 programme.

6. References

[1] Malaysian Innovation Agency (MIA) 2011 National Biomass Strategy 2020: New Wealth Creation for Malaysia's Palm Oil Industry. Available at: http://www.

innovation.my/pdf/1mbas/National_Biomass_Strategy_Nov_2011_FINAL.pdf (accessed January 2013)
[2] Wan Zahari M, Abu Hassan O, Wong H K and Liang J B 2003 *Journal of Anim.* **16** 625-634
[3] Zahari M A K M, Zakaria M R, Ariffin H, Mokhtar M N, Salihon J, Shirai Y and Hassan M A 2012 *Journal of Bioresource Technology* **110** 566–71
[4] Taherzadeh M J and Karimi K 2008 *Int. J. Mol. Sci.* **9** 1621-1651
[5] Zheng J, Choo K, Bradt C, Lehoux R and Rehmann L 2014 *Biotechnology Reports* **3** 99-107
[6] Ehrman T 1994 Standard Method for the Determination of Extractives in Biommass. National Reneawable Energy Laboratory (NREL)
[7] Wise L E, Murphy M and D'Addieco A A 1946 *Paper Trade Journal* **122** 35–43
[8] Teramoto Y, Lee S H and Endo T 2009 *Journal of Bioresource Technology* **100** 4783–4789
[9] Sukri S and Rahman R 2014 *Romanian Biotechnological Letters* **19** 9006–9018
[11] Siti-Normah M D S, Sabiha-Hanim S, Noraishah A, 2012 *Journal of waste.org* **6** 1391-1395
[12] Chen M, Xia L and Xue P 2000 *Int Biodeter Biodegr.* **59** 85–89

Devolatilization Studies of Oil Palm Biomass for Torrefaction Process through Scanning Electron Microscopy

D. Daud, A. Abd. Rahman, A.H. Shamsuddin
Centre for Renewable Energy, Universiti Tenaga Nasional, Putrajaya Campus, Jalan IKRAM-UNITEN, 43000 Kajang, Selangor, Malaysia

E-mail: darsheena@uniten.edu.my

Abstract. In this work, palm oil biomass consisting of empty fruit bunch (EFB), mesocarp fibre and palm kernel shell (PKS) were chosen as raw material for torrefaction process. Torrefaction process was conducted at various temperatures of 240 °C, 270 °C and 300 °C with a residence time of 60 minutes. The morphology of the raw and torrefied biomass was then observed through Scanning Electron Microscopy (SEM) images. Also, through this experiment the correlation between the torrefaction temperatures with the volatile gases released were studied. From the observation, the morphology structure of the biomass exhibited inter-particle gaps due to the release of volatile gases and it is obviously seen more at higher temperatures. Moreover, the change of the biomass structure is influenced by the alteration of the lignocellulose biomass.

1 Introduction

Biomass is widely considered as a very promising alternative source of raw material due to its availability and potential for energy conversion. The main and the most abundant biomass that has been planted in Malaysia came from oil palm. The palm oil industry in Malaysia generates almost 94 % of the biomass feedstock of agricultural and forestry sources such as wood residues, rice and sugar-cane contributes the remaining 6 % [1]. According to the Malaysian Palm Oil Board (MPOB) it is estimated that the oil palm plantation occupies nearly 4.70 million hectares and produce approximately 77.24 million tonnes (dry) of waste [2].

In order to employ biomass as a primary alternative source, advancement and improvement of technology typically in the power generation sector along with the process selection will play a major role in delivering the activities towards renewable energy. The structure of biomass is mainly fibrous and exhibits low energy density and high moisture content. It also pose difficulty on grindability, handling, transportation and storage when compared to coal [3]. Therefore, torrefaction process has been introduced as a one of the techniques that can convert biomass into better quality solid biofuel. Torrefaction involves heating the biomass in the absence of oxygen to a temperature of typically 200 °C to 300 °C [4]. Torrefaction process is used to enhance the structure of biomass until the physical properties becomes similar to coal. The change of structure can be clearly observed through scanning electron microscopy (SEM) images.

2 Materials and Methods

The waste from the oil palm biomass which consists of empty fruit bunch (EFB), mesocarp fibre and palm kernel shell (PKS) were used for this experiment. It is taken from an oil palm mill in Perak, Malaysia. The raw waste underwent a fuel preparation process in order to ensure the samples are monitored especially in term of weight, moisture and particle size before the torrefaction experiment. All samples have prescribed by British standard [5] for moisture content determination. The particle size for all the samples has been processed till the finer size which is 250 μm. Once the samples were ready, it was placed in a vacuum tube furnace with approximately 5 kg per every run for conducting the torrefaction process. The initial temperature was set at ambient temperature, which was 30 °C. The sample was then heated to the drying temperature 105 °C at a rate of 20 °Cmin^{-1} and it was held constant for 10 minutes before proceeding to the next phase. The sample was then heated up again to the desired torrefaction temperature which of 240 °C, 270 °C, and 300 °C and was held constant for 60 minutes of residence time. In order to avoid combustion process from taking place, nitrogen gas was continuously supplied throughout the process. The products of torrefied biomass as well as the raw biomass were then used to observe the pores and surface morphology through the SEM images.

3 Results and Discussion

3.1 SEM images of the raw and torrefied biomass for empty fruit bunch (EFB)

The changes in the morphology structure of raw and torrefied biomass were examined in order to see the correlation between torrefaction temperatures with the volatile gases through the process. The effects of various torrefaction temperatures at 240 °C, 270 °C and 300 °C and a residence time 60 minutes were applied. Figure 1 and Figure 2 shows the SEM images of the raw and torrefied biomass for EFB with magnification of 2000 and 6000 respectively.

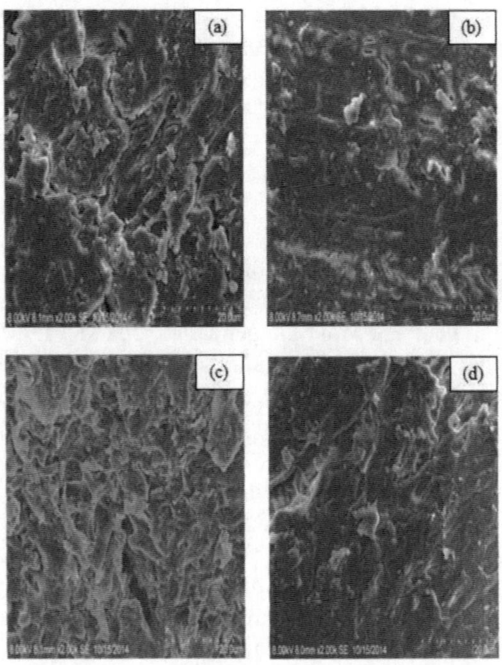

Figure 1: SEM images (2000 x) of (a) EFB raw, (b) EFB torrefied at 240 °C, (c) EFB torrefied at 270 °C and (d) EFB torrefied at 300 °C

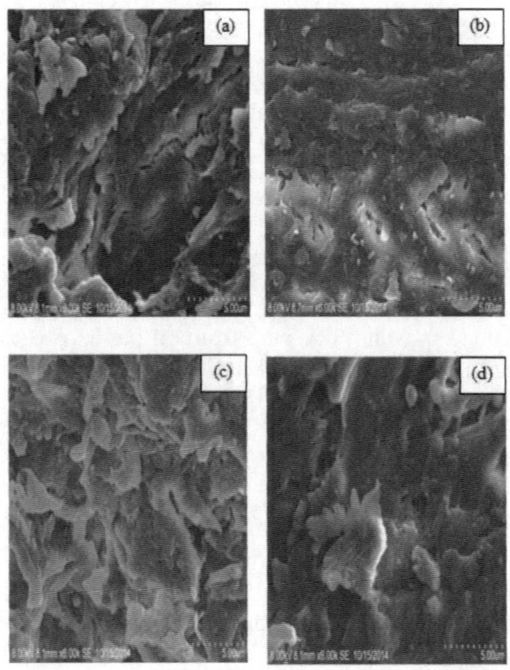

Figure 2: SEM images (6000 x) of (a) EFB raw, (b) EFB torrefied at 240 °C, (c) EFB torrefied at 270 °C and (d) EFB torrefied at 300 °C

From the observation, the SEM images show significant differences between the raw and torrefied biomass. The raw EFB indicated pores which can obviously be seen through a magnification of 6000. According to Sabil et al. (2013) [6] a presence of pores in the surface of EFB verified that EFB is a fibrous material. Furthermore, after the torrefaction process was conducted, the torrefied EFB shows a prominent surface. This may be due to the loss of the moisture and some light organic volatiles under the pre-treatment process. The process may have affected the hemicellulose structure which started to decompose at a lower temperature of 150 °C. Due to this, the biomass slightly shrank, then exhibited the prominent structure at 240 °C.

The observation continues with SEM images of the torrefied biomass at 270 °C. SEM images of the torrefied biomass samples that have been torrefied at 240 °C and 270 °C show dissimilarities. At 270 °C, the torrefied biomass show increase of inter-particle pores in the cell wall. The observation is in agreement with Tsai & Liu (2013) [7] who studied the effect of temperature on thermochemical property and true density of torrefied coffee residue. The increase of inter-particle pores in the cell wall was found to have been caused by the trapping of volatile gases during the devolatilization stages. These gases are released during the process and at the same time the decomposition of hemicellulose is still taking part. At 300 °C, the surface of the torrefied EFB biomass looks damaged and it seems to have melted under the pre-treatment process. According to Stelt et al. (2011) [8] the cohesive failure is due to the fact that the fibre ends and particles stick out of the surface which causes the voids to appear. In addition, lignin and hemicellulose exceed their glass transition temperatures, thus leading to further cracks and crevices.

3.2SEM images of the raw and torrefied biomass for mesocarp fibre

Figure 3 and Figure 4 shows the SEM images of the mesocarp fibre with magnification of 2000 and 6000 at the aforementioned parameters. From the observation, the raw mesocarp fibre shows the inter-particle gaps in the surface structure. The images of the surface structure are slightly changed when the torrefaction process was conducted. The influence of the pre-treatment process changed the internal structure and reduced the amount of gaps at which at this stage, the integrated structure was performed. This observation is similar to the Chen et al. (2014) [9].

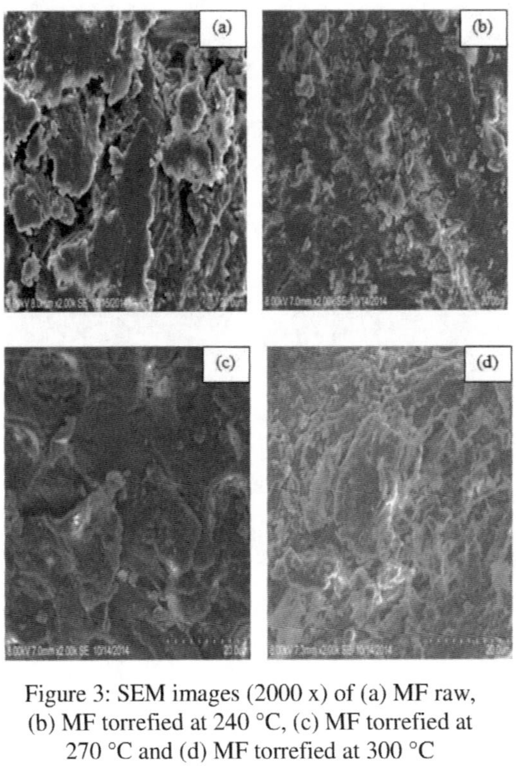

Figure 3: SEM images (2000 x) of (a) MF raw, (b) MF torrefied at 240 °C, (c) MF torrefied at 270 °C and (d) MF torrefied at 300 °C

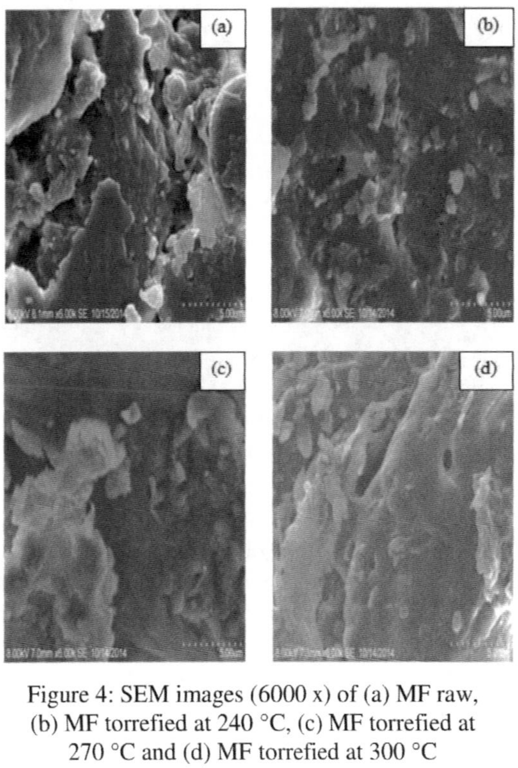

Figure 4: SEM images (6000 x) of (a) MF raw, (b) MF torrefied at 240 °C, (c) MF torrefied at 270 °C and (d) MF torrefied at 300 °C

Meanwhile, the SEM images of the torrefied biomass at 270 °C start to show the pores and cracks. At this temperature the mesocarp fibre reacted with the heat and rearranged the internal structure by losing its fibrous nature. Moreover, lignin started to decompose added to the hemicellulose extensive decomposition as xylan-based hemicellulose also generally starts to decompose around 250 °C – 280 °C. Due to the decomposition of lignocellulose it seems to be easier to obtain a structural shrinkage that makes the structure very brittle or friable. The decomposition of lignocellulose continued until the end of the torrefaction temperature at 300 °C.

3.3 SEM images of the raw and torrefied biomass for palm kernel shell (PKS)

The observation continues with SEM images of PKS at magnification 2000 and 6000 in Figure 5 and Figure 6 respectively. The PKS images show different images between the EFB and mesocarp fibre, especially at the beginning of the process. The raw PKS shows the integrated surface and not much prominent surface is observed. This indicates an image that is quite similar with the torrefied PKS at 240°C. Thus, PKS can be considered as a material that affected lesser at the beginning of the torrefaction temperature. It also shows less fibrous material as compared to EFB and mesocarp fibre.

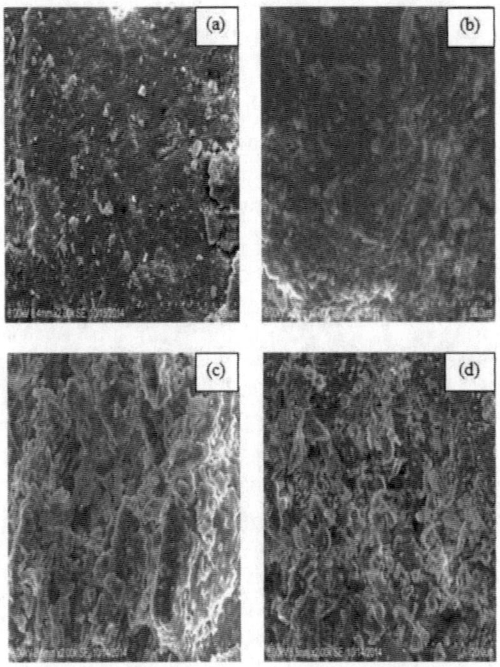

Figure 5: SEM images (2000 x) of (a) PKS raw, (b) PKS torrefied at 240 °C, (c) PKS torrefied at 270 °C and (d) PKS torrefied at 300 °C

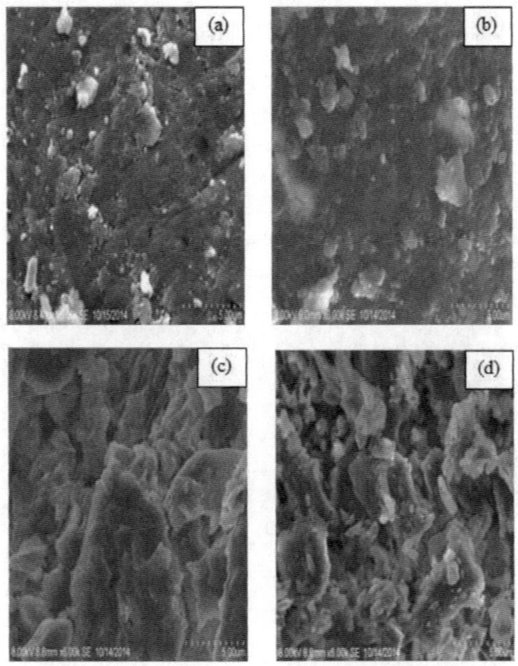

Figure 6: SEM images (6000 x) of (a) PKS raw, (b) PKS torrefied at 240 °C, (c) PKS torrefied at 270 °C and (d) PKS torrefied at 300 °C

Moreover, in the low to the high selection magnification images show cracks and fissures on the torrefied PKS images at 270 °C. It is also creates a more pores and leaving an empty space in between due to the releasing of volatile gases. The alteration of structure increased when the torrefaction temperature was increased until 300 °C. The pores in the torrefied biomass could increase the potential for lignocellulose cracking and fissures in order to produce char. This result is in agreement with the finding of previous studies by Tsai & Liu (2013) [7].

4 Conclusions

Based on the morphology observation of the raw and torrefied biomass it can be concluded at the higher torrefaction temperatures, the structure of the torrefied biomass become brittle and creates more pores and fissures structure. Moreover, the appearance of the inter-particle gap due to the released of the volatile gases is mostly observed at temperature 270°C and above. From the result obtained, the overview change of the biomass structure obviously has correlated with the alterations of the lignocellulose of the biomass. This results in a biomass product that is more homogenous in structure and more brittle or friable in all samples, aiding towards improving its grindability and handling.

Acknowledgments
Authors would like to acknowledge for the financial support given: Long Term Research Grant (01201102 LRGS) from Ministry of Higher Education (MOHE) and Universiti Sains Malaysia (USM).

References

[1] Ng, F. Y., Yew, F. K., Basiron, Y., & Sundram, K. (2011). A renewable future driven with Malaysian palm oil-based green technology. *Journal of Oil Palm & The Environment (JOPE),* 1-7.

[2] *Malaysian Oil Palm Statistics 2009.* (2010). 8-29: Malaysian Palm Oil Board .

[3] Chew, J., & Doshi, V. (2011). Recent advances in biomass pretreatment-Torrefaction fundamentals and technology. *Renewable and sustainable energy reviews*, 4212-4222.

[4] Bergman, P., Boersma, A., Kiel, J., Prins, M., Ptasinski, K., & Janssen, F. (2005). *Torrefaction for entrained-flow gasification of biomass.* Energy Research Centre of the Netherlands (ECN): ECN-C--05-067.

[5] BS EN 14774-2: 2009. *Solid biofuel - Determination of moisture content - Oven dry method.* British Standard. November 2009.

[6] Sabil, K. M., Aziz, M. A., Lal, B., & Uemura, Y. (2013). Effects of torrefaction on the physiochemical properties of oil palm empty fruit bunches,mesocarp fiber and kernel shell. *Biomass and Bioenergy,* 351-360.

[7] Tsai, W. T., & Liu, S. C. (2013). Effect of temperature on thermochemical property and true density of torrefied coffee residue. *Journal of analytical and applied pyrolysis,* 47-52.

[8] Stelt, M. J., Gerhauser, H., Kiel, J. H., & Ptasinski, K. J. (2011). Biomass upgrading by torrefaction for the production of biofuels: A review. *Biomass and Bioenergy,* 3748-3762.

[9] Chen, W. H., Lu, K. M., Lee, W. J., Liu, S. H., & Lin, T. C. (2014). Non-oxidative and oxidative torrefaction characterization and SEM observations of fibrous and ligneous biomass. *Applied Energy,* 104-113.

CFD Investigation of Pollutant Emission in Can-Type Combustor Firing Natural Gas, LNG and Syngas

H Hasini[1], SSA Fadhil, N Mat Zian, NI Om
Centre for Fluid Dynamics, College of Engineering, Universiti Tenaga Nasional, Jalan IKRAM-UNITEN, 43000 Kajang, Selangor

E-mail: hasril@uniten.edu.my

Abstract. CFD investigation of flow, combustion process and pollutant emission using natural gas, liquefied natural gas and syngas of different composition is carried out. The combustor is a can-type combustor commonly used in thermal power plant gas turbine. The investigation emphasis on the comparison of pollutant emission such in particular CO_2, and NO_x between different fuels. The numerical calculation for basic flow and combustion process is done using the framework of ANSYS Fluent with appropriate model assumptions. Prediction of pollutant species concentration at combustor exit shows significant reduction of CO_2 and NO_x for syngas combustion compared to conventional natural gas and LNG combustion.

1. Introduction

Global warming has become an important issue around the world with recent unpredictable weather condition that has affected human lives. This issue is always associated with uncontrolled CO_2 emissions from various hydrocarbon combustion sources. Among the largest contributor for CO_2 and NO_x emission power generation industries. It is known that most of the power generators around the world still depends on hydrocarbon as the main source of combustion. With increasing stringent environment policy to limit pollutant emissions, an alternative cleaner source of fuel is widely researched and studied. Among others, synthetic gas or syngas has becoming a fuel of choice and has the potential to become a major fuel towards zero pollutant emission [1]. Syngas has the potential to replace conventional fuel used in power station gas turbine and attempt has been made to study the combustion behaviour when this fuel is used [2]. In the open literature, many works have been made to investigate the combustion behaviour when hydrogen or syngas is used [3-6]. However, majority of these works were carried out on small or lab-scale at which the combustion characteristics might vary differently with the actual process inside large scale combustor. Thus, this paper attempts to numerically investigate the combustion, flow and emission in full scale gas turbine combustor firing syngas. Emphasis is made to the prediction of pollutant emission, in particular CO_2 and NO_x with comparison made to pollutants from natural gas and LNG.

2. Model Descriptions

The can-type combustor model geometry is obtained by 3D scanning of the actual combustor used in thermal power plant. The cloud data for the combustor is then transferred to a more readable format of

[1] To whom any correspondence should be addressed.

surface data, which is later transformed into CFD model. Detail modelling of the actual combustor model has been thoroughly explain and describes in [7].

3. Model Assumption and Boundary Condition

The calculation of flow and combustion uses pure methane, ethane, propane, liquefied natural gas (LNG) of different composition and syngas the fuel. In reality however, the fuel is made up from several other species and this will be considered into the simulation in an increasing complexity manner. The combustion model was based on eddy dissipation model which compute the rate of reaction under the assumption that the chemical kinetics are fast compared to the rate at which the reactant are mixed by turbulent fluctuations. The mixture fraction model involves the solution for one or two conserved scalars (the mixture fractions). In this approach, the transport equations for individual species are not solved. Instead, the individual component concentrations for the species of interest are derived from the predicted mixture fraction distribution. Physical properties of chemical species and equilibrium data are obtained from the chemical database. The connection of turbulence and chemistry is derived with the probability density function (PDF), which describes the fraction of time that the fluctuating variable, f takes in a value between f and $f+\Delta f$. The mixture fraction/PDF model is computationally efficient because it does not require the solution of a larger number of species transport equations. This approach has an additional benefit of allowing a more accurate estimation of the flow field mean density than is possible using the finite rate formulation. The flow is also assumed to follow the ideal gas behaviour and the specific heat values are assumed to obey the mixing law. These assumptions are general for most gaseous combustion found the open literature

4. Results and Discussion

Two pollutant species are calculated at combustor exit which are CO_2 and NO_x. Detail comparison is made on the emission of these pollutants when fuel compositions are altered. In this case, the CH4 combustion is treated as a reference case.

4.1. CO_2 and NO_x Emission for Natural Gas

The distribution of CO_2 inside the combustor for different natural gases is shown in figure 1. General trend illustrates that the maximum CO_2 concentration is approximately 17% (by mass fraction) and is given by propane, followed by ethane and methane. Theoretically, this agrees well with the theoretical value which indicates that the CO_2 emission is highly dependent fuel carbon number. In this case, propane is expected to produce the highest CO_2 due to its highest carbon number compared to methane and ethane. At the combustor exit however, the trend does not agree with the maximum CO_2 concentration and this is given in figure 2. The emission for all cases shows almost identical trend with only a slight difference of less than 4%. The average emission of CO_2 is approximate 1080 ppm. In terms of NO_x emission methane shows the highest emission at combustor exit with an approximate value of 670 ppm. This is followed by ethane and propane with 545 ppm and 262 ppm respectively.

4.2. CO_2 and NO_x Emission for LNG

In this simulation, the species were varied by retaining CH_4 as the major source of fuel. The other species such as C_2H_6, C_3H_8 and C_4H_{10} are treated as minor and their composition were varied such that the volume fraction does not exceed 5%. Therefore the resulting CO_2 distribution does not differ much as compared to the one produced by pure methane combustion. Detail CO_2 distribution at combustor exit is shown in figure 3. The fraction of CO_2 ranges from 950 ppm to 1117 ppm. No clear trend could be deduced from the comparison of different LNG compositions due to the fact that the majority of component species that make up the fuel consist of CH_4. Thus it is anticipated that all cases will follow the CH_4 combustion emission. The NO_x emission for different LNG composition at combustor exit is shown in figure 4. Due to the uncertain variation of temperature magnitude at combustor exit, similar trend for NO_x is also anticipated. Comparison with methane shows slightly lower NO_x emission for different LNG composition at combustor exit.

4.3. CO_2 and NO_x Emission for Syngas

Figure 5 shows the CO_2 emission of syngas combustion at combustor exit for different syngas composition. It is well acknowledged that coal, natural gas and most hydrocarbon combustion is the biggest CO_2 producer in the world. Thus it is anticipated that CH_4 combustion results in the maximum CO_2. For syngas combustion however, the CO_2 production does not depend on the H_2 composition. The trend shows that it is highly dependant on the composition of CO, i.e. higher CO in syngas results in higher CO_2. Nevertheless, the difference in CO_2 production is very little between different syngas composition thus could be neglected. The distribution of thermal NO_x at combustor exit for different syngas composition is illustrated in figure 6. Compared to CH_4, syngas combustion produces much less NO_x. For syngas, the fuel element consists primarily on fixed CH_4 and varied composition of H2/CO ratio. In this case, the amount of CH_4 is very minimal (fixed to 10% mole fraction). The volume fraction occupied by the syngas within the primary combustion region is higher thus limiting the occupation of N_2 element in air which is known to contribute to the production of NO_x.

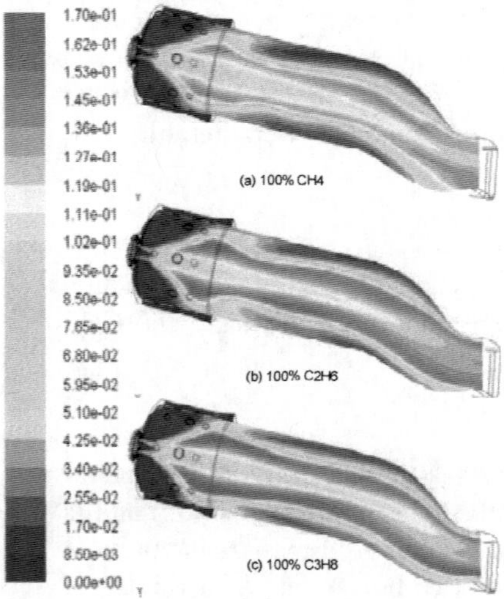

Figure 1. CO2 distribution at combustor mid plane for different Natural Gas

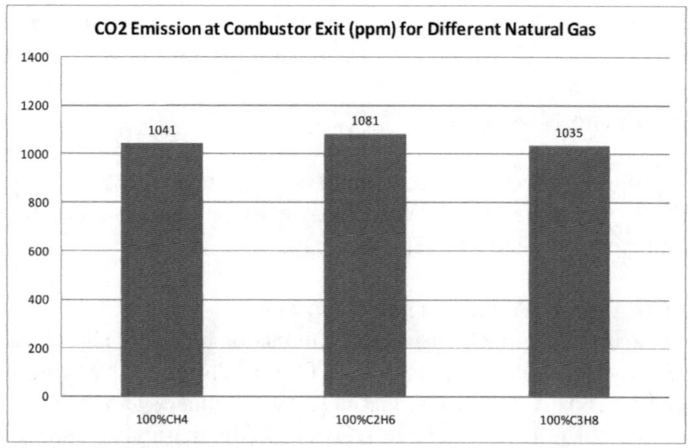

Figure 2. CO2 emission at combustor exit for different natural gas

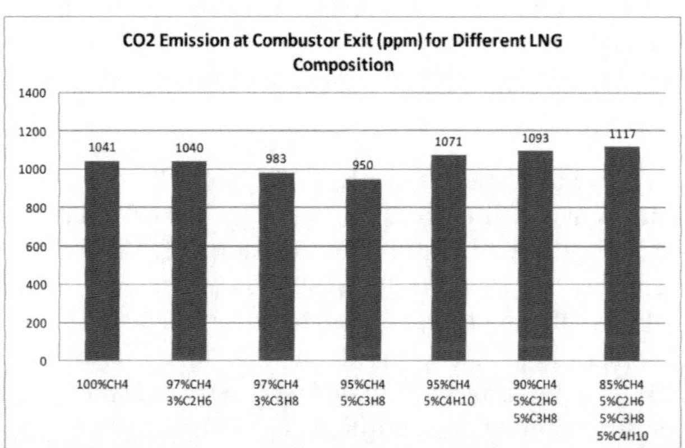

Figure 3. CO2 emission at combustor exit for different LNG compositions

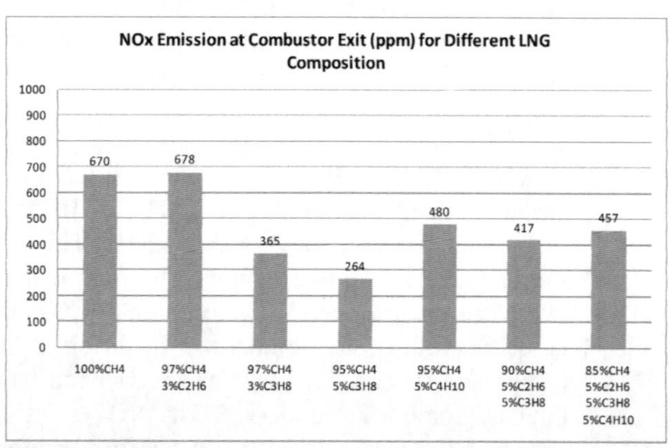

Figure 4. NOx emission at combustor exit for different LNG compositions

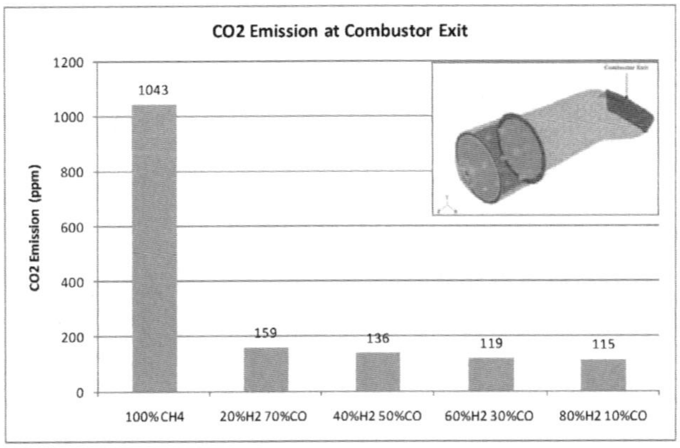

Figure 5. CO2 emission for different syngas composition compared to CH4 combustion (the composition of CH4 for syngas combustion is fixed to10%)

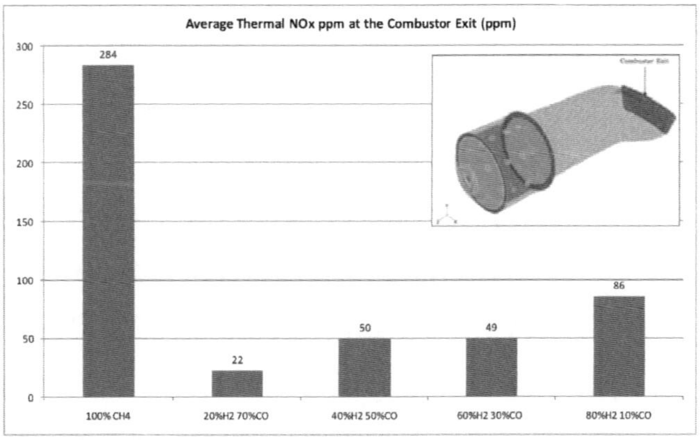

Figure 6. Thermal NOx emission for different syngas composition compared to 100% methane combustion (the composition of methane for syngas combustion is fixed to10%)

5. Conclusions

The prediction of combustion gas temperature for natural gas agrees reasonably well with theoretical stoichiometric value for flame temperature. The emission of CO_2 at combustor exit is highly dependent on the fuel composition. Highest emission of CO_2 at combustor exit is given by natural gas combustion, followed by LNG and syngas. Different compositions in LNG do not significantly influence the CO2 emission as the major species mole fraction was occupied by CH_4. Similarly for syngas, different syngas composition does not significantly influence the CO_2 emission since the controlling parameter was conventional hydrocarbon, C_xH_y. Despite having higher flame temperature, the thermal NO_x generated at combustor exit for syngas is relatively low compared to CH_4. It is approximately 70% lower than the NO_x produced by conventional CH_4 combustion.

Acknowledgments

The authors would like to acknowledge the Ministry of Education Malaysia (MOE) for the financial support through the Fundamental Research Grant Scheme (20150210FRGS). Constructive discussion and suggestion with Tenaga Nasional Berhad (TNB) personnel on gas turbine operation is also appreciated.

6. References

[1] Habib M A, Mokheimer E M A, Sanusi S Y and Nemitallah M A 2014 *Energy Conversion and Management* **83** 296
[2] Kim Y S, Lee J, Kim T S, Sohn J L 2011 *Energy Conversion and Management* **52** 2262
[3] Boivin P, Jimenez A L, Williams F A 2011 *Combustion and Flame* **158** 1059
[4] Ciesa P, Lozza G, Mazzocchi L 2000 *J. Eng. Gas Turbine and Power* **122** 27
[5] Chachartegui R, Torres M, Sanchez D, Jimenez F, Munoz A and Sanchez T 2011 *Fuel Processing Tech.* **92** 213
[6] Delattin F, Di Lorenzo G, Rizzo G, Bram J, De Ruyck J 2010 *Applied Energy* **87** 1441
[7] Hasini H, Shuaib N H and Wan Abdullah W A F 2013 Applied Mechanics and Materials **393** 741

IOP Conf. Series: Earth and Environmental Science **32** (2016) 012074 doi:10.1088/1755-1315/32/1/012074

The Application of Waiting Lines System in Improving Customer Service Management: The Examination of Malaysia Fast Food Restaurants Industry

Zurina Ismail[1], Shahrul Suhaimi AB Shokor[2]
College of Business Management & Accounting,
Universiti Tenaga Nasional, 26700 Muadzam Shah, Pahang,

E-mail: zurina@uniten.edu.my

Abstract. Rapid life time change of the Malaysian lifestyle had served the overwhelming growth in the service operation industry. On that occasion, this paper will provide the idea to improve the waiting line system (WLS) practices in Malaysia fast food chains. The study will compare the results in between the single server single phase (SSSP) and the single server multi-phase (SSMP) which providing Markovian Queuing (MQ) to be used for analysis. The new system will improve the current WLS, plus intensifying the organization performance. This new WLS were designed and tested in a real case scenario and in order to develop and implemented the new styles, it need to be focusing on the average number of customers (ANC), average number of customer spending time waiting in line (ACS), and the average time customers spend in waiting and being served (ABS). We introduced new WLS design and there will be prompt discussion upon theories of benefits and potential issues that will benefit other researchers.

1. Introduction

Queuing theory or WLS is the most important part of operations and can be described as valuable tools for operation managers (OP). Service industry such as retailing, banking and fast food providers continuously looking for an option to reduce the customer's frustration waiting in a slow line. Generally, the retail industry seems to be retaining the multiple line / multiple checkout system, banks and fast food providers had switched over the recent years to the queuing system where customers wait for the next available cashier. The fast food chain is categorized under the quick service food provider based on the business characteristic [1, 2]. For the industry, speed is the factor which provides efficiency of the service operations of each fast food chain [3]. To choose an accurate queue of a server, most of the customer will refer to certain criteria which include waiting time and how long the waiting line is [4]. The reason is that the management eventually avoiding the customer's negative perception while waiting to be served with a good experience [5, 6]. They might lose their customers if they failed to meet the expectation in providing a speedy quality service. The customer will frequently decide to change the queuing system based on the length and the amount of time they have to take to get the service. The terms "balking" is best to describe the customer's rejection, hence the possibilities to withdraw from the queue based on the service speed (queue length and time they must

[1] Marketing & Entrepreneur Department.
[2] Human Resource Department.

IOP Conf. Series: Earth and Environmental Science **32** (2016) 012074 doi:10.1088/1755-1315/32/1/012074

wat to be served) [7]. Either "balking" or "reneging", both actions will affect negatively to the fast food chain. These research papers have the intention to reduce "balking" and "reneging" scenarios by proposing the new WLS. By focusing on queuing method, researchers are continually urged to ensure the current model, including single-queue (SQ), and SSSP are upgraded. It is a challenge to suggest and prove SQ, SSSP and SSMP design since it had a significant relationship in improving the model thus the growing concern with the queuing method in the fast food industry. The research paper provides the interpretation of alternative WLS (multiphase) and at the same time differentiate the effectiveness of the existing WLS (SSQ, SSSP). Its focus mainly on similarity and dissimilarity of data based on the average sum of time customer will spend in lining up until being served for current and proposed WLS.

2. The Queuing System Design

A real case scenario was carried out in this study, which was based on the previous literature of WLS. By using MQ analysis (M/M/1), data gathered were used to design and analyse accurately for the convenience of the research. For sampling, Six (6) fast food restaurants in Klang Valley were identified. The actual WLS processes were physically recorded, then extracted and analysed by using Monte Carlo Simulation and Markovian waiting line analysis technique. This research prepared a conclusion which either the Markovian SQ Single Server of SSMP waiting line design provides a better effect on the efficiency of the WLS for the fast food restaurants in Malaysia. Generally in Malaysia, SQ single server (parallel) WLS were frequently used by most of the fast food restaurant. The effectiveness of the system can only be determined when one server serving one customer at a time. This was overwhelmingly problematic for the fast food restaurant in recent years when the customer had to wait for the next available cashier without any other option in this particular system. Illustration of the SQ single server (parallel) WLS based on this assumption; arrivals are described by a Poisson probability distribution and come from an infinite or very large population, arrivals are independent of preceding arrivals, but the average number of arrivals (arrival rate) does not change over time, arrivals are served on a first-in, first-out (FIFO) basis, and every arrival waits to be served, regardless of the length of the line of the queue, and service times vary from one customer to the next and are independent of one another, but their average rate is known. Measuring the queue's performance by comparing between proposed and current WLS based on the number of phase for customer to get their services completed. The difference will be explained in figure 1 and figure 2 as below:

Figure 1: Current Design, Single Queue, Single Server (Parallel) Waiting Lines

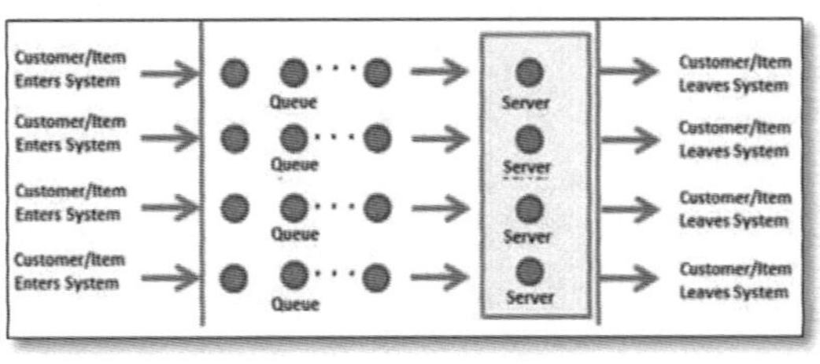

Figure 2: Proposed Design Single Queue, Multi Server Waiting Lines

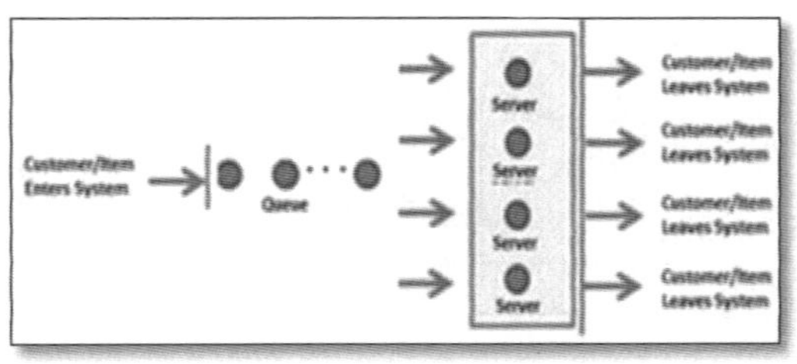

3. Data Analysis

Overall data gathered was based on 1664 customers, who came to the respective fast food restaurants to buy their meals (Table 1), and the fastest interval time between customers arrival was exactly 1 minute and the longest time was 6 minutes. On the other hand, the fastest customer being served was 2 minutes and the longest time was 11 minutes.

Table 1.

Interval Time In	Frequency	Time Serve	Frequency

	1	375	1	0
	2	362	2	42
	3	520	3	114
	4	180	4	681
	5	98	5	478
	6	129	6	207
			7	74
			8	28
			9	7
Total number of customer		1664		1631

Data collected shown the value of λ which is a value for the arrival rate per hour and μ is a calculated value for service rate per hour. Therefore, based on the actual event the value of λ = 11.56 and the value of μ is = 12.9. In order to get the value of λ and μ more accurate, Monte Carlo Simulation was used and the average value of λ and μ both respectively are 11.56 and 12.90 (Table 2).

Table 2.

	Number of simulation	Average Service Time	μ
	1	4.66	12.90324
	2	4.65	12.84796
	3	4.52	12.95897
	4	4.69	12.82050
	5	4.63	12.98702
	6	4.67	12.87553
	7	4.63	12.93104
	8	4.67	12.82050
	9	4.61	12.98700
	10	4.66	12.84798
Total service Time Per simulation		4.652	12.8976784

Software called QM for Windows was used to determine the efficiency for both waiting line designs, since the value of λ and μ was calculated. The result of analysis for each waiting line design can be referred in table 3.

Table 3. The Comparison Analysis between Waiting Lines Design

Parameter	Value	Minutes	Value	Minutes
Average server utilization	.9		.9	
Average number in the queue (Lq)	7.73		6.94	
Average number in the systems (L)	8.63		9.63	
Average time in the queue (Wq)	.67	40.12	.2	12.02
Average time in the systems (W)	.75	44.78	.28	16.67

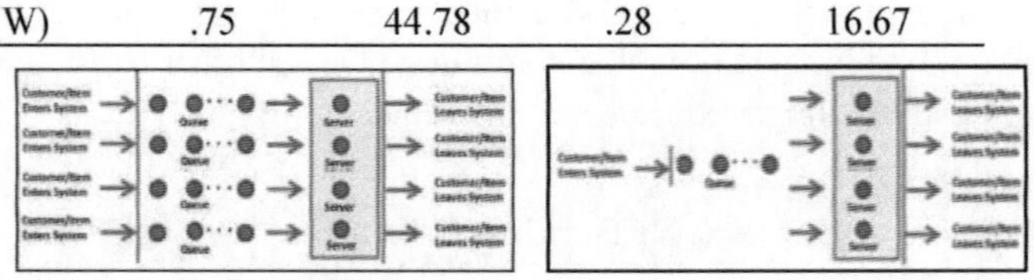

By looking at the table, the info stated the average time for a customer have to wait in a queue using the Single Server (Parallel) design was 40.12 minutes and the average time for a customer has to wait in the system was 44.78 minutes. For Multiple Server design, the average time a customer has to wait in the queue is improving which was 12.02 minutes; saving about 70% of the waiting time in queue and for the system was 16.67 minutes which saved 62.7% of waiting in the system. As referred to the data, the research has the evidence supporting the multiple server design will improve the quality of the services in Malaysia fast food restaurants. A good waiting line design will provide better and more conclusive results on how to attract more customers or even make them faithful to the organization itself. A very good waiting line design will assure an advantage to gain more benefits and compete much better in the industry.

4. Conclusion

There are many research opportunities and upgrades that can be developed with the idea of waiting line system proposed in this paper. To determine the success of organizations or individuals in achieving their objectives and strategies, Performance Management can be used. Reference [9] suggested that we can extend the study to discuss the organization performance base on 4 dimensions, employees' measures, customer measures, financial measures and suppliers' measures. This research proposed the three WLS design which includes Markovian Single Queue, Single Server and Multiphase Queuing will give significant effect on the organization performance positively. The main focus of this research is to justify and to share the idea of effectiveness for improving the queuing design for fast food restaurants. The data analysed includes the average number of customers in system, average number of customers in waiting line, average time a customer spends waiting and being served and average time customer spends waiting in the queue. The study also provides several potential issues related to organizational performance that can be carried out for future study.

5. References

[1] Jackson, J. (2011). Kinds of Food Service in a Restaurant. http://www.ehow.com/info_8570902_kinds-food-servicerestaurant. html (accessed May 4, 2013)

[2] Walker, J. R. (2011). The Restaurant: from Concept to Operation, 6th ed. New Jersey. John Wiley and Sons, Inc.

[3] Drysdale, J. A. and Galipeau, J. A. (2009). Profitable Menu Planning, 4th ed. New Jersey: Pearson International Edition

[4] Adan, I.J.B.F., Boxmal, O.J., Resing, J.A.C. (2000), "Queuing models with multiple waiting lines," Department of Mathematics and Computer Science, Eindhoven University of Technology

[5] Dharmawirya, M & Adi H.O.E (2012), Industrial Engineering Letters, ISSN 2224-6096, Vol 2, No.5, 2012

[6] Zhao, X., Lau, R.S.M, and Lam, K. (2002). "Optimizing the service configuration with the least total cost approach." International Journal of Service Industry Management, Vol. 13 No. 4, pp. 348-361.

[7] Broyles JR, Cochran JK. (2007). Estimating business loss to a hospital emergency department from patient reneging by queuing-based regression. IIE Industrial Engineering Research Conference. Nashville, TN.

[8] Kumar, R & Sharma S.K, An M/M/1/N Queueing Model with Retention of Reneged Customers and Balking, American Journal of Operational Research 2012; 2(1): 1-5 doi: 10.5923/j.ajor.20120201.01

[9] Dimovski, V., Škerlavaj, M. (2005), Performance Effects of Organizational Learning in aTransitional Economy, Problems and Perspectives in Management , No. 4.References to preprints.

Institute of Physics
Dirac House, Temple Back
Bristol BS1 6BE UK

ISSN: 1755-1307
ISBN 978-1-5108-2273-3

9 781510 822733

CURRAN ASSOCIATES INC.
proceedings.com

International Conference "Structural and Phase Transformations in Materials: Theory, Computer Modelling and Experiment" (SPTM 2017)

IOP Conference Series: Materials Science and Engineering
Volume 192

Ekaterinburg, Russia
23 – 27 March 2017

IOP
Institute of Physics